R0006307258

D1064438

Microbiological Control for Foods and Agricultural Products

Analysis and Control Methods for Foods and Agricultural Products
A reference set in four volumes

Volumes in the set:

Analysis of Food Constituents
edited by J.L. Multon

Analytical Techniques for Foods and Agricultural Products
edited by G. Linden

Microbiological Control for Foods and Agricultural Products
edited by C.M. Bourgeois and J.Y. Leveau

Quality Control
edited by J.L. Multon

Microbiological Control for Foods and Agricultural Products

EDITED BY

C. M. Bourgeois and J.-Y. Leveau

TRANSLATED BY

Stephen Davids

EDITOR FOR THE ENGLISH-LANGUAGE EDITION

Daniel Y. C. Fung

Editors:
C. M. Bourgeois
A.D.R.I.A
6 rue de l'Université
F.29191 Quimper Cedex
France

J.-Y. Leveau
E.N.S.I.A
Département de Biotechnologie
F.91305 Massy
France

Translator:
Stephen Davids
880 avenue Casot
Québec G1S 2X9
Canada

Editor of the English-language Edition:
Daniel Y. C. Fung
Dept. of Animal Sciences and Industry
Kansas State University
Manhattan, KS 66506-1600
USA

This book is printed on acid-free paper. ♾

Library of Congress Cataloging-in-Publication Data

[Le contrôle microbiologique, 2nd edition. English.]
Microbiological control for foods and agricultural products/ edited by C.M. Bourgeois and J.Y. Leveau; translated by Stephen Davids; editor for the English-language edition, Daniel Y.C. Fung.
p. cm. -- ([Techniques d'analyse et de contrôle dans les industries agro-alimentaires. English.]
Analysis and control methods for foods and agricultural products/ J.L. Multon, editor.)
Includes bibliographical references and index.
ISBN 1-56081-673-2 (acid-free paper)
1. Food adulteration and inspection. 2. Food--Analysis.
I. Bourgeois, C.M. II. Leveau, J.Y.
TX531.T413 1995
363.19'26--dc20 **95-7026**
CIP

This book was originally published as "Techniques d'analyse et de contrôle dans les industries agro-alimentaires, Volume 3: Le Contrôle microbiologique, Second Edition," by Lavoisier-Tec & Doc, 11, rue Lavoisier, F.75384 Paris Cedex 08, and APRIA, 35, rue du Général Foy, F.75008 Paris, France

Printed in the United States of America

ISBN 1-56081-673-2 VCH Publishers, Inc.

Printing History:
10 9 8 7 6 5 4 3 2 1

Published jointly by:
VCH Publishers, Inc.
220 East 23rd Street
New York, New York 10010

VCH Verlagsgesellschaft mbH
P.O. Box 10 11 61
69451 Weinheim, Germany

VCH Publishers (UK) Ltd.
8 Wellington Court
Cambridge CB1 1HZ
United Kingdom

Preface for the English Edition

This book contains a collection of articles, originally written in French, by well-known food microbiologists and scientists mainly from France. The book is divided into five parts. The first part contains methods and procedures for applied food microbiology. The emphasis is on basic techniques and approaches relative to handling of samples, viable cell count methods, newer methods, immunology, identification of microbes by various methods, and predictive microbiology. The second part deals with specific groups of microbes (mesophiles, psychrotrophs, lactics, etc.) important to food microbiology. The third part is on specific genera of important microbes (*Salmonella, Campylobacter, Listeria, Clostridium*, etc.) in public health and food sanitation. The fourth part concerns microbial problems of various food groups (water, milk, meats, etc.). The book concludes with a short fifth part on starter culture technology and monitoring of microbes in the food industry.

These chapters contain a wealth of information on applied microbiology not readily available in other similar books. Stephen J. Davids translated the work from French to English. I have the honor of editing the English version of the work. A new chapter, entitled "Rapid Methods and Automation in Microbiology: An Update," is added to provide the most current information on this topic. Suggested readings on selected topics also is presented, so that readers can locate updated information published in English, especially from 1990–1994. The purpose of these additions is to inform readers about recent developments in applied food microbiology, molecular biology (e.g., polymerase chain reactions), techiques in nucleic acid probes, and ELISA tests.

I am indebted to Karim Kone and Reneé Hart-Thakur, my graduate students, for assisting me in updating the references.

This book should be a useful reference and a handy manual for food and applied microbiologists. It also can be used as a textbook in food microbiology.

Daniel Y. C. Fung
Manhattan, Kansas
January 1995

Contents

PART I. GENERAL AND BASIC TECHNIQUES

1. Rapid Methods and Automation in Food Microbiology: An Update 3
Daniel Y. C. Fung

1.1 Introduction 3
1.2 Improvements in Sampling and Sample Preparation 4
1.3 Alternative Methods for Viable Cell Count Procedure 6
1.4 New Methods for Estimation of Microbial Populations and Biomass 10
1.5 Miniaturized Microbiological Techniques 14
1.6 New and Novel Techniques 15
1.7 Conclusion 18
References 18

2. Basic Principles of Industrial Microbiological Testing and Uses of its Findings 23
C. M. Bourgeois, J. J. Cleret

2.1 Aims and Requirements of Industrial Microbiological Testing 24
2.2 Implementation of Industrial Testing 24
2.3 Uses of Test Results 28
References 32

3. Sample Removal, Transport, and Preparation 35
C. M. Bourgeois, A. Plusquellec

3.1 Sample Collection 35
3.2 Sample Transportation and Preservation 37
3.3 Grinding of Solid Food Samples 38
3.4 Surface Sampling 41
3.5 Adjusting Microbial Concentrations to Analytical Needs: Dilution and Concentration 42
References 44

4. Rapid Methods of Enumerating Microorganisms 47
C. M. Bourgeois, R. Malcoste

4.1 Microscopic Detection and Enumeration 47
4.2 Estimation of Viable Flora by Culture in or on Solid Media 52
4.3 Methods of Evaluating the Most Probable Number (MPN) of Microorganisms by Statistical Examination of the Numbers of Positive Responses to Growth Tests 58
4.4 Electronic Counting of Cells and Flow Cytometry 66
4.5 Conclusion 68
References 68

5. Evaluation of Microflora by Nonmicrobiological Techniques 73
C. M. Bourgeois, P. Mafart

5.1 General Principles 73
5.2 Detection Targets 74
5.3 Spectroscopy 75
5.4 Electrochemistry 83
5.5 Other Processes 87
5.6 Conclusion 89
References 89

6. Identification 97
Marielle Bouix, J.-Y. Leveau

6.1 The Examination of Cultural Traits 97
6.2 The Examination of Cell Morphological and Structural Traits 98

6.3 The Examination of Sexual Traits 99
6.4 The Examination of Biochemical and Physiological Traits 99
6.5 The Examination of Genetic Traits 104
6.6 The Examination of Immunological Traits 105
6.7 The Examination of Pathogenic Capabilities 106
6.8 Use of Results 106
References 107
Suggested Readings, Chapters 2–6 107

7. Microbiological Applications of Immunology 109
Florence Humbert, Cécile Lahellec

7.1 Introduction 109
7.2 Precipitation 110
7.3 Agglutination 112
7.4 Immunofluorescence 114
7.5 ELISA 115
7.6 Radioimmunology 116
References 117

8. Identification of Microorganisms by Nucleic Acid Probe Hybridization 119
Aline Lonvaud-Funel

8.1 Introduction 119
8.2 The Principle of Molecular Hybridization 120
8.3 Radioactive Probes and Cold Probes: Detection Principles 121
8.4 Principles of Labeled Nucleotide Incorporation 125
8.5 Hybridization Protocols 127
8.6 Technical Developments 132
8.7 Food Applications 133
8.8 Conclusion 136
References 136

9. Mechanization and Automation of Techniques 139
R. Grappin, Christine Piton

9.1 Introduction 139
9.2 The Aim of Automation 140
9.3 The Limitations of Automation 141
9.4 Conclusion 146
References 147

10. Prediction of Product Life Span 151

C. M. Bourgeois

10.1 The Value and Importance of Such Predictions 151
10.2 Predictive Microbiology 152
10.3 Modelizing the Effect of Temperature 153
10.4 Multifactorial Models 157
10.5 Prediction of the Life Span of a Manufactured Product 161
10.6 Prediction of the Potential Life Span of a New Product 164
10.7 Future Possibilities 165
References 165
Suggested Readings, Chapters 7–10 167

PART II. METHODS OF EVALUATION OF VARIOUS TECHNOLOGICALLY SIGNIFICANT MICROFLORA

11. Total Aerobic Mesophilic Microflora 175

C. M. Bourgeois

11.1 Value of the Total Aerobic Mesophilic Microflora Count as an Indicator 175
11.2 Counting Methods 176
11.3 Methods of Overall Evaluation 178
References 179

12. Psychrotrophic Microflora 181

Cécile Lahellec, P. Colin

12.1 Introduction 181
12.2 Techniques for the Examination of Psychrotrophic Bacteria 182
12.3 Appendix 187
References 187

13. The Lactic Microflora 189

J.-Y. Leveau, Marielle Bouix, H. de Roissart

13.1 Definition, Classification, and General Properties 189
13.2 Common Techniques of Examination 192

13.3 The Examination of Different Genera 205
References 221

14. Bacteriophages of Lactic Starters: Detection and Enumeration 227
J.-P. Accolas, Marie-Christine Chopin, G. K. Y. Limsowtin

14.1 Review of the Characteristics of Bacteriophages 227
14.2 Review of Methods: Meaning and Significance of Results 230
14.3 Recommended Methods 238
References 244
Suggested Readings, Chapters 11–14 246

15. The Yeasts 249
Marielle Bouix, J.-Y. Leveau

15.1 Introduction 249
GENERAL CHARACTERISTICS OF YEASTS 250
15.2 Definition and Classification of Yeasts 250
15.3 Biological Characteristics 252
15.4 Physiological Characteristics 253
15.5 Genetic Characteristics 257
15.6 Ecological Characteristics and Consequences of Yeast Development 258
YEAST EXAMINATION TECHNIQUES 260
15.7 Detection and Culture Techniques 260
15.8 Identification Techniques 264
15.9 Fine Differentiation Technique 272
References 274
Suggested Readings 275

16. The Molds 277
C. Moreau

16.1 Molds in the Food Industry 277
16.2 Methods of Analysis 279
16.3 Types of Analyses Performed 283
16.4 Identification of Molds 287
16.5 Expression of Results 287
16.6 Experimental Studies and Prospects for Improved Defense Against Mold Infections 288
16.7 Conclusion 289

PART III. EVALUATION METHODS FOR MICROFLORA OF SANITARY INCIDENCE

17. Indicators of Fecal Contamination 293
M. V. Catsaras

17.1 Basis for the Use of Indicator Microorganisms and the Justification of Their Choice 293
17.2 Coliforms and *E. coli* 294
17.3 The Enterobacteria 297
17.4 The Fecal Streptococci 298
17.5 The Sulfite-Reducing *Clostridium* Species 300
17.6 Conclusion 300
17.7 Appendix 301
References 304
Suggested Readings 305

18. The Genus *Salmonella* 309
J. Gledel, Béatrix Corbion

18.1 Definition 309
18.2 Principal Biochemical Traits 309
18.3 Antigenic Structure 310
18.4 Taxonomy 311
18.5 The Kauffmann White Scheme 313
18.6 Complementary Distinctive Traits 314
18.7 Epidemiology-Pathogenic Potential 315
18.8 Detection of *Salmonella* in Foods 316
References 321
Suggested Readings 323

19. The Genus *Campylobacter* 325
M. Catteau

19.1 Culture Conditions and Identification 325
19.2 Pathogenic Potential 326
19.3 Analysis of *Campylobacter* in Foods 329
19.4 Occurrence in the Environment and in Foods—Epidemiology 330
19.5 Conclusion 332
References 332
Suggested Readings 333

20. The Genus *Yersinia* 335
M. Catteau

20.1 Taxonomy 335
20.2 Pathogenic Potential 338
20.3 Analysis for *Yersinia Enterocolitica* in Foods 341
20.4 Incidence of *Y. Enterocolitica* in Food Products 342
References 344
Suggested Readings 344

21. The Genus *Clostridium* and the Sulfite-Reducing Anaerobes 347
Martine Poumeyrol, J. Billon

21.1 *Clostridium Perfringens* 347
21.2 *Clostridium Botulinum* 353
References 358
Suggested Readings 359

22. Coagulase-Positive Staphylococci 361
Marie-Laure De Buyser

22.1 Taxonomic Position 361
22.2 Significance of the Presence of Coagulase-Positive Staphylococci in Foods 362
22.3 Characterization of Enterotoxigenic Staphylococci 362
22.4 Detection and Enumeration 363
22.5 Identification 366
22.6 Interpretation of Results 367
References 369
Suggested Readings 370

23. The Genus *Listeria* 373
M. Catteau

23.1 Taxonomy-Identification 373
23.2 Serotyping 375
23.3 Lysotyping 375
23.4 Pathogenic Potential 375
23.5 Epidemiology 375
23.6 Contaminated Foods and Multiplication of *L. monocytogenes* 377
23.7 Detection of *Listeria* in Foods 377

23.8 Method of Detection of *Listeria monocytogenes* in Cheeses 378
23.9 Appendix: Formulas of the Media Used 380
References 380
Suggested Readings 381

PART IV. MICROBIOLOGICAL TESTING OF STARTING MATERIALS AND FINISHED PRODUCTS

24. Water 385
A. Plusquellec

24.1 Definition 385
24.2 Regulation of the Bacteriological Quality of Water 386
24.3 Bacteriological Analysis 387
24.4 Interpretation of Results 392

25. Milk and Dairy Products 395
A. Plusquellec

25.1 Testing of Raw Milk at the Point of Collection 395
25.2 Microbiological Testing in Milk Processing 399
25.3 Microbiological Testing of Distributed Dairy Products 406

26. Beer and Soft Beverages 415
A. Plusquellec

26.1 Introduction 415
26.2 Microbial Contaminations of Beer 415
26.3 Microbiological Testing 416
26.4 Analysis of Soft Beverages 418

27. Meat and Meat Products 421
A. Plusquellec

27.1 Introduction 421
27.2 Characteristics of Animal Muscle 422
27.3 Microbiological Characteristics of Meat 422

27.4 Endogenous Infections 422
27.5 Spoilage of Meat 423
27.6 Microbiological Testing 426
27.7 Regulations, Microbiological Standards 427

28. Eggs and Egg Products 431
A. Plusquellec

28.1 Contamination of Eggs 431
28.2 Bacteriological Analysis 432
28.3 Regulation of Bacteriological Quality 434

29. Fisheries Products: Fish, Crustaceans, and Shellfish 437
A. Plusquellec

29.1 Bacteriological Quality of Fish and Fisheries Products 437
29.2 Spoilage Processes of Fisheries Products 439
29.3 Testing 440

30. Vegetable Products 445
A. Plusquellec

30.1 Introduction 445
30.2 Characteristics of Fruits and Vegetables 445
30.3 Spoilage of Fruits and Vegetables 447
30.4 Microbiological Testing 449
30.5 Regulation 450
30.6 Dehydrated Plant Products 450
30.7 Minimally Processed Vegetables 451

31. Prepared Meals 455
A. Plusquellec

31.1 Definition 455
31.2 Microbiological Problems Posed by Prepared Meals 456
31.3 Conditions Required for Good Bacteriological Quality 456
31.4 Microbiological Testing 458
31.5 Regulatory Criteria 459

32. Preserves and Semipreserves 461
A. Plusquellec

32.1 Definitions 461
32.2 Role of the Testing Laboratory 462
32.3 Stability Testing 462
32.4 Sterility Testing 464
32.5 Semipreserves 467

33. Testing of Intermediate Moisture Foods 469
Laurence Lesage, D. Richard-Molard

33.1 Intermediate Moisture Foods 469
33.2 Microbiological Testing of IMFs 471
33.3 Biochemical Tests 478
33.4 Conclusion 480
33.5 Appendix: Culture Media 480
References, Part IV 486
Suggested Readings, Chapters 24–26 489
Suggested Readings, Chapters 27–33 491

PART V. MISCELLANEOUS APPLICATIONS OF MICROBIOLOGICAL TESTING

34. Testing of Starter Cultures for Purity 497
T. Germain, A. Miclo

34.1 General Aspects of Testing 497
34.2 Conventional Methods 498
34.3 Accelerated Conventional Methods 505
34.4 Rapid Techniques 506
34.5 Conclusion 510
References 511

35. Microbiological Monitoring of Factory Equipment, Atmosphere, and Personnel 515
A. Plusquellec, J. Y. Leveau

35.1 Introduction 515
35.2 Microorganisms Analyzed 516
35.3 Testing of Equipment and Work Areas 516

35.4 Atmospheric Testing 525
35.5 Testing at the Level of Personnel 526
35.6 Conclusion 527
References 528
Suggested Readings, Chapters 34 and 35 528
Suggested Readings, General Interest 529

Index 531

Contributors

J. P. Accolas. Chief Scientist, Department of Dairy Microbiology and Food Engineering, *INRA* (National Institute of Research in Agronomy), Jouy-en-Josas, France

J. Billon. Chief Veterinary Inspector, *LCHA* (Central Laboratory for Food Safety), Paris, France

M. Bouix. Professor, ENSIA (National School of Agriculture and Food Studies), Massy, France

C. M. Bourgeois. Professor, University of Western Brittany (*Bretagne occidentale*) and Director, *ADRIA* (Association for the Advancement of Applied Research in the Food Industry), Quimper, France

M.-L. de Buyser. Associate Scientist, *CNEVA* (Aviculture Research Center), *LCHA* (Central Laboratory for Food Safety), Paris, France

M. Catteau. Professor and Director, *SERMHA* (Department of Microbiology and Food Safety), Pasteur Institute, Lille, France

M. Catsaras. Professor, Pasteur Institute, Lille, France

M.-C. Chopin. Chief Scientist and Director, Bioengineering Laboratory, *INRA* (National Institute of Research in Agronomy), Jouy-en-Josas, France

J.-J. Cléret. Professor, University of Western Britanny (*Bretagne occidentale*), Quimper, France

P. Colin. Scientist, *CNEVA* (Aviculture Research Center), Ploufragan, France

B. Corbion. Scientist, *CNEVA* (Aviculture Research Center), and Head of the Research Group in Epidemiology and Advanced Techniques in Microbiology, *LCHA* (Central Laboratory for Food Safety), Paris, France

D. Y. C. Fung. Professor, Kansas State University, Manhattan, Kansas, U.S.A.

J. Gledel. Chief Scientist and Director, *CNEVA* (Aviculture Research Center), France

P. Germain. Director, *LCHA* (Central Laboratory for Food Safety), Paris, France

R. Grappin. Professor, *ENSAIA* (National Polytechnic Institute), Vandoeuvre-lès-Nancy, France

F. Humbert. Associate Scientist, *CNEVA* (Aviculture Research Center), Ploufragan, France

C. Lahellec. Chief Scientist and Head of the Research Group for Safety and Quality of Poultry Products, *CNEVA* (Aviculture Research Center), Ploufragan, France

L. Lesage. Scientist, *INRA* (National Institute of Research in Agronomy), Nantes, France

J.-Y. Leveau. Professor, Department of Biotechnology, *ENSIA* (National School of Agriculture and Food Studies), Massy, France

G. K. Y. Limsowtin. Research Officer, New Zealand Dairy Research Institute, New Zealand

A. Lonvaud-Funel. *Institut d'oenologie* (Wine Study Institute), Talence, France

P. Mafart. Professor, University of Western Brittany (*Bretagne occidentale*), Quimper, France

R. Malcoste. University of Western Brittany (*Bretagne occidentale*), Quimper, France

A. Miclo. Associate Professor, *ENSAIA* (National Polytechnic Institute), Vandoeuvre-lès-Nancy, France

C. Moreau. Chief Scientist, *CNRS* (National Scientific Research Center), University of Western Britanny (*Bretagne occidentale*), Brest, France

C. Piton. Department of Research in Dairy Technology, *INRA* (National Institute of Research in Agronomy), Poligny, France

A. Plusquellec. Associate Professor, University of Western Brittany (*Bretagne occidentale*), Quimper, France

M. Poumeyrolle. Chief Scientist, *CNEVA* (Aviculture Research Center), and *LCHA* (Central Laboratory for Food Safety), Paris, France

D. Richard-Molard. Chief Scientist and Director, Department of Carbohydrate and Protein Research, *INRA* (National Institute of Research in Agronomy), Nantes, France

Guide to Bibliographic Sources

Aerobic: *Total Aerobic Mesophilic Microflora*
Automation: *Rapid Methods and Automation*
Automation: *Mechanization and Automation of Techniques*
Bacteriophages of Lactic Starters
Basic Principles
Beer and Soft Beverages
Campylobacter
Clostridium (*and Sulfite-Reducing Anaerobes*)
Egg Products
Enumerating Microorganisms: *Rapid Methods...*
Fecal Contamination: *Indicators...*
Fisheries Products
General Interest Sources
Identification
Immunology: *Microbial Applications of Immunology*
Intermediate Moisture Foods
Lactic Microflora
Life Span: *Prediction of Product Life Span*
Listeria
Meat
Milk
Monitoring of Equipment, Personnel
Nonmicrobial Techniques: *Evaluation of Microflora...*
Nucleic Acid Probe Hybridization: Identification of Microorganisms...

Prepared Meals
Preserves
Psychrotrophic Microflora
Salmonella
Sample Handling
Staphylococci: Coagulase-positive Staphylococci
Starter Culture: *Testing of Starter Culture*
Vegetable Products
Water
Yeasts
Yersinia

PART I

GENERAL AND BASIC TECHNIQUES

CHAPTER

1

Rapid Methods and Automation in Food Microbiology: An Update

Daniel Y. C. Fung

1.1 Introduction

Rapid methods and automation in microbiology are dynamic fields of study that address the utilization of microbiological, chemical, biochemical, biophysical, immunological, and serological methods for the study of improving isolation, early detection, characterization, and enumeration of microorganisms and their products in clinical, food, industrial, and environmental samples. Medical microbiologists started to develop rapid methods about 25 years ago. In the past 15 years, food microbiologists have started to adapt rapid and automated methods in their laboratories (Figure 1.1). Both groups are moving forward aggressively. Conventional methods of detection, enumeration identification, and characterization of microbes are described in reference books such as *Compendium of Methods for the Microbiological Examination of Foods* (Vanderzant and Splittstoesser, 1992), *Official Methods of Analysis of the AOAC* (AOAC, 1990), *Bacteriological Analytical Manual* (FDA, 1992), *Standard Methods for the Examination of Dairy Products* (APHA, 1985), and *Modern Food Microbiology* (Jay, 1992). The new *Bergey's Manual of Determinative Bacteriology* (9th edition by Holt et al., 1994) is an excellent resource book.

Important publications on the subject of rapid methods for medical specimens, water, food, industrial, and environmental samples are in a series of papers by Fung and colleagues (Fung, 1991, 1992; Fung et al., 1989), and books such as *Mechanizing Microbiology* (Sharpe and Clark, 1978), *Foodborne Microorganisms and Their Toxins: Developing Methodology* (Pierson and Stern, 1986), *Rapid Methods in Food Microbiology* (Adams and Hope, 1989), and

Food Microbiology

- Sample preparation
- Total viable cell count
- Differential cell count
- Pathogenic organisms
- Enzymes and Toxins
- Metabolites and Biomass

Figure 1.1 Development of interests in rapid methods.

Instrumental Methods for Quality Assurance in Foods (Fung and Matthews, 1991). Hartman and co-workers (1992) have an excellent chapter on *Rapid Methods and Automation* in *Compendium of Methods for the Microbiological Examination of Foods* (Vanderzant and Splittstoesser, 1992).

The purpose of this chapter is to review the basic principles and practical applications of a variety of instruments and procedures directly and indirectly related to improved methods for microbiology.

1.2 Improvements in Sampling and Sample Preparation

The six areas of research and development in food microbiology are listed in Figure 1.2. Each area is important and is being studied by various scientists in the field. In order to carry out a microbiological analysis, the scientists must first prepare the food sample properly. One of the most useful instruments developed for sample preparation is the Stomacher (Tekmar, Cincinnati, OH). This instrument is designed to massage food samples in a sterile disposable bag. The weighted food sample (ground beef, flour, milk, fish, etc.) is first placed in the sterile disposable plastic bag, and appropriate sterile diluents are added. The bag with the food is placed in the open chamber. After the chamber is closed, the bag is then massaged by two paddles for a suitable time period, usually from 1 to 5 minutes. No contact occurs between the instrument and the sample. During massaging, microorganisms are dislodged into the diluent for further microbiological manipulation. Massaged slurries are then used for microbiological analysis. Sharpe and Jackson (1975) and Emswiler and co-workers (1977) have shown that satisfactory results can be obtained by this method compared with conventional blending of foods. Food with sharp

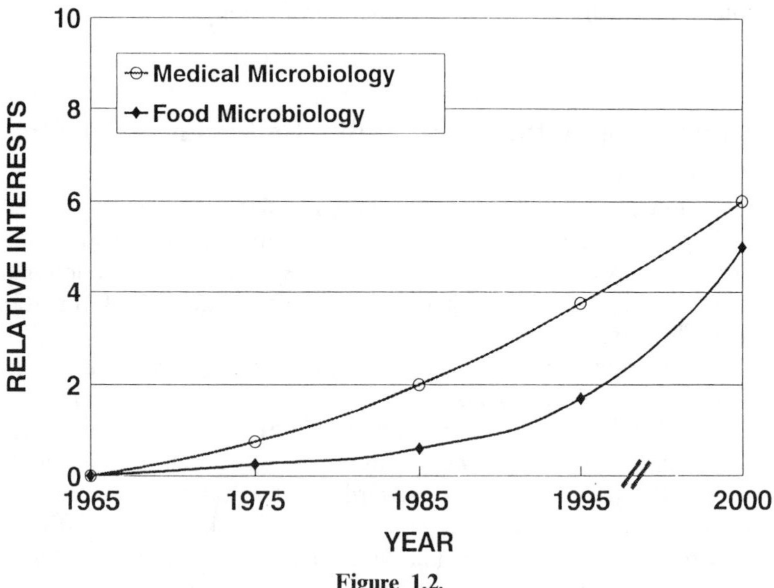

Figure 1.2.

particles such as bone, shells and so forth should not be used in this instrument because the particles may break the plastic bag.

Another new instrument that holds promise for sample preparation is the Gravimetric Diluter (Spiral Biotech, Bethesda, MD). One of the most time-consuming procedures of routine microbiological work is to aseptically measure a sample of food (e.g., 10 g of meat) and then aseptically add an exact amount of sterile diluent (e.g., 90 ml) to make a desired dilution (1:10). With the Gravimeter Diluter, the analyst needs only to aseptically place an amount of food (e.g., 5.3 g) into a pretared Stomacher bag or a sterile blending jar, set a desired dilution (1:10), and set the instrument to deliver the appropriate amount of sterile diluent (e.g., 47.7 ml). Thus, the dilution operation can be done automatically. The dilution factor can be chosen by the analyst to satisfy the need (1:10, 1:50, 1:100, etc.) simply by programming the instrument. Manninen and Fung (1992b) evaluated this instrument and found that depending on the volume tested, the accuracy of delivery for most samples was found to be in the range of 90–100%. A new version of this instrument is called Diluflo and has been in use satisfactorily in the author's laboratory since 1992. The new model can handle presterilized diluent in convenient jars for ease of operation.

A review by Lee and Fung (1986) documented many useful ways to obtain surface samples from meats and other foods. Fung and co-workers (1980) developed an adhesive tape method that can effectively "peel" viable cells from meat surfaces and later deposit them on agar surfaces to obtain viable cell counts of meat surfaces. The same tape can be transferred from agar to agar up to four times without noticeable reduction of colony numbers.

1.3 Alternative Methods for Viable Cell Count Procedure

The conventional viable cell amount or standard plate count method is still the "standard" procedure to obtain cell numbers from food and water, but the procedure is time consuming both in terms of operation and collection of data. Several methods have been explored to improve the efficiency of operation of the viable cell count procedure.

The spiral plating method is an automated system for obtaining viable cell counts (Spiral Biotech, Bethesda, MD). The instrument can spread a liquid sample on the surface of agar contained in a Petri dish in a spiral shape (the Archimedes spiral) with a concentration gradient starting from the center and decreasing as the spiral progresses outward on the rotating plate. The volume of liquid deposited at any segment of the agar plate is known. After the liquid containing microorganisms is spread, the agar plate is incubated overnight at an appropriate temperature for the colonies to develop. The colonies appearing along the spiral pathway can be counted either manually or electronically by laser. The spiral system has been used in the United States for a variety of foods with satisfactory results (Schalkowsky, 1986).

Manninen and colleagues (1991) evaluated the spiral system and laser colony scanner for enumeration of microorganisms and found excellent correlation between pour plate and spiral plate, both using manual and laser counting procedures for bacteria (Table 1.1). Yeast counts were also very comparable but mold counts were not as reliable, especially for fast growing mold such as *Rhizopus oligosporus.* Manninen and Fung (1992b) also evaluated the spiral plater and laser colony scanner for enumeration of microorganisms in meat and devised a convenient protocol combining the conventional aerobic plate count method with the spiral system to evaluate surface microbial loads of pork loins. In this procedure, a surface (25, 50, or 125 cm^2) is swabbed and then placed in 9 ml of sterile buffer. For each sample both the conventional and the spiral plating method is used. For conventional plating 1 ml is used. For spiral plating the liquid is plated without further dilution. In this protocol when the number of cells is too numerous to count for the conventional plating (more than 300 per plate), the spiral plate sample provided the accurate count. Conversely, when the number is very low (1–250 cells per plate), the conventional pour plate sample will provide accurate count while the spiral plated sample will have virtually no colonies. Using 4 plates (2 plates are really needed), a large range of microbial load on meat surfaces can be counted effectively.

Table 1.1 Comparison of pour plate and spiral plate counted manually and by laser for bacterial cultures

	Pour plate		Spiral plate	
Test cultures	Manual	Laser	Manual	Laser
		(Log_{10}cfu/ml)		
Escherichia coli	8.86	8.85	8.73	8.85
Salmonella enteritidis	8.76	8.66	8.78	8.92
Pseudomonas aeruginosa	8.00	8.00	8.00	8.00
Staphylococcus aureus	8.04	7.78	8.18	8.18
Lactobacillus plantarum	9.48	9.40	9.60	9.69
Streptococcus sp.	7.73	7.66	8.00	8.08
Bacillus cereus	7.26	7.15	7.15	7.26
Micrococcus luteus	7.40	7.32	7.51	7.57

Source: Data from Manninen et al. (1991).

A new version of the spiral plater was recently introduced named "Autoplater" (Spiral Biotech, Bethesda, MD). With this instrument, an analyst needs only to present the liquid sample and the instrument completely and automatically processes the sample, including resterilizing the sampling tube (dispensing styles) for the next sample.

One of the problems of the spiral system is clogging of the dispensing stylus by food particles. Konuma and Kurata (1982) modified a Stomacher bag by placing a filter in the bag so the homogenized liquid poured from the bag would be free of particles. Since then, commercial companies have manufactured stomacher bags with filters in place. The liquid, after filtration and presented to the spiral system, will not clog the stylus.

Another alternative method for viable cell count is the Isogrid system (QA Laboratories Ltd., San Diego, CA). This system consists of a square filter with hydrophobic grids printed on the filter to form 1,600 squares for each filter. Food samples are weighed, blended, and enzyme treated (if necessary) before passage through the membrane filter containing the hydrophobic grids by vacuum. The filter is then placed on agar containing a suitable nutrient for growth of the bacterium, yeast, or mold. The hydrophobic grids prevent colonies from growing further than the square grids; thus all colonies have a square shape. This facilitates counting of the colonies both manually and electronically. This method has been successfully used to make viable cell counts for a variety of foods, including milk, meat, black pepper, flour, peanut butter, mushrooms, rice, fish, fish shrimp, and fish oyster (Entis, 1983, 1984, 1986; Entis et al., 1982; Sharpe and Peterkin, 1988). Other applications of the Isogrid system include determination of total coliforms, fecal coliforms, *Escherichia coli, Salmonella* spp., and so forth (Sharpe, 1991).

Recently, an effective television image analyzer was developed for rapid enumeration of colonies grown on the Isogrid. Rehydratable nutrients are embedded into a series of films in the Petrifilm system (3M Co., St. Paul, MN). The top layer of the protective cover is lifted and 1 ml of liquid sample is introduced to the center of the unit, and the cover is then replaced. A plastic template is placed on the cover to make a round mold. The rehydrated medium with nutrient will support the growth of microorganisms after suitable incubation time and temperature. The colonies can be counted directly in the unit. The unit is about the size and thickness of a plastic credit card, thus providing great savings of space in storage and incubation. Petrifilm units have been developed for total bacterial count, coliform count, fecal coliform count, and yeast and mold counts. Petrifilm has been successfully used to count organisms from milk and meat (Ginn et al., 1986; Smith et al., 1986). Fung and associates

Table 1.2 Comparison of the Standard Plate Count Method and the Petrifilm Method for Viable Cell Counts of Various Seafood

		Colony-forming units/g	
Seafood	Sample	Standard Plate Count	Petrifilm SM
Shrimp	1	2.1×10^4	0.9×10^4
	2	9.3×10^3	6.8×10^3
	3	1.5×10^4	1.2×10^4
	4	4.4×10^4	2.0×10^4
	5	2.0×10^4	0.9×10^4
Perch	1	1.4×10^3	1.2×10^3
	2	5.4×10^2	3.7×10^2
	3	8.0×10^2	4.8×10^2
	4	5.0×10^2	3.2×10^2
	5	1.0×10^3	1.2×10^3
Cod	1	1.2×10^2	1.1×10^2
	2	2.2×10^2	2.4×10^2
	3	1.0×10^4	1.0×10^4
	4	2.1×10^4	2.3×10^4
	5	1.2×10^4	1.2×10^4
Whiting	1	3.0×10^2	3.1×10^2
	2	5.8×10^2	5.6×10^2
	3	1.2×10^3	1.1×10^3
	4	5.4×10^2	5.2×10^2
	5	8.8×10^2	8.4×10^2

Source: Data from Fung et al. (1987).
Samples were massaged in a Stomacher for 1 minute in sterile diluent. Viable cell counts were made according to standard method. Incubation time was 48 hours at 32°C. All samples were done in duplicate. Correlation coefficient between the two methods is $r = 0.99$.

(1987) obtained a 0.99 correlation coefficient comparing Petrifilm SM with the conventional viable cell count method for seafood analysis (shrimp, perch, cod, whiting) of mesophiles (Table 1.2). Recently, in conjuction with immunological techniques, the Petrifilm system can help detect *E. coli* 0157:H7.

Redigel system (RCR Scientific, Inc., Goshen, IN) is another convenient viable cell count system. This system consists of sterile nutrients with a pectin gel in a tube. The tube is ready to be used any time and no heat is needed to "melt" agar. A 1 ml food sample is first pipetted into the tube. After mixing, the entire content is poured into a special Petri dish previously coated with calcium. When liquid comes in contact with the calcium, a Ca-pectate gel is formed and the complex swells to resemble conventional agar. After an appropriate incubation time and temperature, the colonies can be counted exactly like conventional standard plate count method. Fung and Chain (1991) compared 17 different foods (pasteurized milk, raw milk, cheddar cheese, chocolate chips, rice, wheat germ, corn meal, whole wheat, flour, peanuts, pecans, ground beef, chicken, ground black pepper, thyme, broccoli, mushrooms, and turkey pot pie; 20 samples each) and obtained a correlation of 0.964. The system also has tubes for coliform, yeast, mold, and other organisms.

The four aforementioned methods have potential as an alternative to the conventional agar pour plate method. Chain and Fung (1991) made a comprehensive analysis of all four methods against the conventional method on seven different foods (skinless chicken breast, fresh ground beef, fresh ground pork, packaged whole shelled pecans, raw milk, thyme, and whole wheat flour; 20 samples each) and showed that the new systems and the conventional method were highly comparable and exhibited a high degree of accuracy and agreement ($r = 0.95+$). It should be noted that these newer methods need some training and experiences before satisfactory results can be obtained consistently. They are good methods if operated carefully. In the same paper, Chain and Fung (1991) also determined that for viable cell counts the alternative methods are less expensive than the conventional method.

In the direct epifluorescent filter technique (DEFT) method, the liquified sample is first passed through a filter that retains the microorganisms, the filter is then stained with acridine orange, and the slide observed under ultraviolet microscopy. "Live" cells usually stain orange-red, orange-yellow, or orange-brown, whereas "dead" cells fluoresce green. The slides can be read by the eye, or by a semiautomated counting system marketed by Bio-Foss. A "viable cell count" can be made in less than 1 hour. With the use of an image analyzer, an operator can count 50 DEFT slides per hour (Pettipher, 1986). This method has been used satisfactorily for counting cells in milk and other food samples (Pettipher, 1989; Pettipher and Rodrigues, 1980, 1982; Pettipher et al., 1980). Recently, the Nordic countries have begun using this method for quality assurance of ground beef.

These alternative methods for viable cell counts of microorganisms in foods provide an intermediate step toward replacing the cumbersome conventional method. These methods need to be tested in more laboratories with more classes of foods before they can completely replace the time-honored standard plate count method.

1.4 New Methods for Estimation of Microbial Populations and Biomass

Many methods have been developed and tested in recent years to estimate the total number of microorganisms by parameters other than the viable colony count. For a new method to be acceptable, it must have some direct correlation with the total viable cell count. Thus, standard curves correlating parameters such as adenosine triphosphate (ATP) level, detection time of electrical impedance or conductance, generation of heat, radioactive CO_2, and so forth with viable cell counts of the same sample series must be made. In general, the larger the number of viable cells in the sample, the shorter the detection time of these systems. A scattergram is then plotted and used for further comparison of unknown samples. The assumption is that as the number of microorganisms increase in the sample, these physical, biophysical, and biochemical events will also increase accordingly.

Theoretically, these methods can detect as low as one viable cell in the sample if the incubation period is long enough (days or weeks). On the practical side, usually the limit is 10^{4-5} cells/ml. When a sample has 10^6–10^7 organisms/ml, detection can be achieved in about 4–6 hours.

All living things utilize ATP. In the presence of a firefly enzyme system (luciferase and luciferin system), oxygen, and magnesium ions, ATP will facilitate the reaction to generate light. The amount of light generated by this reaction is proportional to the amount of ATP in the sample; thus, the light units can be used to estimate the biomass of cells in a sample. The light emitted by this process can be monitored by a variety of fluorimeters. These procedures can be automated for handling of large numbers of samples. Some of the instruments can detect as little as 10^2–10^3 fg. The amount of ATP in one colony-forming unit has been reported as 0.47 fg with a range of 0.22–1.03 fg.

Using this principle, many researchers have tested the efficacy of using ATP to estimate microbial cells in foods and beverages. Littel and co-workers (1986) indicated that the ATP procedure was able to predict bacterial levels within $0.5 \log_{10}$ of the actual count for beef and chicken samples. Minimum sensitivity is 5×10^4 colony-forming units/g of meat sample. Ward and colleagues (1986) also found positive correlation between the ATP method and the conventional method in evaluating fish samples.

Lumac (Landgraaf, the Netherlands) markets several models of ATP instruments and provides customers with test kits with all necessary reagents, such as a fruit juice kit, hygiene monitoring kit, and so forth. The reagents are injected into the instrument automatically and readout is reported as relative light units (RLUs). By knowing the number of microorganisms responsible for generating known RLUs, the number of microorganisms in the food sample can be estimated. In some food systems, such as wine, the occurrence of any living matter is undesirable; thus, monitoring of ATP can be a useful tool for quality assurance in the winery. Recently, much interest has been expressed by companies such as BioTrace (Plainsboro, NJ) and Lumac in using ATP estimation not for total viable numbers but as a sanitation check.

A new procedure named Enliten (Promega) utilizes centrifugation to concentrate cells from milk and then lyse the cells for ATP determination. It can provide estimation of bacteria in milk in about 1 hour.

As microorganisms grow and metabolize nutrients, large molecules change to smaller molecules in a liquid system and cause a change in electrical conductivity and resistance in the liquid as well as at the interphase of electrodes. By measuring the changes in electrical impedance, capacitance, and conductance, scientists can estimate the number of microorganisms in the liquid because the larger the number of microorganisms in the fluid, the faster the change in these parameters which can be measured by sensitive instruments. A detailed analysis on the subject of impedance, capacitance, and conductance in relation to food microbiology has been made by Eden and Eden (1984).

The Bactometer (bioMerieur Vitek, Inc., Hazelwood, MO) is an instrument designed to measure impedance changes in foods. Samples are placed in the wells of a 16-well module. After the module is completely or partially filled, it is plugged into the incubator unit to start the monitoring sequence. At first, there is a stabilization period for the instrument to adjust to the module, then a base line is established. As the microorganisms metabolize the substrates and reach a critical number (10^{5-6} cells/ml), change in impedance increases sharply, and the monitor screen shows a slope similar to the log phase of a growth curve. The point at which the change in impedance begins is the *detection time*, and this is measured in hours from the start of the experiment. The detection time is inversely proportional to the number of microorganisms in the sample. By knowing the number of microorganisms per milliliter in a series of liquid samples and the detection time of each sample, a standard curve can be established. The cutoff points to monitor certain specifications of the food products can be decided from this curve. For example, if meat having 10^6 organisms/g when applied to the instrument results in a detection time of 5 hours, then 5 hours can be determined as a cutoff point for an indicator that meat samples have fewer or more than 10^6 organisms/g. A similar meat sample with a detection time of 3 hours would indicate that the meat has more than 10^6 organisms/g. Conversely, a similar meat with a detection time of 7 hours would indicate that the meat has fewer than 10^6 organisms/g. In conjunction with food laboratories, the instrument company helps to establish "spec" for a particular food. For example, if a food is "within spec," the screen will give a "green" signal. When a food is "out of spec," the screen will give a "red" signal. A "yellow" signal indicates a food has questionable quality.

Impedance methods have been used to estimate bacteria in milk, dairy products, meats, and other foods (Bishop and White, 1985; Eden and Eden, 1984; Waes and Bossuyt, 1984; Zindulis, 1984). Of particular interest is the application of this method for determining the shelf life potential of pasteurized whole milk by Bishop and co-workers (1984).

The Malthus system (Crawley, UK) works by measuring the conductance of the fluid as the organisms grow in the system. It also generates a conductance curve similar to the impedance curve of the Bactometer, and it also uses detection time in monitoring the density of the microorganisms in the food. The major difference between the two systems, besides the scientific principle (impedance versus conductance), is the incubation units. In the Bactometer system, the size of the well (about 2 ml in capacity) in the 16-well module is fixed. No modification is possible because the module is designed to fit into the incubator chamber. The unit (with two separate chambers) can be operated in two different temperatures at the same time. The temperature is controlled by heating air. The Malthus system, however, allows analysts to choose three sizes ranging from 2 to 100 ml samples, depending on the sample involved. The temperature unit is controlled by heating water. Another important difference is that the modules of the Bactometer are disposable, whereas the jars, tubes, and electrodes of the Malthus system are autoclavable and reusable. In terms of performance, the systems are equivalent in sensitivity and detection time.

The Malthus system has been used for microbial monitoring of brewing liquids (Evans, 1985), milk (Visser and de Groote, 1984), and hygiene monitoring (McMurdo and Whyward, 1984). Gibson and colleagues (Gibson and Hobbs, 1987; Gibson and Ogden, 1980; Ogden, 1986) have done a considerable amount of work using the Malthus system to study seafood microbiology.

Besides estimating viable cells in foods, both the Bactometer and the Malthus systems can detect specific organisms by the use of selective and differential liquid media. The Malthus system was approved recently for the detection of *Salmonella*. New developments of these two systems are constantly being made. In fact, a tube system has been developed for the Malthus system to detect CO_2 production by yeast using indirect conductance measurements. Disposal units in the system have also been introduced.

A new instrument called the "Omnispec bioactivity monitor system" (Wescor, Inc., Logan, UT) is a tristimulus reflectance colorimeter that monitors dye pigmentation changes mediated by microbial activity. Dyes can be used that produce color changes as a result of pH changes, changes in the redox potential of the medium, or the presence of compounds with free amino groups. Samples are placed in microtiter wells or other types of containers and are scanned by an automated light source with computer interference during the growth stages (0–24 hours). The change of color of hue (a*, b*, L) can be monitored similar to an impedance curve or conductance curve. Manninen and Fung (1992c) evaluated this system in a study of pure cultures of *Lactobacillus monocytogenes* and food samples, and found high correlation coefficients (r) of 0.90–0.99 for pure bacterial cultures and 0.82 for minced beef between the colony counts predicted by the colorimetric technique and the results of the traditional plate count method. They also showed that detection times for bacterial cultures such as *Enterobacter aerogenes*, *E. coli*, *Haffnia alvei*, and several strains of *L. monocytogenes* were substantially (2–24 hours) shorter using the instrument than using the traditional method, and concluded that the colorimetric detection technique employed by the Omnispec system simplifies the analyses, saves labor and materials, and provides a high sampling capacity. Tuitemwong (1993) recently completed an extensive study using Omnispec 4000 to monitor growth responses of food pathogens in the presence or absence of membrane bound enzymes. This instrument is highly efficient in large-scale studies of microbial interaction with different compounds in liquid and food.

The catalase test is another rapid method for estimation of microbial populations in certain foods. Catalase is a very reactive enzyme. Microorganisms can be divided into catalase-positive and catalase-negative. Both groups are important in food microbiology; however, under certain food-storage conditions, a certain group predominates. Most perishable foods (commercial as well as domestic) are cold stored under aerobic conditions. The organisms causing spoilage of these foods are psychrotrophs. The predominated psychrotrophic bacteria are *Pseudomonas* spp., which are strongly catalase-positive. Other important psychrotrophs such as *Micrococcus*, *Staphylococcus*, and a variety of enterics are also catalase-positive. Thus, one can make use of the presence of catalase to estimate the bacterial population. Catalase activity can be deleted by a simple capillary tube method (Fung, 1985).

Another way to use the catalase test is as an index of cleanliness of meat-processing areas. Preliminary data from the author's laboratory indicate that a simple swab-catalase test can determine the degree of cleanliness after cleanup of a meat-processing plant. A moist swab was applied to a 2 × 2 in. area and then the swab was placed in 2 ml of 3% H_2O_2 in a test tube. The amount of bubbles generated in the tube can be scaled as 0, 1+, 2+, 3+, 4+, and 5+ depending on the activity of sample. A bacterial count of the immediate adjacent area was also taken. The results showed that prior to slaughter, the room had very low catalase activity, which indicated that bacteria as well as blood or other particles that may carry catalase activity were present in low number. During the slaughter operation, the catalase activity was high because of meat particles, blood, and bacteria in the environment. After proper cleanup, the catalase activity in the area fell to the original level. Bacterial counts did not directly correlate with the catalase activity. This is to be expected because the swab picks up material other than bacteria that can generate bubbles in the presence of H_2O_2. Further work is in process to ascertain the value of this simple test for environmental sanitation monitoring.

Another exciting development of the catalase test is to use it to monitor end point heating temperature of food. Ang and associates (1993) recently showed that heating poultry meat to 71°C (a legal requirement for these products) will destroy both bacterial and animal catalase. The test is 99% accurate and is simple and inexpensive to perform. Similar tests can be developed for beef, fish, pork meat, and so on. The key to success of this test is to separate bacterial catalase from catalaselike enzyme from nonbacterial source. However, above 71°C both bacterial catalase and other catalaselike enzymes will be destroyed.

1.5 Miniaturized Microbiological Techniques

Identification of microorganisms is an important part of quality assurance and control programs in the food industry. In the past 25 years, the author has developed many miniaturized methods to reduce the volume of reagents and media (from 5 to 10 ml to about 0.2 ml) for microbiological testing in microtiter plates. The basic components of the miniaturized system are the microtiter plates for test cultures, a multiple inoculation device, and containers to house solid media (large Petri dishes) and liquid media (another series of microtiter plates). The procedure involves placing liquid cultures (pure cultures) to be studied into sterile wells of a microtiter plate to form a master plate. Each microtiter plate can hold up to 96 different cultures, 48 duplicate cultures, or varius combinations as desired. The cultures are then transferred by a sterile multipoint inoculator (96 needles protruding from a template) to solid or liquid media. Sterilization of the inoculator is by alcohol flaming. Each transfer represents 96 separate inoculations in the conventional method. After incubation at an appropriate temperature, the growth of cultures on solid media or liquid media can be observed and recorded, and the data can be analyzed. These miniaturized procedures save considerable time in operation, effort in manipulation, materials, labor, and space. These methods are ideal for studying large numbers of isolates or for research involving challenging large numbers of microbes against a host of test compounds.

The miniaturized methods have been used to study large numbers of isolates from foods (Fung and Hartman, 1975; Lee et al., 1982, 1985) and to develop bacteriological media and procedures (Chen and Fung, 1991). Miniaturized methods for studying food yeast were also developed in the author's laboratory (Fung and Liang, 1989; Lin and Fung, 1985, 1987). Currently, these miniaturized methods are being used to study food mycology (Hart and Fung, 1990, Hart et al., 1991).

On the commercial side, many diagnostic kits to identify microorganisms have been developed and marketed since the 1970s. Currently, API, Enterotube, R/B, Minitek, MicroID, and IDS are available. Most of these systems were first developed for the identification of enterics (*E. coli*, *Salmonella*, *Shigella*, *Proteus*, *Enterobacter* spp., etc.). Later, many of the companies expanded their capacity to identify nonfermentor, anaerobes, gram-positive organisms, and even yeasts and molds. Most of the early comparative analyses centered on evaluation of these kits for clinical specimens. Cox et al. (1977) and

Fung and Cox (1981) studied these systems from the standpoint of food microbiology and concluded that these systems generally provide 90–95% accuracy when compared with conventional methods. Comparative analysis of diagnostic kits and selection criteria for miniaturized systems were made by Fung et al. (1984) and Cox et al. (1984). They concluded that these miniaturized systems are accurate, efficient, labor saving, space saving, and cheaper than the conventional procedure. Their usefulness in clinical and food microbiological laboratories will continue to be important.

One recent development in miniaturized microbiology is the Biolog system (Hayward, CA). A pure culture in liquid form is placed in a mitrotiter plate which has 95 substrate wells and a control well. After growth, the pattern of growth in the wells can be electronically read. Identification is made by comparing the growth pattern with patterns in the database. Hundreds of bacterial genera and species can be identified by this system.

1.6 New and Novel Techniques

More sophisticated instruments have been developed to identify isolates from clinical specimens such as Sensititre (Radiometer Amer, Westlake, OH). One of the most automated systems for the identification of isolates (clinical and foods) is the Vitek system. This system depends on the growth of target organisms in specially designed media housed in tiny chambers in a plastic "card." The card is then inserted into the incubation chamber. The instrument periodically scans the wells of the cards and sends information to the computer, which then matches the database and identifies the unknown cultures in the cards. The system is entirely automated and computerized and provides hard copies for record keeping. Most evaluations of the usefulness of the Vitek system had previously used clinical specimens. Bailey and co-workers (1985) were successful in utilizing the Vitek system to identify Enterobacteriaceae from foods. The system is capable of identifying enterics, yeast, *Bacillus*, selected gram-positive pathogens, and other organisms.

Concepts and applications of miniaturized kits, immunoassays, and DNA hydribization for recognition and identification of foodborne bacteria are reviewed by Cox et al. (1987). The DNA probe (Genetrak, Framingham, MA) is a sensitive method to detect pathogens such as *Campylobacter*, *Salmonella*, *Listeria*, and *E. coli*. At first, the system utilized radioactive compounds for assay. The second generation of probes uses enzymatic reactions to detect the presence of pathogens. Another major change in this area is the development of probes to detect target RNA. In a cell there is only 1 copy of DNA; however, there may be 1,000–10,000 copies of ribosomal RNA. Thus the new generation of probes is designed to probe target RNA. Currently, kits are available for *Salmonella*, *Listeria*, *Campylobacter*, *Yersinia*, and the like. As the need arises more organisms will be added to the list. A semiautomated robotic system is currently under development for use in DNA probe systems.

The enzyme-linked immunosorbent assay (ELISA) systems method commercialized by Organon Teknika (Durham, NC) utilizes two monoclonal antibodies specific for *Salmonella* detection. In a comparative study involving 1,289 samples, Eckner and associates (1987) found that there was no significant difference between the conventional method and the ELISA method for food samples except cake mix and raw shrimp. Another ELISA system, the Tecra system (International BioProducts, Redmond, WA), was developed in Australia which uses polyclonal antibodies to detect *Salmonella*. These methods have also been used to detect *Listeria* and *E. coli*. Many companes are providing a host of monoclonal and polyclonal antibodies for a variety of diagnostic tests, some including food pathogens (Fung et al., 1988). Automated systems for ELISA tests are now being developed and marketed.

The VIDAS™ (Vitek immunodiagnostic assay system), is a multiparametric immunoanalysis system. The system utilizes the enzyme-linked fluorescent immunoassay (ELFA) method. According to the manufacturer, "The end result of the test protocol is a fluorescent product and the VIDAS reader utilizes a special optical scanner that measures the degree of fluorescence. From the moment the solid phase receptacles and the reagent strips are placed in the instrument, the VIDAS is fully automated." This is a revolutionary development because one of the drawbacks of the ELISA test is the many steps necessary for adding reagents and washing test samples. In the VIDAS system, all intermediate steps are automated.

One of the newest developments in rapid methods is polymerase chain reaction (PCR) technology (McNamara, 1993). Theoretically, with this technology, one piece of target DNA can be amplified to many millions of copies in several hours, thus enabling rapid detection of target microorganisms. For a sucessful technique, the sequence of two known regions of the target DNA must be recognized and two oligonucleotides must serve as "primers" for a series of polymerization reactions that will result in millions of copies of the DNA. Typical PCR protocol involves denaturation of DNA at 96°C (30 seconds to several minutes), primer annealing at 55°C (30 seconds), and primer extension at 72°C (1.5 minutes). The reaction components include target DNA, bases, DNA polymerase (TAQ or other thermo-stable enzymes), Mg^{++} or Mn^{++}, primers, buffer (pH 8.0–8.3), and mineral oil. The PCR products can be detected by gel electrophoresis, Southern blot, dot-plot, or ELISA tests. Automation and improvements of the entire system are constantly being made, especially by Perkin Elmer Cetus (Norwalk, CT). For example, in the most recent model of the thermocycler, mineral oil is no longer needed due to the design of new tube assemblage in the instrument. This highly sophisticated procedure started in research laboratories for molecular biology and is now slowly finding its way to applied microbiology. Ferreira and co-workers (1992) used PCR for improved assay for identification of Type A *Clostridium botulinum*. Certainly, PCR will be used in all areas of applied microbiology, including food microbiology, in the near future.

Motility enrichment is a very useful concept in rapid isolation and identification of food pathogens. Fung and Kraft (1970) described a motility flask system for rapid detection and isolation of *Salmonella* spp. from mixed cultures and poultry products. The system involves a flask with a side arm that contains several agar layers. Lactose broth is placed in the flask, and then a sample (with or without salmonellae) is inoculated into the lactose broth. When salmonellae are present, they swim through the first level of agar, which contains selenite cysteine and sodium lauryl sulfite which inhibit other organisms but allow salmonellae to pass through. Once salmonellae pass through the first layer, they can grow and metabolize compounds in the second and third layers. The color changes in the second and third agar layers verify that salmonellae were in the sample that was put into the lactose browth. Further serological tests can then be performed to confirm the presence of *Salmonella*.

In more recent years, a commercial system called *Salmonella* 1-2 test (BioControl, Bothwell, WA) was developed that utilizes motility as a form of selection. The food sample is first preenriched for 24 hours in lactose broth, and then 0.1 ml is inoculated into one of the chambers in an L-shaped system. The chamber contains selective enrichment liquid medium. There is also a small hole connecting the liquid chamber with the rest of the system, which has a soft agar through which salmonellae can migrate. An opening on the top of the second chamber allows the analyst to deposit a drop of polyvalent anti-H antibody. If the sample contains salmonellae from the lower side of the L unit, salmonella will migrate through the hole and up the agar column. Simultaneously, the antibody against flagella of salmonellae will move downward by gravity. When the antibody meets the salmonellae they will form a visible "immunoband." The presence of an immunoband in this system is a positive test for *Salmonella* spp. The system has gained popularity because of its simplicity.

"Oxyrase" (Mansfield, OH), a membrane fraction of *E. coli*, was found in the author's laboratory to stimulate the growth of a large number of important facultative anaerobic food pathogens. In the presence of a hydrogen donor such as lactate, "Oxyrase" can convert O_2 to H_2O, thereby reducing the oxygen tension of the medium and creating anaerobic conditions that favor the growth of facultative anaerobic organisms. In a medium containing 0.1 units/ml of the enzyme, the growth of *Listeria monocytogenes*, *E. coli* 0157:H7, *Salmonella typhimurium*, *Streptococcus faecalis*, and *Proteus vulgaris* were greatly enhanced; colony counts were greater by 1–2 log units, depending on the initial count and the strain studied, after incubation in the presence of the enzyme for 5–8 hours at 35–42°C compared with control without oxyrase.

By combining the oxyrase enzyme and a unique "U"-shaped tube, Yu and Fung (1991a, b, 1992) developed an effective method to detect *Listeria monocytogenes* and *Listeria* spp. from laboratory cultures and meat systems. Currently, much work is being done in the author's laboratory to stimulate the growth of *Campylobacter* from foods. Furthermore, membrane fractions from food-grade bacteria such as *Acetobacter* and *Gluconobacteria* have been isolated in the author's laboratory. These membrane fractions can stimulate fermentation as well.

There are many other systems that involve modern biochemistry, chemistry and immunology. For example, one can use protein profiles for microbial "finger painting" (AMBIS system, San Diego, CA), or cell composition as a way to identify bacterial cultures (Hewlett-Packard, Palo Alto, CA).

1.7 Conclusion

This chapter describes a variety of methods that are designed to improve current methods, explore new ideas, and develop new concepts and technologies for the improvement of applied microbiology. Although many of these methods were first developed for clinical microbiology, they are now being used for food microbiology. A new journal named *Journal of Rapid Methods and Automation* was started in 1992 by the author to encourage quick dissemination of information concerning current developments in rapid methods and automation. This field will certainly grow and many food microbiologists will find these new methods very useful in their routine work in the immediate future. In fact, many methods described here are already being used by applied microbiologists nationally and internationally.

Acknowledgment

Contribution No. 95-81-B, Agricultural Experiment Station, Kansas State University, Manhattan, Kansas 66506. This material is based on work supported by the Cooperative State Research Service, U.S. Department of Agriculture, under agreement No. 8890341874511.

References

Adams M.R. and Hope C. F. A., 1989. *Rapid methods in food microbiology.* Elsevier, Amsterdam.

Ang C. Y. W., Liu F., Townsend W. E., Fung D. Y. C., 1993. Determination of end-point temperature of cooked chicken meat by catalase test. *IFT Annual Meeting Technical Program: Book of Abstracts* #844.

AOAC, 1990. *Official methods of analysis of the AOAC (Association of Official Analytical Chemists)*, vols. 1 and 2. Association of Official Analytical Chemists, Arlington, VA.

ALPHA, 1985. *Standard methods for the examination of dairy products.* American Public Health Association, Washington, D.C.

Bailey J. S., Cox N. A., Thomson J. E., Fung D. Y. C., 1985. Identification of Enterobacteriaceae in foods with the automicrobic system. *J. Food Protect.*, **48**, 147.

Bishop J. R. and White C. H., 1985. Estimation of potential shelf-life to cottage cheese utilizing bacterial numbers and metabolites. *J. Food Protect.*, **48**, 663.

Bishop J. R., White C. R., Firstenberg-Eden R., 1984. A rapid impedimetric method for determining the potential shelf-life of pasteurized whole milk. *J. Food Protect.*, **47**, 471.

Chain V. S. and Fung D. Y. C., 1991. Comparison of Redigel, Petrifilm, Spiral Plate System, ISOGRID and standard plate count for the aerobic count on selected foods. *J. Food Protect.*, **54**, 208.

Chein S. P. and Fung D. Y. C., 1991. Acriflavin violet red bile agar for the isolation and enumeration of *Klebsiella pneumoniae*. *Food Microbiol.*, **7**, 73.

Cox N. A., McHan F., Fung D. Y. C., 1977. Commercially available minikits for the identification of Enterobacteriaceae: A review. *J. Food Protect.*, **40**, 866.

Cox N. A., Fung D. Y. C., Goldschmidt M. S., Bailey J. S., Thomson J. E., 1984. Selecting a miniaturized system for identification of *Enterobacteriaceae*. *J. Food Protect.*, **47**, 74.

Cox N. A., Fung D. Y. C., Bailey J. S., Hartman P. A., Vasavada P. C., 1987. Miniaturized kits, immunoassays and DNA hydribization for recognition and identification of foodborne bacteria. *Dairy Food Sanitation*, **7**, 628.

Eckner K. F., Flowers R. S., Robinson B. J., Mattingly J. A., Gabis D. A., Silliker J. A., 1987. Comparison of *Salmonella* Bio-EnzaBead immunoassay method and conventional culture procedure for detection of *Salmonella* in foods. *J. Food Protect.*, **50**, 379.

Eden R. and Eden G., 1984. *Impedance Microbiology*. Letchworth: Research Studies Press.

Emswiler B. S., Pierson C. J., Kotula A. W., 1977. Stomaching vs. blending. *Food Technol.*, **31**, no. 10, 40.

Entis P., 1983. Enumeration of coliforms in nonfat dry milk and canned custard by hydrophobic grid membrane filter method: Collaborative study. *J. Assoc. Off. Anal. Chem.*, **66**, 897.

Entis P., 1984. Enumeration of total coliforms and *Escherichia coli* in foods by hydrophobic grid membrane filters: Collaborative study. *J. Assoc. Off. Anal. Chem.*, **67**, 812.

Entis P., Brodsky M. H., Sharpe A. N., 1982. Effect prefiltration and enzyme treatment on membrane filtration of foods. *J. Food Protect.*, **45**, 8.

Evans H. A. V., 1985. A note on the use of conductivity in brewery microbiology control. *Food Microbiol.*, **2**, 19.

FDA, 1992. *Bacteriological Analytical Manual*, 6th ed. Association of Official Analytical Chemists. Food and Drug Administration, Arlington, VA.

Ferreira J. L., Hamdy M. K., McCay S. G., Baumstark B. R., 1992. *J. Rapid Methods Automation Microbiol.*, **1**, 29.

Fitts R., 1985. Development of a DNA-DNA hybridization test for the presence of *Salmonella* in foods. *Food Technol.*, **39**, no. 3, 95.

Flowers R. S., 1985. Comparison of rapid *Salmonella* screening methods and the conventional culture method. *Food Technol.*, **39**, no. 3, 103.

Fung D. Y. C., 1985. Procedures and methods for one-day analysis of microbial loads in foods. *In:* Habermehl K.-O. (ed). *Rapid Methods and Automation in Microbiology and Immunology*, 656–664, Springer-Verlag, Berlin.

Fung D. Y. C., 1991. Rapid methods and automation for food microbiology. *In*: Fung D. Y. C., Matthews R. F. (eds). Marcel Dekker, Inc., New York.

Fung D. Y. C., 1992. Historical development of rapid methods and automation in microbiology. *J. Rapid Methods Automation Microbiol.*, **1**, no. 1, 1.

Fung D. Y. C. and Chain V. S., 1991. Comparative analysis of Redigel and aerobic plate count methods for viable cell count of selected foods. *Food Microbiol.*, **8**, 299.

Fung D. Y. C. and Cox N. A., 1981. Rapid identification systems in the food industry: Present and future. *J. Food Protect.*, **44**, 877.

Fung D. Y. C. and Hartman P. A., 1975. Miniaturized microbiology techniques for rapid characterization of bacteria. *In*: Heden C. G., Illeni T. (eds). *New Approaches to the Identification of Microorganisms*, 347–370, Wiley, New York.

Fung D. Y. C. and Kraft A. A., 1970. A rapid and simple method for the detection and isolation of *Salmonella* from mixed cultures and poultry products. *Poult. Sci.*, **49**, 46.

Fung D. Y. C and Liang C., 1989. A new fluorescent agar for the isolation of *Candida albicans. Bulletins d'information des Laboratoires des Service Vétérinaires* (France), no. 29/30, 1–2.

Fung D. Y. C. and Matthews R. F., eds., 1991. *Instrumental methods for quality assurance in foods.* Marcel Dekker, Inc. New York.

Fung D. Y. C., Lee C. Y., Kastner C. L., 1980. Adhesive tape method for estimation of microbial load on meat surface. *J. Food Protect.*, **43**, 295.

Fung D. Y. C., Goldschmidt M. C., Cox N. A., 1984. Evaluation of bacterial diagnostic kits and systems at an instructional workshop. *J. Food Protect.*, **47**, 68.

Fung D. Y. C., Hart R. A., Chain, V., 1987. Rapid methods and automated procedures for microbiology evaluation of seafood. *In*: Kramer D. E., Liston J. (eds). *Seafood Quality Determination*, 247–253, Elsevier, Amsterdam.

Fung D. Y. C., Bennet R., Lehleitner G. C., 1988. Rapid diagnosis in bacteriology: Contribution of polyclonal and monoclonal antibodies. *In*: Gaeteau M. M., Henry, J., Siest G. (eds). *Biologie Prospective, le Colloque de Pont-e-Mousson*, 21–26, John Libbey, London.

Fung D. Y. C., Cox N. A., Goldschmidt M. C., Bailey J. S., 1989. Rapid methods and automation: A survey of professional microbiologists. *J. Food Protect.*, **52**, 65.

Ginn R. E., Packard V. S., Fox T. L., 1986. Enumeration of total bacteria and coliforms in milk by dry rehydratable film method: Collaborative study. *J. Assoc. Off. Anal. Chem.*, **69**, 527.

Hart R. A. and Fung D. Y. C., 1990. Evaluation of dye media selective for *Aspergillus* and/or *Penicillium.* In: *Proc. Ann. Mtg. Am. Soc. Microbiol.* Anaheim, CA, May 13–17, 279.

Hartman P. A., Curiale M. S., Sharpe A. N., Swaminathan B., Eden R., Cox N. A., Goldschmidt M. S., Fung D. Y. C., 1992. Rapid methods and automation. *In*: Vanderzant C., Splittstoesser D. (eds). *Compendium of Methods for the Microbiological Examination of Foods.* American Public Health Association, Washington, D.C.

Holt J. G., Krieg N. R., Sneath P. H. A., Staley J. T., Williams S. T., 1994. *Bergey's manual of determinative bacteriology*, 9th ed., Williams Wilkins, Baltimore.

Jay J. M., 1992. *Modern Food Microbiology*, 4th ed., Van Nostrand Reinhold, New York.

Konuma H. A. and Kurata H., 1982. Improved Stomacher 400 bag applicable to the spiral plate system for counting bacteria. *Appl. Environ. Microbiol.*, **44**, 765.

Lee C. Y. and Fung D. Y. C., 1986. Surface sampling techniques for bacteriology. *J. Environ. Health*, **48**, 200.

Lee C. Y., Fung D. Y. C., Kastner C. L., 1982. Computer-assisted identification of bacteria on hot-boned and conventionally processed beef. *J. Food Sci.*, **47**, 363.

Lee C. Y., Fung D. Y. C. Kastner C. L., 1985. Computer-assisted identification of Microflora on hot-boned and conventionally processed beef: Effect of moderate and slow initial chilling rate. *J. Food Sci.*, **50**, 553.

Lin C. C. S. and Fung D. Y. C., 1985. Effect of dyes on growth of food yeast. *J. Food Sci.*, **47**, 770.

Lin C. C. S. and Fung D. Y. C., 1987. Critical review of conventional and rapid methods for yeast identification. CRC Crit. Rev. *In: Microbiol.*, **14** no. 4, 273.

Littel K. J., Pikelis S., Spurgash A., 1986. Bioluminescent ATP assay for rapid estimation of microbial numbers in fresh meat. *J. Food Protect.*, **49**, 18.

McMurdo I. H. and Whyward S., 1984. Suitability of rapid microbiological methods for the hygiene management of spray drier plant. *J. Soc. Dairy Technol.*, **34**, no. 1, 4.

McNamara A. M., 1992. Polymerase chain reaction: Techniques and application. *In*: Fung D. Y. C. (ed). *Handbook for Rapid Methods and Automation in Microbiology Workshop*, July 16–23, 1993, Kansas State University, Manhattan, KS.

Manninen M. T. and Fung D. Y. C., 1992a. Estimation of microbial numbers from pure bacterial cultures and from minced beef samples by reflectance colorimetry with Omnispec 4000. *J. Rapid Methods Automation Microbiol.*, **1** no. 1, 41.

Manninen M. T. and Fung D. Y. C., 1992b. Use of the Gravimetric diluter in microbiological work. *J. Food Protect.*, **55**, 59.

Manninen M. T. and Fung D. Y. C., 1992c. Use of spiral plater and laser colony scanner for enumeration of microorganisms in meat. *J. Rapid Methods Automation Microbiol.*, **1** no. 2, 117.

Manninen M. T., Fung D. Y. C., Hart R. A., 1991. Spiral system and laser colony scanner of enumeration of microorganisms. *J. Food Safety.*, **11**, 177.

Pettipher, G. L., 1986. Review: The direct epifluorescent filter technique. *J. Food Technol.*, **21**, 535.

Pettipher G. L., 1989. The direct epifluorescent filter technique. *J. Food Technol.*, **21**, 535.

Pettipher G. L. and Rodrigues V. M., 1980. Rapid enumeration of bacteria in heat treated milk and milk products using a membrane filtration-epifluorescent microscopy technique. *J. Appl. Bacteriol.*, **50**, 157.

Pettipher G. L. and Rodrigues V. M., 1982. Rapid enumeration of microorganisms in foods by the direct epifluorescent filter technique. *Appl. Environ. Microbiol.*, **44**, 809.

Pettipher G. L., Mansell R., McKinnon C. H., Cousins C. M., 1980. Rapid membrane filter epifluorescent microscopy technique for the direct enumeration of bacteria in raw milk. *J. Appl. Environ. Microbiol.*, **39**, 423.

Pierson M. D. and Stearn N. J., 1986. *Foodborne microorganisms and their toxins: Developing methodology.* Marcel Dekker, New York.

Schalkowsky S., 1986. Plating systems. *In*: Pierson M. D., Stern N. J. (eds). *Foodborne microorganisms and their toxins: Developing methodology*, 16–26, Marcel Dekker, New York.

Sharpe W. V., 1991. Rapid methods: Consideration to adoption by regulatory agencies. *Proc.* 105 *AOAC Ann. Intern. Mtgs., Aug.* 12–15, 1991, Phoenix, AZ.

Sharpe A. W. and Clark D. S. (eds), 1978. *Mechanizing microbiology.* Thomas, Springfield, IL.

Sharpe A. W. and Jackson A. K., 1975. Automation requirements in microbiological quality control of foods. *In*: Heden C. G., Illeni T. (eds). *Automation in microbiology and immunology*, 117–124, Wiley, New York.

Sharpe A. W. and Peterkin P. I., 1988. *Membrane filter food microbiology*, Letchworth, Research Studies Press.

Tuitemwong K., 1993. "Characteristics of food grade membrane bound enzymes and applications in food microbiology and food safety." Ph.D. Diss., Kansas State University Library, Manhattan, KS.

Vanderzant C. and Splittstoesser D. (eds), 1992. *Compendium of Methods for the Microbiological Examination of Foods.* American Public Health Association, Washington, D.C.

Visser I. J. R. and deGroote J., 1984. Prospects for the use of conductivity as an aid in the bacteriological monitoring of pasteurized milk. *Antonin van Leewenhoek.*, **50**, 202.

Waes G. M. and Bossuyt R. G., 1984. Impedance measurements to detect bacteriophage problems in cheddar cheese. *J. Food Protect.*, **47**, 349.

Ward D. R., La Rocco K. A., Hopson D. J., 1986. Adenosine triphosphate bioluminescent assay to enumerate bacterial numbers on fresh fish. *J. Food Protect.*, **49**, 647.

Yu L. S. L. and Fung D. Y. C., 1991a. Effect of oxyrase enzyme in *Listeria monocytogenes* and other facultative anaerobes. *J. Food Safety*, **11**, 163.

Yu L. S. L. and Fung D. Y. C., 1991b. Oxyrase enzyme and motility enrichment Fung–Yu tube for rapid detection of *Listeria monocytogenes* and *Listeria spp. J. Food Safety*, **11**, 149.

Yu L. S. L. and Fung D. Y. C., 1992. Growth kinetics of *Listeria* in the presence of oxyrase enzyme in a broth model system. *J. Rapid Methods Automation Microbiol.*, **1** no. 1, 15.

Zindulis J., 1984. A medium for the impedimetric detection of yeasts in foods. *Food Microbiol.*, **1**, 159.

CHAPTER

2

Basic Principles of Industrial Microbiological Testing and Uses of its Findings

C. M. Bourgeois, J. J. Cleret

The bacteriological quality of a food product comprises two elements:

- The hygienic quality, which determines the risk to the health of the consumer. A product is of unacceptable quality if it contains sufficient numbers of pathogenic microorganisms or toxins to pose a real or even a potential danger to the consumer.
- The commercial quality, which is associated with the existence of or risk of product spoilage. A product is of unacceptable quality if it contains sufficient numbers of microorganisms to decrease product appeal before the date by which it should normally be consumed.

Managing this quality implies good manufacturing, storing, and distribution practices. The industry is particularly preoccupied with good manufacturing practices or "GMPs" which consist essentially of the following:

- imparting to foods an intrinsic protection against microbial proliferation;
- reducing as much as possible the level of contamination of the finished product by judicious choice of starting materials and by constant monitoring of the manufacturing process.

Herein lies the role of industrial microbiological testing.

2.1 Aims and Requirements of Industrial Microbiological Testing

2.1.1 Aims

Generally, the goal of microbiological testing is to guarantee hygienic safety as well as some predetermined level of organoleptic quality insofar as these depend on microorganisms. The difficulties in achieving this goal are due to the complexity of the evaluation procedures required in order to certify the bacteriological quality of products. They are slow to be carried out as well as being costly and of limited safety. The delay in obtaining test results, already a nuisance in regulatory testing, is much more so in the case of industrial testing. The analysis of the finished product represents only a terminal inspection which allows the detection of anomalies and the exclusion of defective products but does not remedy manufacturing mishaps within a useful time frame.

What is required in the food industry is a system that can monitor product quality during manufacturing and ensure early detection of anomalies in order to prevent product defects. In general, quality must be designed into the product at the same time as the product itself is developed, rather than settling for simple testing of this quality. This is achieved by preventive means through monitoring of all inputs (starting materials, adjuncts, processing steps) and by responding quickly to any anomalies detected (Briskey, 1978). The aim of testing is therefore not to monitor the quality of the food but rather to monitor the process by which it is manufactured.

2.1.2 Requirements

The principal requirements of this type of industrial process monitoring are speed and low cost, the former in order to be able to respond quickly to any anomalies and the latter in order to allow sufficient diversity of analyses to provide close surveillance of production without becoming a financial burden.

It is not essential, however, that such internal monitoring be carried out according to official methods. It may even be preferable to switch to less proven newer techniques in order to satisfy the principal requirements, so long as they correlate sufficiently with the conventional methods and provide supervisors with a clear picture of the evolution of their manufacturing practices (Insalata, 1972).

2.2 Implementation of Industrial Testing

Testing is best considered part of a regulating system. Its function is to indicate as early as possible failure or any undesirable tendencies in the production system to allow a quick and simple intervention reversing any undesirable trends in quality-determining events. Each type of test therefore plays the role of a sensor which gathers data characterizing product quality in such a way as to allow comparison to reference values, such comparisons leading to appropriate responses.

2.2.1 Levels of Testing

Limiting testing to the final production stage carries the risk of rejecting the product for undetermined causes, which does not improve the process. Extensive analysis at the final stage with identification of microbial species can sometimes result in the pinpointing of an anomaly—for example, the presence of coliforms in a heated product suggesting subsequent recontamination—but it is time-consuming and prevents prompt responses. It is therefore preferable to implement a regulating loop at each critical level (Figures 2.1 and 2.2).

2.2.2 Selection of Criteria and Parameters to be Measured

In automated industrial testing, where the primary aim is efficiency, the choice of a limited number of sufficiently informative analyses as indicators of quality is imperative.

2.2.2.1 Indicators of Hygienic Quality

Although none of the proposed indicators is well correlated with sanitary risk (Miskimin et al., 1976), those that seem to be the most relevant are as follows:

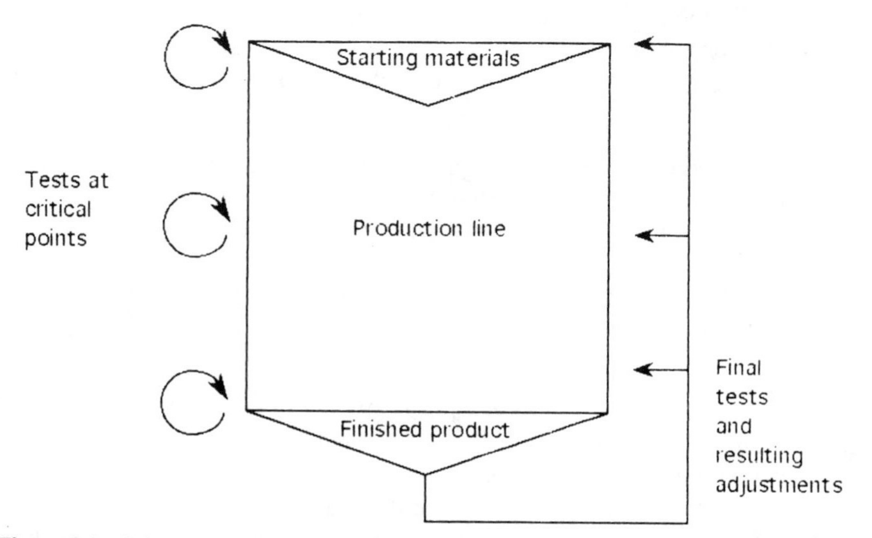

Figure 2.1 Schematic overview of a factory monitoring system (adapted from Corlett, 1974).

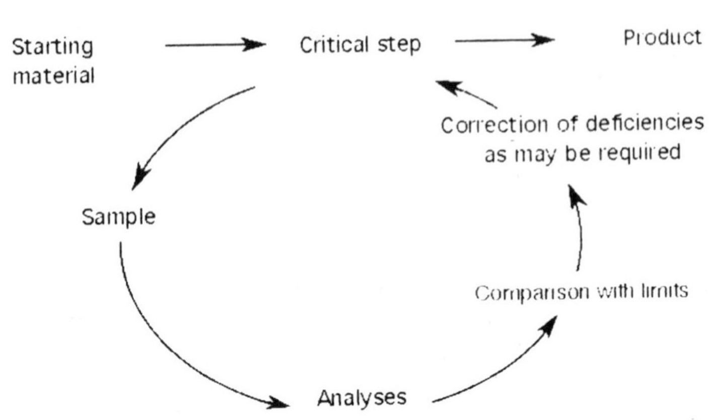

Figure 2.2 Monitoring of critical points in a manufacturing operation (adapted from Corlett, 1974).

- *Total mesophilic microflora.* Generally evaluated by counting colonies visible in agar after 3 days at 30°C, this provides some information about the safety of the product. Practically speaking, even in the absence of a correlation between the number of mesophiles and the number of pathogens, it is generally recognized that pathogens may occur with high total floral counts. In foods suspected of having caused a food intoxication, the mesophilic count is rarely below 10^5/g, the usual value being from 10^6 to 10^7/g (Hobbs and Gilbert, 1974).

Moreover, in foods that have caused a food intoxication, the microorganisms directly responsible are themselves present in very high numbers, i.e., generally at least 10^5/g, except in the case of salmonella, of which the harmful threshold may be as low as one or a few units.

- Enterobacteria. The total number of enterobacteria, the number of coliforms, the number of fecal coliforms, or, more routinely, the number of *Escherichia coli* may be used as indicators. Direct analysis for salmonella may be necessary for some products.
- Staphylococci. If a food is predisposed to the development of these presumably pathogenic bacteria, it is desirable and relatively easy to enumerate them.

According to Miskimin and colleagues (1976), a good correlation exists between total mesophilic flora, the number of coliforms, and the number of *E. coli* such that any of these three indicators may be used at least for monitoring good manufacturing practices if not for the evaluation of sanitary risks. The total mesophilic flora enumeration procedure seems to be the most well adapted to industrial monitoring.

2.2.2.2 Indicators of Organoleptic Quality and Shelf Life

With respect to organoleptic quality and the foreseeable duration of product storage, the total flora also provides some useful information. Product spoilage appears only when the total flora reaches levels of 10^6–10^8/g, with the bulk of the flora being composed of the microbial agent responsible for the spoilage.

In any event, such overall evaluations of microflora must be done under conditions that allow reliable determination of the spoilage causing flora, that is, under conditions similar to those normally experienced by the product (Mossel, 1976). For example, in the case of food stored under refrigeration, the evaluation of psychrotrophic flora is more pertinent. Similarly, for foods that have undergone thermal treatment and that risk spoilage by clostridia, the significant flora is that enumerated following anaerobic incubation after heating for 10 minutes at 80°C.

2.2.2.3 Criteria

Microbiological criteria (Fruin, 1978) include standards, specifications, and recommendations or guidelines. A *standard* is a regulatory precept indicating the maximum tolerable number of microorganisms of a given type, enumerated by a defined method in a specified food. A *specification* is a precept similar in nature to a standard but defined by contractual agreement between a purchaser and a supplier. A *recommendation* is a guide similar in nature to the other two except without any legal or contractual status and therefore is not as binding. It may be a preparative stage in the definition of a standard. A *limit* is a guide similar in nature but defined by the applicable department within the company itself.

2.2.3 Methods of Analysis

Conventional techniques are long and tedious and consequently poorly adapted to present needs. Their progress has been slow since their underlying principle, which consists essentially of evaluating the number of microorganisms in a sample, does not readily lend itself to improvements in terms of acceleration and mechanization. This has led many to wonder whether or not other types of measurements may have at least as much significance. Practically speaking, the likelihood of an initially safe product becoming spoiled or dangerous does not depend only on numbers of bacteria but also on their key characteristics such as growth rate, speed of toxigenesis and proteolysis, as well as on environmental conditions. For this reason, the estimation of the quality of a food may be based on the aptness of the microflora or the time they require to bring about biochemical changes, in particular those changes which render the food unfit for consumption (Sharpe, 1976). The conventional methylene blue method is an example of such a test, although choosing a parameter that directly reflects the undesirable transformation would be even more valuable—for example, enterotoxin production by pathogens, ammonia, acetoin or pyruvate production, modifications of color, or viscosity by spoilage microorganisms.

2.3 Uses of Test Results

The results of analyses must be objectively interpreted, that is, translated into indicators useful for objective comparisons. These indicators must therefore be associable with random variables governed by known laws and capable of offering theoretical support. The quantification resulting from these analyses and measurements may thus form the basis of an objective estimation of the risks implicit in various final decisions downstream from the process being monitored. Once such risk assessments have been made, additional information may be acquired for increasing testing sensitivity and precision. For this purpose, monitoring results will be compared to reference values which are essential elements of the process testing plan.

2.3.1 Choice of Reference Values

Whether for regulatory development or for in-house monitoring, references must meet the following conditions (Corlett, 1974):

Pertinence

- Definition and description of product manufacturing and handling conditions
- Effective implementation of "good manufacturing practices" as defined by specialists

Representivity

- Sampling according to a suitable and totally defined procedure

Reliability

- Analysis of reference samples by standardized methods to be used for later analytical testing

Effectiveness

- Determination of the parameters influencing the distribution of measurements obtained under previous conditions

At this stage, development of the references translates into a number ϕ of colony-forming units (CFUs) per gram, determined statistically and generally representing the number which is exceeded not more than 5 times out of 100 under good manufacturing practice conditions.

In the context of manufacturing and risk thus defined, ϕ depends on the microbiological methods implemented, notably on their precision (reproducibility of the figures obtained), on their accuracy (experimentally determined by deliberate contamination of a known degree), and on their sensitivity (the smallest number detectable under ordinary circumstances).

2.3.2 Choice of Testing Thresholds

Realistic quality control objectives require reconciliation of the statistical constraints of the previous references with economic, commercial, and sanitary constraints. As a whole, these constraints allow the optimization of the chosen limits within the framework of a sampling plan intended for routine monitoring.

The system, most often based on estimation of the total flora, recognizes three zones between which two limits are fixed. Figure 2.3 is applicable to cases where the value of ϕ is lower than the value of M, the threshold for rejection being defined by sanitary regulations (minimum infectious dose) or by commercial quality criteria (maximum tolerable level of physicochemical spoilage). Obviously if ϕ were greater than M, no margin for decision would be available, the product being unacceptable. The limits n and N therefore delimit three zones:

- a green zone corresponding to an acceptable product;
- a yellow zone corresponding to a marginally acceptable product within limits of tolerance. Tolerance may be defined in terms of the maximum tolerable proportion of samples having an index greater than n, for example;

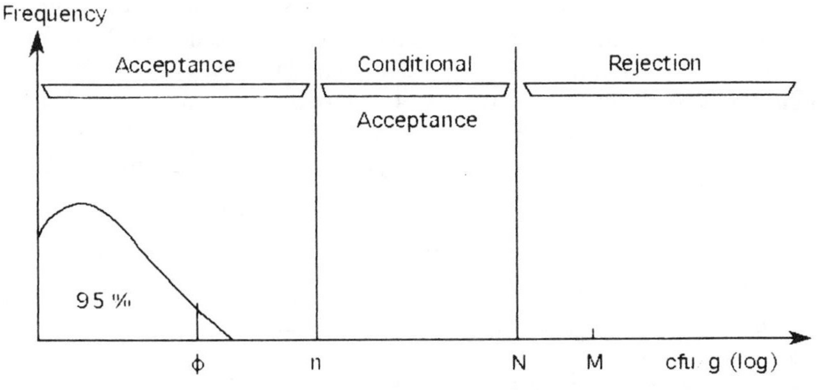

Figure 2.3 Positioning of the limits of a three-zone system (adapted from Mossel, 1974). ϕ = maximum number of colonial units/g for 95% confidence under GMP conditions; n = desired maximum contamination threshold; N = tolerated maximum contamination threshold; M = threshold for rejection.

- a red zone corresponding to product representing a risk sufficient to warrant rejection.

The choice of n and N is established for each product by taking into consideration the following constraints:

- technical and commercial (food properties, manufacturing conditions, product conditioning, distribution, delays prior to consumption);
- statistical (distribution characteristics of the contamination index, coefficient of variation of the analysis method, sampling plan) (Kilsby and Pugh, 1981; Jarvis, 1989);
- sanitary and technical (margin between ϕ and M, risk-multiplying factors).

2.3.3 Data Interpretation

Given sampling and measurement fluctuations, comparison of data with reference parameters requires a concentration of information to be able to make a clear and objective decision to accept or reject the sampled lot. For this purpose, indexes are devised based on the effective distribution of the measurements. For example, in the method of Dyett (1970), the ranking of samples in four zones is used as the basis for the calculation of one of these indexes. The limits n_1, n_2, n_3 are established empirically from the references and in such a way that 75% of the reference samples are assigned numbers lower than n_1, 95% numbers lower than n_2, and 100% numbers lower than n_3. A score is attributed to each sample according to an arithmetic scale applied to the classification. The sum of the scores for a sample or samples from a given period yields the contamination index.

All of the principles based on nonparametric test score sheets may thus be applied (six consecutive index values lower than the reference, six consecutive increases or decreases are significant, seven are very significant, etc.). Tests of frequency analysis establish alarm and action thresholds for the index serving the same purpose as the equivalent parametric based thresholds (Dyett, 1970). The derivation of such an index from empirical data gives it great flexibility and facilitates its adaption to a variety of situations. The limits for the various categories must be reviewed periodically in order for the index to retain its usefulness. This may be implemented by integrating testing results into the

	Category	Score 3 ($n<n_1$)	Score 2 ($n_1<n<n_2$)	Score 1 ($n_1<n<n_2$)	Score 0 ($n<n_3$)	Total (index)
Reference	No. of samples	75	20	5	0	100
	Weighted scores	225	40	5	0	270
Test example	No. of samples	78	18	3	1	100
	Weighted scores	234	36	3	0	273

Figure 2.4 The various index thresholds.

references or by performing new reference sampling. Revision of the references obviously becomes necessary if the product bacteriological quality improves sufficiently to allow the index to approach its maximum value to a significant degree. Such an index upgrades the value of all tests. Aside from and independent of qualitative decisions associated with quality control objectives, the information is used to detect significant process tendencies, thereby allowing early intervention before product characteristics have gone beyond acceptable limits.

2.3.4 Overall Interpretation

Such systems are applicable to different points in the manufacturing process, notably to starting materials, critical manufacturing steps, finished products leaving the production line and after significant time in storage.

Generalizing the same type of index to different points allows their comparison as a function of time, consequently facilitating the rooting out of possible anomalies. Direct superimposing of graphs can thus be obtained, since all of the various distributions have been in principle "reduced" to the same frequency zones. Furthermore, it is not necessary for the sampling plans to be the same at the various monitoring/control points. These systems, therefore, are able to detect significant deviations. The task of centralizing all other information remains, which allows the causes of observed deviations to be determined.

2.3.5 The HACCP System (ICMSF 1988)

The hazard analysis critical control point or HACCP concept is a systematic approach to the identification, evaluation, and control of risks first proposed in 1971 during an American congress on food safely (APHA, 1972). Emphasizing key factors of food safety and quality, this system offers greater cost-effectiveness than the traditional system.

The following list describes the general features of the system as presented in the ICMSF manual. The reader is referred to this excellent reference for the application of the system. The HACCP system includes sequentially the following steps:

1. Identification of risks and evaluation of their severity and probability (risk analysis), for each food production, processing, distribution, and preparation step. Risk refers to any unacceptable contamination, any unacceptable growth and/or survival of microorganisms that may have an impact on the safety and/or degree of deterioration as well as any unacceptable production or persistence of undesirable microbial metabolites (toxins, enzymes, biogenic amines) in the product.
2. Determination of the points (places, practices, procedures, or processes) where risks may be controlled (critical control points or CCPs) by appropriate actions with respect to one or several risk factors. Two types of CCPs may be distinguished: those for which the associated risk may be eliminated and those for which the associated risk may only be minimized.
3. Specification of criteria which indicate whether or not the risk associated with a given operation is being controlled correctly at the level of a given CCP. These criteria are specified limits for parameters of a physical nature (e.g., temperature), chemical nature (e.g., salt concentration), or biological nature (e.g., concentration of microorganisms of a given type).
4. Establishment and implementation of procedures which yield verification that the operation carried out at the level of each CCP is correctly controlled. These procedures must be chosen such that corrective measures which may become necessary can be taken either before or during the operation in question.
5. Implementation of any corrective action that may be necessary when test results indicate faulty control at a given CCP level.
6. Verification, that is, use of additional information acquired in order to ensure that the HACCP system is functioning effectively.

2.3.6 Conclusion

The generalization of these testing systems throughout the food industry should reduce the burden of regulatory testing as well as promote regulatory services focused less on the monitoring of products themselves and more on the supervision of the systems installed to guarantee product quality and integrity.

References

Briskey E. J., 1978. Management's view of quality assurance—Aspiration and requirements. *Food Technol.,* October, 43–45.

Corlett, D. A., 1974. Setting microbiological limits in the food industry. *Food Technol.*, October, 34–39.

Dyett E. J., 1970. Microbiological standards applicable in the food factory. *Chemistry and Industry*, February 7, 198–192.

Fruin J. T., 1978. Microbiological criteria for food. *J. Food Prot.*, **41**, 481–482.

Hobbs B. and Gilbert J., 1974. Microbiological counts in relation to food poisoning in IUFOST. Proceedings of the International Congress in Food Sciences and Technology, vol. 3, 159–169, Madrid.

ICMSF, 1988. Microorganisms in foods. *1: Application of the hazard analysis critical control point (HACCD) system to ensure microbiology safety and quality.* Blackwell Scientific Publications.

Insalata N. F., 1972. The technical microbiological problems in intermediate moisture products. *Food Product Dev.*, August–September, 72–76.

Jarvis B., 1989. Statistical aspects of the microbiological analysis of foods. *In: Progress in industrial microbiology*, vol. 21, Elsevier.

Kilsby D. C. and Pugh M. E., 1981. The relevance of the distribution of micro-organisms within batches of food to the control of microbiological hazards from foods. *J. Appl. Bacteriol.*, **51**, 345–354.

Miskimin D. K., Berkowitz K. A., Solberg M., Riha W. E. Jr, Franke W. C., Buchanan R. L., O'Leary V., 1976. Relationships between indicator organisms and specific pathogens in potentially hazardous foods. *J. Food Sci.*, **41**, 1001–1006.

Mossel A., 1976. *Facts and fallacies on the microbiological quality of foods.* Gordian, **1**, 10–16.

Sharpe A. N., 1976. Viewpoint from food microbiology. *Symposium rapid methods and automation in microbiology.* Cambridge.

CHAPTER

3

Sample Removal, Transport, and Preparation

C. M. Bourgeois, A. Plusquellec

The value of the results obtained from bacteriological analysis is determined by two major factors:

- Since the analysis performed on a sample must provide a correct evaluation of the quality of an entire lot, the sampling procedure must be carried out according to statistically valid sampling criteria.
- The test sample used for the analysis must not have undergone any bacteriological changes relative to the food whose quality it is to reflect. This implies that operations prior to the actual analysis must not have any influence on the microflora either qualitatively or quantitatively. The purpose of this chapter is to examine the conditions that enable these ideal conditions to be attained in practice.

3.1 Sample Collection

Whenever possible, samples collected for analysis should be left in their original packaging under refrigeration until arrival at the analytical laboratory. This is not always possible, however, particularly in the case of starting materials that are normally delivered as bulk shipments (milk in tank trucks or milk cans, vegetables on trucks) or for masses of material too large to be treated as a single sample (beef carcasses). In such cases, the sample must be taken with particular attention to two factors:

- statistical considerations of taking a sample which is representative of the lot being examined;
- bacteriological considerations of preserving the product microflora and especially of not introducing any foreign microorganisms.

We will examine here the case of beef carcasses and of a liquid food product.

3.1.1 The Case of Beef Carcasses

The mass of such an item prevents sampling as a whole, thus limiting the procedure to a small quantity of meat or to a sampling of the surface flora.

The AFNOR reference procedure consists of using a scalpel to remove a thin slice of meat by following the contour of a metal frame delimiting a surface having an area of $25\,cm^2$.

Nevertheless, it is often sufficient to sample microorganisms by detaching them from the carcass using an appropriate procedure. Clark (1965a, b), followed by Baumgart and Kussman (1975) and finally Thran (1979) described processes based on the use of pistollike instruments that project a sterile solution against the meat under pressure sufficient to rinse a defined surface while recovering all of the solution for analysis. These methods have the advantage of being easy to perform on suspended carcasses.

3.1.2 The Case of Liquid Food Products

Sampling is most often carried out by withdrawing a sufficient quantity of liquid into a sterile flask. It is obviously necessary to clean the sample spout, preferably by flaming, and to let run some quantity of liquid to be discarded before taking the sample. It is also important to take reasonable precautions against the risk of contamination from the ambient air.

Various systems have been developed to take samples under suitable conditions, especially in brewing. The opening of a sampling port may be covered by a permanent septum (a rubber disk) which makes it possible to do direct sampling into a sterile syringe using a needle after disinfecting the rubber surface.

It is also possible to sample continuously by installing a filtration assembly in a line branched off a sampling port. All microorganisms present in the filtered volume of liquid during the sampling cycle will be retained on the filtering membrane for analysis.

Finally, in the Portno and Molzahn (1977) system, beer is deposited drop by drop onto a sterile filtering ribbon which scrolls at a constant speed while microorganisms are subsequently enumerated directly on the filter after staining. These latter two systems thus ensure sample withdrawal and preparation at the same time.

3.2 Sample Transportation and Preservation

From the time of sample removal until processing of the sample, all possible precautions must be taken to stabilize both qualitatively and quantitatively the flora present at the moment of sampling. The more successful precautions are as follows:

- as short a delay as possible between sample removal and processing, i.e., quick transportation and brief holding;
- storage at low temperature throughout the duration of transportation and subsequent holding.

Storage at a low temperature above freezing (between 0 and 5°C) stabilizes the microflora quite well, although it will allow psychrotrophic bacteria to multiply slowly.

Storage at temperatures below freezing (frozen) has a definite selective bactericidal effect leading to qualitative and quantitative modifications of the microflora (Lebert and Rosset, 1970). This effect is less pronounced, however, when freezing is very rapid (deep-freezing) and storage is at a sufficiently low temperature (at least −18°C). Furthermore, it is esential that the food remain frozen until analysis, which implies transportation in low-temperature containers (e.g., coolers containing dry ice) and that thawing be done quickly.

Kotula and co-workers (1979) investigated chemical agents capable of ensuring the protection of the microflora during the transportation of frozen chopped steaks. Samples were either frozen at −11°C for 24 hours and then stored at −3°C for 72 hours or frozen in dry ice (−55°−−60°C) and then stored therein for 72 hours, having been mixed with buffered solutions of dimethyl sulfoxide (DMSO) or glycerol prior to freezing. These additives, used at various concentrations, improve the survival of *Clostridium perfringens* and coliforms in both cases. The addition of an equal volume of 10% DMSO ensures some protection of *Escherichia coli* in dry ice flakes, and total protection of *Staphylococcus aureus* and aerobic flora in the −11°C/−3°C system.

It must be noted also that the effect of coldness on bacteria, without being lethal, may bring metabolic stresses (cell injury) which create specific nutritional requirements, particularly sensitivity to inhibitors in selective media, or a lengthening of the lag phase. The effects of this stress are simply added to those of the industrial processing undergone by the food, that is, the action of heat or cold, acidification, dehydration, addition of inhibitory agents, and the like.

In practical terms, these stress effects result in microorganisms that are viable but injured, that is, able to multiply in the food given a sufficient recovery time, but which will not necessarily be detected on the selective medium that usually detects them.

To ensure that injured cells are counted, the sample may be left to stand at ambient temperature for a resuscitation period of a half hour to an hour, or it may be plated initially on a rich medium followed by a selective medium after a few hours of preincubation, or a less selective medium may be used.

3.3 Grinding of Solid Food Samples

Microbial analyses of food samples are almost always performed on liquid suspensions. The blending of solid foods in a sterile diluent provides a liquid suspension that may be subsequently diluted as desired. A quantitative procedure is used to express numerical results in terms of a given mass of food (generally 1 g). The original solutions are generally prepared at a dilution of 1:10 or 1:5. The dilution of the suspension is expressed as in the following ratio:

$$\frac{\text{Weight of food}}{\text{Total volume (diluent + food)}}$$

In the case of liquid foods, the specific gravity is considered to be close to unity and 1 g of food represents 1 ml. A 1:5 solution is thus obtained by adding 50 g of food to 200 ml of diluent. Results expresed per gram of food are then calculated by multiplying by 5, the figure reported for 1 ml of suspension. This is the dilution factor.

In the case of dehydrated foods and of soluble foods, the procedure is different, a certain quantity of food being weighed (in grams) and the final volume adjusted in order to obtain the desired dilution.

3.3.1 The Test Sample

The fraction of actual test material taken for the analysis must be as representative as possible of the lot being analyzed. If a food is composed of several elements, each element will be sampled in a proportion substantially equivalent to that which it represents in the food as a whole. The test sample must also include surface material and central material in the same proportions as they occur in the food, except when recommendations specify otherwise.

In general, for the blending process, it is worthwhile to take a rather large test sample or to test an entire unit of consumption, for example, a slice of meat for microbial analysis.

Removal of the test sample must be performed aseptically and in the immediate proximity of a flame. Depending on the nature of the product, the handler may use sterile tweezers, scalpels, or spatulae for this purpose. These tools must be of stainless steel and sterilizable by flaming or in a Pasteur oven. For each type of food, appropriate apparatuses frequently have been developed which enable the test sample to be taken under defined conditions.

3.3.2 Weighing

It is possible to aseptically weigh an exact quantity of a food into a container already holding a given quantity of diluent, for example, weighing 10 g into a container holding 90 ml of diluent. It is sometimes difficult, however, to maintain asepsis on and around a balance, and even more so given that adjusting to an exact quantity must lengthen the time required for weighing and consequently the risk of contamination.

The simplest technique for obtaining a desired dilution consists of the following:

- taring the sterile container to be used for homogenizing (Erlenmeyer flask or special homogenizing container);
- aseptically introducing some undetermined quantity of food sample (greater than the quantity necessary for the analysis);
- determining by means of a second weighing, the weight in grams of the food thus taken;
- adding by means of a sterile graduated pipette the volume of sterile diluent necessary to obtain the desired dilution; for example, if the original suspension is to be diluted to 1:5, × grams of food product will be diluted with 4 times × ml of sterile diluent.

Equipment (such as Diluflo [Spiral Biotech Inc., Bethesda, MD]) is now available that automates the operations subsequent to the introduction of the test sample into its container.

3.3.3 Diluent

The diluent used to prepare the liquid suspension is generally the same as that used to make the decimal dilutions. It must not induce any qualitative or quantitative variations in either direction in the microbial flora present, that is, it must ensure the survival of all microorganisms but not promote their multiplication. The most commonly used diluents in food microbiology are tryptone salt solution, one-quarter strength Ringer's solution, and phosphate buffer solution.

For cases in which the examination includes tests for *Salmonella*, the liquid suspension may be prepared using buffered peptone water, this being the preparation that is incubated for the preenrichment step.

3.3.4 Homogenizing

Homogenizing is an important step. It must be efficient enough to permit homogenization of a wide variety of foods. It is essential, however, that there be no detrimental effects on the microorganisms present.

Several types of homogenizing methods are used in food microbiology:

- *Grinder-blenders.* The food sample and diluent are introduced ino a special jar equipped with propeller blades. The jar is then placed on its motor shaft connection which drives the propeller at high speeds. The jar must be sterilizable (Pyrex and/or stainless steel).
- *Agitator shafts.* Homogenizing is achieved by introducing a shaft equipped with propeller blades into the food-diluent mixture. These grinders may or may not require the use of a special container. High-speed revolution of propellers may bring about heating of the mixture, which necessitates either cooling or operation for short periods of time.
- *Paddle massagers (Stomacher type).* Unlike the previous two types, this device functions not by shearing but by impact. The food-diluent mixture is sealed into a sterile plastic bag and placed inside the device where it is subjected to repeated beatings with two paddles. The physical massaging of the product releases the microorganisms into the suspension.

The stomacher instrument provides cleanliness and sterility of the homogenizing container as well as saves time, reduces noise, and increases negligible temperatures. Many researchers have indicated that total cell count and coliform counts do not differ significantly from those obtained by conventional homogenizing methods for a variety of food products such as meats and meat products, fish, raw and cooked vegetables, and fruits (Sharpe and Jackson, 1972; Belica, 1973; Tuttlebee, 1975; Lahellac et al., 1976). In comparison with propeller devices, stomachers give slightly lower results for products with a high fat content (Sharpe and Jackson, 1972) and slightly higher for cooked foods (Tuttlebee, 1975). In examining specific organisms such as *Clostridium perfringens, Staphylococcus aureus, Streptococcus faecalis, Escherichia coli,* Sharpe and Harshman (1976) did not find any significant differences between conventional techniques and stomaching except for fat contents exceeding 80%, for which recovery was slightly lower by the stomacher method, a difference correctable by adding Tween 80 to the diluent at a concentration of 1%. Finally, lethal effects (such as heat) associated with conventional homogenizing methods are reduced in the stomacher procedure. Due to these advantages, the stomacher has gained widespread usage.

The consistency and other properties of liquid suspensions for microbiological analysis depend at the same time on the food and the type of homogenizing method used. Regardless of the conditions, homogenizings tend to be nonuniform, resulting in difficulty in pipetting or poor spreading over the plating medium.

Newer techniques such as the so-called spiral method, the HGMF technique, or techniques involving filtration (DEFT) require that a relatively clear suspension be obtained, and the development of new processes therefore tends to be aimed at obtaining suspensions free of food debris. Peterkin and Sharpe (1981) proposed for this purpose pipette tips fitted with a polyethylene mesh of 111 μm.

Another possibility is provided by the use of a modified stomacher bag (Konuma et al., 1982). The bag is separated into two compartments equipped with a Teflon filter. The food and the diluent are introduced into the large compartment and a relatively clear and homogeneous suspension collects in the smaller compartment through the filter during stomaching for subsequent sampling. This process is well suited for sample preparations to be applied to the spiral method.

After homogenizing, the rest of the analytical procedure, which usually involves further dilution and plating onto agar-based media, must be done quickly to avoid the multiplication of microorganisms in the dilution liquid which contains nutrients carried over from the food product.

3.4 Surface Sampling

At times the food product often may be contaminated only on the surface. Also, the surface microflora is sometimes of particular interest for product-testing purposes. In such cases no homogenizing is done. The detachment of the microorganisms from the surface and their collection in a suspension for further evaluation is sufficient. The procedures used for this purpose are similar to those used for the testing of glass or plastic container surfaces. The results obtained may be expressed either in microbial counts/g or cm^2.

Usually, the surface to be analyzed is swabbed and then placed in an appropriate diluent, which often contains a nongermicidal surface active agent. Some type of agitation is also applied. A microbial count is then performed using the suspension. This procedure is applicable to testing contamination levels of vegetables (to be deep-frozen) and meat samples.

In the case of meats, several procedures and apparatuses have been developed to allow sampling under proper conditions:

- *Swabbing.* The microorganisms are obtained by a cotton or alginate swab, then suspended in a diluent for further microbial count, or the swab can be directly streaked onto agar-based medium (Ølgaard, 1977).
- *Membrane contact.* A pre-wetted filtration membrane can be pressed against the surface of the meat to obtain the microorganisms (Silliker et al., 1957).
- *Agar sausage.* A sterile agar surface can be applied directly to a surface to be analyzed (Cate, 1965). The agar is then incubated for colony development.
- By applying to the surface a short metal cylinder, of which the inside surface is scraped using a spatula after having poured a diluent (Williams, 1967).
- *Filter paper.* A filter paper impregnated with a redox indicator (e.g., resazurin) can be used to contain a surface to be analyzed (Tompkin, 1970).

When direct contact with agar sausages or filtration membranes are applied to a meat surface, development of the microbial colonies can be seen on agar or membrane after growth occurs. These colonies may be the outgrowth of a single cell or from members of microcolonies on the surface. The microbes collected on the agar or membrane may be resuspended in an appropriate diluent and further diluted and plated for a more exact enumeration.

Finally, it is possible, without detaching the microbes from the meat, to visualize them in situ by means of an appropriate staining reaction, for example, redox. Bradshaw and colleagues (1961) proposed to do this by pulverizing 5-phenyl-2-(p-2 iodophenyl)-3-(p-nitrophenyl) tetrazolium chloride on the meat surface, incubating the meat in a plastic bag kept at 37°C for 15–30 minutes, and looking for the color reaction. Emswiler and associates (1976) suggested a built-in testing of packaged ground meat by placing indicator disks impregnated with methylene blue on its surface just prior to wrapping. Growth of microbes change the color of the indicator.

3.5 Adjusting Microbial Concentrations to Analytical Needs: Dilution and Concentration

Bacteriological analytical procedures are effective only within a narrow range of bacterial concentrations. Counting of colonies on agar plates, for example, may be done properly only when the number of colony-forming units (CFUs) per plate is between 30 and 300, which for a test sample of 1 ml corresponds to a concentration between 30 and 300 cells or cell aggregates/ml. This range has been recently changed to 25–250/plate. To obtain samples within this narrow range of microbial concentrations, a sample can either be diluted or concentrated.

3.5.1 Modifying the Test Sample

The volume of the test sample may be varied in order to fall into the countable range of 30–300 (or 25–250). Various automated systems are used in the dairy industry for this purpose. The technique of Thompson and co-workers (cited in Posthumus et al., 1974) consists of removing 1 μl of milk with a calibrated loop and rinsing off the 1 μl of milk with 1 ml of water into a Petri dish before pouring the agar-based medium.

To verify that a product meets a specific standard (e.g., 100,000 CFU/ml), a suitable quantity can be used for direct plating. For example, if the standard for a liquid food is 100,000/ml, depositing 1 μl for plating will result in 100 CFU/agar plate, which falls in the readable zone of 30–300 colonies/plate.

These procedures appear best suited for liquid foods. In the case of test suspensions obtained from solid products, food particles may prevent the removal of such small volumes for testing.

3.5.2 Dilution

Usually a tenfold serial dilution is made from solid foods. The first suspension is usually a 1:10 dilution of the food (i.e., 1 g of food into 9 ml of diluent or 5 g of food into 45 ml of diluent, etc.). This 1:10 suspension is further diluted using the same diluent.

Dilutions are made as follows: Diluent is aseptically distributed in 9 ml volumes to a series of tubes. After shaking to homogenize the 1:10 suspension, 1 ml is aseptically transferred (manually or automatically by a machine) to tube no. 1 to make a 1:100 dilution of the sample. After the 1:100 sample dilution, 1 ml is aseptically transferred to tube no. 2 to make a 1:1,000 dilution. This procedure is continued using a fresh pipette each time until the desired dilution is reached (i.e., a 1:1,000,000 dilution, etc.) These operations can be entirely automated (Trinel et al., 1983). Next, from each dilution, two 1 ml samples are placed in Petri dishes and appropriate agar is poured for colony development. In principle, only the plates containing 30–300 colonies are read. The results are then averaged and the count of the original sample can be obtained by multiplying this number by the dilution factor. Plates containing fewer than 30 colonies may be usable when the purpose of the test is simply to verify that the number of bacteria does not exceed a certain acceptable level.

3.5.3 Concentration onto Filtration Media

In some foods such as beer, wine, sweet juice, or water samples, the contamination level is too low to use 1 ml for sampling. In these cases concentration of microbes can be done by filtering hundreds of milliliters of sample through a membrane filter with a diameter of 47 mm. The number of colonies developed on the surface of the filter after incubation of the filter on an appropriate agar can be enumerated and then reported as the number of colonies/100 ml or 1,000 ml of sample.

In this process, the membrane has three functions: (1) to concentrate the microorganisms, (2) to remove them from the medium in which they are suspended, and (3) to provide a culture support.

The first function of the membrane filter (retain and concentrate) is theoretically attainable when the average pore diameter of the membrane is smaller than the diameter of the cells. In practice, however, two phenomena complicate the process. First, if the electrical charge of the filter is the same sign as that of the microorganisms, the microorganisms will not pass through even if the pore size is larger than the diameter of the microorganisms. Second, food particles suspended in the liquid sample can clog the filter.

The pore size of the membrane filter is usually 0.45 μm in diameter; occasionally 0.8 μm or even larger are used depending on the electrostatic properties of the membrane (Zierdt, 1979). Polycarbonate membranes of 2 μm pores have thus been used to enumerate yeast cells in beverages (La Rocco et al., 1985). In addition, new membrane filters having suitable charges have been designed in order to maximize the benefits of this effect.

Clogging of membrane filters by food particles severely limits the filterable volume. This can be remedied by (1) increasing the membrane porosity as much as possible, (2) prefiltering to selectively eliminate larger particles, (3) subjecting the suspension to an enzymatic pretreatment or, (4) filtering using hyperbaric pressures. Processes such as these have been successfully applied to the analysis of a variety of food products (Sharpe and Peterkin, 1979), notably meats (Standard and Wood, 1983) and milk (Bourgeois et al., 1984).

The second function of the membrane filter, which is to separate the cells from their original medium, allows the study of the flora surviving an antiseptic treatment or the enumeration of microorganisms living in a product containing antimicrobial agents. Here one must be aware of the adsorption properties of the membrane filter which may enable it to retain key molecular substances, that is, those from whose influence one specifically wishes to protect the microbial cells for the purposes of the analysis.

Finally, the third function of the membrane filter is to act as a culture support, which enables the colonies to develop by drawing the nutrients they require from the agar-based medium through the filter by capillary action. Substantial differences occur among membranes of different types and brands with respect to the counts yielded and the growth rates of the colonies (Mafart et al., 1985).

Filtration as a concentrating process is essentially limited to liquid foods, although other applications of filtration have found use in various recent analytical techniques (e.g., the DEFT method, hydrophobic grid membranes, etc.) which also deal with solid foods. The filtration of suspensions obtained by the homogenizing of solid foods may be possible.

According to Sharpe and co-workers (1979), suspensions prepared from numerous food products may be filtered and quantities of food as large as 0.5–1 g may be thus deposited onto 10 cm^2 filters. Given that the incorporation into agar of 1 ml of a 1/10 original suspension represents 0.1 g of product, it is immediately apparent that this method gives a five- to tenfold concentration, which may provide interesting possibilities in certain cases. Sharpe et al. (1979) also made an in-depth study of the effects of different factors on filterable quantity, concluding that the application of the hydrophobic grid filter method to solid food analyses (see Chapter 2) does not present any special difficulties. Filtration is thus a preparative procedure offering a wide range of possibilities for the microbiological analysis of foods.

References

Baumgart J. and Kussmann H., 1975. Eine Sprühmethode zur Ermittlung des Oberflächenkeingehaltes bei Schlachtschweinen. *Fleischwirtschaft*, **55**, 1113–1114.

Belica J., 1973. Katalase-Schnelltest zur Hygienekontrolle in Schlacht- und Verarbeitungsbetrieben. *Fleischwirtschaft*, **53**, no. 11, 1600–1601.

Bourgeois C., Le Parc O., Abgrall B., Cleret J. J., 1984. Membrane filtration of milk for counting spores of *Clostridium tyrobutyricum*. *J. Dairy Sci.*, **67**, 2493–2499.

Bradshaw N. J., Dyett E. J., Herschdoerfer S. M., 1961. Rapid bacteriological testing of cooked or cured meats, using a tetrazolium compound. *J. Sci. Food Agric.*, **12**, April, 341–344.

Cate Ten L., 1965. A note on a simple method of bacteriological sampling by means of agar sausage. *J. Appl. Bacteriol.*, **28**, 221–223.

Clark D. S., 1965a. Improvement of spray gun method of estimating bacterial populations on surfaces. *Can. J. Microbiol.*, **11**, 1021–1022.

Clark D. S., 1965b. Method of estimating the bacterial populations on surfaces. *Can. J. Microbiol.*, **11**, 407–413.

Emswiler B. S., Kotva A. W., Chesnut C. M., Young E. P., 1976. Dye reduction method for estimating bacterial counts in ground beef. *Appl. Environ. Microbiol.*, **31**, no. 4, 618–620.

Konuma H., Suzuki A., Kurata H., 1982. Improved Stomacher 400 bag applicable to the Spiral Plate System for counting bacteria. *Appl. Environ. Microbiol.*, **44**, 765–769.

Kotula A. W., Pierson M. D., Emswiler B. S., Guilfoyle J. R., 1979. Effect of sample transport systems on survival of bacteria in ground beef. *Appl. Environ. Microbiol.*, **38**, no. 5, 789–794.

Lahellec C. and Colin P., 1976. Influence du mode de broyage sur les résultats d'analyses bactériologiques. Bulletin d'information, Station Expérimentale d'Aviculture de Ploufragan, **16**, 141–142.

La Rocco K., Galligan P., Littel K., Spurgash A., 1985. A rapid bioluminescent ATP method for determining yeast contamination in a carbonated beverage. *Food Technol.*, **39**, 49–52.

Lebert F. and Rosset R., 1976. Incidence de la congélation sur la microflore des plats cuisinés et des viandes hachées. *Bull. l'Académie Vétérinaire de France*, **49**, 285–301.

Mafart P., Cleret J. M., Bourgeois C., 1985. Cinétique de croissance de *Escherichia coli* et *Saccharomyces cerevisiae* en culture sur membranes filtrantes. *Ann. Inst. Pasteur*, **136A**, 246–259.

Ølgaard K., 1977. Determination of relative bacterial levels on carcasses and meats—a new quick method. *J. Appl. Bacteriol.*, **42**, 321–329.

Peterkin P. I. and Sharpe A. N., 1981. Filtering out food debris before microbiological analysis. *Appl. and Environ. Microbiol.*, **42**–1, 63–65.

Porto A. and Molzahn S., 1977. New methods for the detection of viable microorganisms. *Brew. Digest*, **52**, no. 3, 44–50.

Posthumus G., Klijn C., Giesen T., 1974. A mechanized loop method for total count of bacteria in refrigerated suppliers' milk. *Nederlands Melk en Zuiveltijdschrift*, **28**, no. 2, 79–92.

Sharpe A. and Harshman G., 1976. Recovery of *Clostridium perfringens*, *Staphylococcus aureus* and molds from foods by the Stomacher: Effect of fat content, surfactant concentration and blending time. *Journal de l'Institut Canadien de Science et Technologie Alimentaire*, **9**, no. 1, 30–34.

Sharpe A. and Jackson A., 1972. Stomaching: A new concept in bacteriological sample preparation. *Appl. Microbiol.*, **24**, no. 2, 175–178.

Sharpe A. N., Peterkin P. I., and Dudas I., 1979. Membrane filtration of food suspensions. *Appl. Environ. Microbiol.*, **37**, no. 1, 21–35.

Sharpe A. N. and Peterkin P. I., 1979. Membrane filtration of food suspensions. *Appl. Environ. Microbiol.*, **37**, 21.

Silliker J. H., Andrews H. P., Murphy J. F., 1957. A new non-destructive method for the bacteriological sampling of meats. *Food Technol.*, June, 317–318.

Standard C. and Wood J., 1983. The rapid estimation of microbial contamination of raw meat by measurement of adenosine triphosphate (ATP). *J. Appl. Bacteriol.*, **55**, 429–438.

Thran V., 1979. Mikrobiologische Untersuchung von Oberflächen-ein Probennahmegerät. *Fleischwirtschaft*, **59**, no. 7, 950–953.

Tompkin R. B., 1970. Microbiological control systems. Communication at a series of Plant Sanitation and Consumer Protection Workshops sponsored by the American Meat Institute.

Trinel P. A., Bleuze P., Leroy G., Moschetto Y., LeClerc H., 1983. Automatic diluter for bacteriological samples. *Appl. Environ. Microbiol.*, **45**, 451–455.

Tuttlebee J., 1975. The Stomacher—its use for homogenization in food microbiology. *J. Food Technol.*, **10**, no. 2, 113–122.

Zierdt C. H., 1979. Adherence of bacteria, yeast, blood cells, and latex spheres to large-porosity membrane filters. *Appl. Environ. Microbiol.*, **38**, no. 6, 1166–1172.

CHAPTER

4

Rapid Methods of Enumerating Microorganisms

C. M. Bourgeois, R. Malcoste

In all of the techniques mentioned previously the number of microorganisms is evaluated by counting. Evaluation of total microbial counts, without any distinction between those capable of multiplication and those which are dead, is done by means of optical processes such as microscopy. In order to distinguish viable flora, biological methods are employed which take into consideration the multiplication of the microorganisms.

4.1 Microscopic Detection and Enumeration

The minute size of microorganisms makes direct observation possible only with the aid of a microscope. The use of high magnification (× 1,000), however, reduces the area of the field of observation, which therefore reduces the possibilities of detection of the microbes in the specimen. This problem is overcome somewhat by using suitable stainings to make the microorganisms more visible and thereby make the search for them possible at lower magnifications.

The two stainings most commonly used are the simple methylene blue stain and the double Gram stain. The latter offers the advantage of grouping the bacteria into two categories known as *gram-positive* and *gram-negative*, which is the first step in their identification. Processes using fluorescent markers (the DEFT technique) are gaining popularity.

4.1.1 Counting Chambers

Enumerations may be done using special slides (hemacytometer type) which are also known as counting chambers. These are glass slides 2–3 mm thick, with a portion of the upper surface marked into quadrants such that when the slide is covered with an optically planar cover slip, a known volume of microbial suspension is entrapped. The Thomas chamber, for example, has a cavity 0.1 mm in depth and 1 mm^2 in surface area, the bottom of which is divided into 400 squares. It is used in the following manner:

- The microbial suspension is diluted in order to obtain from 5 to 10 microbes per small square, using 10% formol solution to immobilize the microbes if necessary, and possibly incorporating a few drops of methylene blue to aid observation.
- A small volume of the diluted suspension is deposited on the counting grid area, without overflowing, and covered with an optically planar cover slip. The slide is then allowed to sit undisturbed for a moment.
- The slide is examined under microscope with reduced illumination or dark field. Microorganisms are counted in 50–100 squares selected randomly such that the total count is around 500.
- The average per square is taken and multiplied by 4×10^6 as well as by the dilution factor to obtain the number of organisms/ml. It is a good idea to repeat the operation twice and to take the average of the three counts.

This method is applicable to all types of testing but becomes tiresome with bacteria because of their small size and is therefore better suited to organisms which are at least the size of yeasts.

4.1.2 The Breed Method

The Breed method employs slides with a 1 cm^2 marked area on which 0.01 ml of liquid is placed for examination. After drying the liquid sample and staining with methylene blue or by the Gram method, the microorganisms are counted in 30–50 microscopic fields. Once field diameter is known (or deduced from the magnification or measured by means of an objective micrometer), the number of microbes/ml may be calculated using the following formula:

$$N = \frac{n \times S}{s \times v}$$

where:

n is the average number of cells per field
S is the total surface area of the sample, i.e., 1 cm^2
s is the surface area of the field
v is the sample volume, i.e., 0.01 ml

It is noteworthy that if $n = 1$ and $s = 10^{-3}\,cm^2$, N is equal to 10^5, which practically speaking constitutes the detection level of the system.

In spite of its low sensitivity, the Breed method has been applied to dairy testing, infant foods, fish fillets, and other products since it is a very rapid procedure which generally correlates well with the spread plate method. It is also useful for repetitive testing such as on production lines. It should be noted, however, that visual fatigue limits the number of tests that may be performed by the same person in 1 day, unless an image analyzer is used.

4.1.3 Epifluorescent Microscopy (DEFT: direct epifluorescent filter technique)

Various modifications to the Breed method have brought about considerable improvements in performance resulting in a very efficient procedure known as the DEFT method, which has been automated. Among the innovations underlying this evolution are the following:

- The use of fluorochromes for staining, that is, stains which when illuminated with ultraviolet (and therefore invisible) light give off a portion of the absorbed energy in the form of transmitted light at a longer and hence visible wavelength. Of particular interest is the colored appearance of the microorganisms against a black background, enabling them to be detected at moderate magnification, thereby increasing the field area that may be observed in a given time and lowering the practical detection threshold. Various stains are used, among the four tested by Afifi and Muller (1975), that is, primuline, auramine, acridine orange, and acridine yellow; the lattermost provided the best results, although acridine orange is in fact the most widely used.
- The concentration of microorganisms by filtration onto a filtering membrane, allowing direct enumeration, has been applied especially to milk testing (Godberson, 1973; Kynast, 1973). This may be followed by direct staining on the membrane, dehydration and rendering transparent for microscopic examination using the oil-immersion objective. This procedure has been recommended for beer testing among others, using Loeffler blue or fuchsin-methylene blue as stains. If 100 ml of beer is filtered onto a 47 mm (surface area of 10 cm^2) filter, the concentration factor with respect to a 0.01 ml smear over 1 cm^2 is 1,000, and the detection threshold is reduced by a similar factor. This method generates some interfering stains, however. The current trend favors the use of fluorochromes, which implies selection of appropriate membranes. Hobbie and colleagues (1977) have demonstrated that Nucleopore™ polycarbonate membranes are preferable to cellulose filters. Their uniform porosity and even surface allow them to retain all bacteria on the surface whereas about 50% of the cells escape detection on cellulose by penetrating into the depths of the filter. These membranes must nevertheless undergo a pretreatment in order to eliminate autofluorescence.
- The use of the epifluorescence microscope, which projects incident light onto the membrane surface rather than having it cross the membrane from below as is the case with conventional microscopes, significantly improves contrast.

The application of these three innovations has led to the development of the DEFT method, notably by Pettipher and colleagues (Pettipher et al., 1980; Pettipher and Rodrigues, 1981, 1982a, b; Pettipher, 1983; Rodrigues and Kroll, 1985).

Applied to milk, this method consists of a pretreatment with a proteolytic enzyme and a surfactant followed by filtering 2 ml of the milk onto a polycarbonate filtering membrane (25 mm diameter and 0.6 μm pore size) and rinsing, after which the membrane is stained with acridine orange, washed with citric acid, made transparent using isopropyl alcohol, and observed using an oil-immersion lens. Metabolically active bacteria appear orange colored (staining associated with a high RNA content) whereas inactive bacteria are stained green (due to the binding of the stain to DNA). The entire series of steps requires approximately 25 minutes.

The method subsequently underwent various improvements which increased its sensitivity (Rodrigues and Kroll, 1985), and its applications were broadened to include a variety of analyses and products, for example, monitoring of dairy equipment hygiene (McKinnon and Mansel, 1981; Hunter and McCorquodale, 1983), predicting the shelf life of pasteurized milk (Rodrigues and Pettipher, 1984; Kroll and Rodriques, 1986), solid foods after grinding (Pettipher and Rodrigues, 1982b), meat and meat products (Qvist and Jakobsen, 1985; Shaw et al., 1987), enumeration of yeasts in beverages (Rodrigues and Kroll, 1986), and examination of dry foods (Oppong and Snudden, 1987).

Evaluations of the method have generally confirmed its advantages (Pettipher et al., 1983; Abgrall and Bourgeois, 1989), with reduction of time of operation and high correlation with enumeration on solid media, although the numbers counted are generally higher than for plate counts due to the separation of cell aggregates by the pretreatment step. The drawbacks of the method are its relatively low sensitivity and especially the tediousness of the microscopic examination.

In order to lessen the burden of the visual task, efforts have been devoted since 1982 to automating the reading using image analyzers (Pettipher and Rodrigues 1982). These have resulted in the development of efficient systems such as the Bactoscan by Foss Electric, which renders the DEFT method practicable on a routine basis (Jaeggi et al., 1989; Abgrall and Bourgeois, 1989) and makes it one of the most interesting methods to be proposed in recent years.

4.1.4 Microscopic Enumeration after a Multiplication Phase

Epifluorescent microscopy alone, in spite of its remarkable performances, is not able to detect low level contaminations. The difficulty lies with the requirement for high magnification due to the small size of bacteria, which imposes very narrow fields of view for microscopic examination. Under these conditions it becomes necessary to scan a large number of fields, which makes the visual examination very tedious. If the examination is done by optoelectronic means, there is the risk that small numbers of bacteria may be difficult to distinguish from the background noise level.

It is therefore desirable to proceed using biological amplification, that is, allowing the microorganisms to multiply to a level which renders them more accessible to observation. This may be done either on solid supports or in liquid media. There are two different modes of biological amplification, one being to increase the size of the object, the other being to increase the number such that the probability of their presence in a field of view at a specified magnification is increased. In both cases, the result is that lower magnification may be used, thereby widening the field and increasing the surface area scanned.

Culture on solid support prior to microscopic enumeration is the basis of the microcolony method which will be examined in Section 4.2.

Culture in liquid media is usually done using an enrichment medium selectively favoring a specified type of microorganism, therefore increasing its portion of the total population. The main disadvantge of this method is that it is not quantitative. Nevertheless, if for a given species the incubation is carried out under rigorously standardized conditions, there may be some consistent correlation between the number of microbes present at the beginning and at the end of the incubation. The final number may then provide a semiquantitative indication of the initial number. This method has long been used in the brewing industry to search for lactobacilli contaminations in starter cultures.

Enrichment in liquid media is also the first step in searching for various pathogenic or indicator microbes, in particular for *Salmonella.* In the case of *Salmonella,* the enrichment steps are generally followed by plating onto selective media, although detection may also be done microscopically in conjunction with a special marking procedure known as immunofluorescence.

4.1.5 Field of Application and Advantages of Microscopic Methods

Microscopic methods are practically the only methods usable when enumeration of all microorganisms living and dead is required.

For the specific enumeration of viable microorganisms, the major competing methods are culture methods. Microscopic methods have the advantage of avoiding delays for incubation, the amplification essential for the detection of the microorganism being of an optical nature. Their drawbcks are low specificity and, even with the improvements afforded by epifluorescence, a rather low practical sensitivity in the order of 10^6/g (Abgrall and Bourgeois, 1989).

4.2 Estimation of Viable Flora by Culture in or on Solid Media

Methods to detect or enumerate living microorganisms are of more frequent use and greater value in food microbiology than are estimations of total flora. Being of a biological nature, they depend on the development of the microorganisms.

4.2.1 Enumerating in Solid Medium

4.2.1.1 Standard Plate Count Technique

Although standard plate count technique is not a rapid method, the principle is given here for reference. Two plates per dilution are generally poured according to the following protocol:

- Using aseptic technique, 1 ml of each dilution is deposited per plate, paying particular attention to the homogeneity of the diluted suspension (shaking immediately prior to pipetting is normally adequate) and using a new sterile pipette with each dilution. In practice, the pipette used to transfer 1 ml of a given dilution to the next tube in the series may also be used to inoculate its corresponding plates in order to minimize the number of pipettes used.
- Pour into each plate the contents of one tube of sterile plate count agar medium (15 ml) kept liquified at 48°C in a water bath.
- Mix the contents of each plate according to a standardized technique. The recommended technique is as follows: keeping the covered plate flat on the table surface, it is slid in a circular motion over an area about 6 inches in diameter first clockwise six times, then counterclockwise six times, back and forth six times and finally left to right six times, taking care to not slop inoculated medium over the sides of the plate. Allow the plate to cool and then incubate in the inverted position.

The temperature and time of incubation vary according to the product being examined and the type of microbe of which the population is being estimated. Temperatures are generally 22°, 30°, or 37°C and incubation times 24, 48, or 72 hours.

For correct interpretation of results, only those plates which contain between 30 and 300 (or 25 and 250) colonies (surface and embedded) are taken into consideration.

Enumeration may be facilitated by the use of colony counters. The average obtained from the two plates corresponding to each dilution is taken and multiplied by its dilution factor. Only the first two significant digits are retained, 262 for example being reported as 260.

In cases where two consecutive dilutions give numbers beween 30 and 300 (or 25 and 250) colonies, the average of these may be taken provided that the larger number is not greater than twice the smaller number; otherwise the smaller number is used as such.

This process may be enhanced by some degree of automation at two stages, that is, at the inoculating stage and at the counting stage. These possibilities have been exploited to a large degree in the dairy industry.

Sometimes the development of surface colonies is to be avoided, in which case the double agar layer method may be used. The procedure initially follows the standard plating method, but pouring only about 3/4 of the tube of agar-based medium. The remainder is returned to the water bath to be kept liquified and is poured onto the first layer only after the latter has completely cooled and solidified. Enumeration and counting are done the same way as for the standard method.

4.2.1.2 Agar Drop Culture; Microcolonial Enumeration

Sharpe and colleagues (Sharpe and Kilsby, 1971; Sharpe et al., 1972) proposed a miniaturized technique using culture in agar drops. The sample, any food product homogenate, is diluted in liquified agar at 45°C which may be distributed as 0.1 ml droplets on the bottom of a Petri plate by means of a diluter-dispenser, diluting tenfold or a hundredfold if need be. About 20 such droplets may be deposited per plate, for example, five droplets from four successive dilutions. Following incubation, a special projector enlarges the diameter of the droplets to that of a Petri plate, thus allowing easy counting of still very small microcolonies. An apparatus known as the Droplette system conveniently carries out the dilution, droplet dispensing, and counting steps.

The development of aerobic microbes is favored by the high surface-to-volume ratio of the droplets although the method may be adapted to anaerobes (Sharpe et al., 1976). In the case of mesophilic aerobes, enumeration can be done after 24 hours of incubation at 30°C instead of 48–72 hours as is required for the conventional method. This is also the case for staphylococci on Baird-Parker medium, although 48 hours are necessary for anaerobes as for the conventional method.

This method therefore regroups the advantages of considerable acceleration in detection and savings in time (by reducing number of manipulations) and materials (Petri plates, agar-based medium). The increase in sensitivity which allows rapid detection is, however, essentially the result of the optical magnification of the droplets. In other words, the procedure approaches microscopic techniques. The Droplette system does not appear to have met with the widespread acceptance expected.

4.2.2 Enumeration on Solid Medium Surface

4.2.2.1 Spreading on Agar Plates, Surface Counts

In some cases (e.g., for strict aerobes), it is preferable to employ surface counting. For this purpose previously poured and carefully dried plates are used. A known volume of inoculum (e.g., 0.1 ml), measured using a calibrated loop or micropipette, is deposited on the solid medium surface, and spread evenly over the entire surface using a glass rod spreader. Following incubation for the appropriate time, colonies that have developed on the surface are counted. The rules for interpretation are the same as for the preceding method.

This process has been automated for dairy applications by Thompson and co-workers (1960) and numerous variations have been devised since. The factors influencing the performances of these systems have been analyzed by King and Mabbit (1984).

A semiautomated system for surface inoculating in a spiral pattern (spiral plate count method) was proposed by Trotman (1971) and later by Gilchrist and colleagues (1973). In this system, a machine deposits a small sample volume (about 35 μl) onto the surface of a rotating Petri plate, the dispensing tip moving gradually from the center toward the sides of the plate, thus tracing a spiral. The progressive spacing of the cells as the distance from the center increases serves as a dilution, colonies occurring at confluent density at the spiral origin and then spreading out at some peripheral point which varies as a function of the microbial concentration in the sample. Counting of colonies may be done with the naked eye, with the aid of a standard electronic counter equipped with a special grid, with a laser counter, or even with a video camera image analyzer which enumerates the colonies following the spiral beginning at the periphery. This system in its commercial form has been the subject of numerous publications dealing with its application to milk (Donnelly et al., 1976), prepared dishes (Catteau et al., 1981), and cereal products (Bassfeld and Lambert, 1988). This method, having the advantage of avoiding tedious dilutions, greatly facilitates laboratory work. The primary problem of blockage of the dispensing tip by food particles may be overcome using appropriate prefiltration systems (Konuma et al., 1982, see Chapter 3).

4.2.2.2 Drop Count Method

Known by name as the Miles and Misra method, the drop count method consists of depositing small calibrated drops of the material being examined onto a solid medium surface. The colonies that develop in the inoculated zones are counted by means of a magnifying system allowing an appropriate incubation period. The drops are delivered using special Pasteur pipettes giving 50 drops/ml. It is also possible to use unbeveled hypodermic needles delivering droplets of similar size. The plates must be well dried prior to use. At least five droplets of each dilution are dropped onto each plate carefully avoiding splashes. After incubation, a plate showing 10–20 colonies/drop is chosen. These are counted, the average is taken, and the number obtained is multiplied by 50 to transpose to 1 ml and by the dilution which provided the count.

The system proposed by Fung and Kraft (1968) and cited by Fung and Lagrange (1969), stems from the same principle but uses a microtitration system in series. Inoculating is done by depositing droplets of successive dilutions onto a solid and dried agar-based medium surface. The dilution is created using microtitration plates with 96 wells (8 × 12) resulting in the simultaneous dilution of 12 samples. Each well receives 0.225 ml of medium and into each of the 12 wells of the first series, 0.025 ml of the milk to be analyzed is introduced using a calibrated loop, rotating the loop to homogenize the well contents. Subsequent transfer of 0.025 ml of these 1/10 original solutions into the 12 wells of the following series and so on thereafter up to a dilution of 10^{-8} completes the dilution process. Next, two drops of each dilution (0.025 ml drops) are deposited onto solid medium, four to eight drops per plate. Finally, after 15–20 hours of incubation at 32°C, the counting is done under the microscope at a magnification of 10 in the spots containing 10–100 colonies.

Fung and associates (Fung and Lagrange, 1969; Fung et al., 1976), have found this method to be statistically comparable to the conventional technique, as well as multiplying by 12 the efficiency of the technical work. It consumes ten times less product and reduces by one-half to two-thirds the delay in detection. It has not, however, met with widespread acceptance in the workplace.

These latter two techniques are based on the observation of microcolonies, which will be described further in the following section.

4.2.3 Surface Enumeration After Culturing on Filtering Membrane

In order to benefit from the concentrating effect of microorganisms afforded by filtering membranes for liquid foods, the development of the colonies on the membrane itself may be done by incubating it on the appropriate solid medium or even on an absorbent pad soaked in liquid nutrient medium. In the latter case, it is necessary to take precautions against rapid dehydration of the pad by using sealed plates or by sealing ordinary plates using adhesive tape or hydrocarbon film. Colonies are in this case counted on the membrane as they would normally have been on an agar plate, except that unlike for agar plates on which up to 300 colonies may be counted without much risk of confluence, the number of colonies on a 47 mm membrane must not exceed 60–80.

The procedure of Sharpe and Michaud (1978), using hydrophobic grid membrane filters (HGMF), seems to offer some interesting possibilities. This will be described further in a later section along with data handling methods.

Yet another possibility offered by membranes is the early microscopic detection directly on the membrane following a brief incubation to allow the development of microcolonies. The principle is a simple one. Cells are placed under conditions which allow their multiplication and examination takes place at a moment when the colonies, yet imperceptible to the naked eye, are detectable under low magnification, for example, with a comfortably large field surface area and reduced examination time. After a few hours of growth on a nutrient medium, the microcolonies present are given the appropriate treatment to stain the cells and render the membrane transparent, thereby allowing microscopic examination at low magnification.

This process has been applied to testing beer sterility by Richards (1970), using an incubation period of 5 hours. It is possible to detect yeasts in this manner beginning at a contamination level of 500 cells/filter. After 10–20 hours, examination may be carried out under a magnification of only 100, resulting in a substantial improvement of the detection threshold.

This process is also applicable to testing surface contamination of solid foods such as vegetables (Winter et al., 1971) or meat (Baumgart and Niemann, 1974). The surface microbes are simply suspended in a sterile solution which is then filtered into a membrane.

Winter and colleagues (1971) demonstrated the possibility of using this process to test the level of contamination of vegetables to be deep-frozen. They obtained satisfactory results for incubation times of 4–6 hours by staining the cells with Janus Green and examining at a magnification of 80 and also observed a satisfactory correlation with the conventional method. They suggested that their method was useful for the rapid classification of products into three categories: $<50{,}000/g$; $50{,}000-100{,}000/g$; and $>100{,}000/g$.

The examination of fluorescent microcolonies, which further increases the sensitivity of the microcolony method, has been used successfully in brewing by Harrison and Webb (1971), Kirsop and Dolezil (1973), and Haikara and Boije-Backman (1982).

More recently, Rodrigues and Kroll (1988, 1989) have applied the DEFT method to detect various food product contaminants at the microcolony stage after 6 hours of incubation. Their investigations, which dealt with 59 microbial cultures and eight media, led to the selection of three media for the enumeration of coliforms, pseudomonads, and staphylococci in various fresh, frozen, and heat-treated foods. This method, which gives satisfactory results, has the advantage over the direct DEFT method of enumerating only viable microorganisms as well as enumerating a particular type of flora, given the use of specific media. In the case of heat- or cold-treated products, however, the microorganisms having undergone "stress" require a 3–5 hour revitalizing phase before filtration, which leads to a total analysis time of slightly less than 12 hours.

4.2.4 Dip Slides

Dip slides are glass slides coated on each side with a layer of dried agar-based medium, sold ready to use in sterile protective tubes. These are available especially for milk testing.

At the moment of use, the slide is removed from the tube, plunged into the liquid to be analyzed, freed of excess liquid, and replaced in its tube which is then incubated. Following a delay of 16–24 hours at 30°C, the slide on which colonies have appeared is examined in comparison with control slides, thereby providing the estimation of the number of colonies. When the absorption capacity of the slide is known, it is possible to convert the number of colonies observed into a number of cells/ml.

The results obtained by this method have only a weak correlation with those of the plating technique, however (Macmeekin, 1976), and the delay in detection is not significantly reduced. On the other hand, it has the advantages of being simple to implement and being usable outside the laboratory.

Slides for this method are also available for testing of surfaces.

In a recent development, the support for the agar-based medium is not a slide but a film, providing a system that is particularly easy to transport and handle.

A variation of this process uses samplers consisting of a filtering membrane bonded to an absorbent pad. When the sampler is plunged into the sample, the pad absorbs 1 ml of liquid across the filter which retains the microorganisms. During incubation, the microorganisms absorb nutrients from the pad giving rise to colonies that become visible under low magnification after about 24 hours.

The method of counting on impregnated paper strips uses a paper strip impregnated with nutrient medium, dried and sealed in a plastic envelope. At the moment of use, the sterile strip is removed from its envelope, also sterile, and dipped into the liquid to be analyzed from which a predetermined volume of liquid is absorbed (0.1 or 1.0 ml) which also rehydrates the medium. The strip is then reenclosed in its sterile envelope and incubated for 8–15 hours. Following this incubation time, colonies appear as colored spots. The acceleration of the analysis results from the sensitivity of the reagent, and also possibly from the structure of the paper being more favorable to color expression than is the structure of agar.

These simplified techniques have been the subjects of numerous evaluations (Richards, 1979; Silley, 1985; Goldrick et al. 1986). Without going into the results in detail, it may be said in summary that the values obtained are generally lower and more uncertain than those obtained with the conventional Petri plate technique, but that the simplified techniques nevertheless enable testing to be carried out in production areas, and even in poorly equipped environments.

4.2.5 Range of Application and Value of Solid Media Culture Methods

These easily implemented solid media culture methods are based on the postulate that suggests that a single colony arises from a single microbial cell. This is not always the case. Bacteria are not necessarily isolated and homogeneously distributed. On the contrary, they tend to form clumps and the colonies develop from these clumps, of which the size varies from one sample to another and from one bacterial type to another. It is therefore preferable to state the results in terms of CFUs (number of colony forming units)/ml.

Since sample preparation, dilution procedures, agitation, and so forth all influence the size and distribution of the clumps which will each give rise to a single colony, it is difficult to obtain reproducible results.

Given the imperfections of these methods, correct interpretation of the results is imperative in order to not give the numbers obtained more importance than is justifiable by the technical or statistical precision of the method. It should be noted that a great degree of precision is not often necessary. If, for example, an official standard for a given product indicates that it must not contain more than 1,000 bacteria/ml, the exact number that it contains has little importance. The method need only confirm that the product contains either more than or less than 1,000. This example illustrates that it is not always necessary to use very elaborate techniques that provide a precision having no relationship to the requirements of the circumstances and that involve costly and complex implementation. The chosen method will depend on local conditions of personnel, space, materials, and time which may be mobilized in order to obtain a result. It may be preferable at times to analyze a large number of samplings by means of a slightly crude technique rather than a small number by a very elaborate method.

4.3 Methods of Evaluating the Most Probable Number (MPN) of Microorganisms by Statistical Examination of the Numbers of Positive Responses to Growth Tests

4.3.1 Principle of the Methods

The most commonly known method is that referred to as MPN (most probable number) which represents one of the oldest applications of statistics to microbiology, having been the subject of publications as early as 1908 (Phelps) and 1915 (McCrady).

These methods, which are used to estimate cell concentrations (number of microorganisms per unit of volume) in a microbial suspension, are based on the postulate suggesting that in such a suspension, the microbial cells are dispersed in a totally random fashion. Under these conditions, if the cell concentration is near one per milliliter, each milliliter will contain a number of cells which may be 0, 1, 2, 3, 4..., but with different probabilities defined by a Poisson distribution.

If each of the samplings, for example, among a group of ten, is placed in conditions which allow the growth of the microorganisms present, only two outcomes are possible: nongrowth (absence of microorganisms) or growth (initial presence of 1, 2, 3, 4..., microorganisms).

Poisson's law of rare events, given some number of positive outcomes (growth), allows the average number of cells or more precisely the most probable number of cells per milliliter to be determined. However, since the initial number of cells is unknown and the method works only for concentrations around one per milliliter, it becomes necessary to use several dilutions and to cary out the growth tests on samplings taken from several successive dilutions. MPN can therefore be determined from results obtained from several of these dilutions.

The methods derived from this principle differ essentially by the following:

- the number of levels of dilution taken into consideration for the estimation (4, 3, 2, or 1 in general);
- the number of tests performed and hence the number of elementary results acquired (from one test to several thousand);
- the volume of the samplings; the exmple of 1 ml has been given here, but this volume may vary from a few μl to 50 ml or so.

The validity of the process assumes that a certain number of conditions are in effect:

- that the microorganisms are randomly distributed throughout the sample and consequently that they do not exercise any attractive (clumping, flaking) or repulsive forces on each other;
- that each sampling containing one or several cells gives a positive response; and
- that no sampling initially containing no cells gives a positive response (the recognition of accidental contamination is given special attention).

Haas and Heller recently (1988) proposed a test to assess the degree to which Poisson's law of rare events effectively applies to the distribution of microorganisms in a given product.

4.3.2 The Standard MPN Method

4.3.2.1 Description of protocol and determination of the MPN

The principle of the MPN method is reviewed here as a reference. The following description applies to the "three tubes per dilution" method. After homogenizing the original suspension, several successive decimal dilutions of it are made. For each of these dilutions as well as for the original suspension, three tubes of medium are inoculated with 1 ml per tube. After incubation, the positive tubes are counted (appearance of turbidity, change of color of the indicator, gas production) in the successive dilutions and a characteristic number is noted which consists of the three digits in increasing order of dilution beginning with the number corresponding to the highest dilution for which all the tubes are positive, the first digit being in principle a 3. The characteristic number is then compared to a table which indicates the number of bacteria present (MPN) in the sample corresponding to the lowest dilution considered. Finally, the cell concentration in the initial suspension is calculated by factoring in the dilutions made.

Example
10^{-1} dilution: 3 positive tubes
10^{-2} dilution: 3 positive tubes
10^{-3} dilution: 1 positive tube

Characteristic number: 331
MPN in 10^{-1} dilution: 50
MPN in the original suspension: 500/ml

The determination of the characteristic number is not always this straightforward.

When more than three dilutions are used, the number must be chosen as follows (Table 4.1):

1. Take the highest dilution for which all three tubes are positive plus the next two (as for "a").
2. If a positive result is noted for a dilution higher than the last one thus taken, it must be added to the value of the last one (as for "b");
3. If no dilution gives three positive tubes, take the last three dilutions corresponding to positive tubes (as for "c").
4. Finally, if there is only a single dilution with positive tubes, choose a characteristic number such that the positive dilution digit occupies the tens column (as for "d").

Moreover, it is possible to modify this protocol to a large degree by varying the dilution factors, the numbers of tubes, and the volumes of the samplings

Table 4.1 Determination of the Characteristic Number as a Function of the Number of Positive Tubes for Series of Three Tubes per Dilution

	No. of Positive Tubes per Dilution					Characteristic No.
	10^{-1}	10^{-2}	10^{-3}	10^{-4}	10^{-5}	
	3	3	3	3	3	333
	3	3	3	1	0	310
a	3	3	2	0	0	320
	3	3	2	0	0	321
	3	0	1	0	0	301
b	3	3	2	2	1	323
	3	2	1	1	0	322
c	2	2	2	2	0	222
d	0	1	0	0	0	10

corresponding to the different dilutions. When a protocol has been chosen, the corresponding table must be used. Such tables are found in most of the publications cited in the following section.

4.3.2.2 Confidence limits of the MPN and probabilities of different characteristic numbers

Although it is not necessary to understand the mathematical developments which led to the formulation of the tables or the calculation programs, it is essential to know how to use the tables correctly and how to interpret the results. It should be understood beforehand that the precision of the tables is seldom reflected by the results themselves. For example, the characteristic number of 331 for the three-tube protocol from the table of De Man (1975) corresponds to an MPN of 50, but with 95% confidence limits of 20 and 240. The MPN therefore provides only a very rough estimate of the cell concentration.

It is very useful, in order to establish the significance that should be attached to results acquired by this method, to not simply read the MPN but to also take into consideration the confidence limits from the appropriate tables.

Given the diversity of possible conditions under which the procedure may be used, tables that provide the MPN and confidence limits are not always available for a given protocol. In such cases it is necessary to calculate them using a computer (Finney, 1951; De Man, 1975) or a programmable pocket calculator (Koch, 1982).

The range of the confidence limits may be reduced by increasing the number of tubes inoculated for each dilution, although this has the drawback of increasing the cost of the analysis. Nevertheless, it is possible to improve the precision for a given total number of tubes by optimizing their distribution among the various levels of dilution. Koch (1982) has suggested, citing Finney (1964), that the number of levels and the factors of dilution must vary inversely with the precision with which the result can be predicted. If good precision is possible, it is preferable to use fewer levels and lower dilution factors. Maximum precision is obtained when the number of organisms per tube is around 1.6 (Cochran, 1950; Aspinall and Kilsby, 1979). Furthermore, when the procedure is used to determine whether or not the number of bacteria exceeds some specified critical level rather than to evaluate the number of bacteria per gram, the optimal solution is to use a single dilution, chosen so that if the number of bacteria is equal to the critical number, the average number per inoculated tube will be around 1.6 (Prost, 1974). This method of optimizing the analytical scheme has also been studied by Aspinall and Kilsby (1979).

Note: Testing for a single bacterial cell in a given volume in order to determine if the product meets a standard expressed as "absence of... in ..." (e.g., absence of *Salmonella* in 25 g) is a procedure similar to the MPN method with all of the implications stemming therefrom with respect to the confidence to be associated with the result.

A final point worthy of consideration during the transformation of the characteristic number into an MPN is the probability of obtaining a given result, notably as studied by Halvorsen and Moeglein (1940), Woodward (1957) and De Man (1975).

Consider the extreme case cited by De Man. The calculated MPN from the result 0010 of a ten tube test with decimal dilution tubes is 0.9 with 95% confidence limits of 0.5 and 1.7, whereas the probability of obtaining the characteristic number 0010 when there are 0.9 bacterial cells/g is 0.17×10^{-24}. Obviously such a result must be rejected as being too improbable and the analysis must be repeated. De Man, with the aid of a computer, has applied the method of Parnow (1972) to calculating the probabilities associated with different characteristic numbers and has constructed tables giving the following figures:

- the MPN;
- the 99 and 95% confidence limits;
- the probability category, that is, normal results obtained in 95% of cases distinguished from less probable results obtained in 4% of cases not to be taken into consideration for important decisions (results which are even less probable are not listed in the table and are to be always considered unacceptable)

AFNOR standard NF-V-08-016 (November 1978) for the enumeration of coliforms by the MPN technique gives tables constructed according to this principle and suggests interpreting such improbable results as the product of errors, technical imperfections (contamination), or the presence of a bacteriostatic substance in the sample. Taking into consideration the probability of results may thus lead to questioning of the conditions under which the analytical technique is being implemented.

Finally, Hurley and Roscoe published in 1983 a program that can determine the MPN, the standard deviation, and the confidence interval as well as test the homogeneity of the data, that is, to detect possible errors of manipulation.

4.3.2.3 Advantages and Disadvantages of the MPN Procedure (Oblinger and Koburger, 1975)

The disadvantages of the MPN procedure are quite obvious. First, while the number of bacteria per gram is a linear function of the number of colonies counted on a plate, the conversion of raw MPN data into numbers of bacteria per gram requires complex calculations or careful scrutiny of adapted tables. Second, the extensive labor, large incubation capacity, and quantity of medium required tend to limit the number of tubes used, thus limiting the method to a low level of precision.

The procedure is used in certain cases, however, because it does offer some indisputable advantages. It readily lends itself to the analysis of large sample volumes when the concentration of microorganisms is too low to be measured by plating methods, and is therefore an alternative to filtering membrane techniques which, in spite of their attractiveness, are not without their disadvantages (Schaeffer et al., 1974). In addition, sometimes the microorganism of concern simply grows better in liquid media than on agar or on membranes.

4.3.2.4 Applications

In spite of its archaic aspect, the MPN method continues to be used for the detection of small numbers of microorganisms or of those which have undergone stress precluding their growth on selective agar-based medium.

In the case of water, the microorganisms of interest are generally fecal coliforms (Munoz and Silverman, 1979).

In the case of milk, the procedure is applicable to the detection of coliforms present following pasteurization (Haussler, 1982; AFNOR standard NF-V-08-016 [Nov. 1978]) as well as to the estimation of numbers of *Clostridium tyrobutyricum* (Bryant and Burkey, 1956; AFNOR standard NF-V-08-016).

It is also usable for the detection of *Escherichia coli* biotype I in raw meat, although interlaboratory comparative tests (Rayman et al., 1979) suggest grounds for preferring the Anderson-Baird-Parker method. In general, this technique is giving up ground to micromethods and methods using concentration by filtration.

4.3.3 Micromethods Based on the MPN Principle

Two methods orient the procedure toward miniaturization with all of the advantages resulting therefrom.

4.3.3.1 Microtiter System

Since 1969, Fung and Kraft have advocated the use of 96-well microtitration plates, applying the procedure to milk. Each well receives 0.225 ml of culture medium. Into each of the 12 wells of the first row, 0.025 ml of the milk to be analyzed is introduced using a calibrated loop with twisting to homogenize the inoculum. Each of these 12 original suspensions is then diluted tenfold by transferring 0.025 ml of the next row and so on up to a dilution of 10^{-8}. Three adjacent columns of eight dilutions are thus used for each milk sample (triplicate). The plate is then covered with a sterile glass cover and incubated at 37°C. After 6–12 hours of incubation, the turbid wells are counted as positive and the results are interpreted with the aid of MPN tables for three tubes per dilution. The results have been found by the authors to provide a good correlation with those obtained by conventional plating methods.

Microfiltration plates have the advantage of providing great flexibility in the choice of analytical protocols (number of levels of dilution, number of wells per dilution), the 96-well plates being usable in a variety of ways, which facilitates the optimization of the confidence limits (Koch, 1982).

Various application modes have since been proposed, for example, by Rowe and co-workers (1977). These authors suggest a method involving 12 one-half dilutions using eight wells per dilution and treating a single sample per plate. They provide an adapted MPN table accompanied by a special method for obtaining a characteristic number composed of three digits, P_1, P_2, P_3. P_1 is the number of positive wells in the least concentrated dilution for which this number is maximum; P_2 and P_3 are the numbers of positive wells in the next two dilutions.

This technique, which gives higher MPN values than those obtained by the standard tube technique, saves time, materials, and product.

For the counting of coliforms in river water, Block and co-workers (1982) proposed inoculating 96 wells per dilution. Under these conditions the authors found that the logarithm of the MPN is normally distributed and that the theoretical and experimental precision is good and practically independent of the bacterial density. The analysis of variance can be implemented in nearly all cases.

Micromethods offer the obvious practical advantage of readily lending themselves to automation and improvements in terms of time and reagents consumed, although the dilutions must be done very carefully since the slightest anomaly or contamination will lead to an aberrant result.

4.3.3.2 Hydrophobic Grid Membrane Filter (HGMF)

The HGMF procedure has been the subject of numerous publications by Sharpe and co-workers since 1975, and notably of a more recent work (Sharpe and Peterkin, 1988).

An HGMF membrane is a cellulose acetate membrane on which a hydrophobic grid has been printed delimiting the membrane surface into numerous growth compartments. In one currently commercialized system, the "Isogrid," the membrane comprises 1,600 compartments, the hydrophobic lines constituting barriers which confine growing bacteria to different boxes. Each box is thus a unit of response to either the presence and hence growth or the absence and hence nongrowth of one or more bacterial cells.

This process reduces or eliminates the dilution step and the direct evaluation of numbers of cells as high as 6,000. In effect, as long as the number of cells on the membrane is low, the response may be read directly, as for routine readings of agar plates or ordinary membranes. For higher numbers, when the probability of the initial presence of several cells per box increases, the raw data are processed by a formula which expresses the most probable number as a function of the number of positive responses x obtained on a grid of N boxes:

$$\mathrm{MPN} = N \log_e \frac{\mathrm{N}}{(N - \mathrm{x})}$$

Figure 4.1 shows the relationships between the number of cells and the number of positive growth responses for the following cases:

- for an ideal system allowing the detection of each colony, the relationship remaining linear indefinitely;
- on an agar plate, linearity being observed for up to 300 colonies, at which point diminishing occurs due to overlapping of colonies;
- on an ordinary 47 mm membrane ($10\,cm^2$), the loss of linearity occurring at about 60–80 colonies;
- On an HGMF membrane, the loss of linearity occurring much later and more so the higher the number of boxes in the grid.

This procedure has various advantages such as the elimination of tedious dilutions, more rapidly detectable colonies (due to enhancement of vertical growth combined with limitations on horizontal surfce growth), and electronic counting facilitated by predetermined positioning of colonies. Sharpe and associates (1979) have applied this method to the detection of coliforms and of *E. coli* in foods.

The application of the HGMF system obviously raises the question of filterability. Various authors have addressed this problem, notably Sharpe and co-workers (1976) with specific reference to various food product suspensions and milk on HGMF membranes.

The procedure has in recent years been the subject of various publications which demonstrate its potential value for the enumeration of total aerobic mesophilic microflora (Entis and Boleszczuk, 1986) as well as for the analysis of coliforms in water (McDaniels et al., 1987) and in various food products (Peterkin et al., 1989).

An automatic counter is now commercially available to facilitate the application of this technique (Sharpe and Peterkin, 1988), as well as a single-use disposable filtration unit (Tsuji and Bussey, 1986).

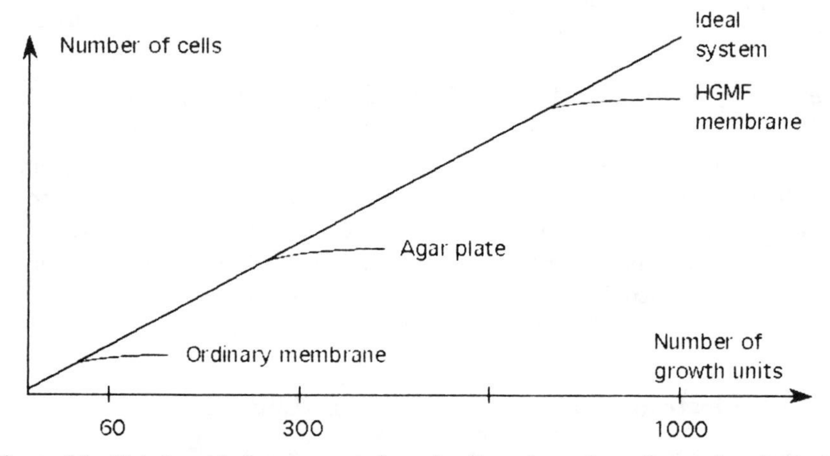

Figure 4.1 Relationship between number of cells and number of growth units (colonies) for various systems (adapted from Sharpe and Michaud, 1978).

4.3.4 Fields of Application and Interest For Methods Based on MPN Estimation

These tests, which are based on the statistical evaluation of responses to multiple binary tests, have been used for a long time, but continue to develop.

The conventional MPN technique is applied principally to the detection and enumeration of coliforms in water but may be used for other microbes, provided that their growth may be easily observed. It can thus be applied to the enumeration of yeasts or molds in fruit juices, to anaerobes (the so-called black tube technique), or to the genus *Clostridium* in dairy products.

The method is particularly useful when the number of microorganisms is low, for example, below one organism/ml, and is at times the only usable method in this case since counts in solid media are inapplicable, whereas filtration onto membranes, although a possible alternative, can only be used for liquids having a low organic matter and suspended particle load.

The main objections to counting methods using liquid media have to do with the time, space, and quantity of supplies necessary for their proper implementation. They remain useful nevertheless for several reasons:

- a large number of dilutions may be used;
- the precision may be increased by increasing the number of tubes inoculated per dilution;
- large sample volumes may be used, which is convenient when the number of microbes is low, the sensitivity and precision in this case being better than for plate counts;
- the survival of bacteria may be better in liquid media.

The miniaturized systems and especially the HGMF system, on the other hand, may give rise to very valuable techniques due to their potential for automation as well as their flexibility and the simplicity of data processing now afforded by desktop computing.

4.4 Electronic Counting of Cells and Flow Cytometry

In both electronic counting and flow cytometry, the microbial cells are counted in a liquid stream but the detection system differs. In the case of electronic counting, variations in the resistivity of a circuit caused by the passage of the cells through a capillary orifice are detected. In the case of flow cytometry, fluorescence emitted by a fluorochrome previously absorbed by the cells is measured under the excitatory effect of a laser ray.

4.4.1 Electronic Counters of Particles (Coulter Type)

These Coulter-type devices are capable of enumerating all types of particles, some versions being adapted to the enumeration of microorganisms. Although they have been available for at least 40 years, their use in microbiology has remained relatively limited.

Microorganisms are suspended in an electrolyte solution which is drawn at known flow rate through a narrow orifice (70 or 80 μm diameter). The solution acts as a conductor between two electrodes at a certain potential positioned on opposite sides of the orifice. When a cell suspended in the electrolyte passes through the orifice of the tube, its presence causes a transient increase in the resistance of the circuit. The instrument may be set to detect increases exceeding a specified threshold such that each passage of a certain type of cell is electronically detected and recorded. This is actually an enumeration process rather than a detection process, which makes it useful for the rapid enumeration of cells in industrial starter cultures, for microbial growth studies (Deysson and Nguyen Thi Lau, 1968), and possibly also for studying the composition of heterogeneous starter cultures (Claveau et al., 1968; Drake and Tsuchiya, 1973).

The evaluation of total microflora in raw milk represents a very unique application of this technology (Paul and Nitsch, 1974; Catoir et al., 1974). The milk sample is diluted automatically in physiological buffer and then in a melted gelatin medium. After 24 hours of incubation at 21°C, addition of an HCl-formalin mixture and sitting at 35°C for 30 minutes liquify the gelatin while hardening the microcolonies, the latter then behave as particles which are enumerated using the counter with a 280 μm orifice.

Catoir and co-workers (1974) found this system to be economical when applied to over 250 analyses/day, giving results at least as reliable as the plate count method and appearing to be perfectible.

4.4.2 Flow Cytometry

Flow cytometry, initially used for animal cell studies, is very useful for studying and sorting hybridomas intended for monoclonal antibody production. Progress in instrumentation has imparted sufficient sensitivity to the process for its application to microbiology, although this aspect of flow cytometry is relatively new. A general outline of the process is given later (the reader is referred to the review of Hadley et al. [1985] for more detailed information).

Microbial suspensions must first of all be subjected to an enzymatic treatment for the purpose of disrupting aggregates and separating the cells. The cells are subsequently contacted with fluorochromes which bind to nucleic acids, proteins, or other molecules. The fluorochromes are used at saturation levels so that binding to cell components is stoichiometric and subsequent measurements are quantitative. Next, the fluorescent marked suspended cells are focused into a narrow flow path delimited by a liquid sheath which enters the detection system at high speed. The detection system consists of a fine laser brush which spans the liquid stream. If a cell passes through this brush, the bound fluorochromes become excited and produce a detectable and measurable fluorescent emission. In practice it is possible to use several fluorochromes simultaneously and effect several simultaneous measurements on a cell, in order to characterize it in detail. Thus, each cell may be analyzed individually by computerized processing of a variety of data allowing enumeration of several cell categories present in a single suspension.

This system of analysis, which is capable of examining at least 1,000 cells/sec., is therefore very rapid, but since the liquid flow stream is by necessity very fine, the detection threshold is relatively high. Some of the speed of the analysis may be sacrificed in favor of a preincubation allowing the microorganisms to multiply.

This technique is currently more of a research tool, but interesting industrial developments can be expected to arise over the next few years.

4.5 Conclusion

In summary, the first two methods presented in this chapter are very similar. In both cases, estimation of the microbial population is done directly by counting after an amplification step. In the case of microscopic examination, the amplification is optical while in the case of plating methods, the amplification is biological, the viable units being rendered visible to the eye as a result of growth. Furthermore, these two types of amplification may be combined.

The evaluation of microflora by the MPN method following culture in liquid medium is characterized by more originality, notably with respect to the type of information generated and the information processing method used. Since the MPN method consists of analyzing a relatively high number of repetitive subsamples of the initial sample, of obtaining from each subsample either a positive or a negative outcome and deducing the cell concentration in the sample by a statistical technique, it obviously lends itself readily to both automation and computerization.

References

Abgrall B. and Bourgeois C. M., 1989. Dénombrement de la flore totale de produits alimentaires par la technique DEFT. *Sci. des Aliments* (in press).

Afifi S. and Muller G., 1975. Fluoreszenz-Bacterioskopie eine direkte Methode zur bakteriologischen Lebensmitteluntersuchung. *Die Nahrung*, **19**, no. 7, 556–567.

AFNOR. Directives générales pour le dénombrement des coliformes. Technique du nombre le plus probable après incubation à 30°C. Standard NF-V-08-016.

Aspinall L. J. and Kilsby D. C., 1979. A microbiological quality control procedure based on tube counts. *J. Appl. Bacteriol.*, **46**, 325–329.

Bassfeld I. and Lambert D., 1988. L'utilisation d'un procédé automatique d'ensemencement, le système "spiral" pour le dénombrement de la flore aérobie mésophile et fongique des produits de cuisson des céréales et de leurs matières premières. *IAA*, June, 495–499.

Baumgart J. and Niemann H., 1974. Mikrobiologische Kontrolle von Frischfleisch mit Hilfe von Schnellverfahren. *Fleischwirtschaft*, **54**, no. 9, 1497–1500.

Block J.-C., Maul A., Dollard M.-A., 1982. Estimation du nombre le plus probable (NPP) des coliformes en rivière en réalisant 96 ensemencements par niveau de dilution. *Revue Française des Sciences de l'Eau*, **1**, 387–399.

Bourgeois C. M., 1968. Les analyses microbiologiques en brasserie. *Biotechnique*, no. 3, 2–39.

Bryant M. P. and Burkey L. A., *J. Bacteriol.*, **71**, 43.

Catoir M., Gerber R., Goy A., 1974. Essais du "Coulter Counter" pour le dénombrement de la flore totale du lait cru. Le lait. *Revue Générale des Questions Laitières*, **54**, no. 531–532, 22–30.

Catteau M., Auby J. C., Catsaras M., 1981. Système spiral et analyse bactériologique des plats cuisinés. *Rec. Méd. Vét.*, **157** (10).

Claveau J., Scriban R., Strobbel B., Carpentier Y., 1968. Etude biométrique des levures en brasserie. Cas particulier: Biométrie et microbiologie de quelques levures d'infection. *Brasserie*, **23**, no. 2, 77–91, no. 3, 141–159.

Cochran W. G., 1950. Estimation of bacterial densities by means of the most probable number. *Biometrics*, **6**, 105–116.

De Man J., 1975. The probability of most probable numbers. *Europ. J. Appl. Microbiol.*, **1**, 67–78.

Deysson G. and Thi Lau N., 1963. Utilisation d'un compteur électronique de particules pour l'étude de la croissance des microorganismes. *Annales Pharmaceutiques Françaises*, **21**, no. 4, 275–285.

Donnelly C., Gilchrist J., Peeler J., Campbell J., 1976. Spiral plate count method for the examination of raw and pasteurized milk. *Appl. Environ. Microbiol.*, **32**, no. 1, 21–27.

Drake J. and Tsuchiya H., 1973. Differential counting in mixed cultures with Coulter Counters. *Appl. Microbiol.*, **26**, no. 1, 9–13.

Entis P. and Boleszczuk P., 1986. Use of fast green FCF with tryptic soy agar for aerobic plate count by the hydrophobic grid membrane filter. *J. Food Prot.*, **49**, no. 4, 278–279.

Finney D. J., 1951. The estimation of bacterial densities from dilution series. *J. Hyg.*, **49**, 26–35.

Finney D., 1964. *Statistical methods in biological assay*, 2d ed., 570–586. Hafner Press, New York.

Fung D. and Lagrange W., 1969. Microtiter method for enumerating viable bacteria in milk. *J. Milk Food Technol.*, **32**, no. 4, 144–146.

Fung D. Y. C. and Kraft A. A., 1969. Rapid evaluation of viable cell counts by using the microtiter system and MPN technique. *J. Milk Food Technol.*, **32**, no. 10, 408–409.

Fung D. Y. C., Donahue R., Jensen J. P., Ullmann W. W., Hausler W. J., Lagrange W., 1976. A collaborative study of the microtiter count method and standard plate count method for viable cell count of raw milk. *J. Milk Food Technol.*, **39**, no. 1, 24–26.

Godbersen G., 1973. Die Instantkeimzahlmethode. *Deutsche Milchwirtschaft*, **24**, no. 5, 146–148.

Haas C. N. and Heller B., 1988. Test of the validity of the poisson assumption for analysis of most-probable-number results. *Appl. Environ. Microbiol.*, vol. 54, no. 12, 2996–3002.

Hadley W. K., Waldman F., Fulwyler M., 1985. Rapid microbiological analysis by flow cytometry *In:* Nelson W. H. (ed). *Instrument Methods for Rapid Microbiological Analysis*, 67–89, VCH.

Haikara A. and Boije-Backman S., 1982. The use of optical brighteners in the detection of bacterial and yeast contaminants in beer. *Brauwissenschaft*, **35**, 113–117.

Halvorson H. O. and Moeglein A., 1940. Application of statistics to problems in bacteriology. V: The probability of occurrence of various experimental results. *Growth*, **4**, 157–165.

Harrisson J. and Webb T. J. B., 1971. Recent advances in the rapid detection of brewery microorganisms and development of a microcolony method. *J. Inst. Brew.*, **85**, 231–234.

Hausler W. J., 1972. Standard methods for the examination of dairy products. 13th ed. *Am. Public Health Assoc.*, 88–89.

Hobbie J., Daley R., Jasper S., 1977. Use of nucleopore filters for counting bacteria by fluorescence microscopy. *Appl. Environ. Microbiol.*, **33**, no. 5, 1225–1228.

Hunter A. C., and McCorquodale R. M., 1983. Evaluation of the direct epifluorescent filter technique for assessing the hygienic condition of milking equipment. *J. Dairy Res.*, **50**, 9–16.

Hurley M. A. and Roscoe M. E., 1983. Automated statistical analysis of microbial enumeration by dilution series. *J. Appl. Bacteriol.*, **55**, 159–164.

Jaeggi N. E., Simes V., Hughest D., 1989. Evaluation of a television image analyser as an aid to estimation of microbial numbers in food using the direct epifluorescent filter technique. *Food Microbiol.*, **6**, 85–91.

King J. S. and Mabbitt L. A., 1984. Factors affecting the volume of milk delivered by a standard loop in the plate loop method for bacterial count. *J. Dairy Res.*, **51**, 317–324.

Kirsop B. H. and Dolezil L. Detection of lactobacilli in brewing. *In:* Carr J. G., Catting C. V., Whiting G. C. (eds). *Lactic acid bacteria in beverages and food*, 159–164. Academic Press, London.

Kiyosh T. and Bussey D. M., 1986. Automation of microbial enumeration: Development of a disposable hydrophobic grid-membrane filter unit. *Appl. Environ. Microbiol.*, **52**, no. 4, 857–860.

Koch A. L., 1982. Estimation of the most probable number with a programmable pocket calculator. *Appl. Environ. Microbiol.*, **43**, no. 2, 488–490.

Konuma H., Suzuki A., Kurata H., 1982. Improved Stomacher 400 bag applicable to the spiral plate system for counting bacteria. *Appl. Environ. Microbiol.*, **44**, no. 3, 765–769.

Kroll R. G. and Rodrigues U. M., 1986. The direct epifluorescent filter technique cytochrome c oxidase test and plate count method for predicting the keeping quality of pasteurized cream. *Food Microbiol.*, **3**, 185–194.

Kynast S., 1973. Uber eine neue Schnellmethode sur direkten Keimzahlbestimmung in Milch. *Deutsche Molkerei Zeitung*, **94**, no. 49, 2004–2005.

McCrady M. H., 1915. The numerical interpretation of fermentation tube results. *J. Infect. Dis.*, **17**, 183–212.

McDaniels A., Borner R. H., Menkedick J. R., Weber C. I., 1987. Comparison of the hydrophobic-grid membrane filter procedure and standard methods for coliforme analysis of water. *Appl. Environ. Microbiol.*, **53**, no. 5, 1003–1009.

McGoldrick K. F., Fox T. L., McAllister J. S., 1986. Evaluation of a dry medium for detecting contamination on surfaces. *Food Technol.*, April, 77–80.

McKinnon C. H. and Mansell R., 1981. Rapid counting of bacteria in rinses of milking equipment by a membrane filtration-epifluorescent microscopy technique. *J. Appl. Bacteriol.*, **51**, 363–367.

Macmeekin T., 1976. Potential use of dip slides for microbiological quality control in the food industry. *Food Technol. Australia*, **28**, no. 4, 129–156.

Munoz E. F. and Melvin P. S., 1979. Rapid, single-step most-probable-number method for enumerating fecal coliforms in effluents from sewage treatment plants. *Appl. Environ. Microbiol.*, **37**, no. 3, 527–530.

Oblinger J. L. and Koburger J. A., 1975. Understanding and teaching the most probable number technique. *J. Milk Food Technol.*, **38**, no. 9, 540–545.

Oppong D. and Snudden B. H., 1988. Comparison of acacridine orange using fluorescence microscopy with traditional methods for microbiological examination of selected dry food products. *J. Food Protect.*, **51**, no. 6, 485–488.

Parnow R. J., 1972. Computer program estimates bacterial densities by means of the most probable numbers. *Food Technol.*, **7**, 56–62.

Paul S. and Nitsch B., 1974. Die Anwendung der elektronischen Mikrokoloniezählung, EMZ, zur Qualitätskontrolle von Rohmilch. *Deutsche Molkerei Zeitung*, **95**, no. 32, 1164–1170.

Peterkin P. I., Conley D., Foster R., Lachapelle G., Milling M., Pruvis U., Sharpe A. N., Malcolm S., 1989. A comparative study of total coliform recovery from foods by most probable number and hydrophobic grid membrane filter methods. *Food Microbiol.*, **6**, 79–84.

Pettipher G. L., 1983a. *The direct epifluorescent filter technique.* Letchworth, Research Studies Press.

Pettipher G. L., 1983b. La microscopie par épifluorescence. *Technique Laitière*, **974**, 37–43.

Pettipher G. L., Mansell R., McKinnon C. H., Cousins C. M., 1980. Rapid membrane filtration-epifluorescent microscopy technique for direct enumeration of bacteria in raw milk. *Appl. Environ. Microbiol.*, **39**, 423–429.

Pettipher G. L., and Rodrigues U. M., 1981. Rapid enumeration of bacteria in heat-treated milk and milk products using a membrane filtration-epifluorescence technique. *J. Appl. Bacteriol.*, **50**, 157–166.

Pettipher G. L. and Rodrigues U. M., 1982a. Rapid enumeration of microorganisms in foods by the Direct Epifluorescent Filter Technique. *Appl. Environ. Microbiol.*, **44**, 809–913.

Pettipher G. L. and Rodrigues U. M., 1982b. Semi-automated counting of bacteria and somatic cells in milk using epifluorescence microscopy and television image analysis. *J. Appl. Bacteriol.*, **53**, 323–329.

Pettipher G. L., Fulford R. J., Mabbitt L. A., 1983. Collaborative trial of the direct epifluorescent filter technique (DEFT), a rapid method for counting bacteria in milk. *J. Appl. Bacteriol.*, **54**, 177–182.

Phelps E. B., 1908. A method for calculating the number of *E. coli* from the results of dilution tests. *Am. J. Pub. Hyg.*, **4**, 141–145.

Prost M., 1974. Tests statistiques pour le contrôle bactériologique en milieux d'ensemencement liquide. *Ann. Fals. Exp. Chim.*, **67**, no. 716, 157–169.

Qvist S. H. and Jakobsen M., 1985. Application of the direct epifluorescent filter technique as a rapid method in microbiological quality assurance in the meat industry. *Int. J. Food Microbiol.*, **2**, 139–144.

Rayman M. K., 1979. ICMSF methods studies. XIII: An international comparative study of the MPN procedure and the Anderson-Baird-Parker direct plating method for the enumeration of *Escherichia coli* biotype I in raw meats. *Can. J. Microbiol.*, **25**, 1321–1327.

Richards G. P., 1979. Evaluation of millipore coli-count samplers for monitoring fecal coliforms in the blue crab, *Callinectas sapidus. Appl. Environ. Microbiol.*, **38**, no. 2, 341–343.

Richards M., 1970. *Detection of yeast contaminants in pitching yeast.* Wallerstein Laboratories Communications, **110**, 11–15.

Rodrigues U. M., Kroll R. G., 1985. The direct epifluorescent filter technique (DEFT): Increased selectivity, sensitivity and rapidity. *J. Appl. Bacteriol.* **59**, 493–499.

Rodrigues U. M. and Kroll R. G., 1986. Use of the direct epifluorecent filter technique for the enumeration of yeasts. *J. Appl. Bacteriol.*, **61**, 139–144.

Rodrigues U. M. and Kroll R. G., 1988. Rapid selective enumeration of bacteria in foods using a microcolony epifluorescence microscopy technique. *J. Appl. Bacteriol.*, **64**, 65–78.

Rodrigues U. M. and Kroll R. G., 1989. Microcolony epifluorescence microscopy for selective enumeration of injured bacteria in frozen and heat-treated foods. *Appl. Environ. Microbiol.*, **55**, no. 4, 778–787.

Rodrigues U. M. and Pettipher G. L., 1984. Use of the direct epifluorescent filter technique for predicting the keeping quality of pasteurized milk within 24 hours. *J. Appl. Bacteriol.*, **57**, 125–130.

Rowe R., Todd R., Waide J., 1977. Microtechnique for most probable number analysis. *Appl. Environ. Microbiol.*, **33**, no. 3, 675–680.

Schaeffer D. J., Long M. C., Janardin K. G., 1974. Statistical analysis of the recovery of coliform organisms on Gelman and Millipore membrane filters. *Appl. Microbiol.*, **38**, 605–607.

Sharpe A., Dyett E., Jackson A., Kilsby D., 1972. Technique and apparatus for rapid and inexpensive enumeration of bacteria. *Appl. Microbiol.*, **24**, no. 1, 4–7.

Sharpe A. N. Kilsby D. C., 1971. A rapid, inexpensive bacterial count technique using agar droplets. *J. Appl. Bacteriol.*, **34**, no. 2, 435–440.

Sharpe A. N. and Michaud G. L., 1975. Enumeration of high numbers of bacteria using hydrophobic grid-membrane filters. *Appl. Microbiol.*, **30**, no. 4, 519–524.

Sharpe A. N. and Michaud G. L., 1978. Enumeration of bacteria using hydrophobic grid-membrane filters *In:* Sharpe A. N., Clark D. S., Charles C. (eds). *Mechanizing microbiology*, 141–153. Thomas Publisher.

Sharpe A. N., Peterkin P. I., Dudas I., 1976. Membrane filtration of food suspensions. *Appl. and Environ. Microbiol.*, **37**, no. 1, 21–35.

Sharpe A. N., Peterkin P. I., Malik N., 1979. Improved detection of coliforms and *Escherichia coli* in foods by a membrane filter method. *Appl. Environ. Microbiol.*, **38**, no. 3, 431–435.

Sharpe A. N. and Peterkin P. I., 1988. *Membrane filter for food microbiology*. Letchworth, Research Studies Press Ltd.

Shaw B. G., Harding C. D., Hudson W. H., Farr L., 1987. Rapid estimation of microbial numbers on meat and poultry by the direct epifluorescent filter technique. *J. Food Prot.*, **50**, no. 8, 652–657.

Silley P., 1985. Evaluation of total-count sampler against the traditional pour plate method for enumeration of total viable counts of bacteria in a process water system. *Letters Appl. Microbiol.*, **1**, 41–43.

Thompson D. I., Donnelly C. B., Black L. A., 1960. A plate loop method for determining viable counts of raw milk. *J. Milk Food Technol.*, **23**, 167–161.

Trotman R. E, 1971. The automatic spreading of bacterial culture over a solid agar plate. *J. Appl. Bacteriol.* **34**, no. 3, 615–616.

Winter F., York G., El-Nakhal H., 1971. Quick counting method for estimating the number of viable microbes on food and food processing equipment. *Appl. Microbiol.*, **22**, no. 1, 89–92.

Woodward R. L., 1957. How probable is the most probable number? *J. Am. Water Works Assoc.*, **49**, 1060–1068.

CHAPTER

5

Evaluation of Microflora by Nonmicrobiological Techniques

C. M. Bourgeois, P. Mafart

5.1 General Principles

In order to become practical in industrial laboratories and ultimately on the production floor itself, microbiological testing must develop in such a way as to meet the following demands:

- rapidness of response, indispensable in order to allow any manufacturing anomalies to be corrected within a useful time frame and thereby prevent the distribution of products which do not meet regulatory requirements or company standards;
- low cost, necessary in order to allow close monitoring of the manufacturing process by large numbers of analyses without excessive financial burden, which implies that the technique must be easy to implement and preferably amenable to automation;
- sufficiently pertinent to allow a correct and precise evaluation of the potential risks.

These demands are only partially satisfied by enumeration precedures (see Chapter 4). For this reason, efforts are currently directed toward developing procedures based on evaluation of the microbial population by measurements detecting signals linked to the presence and activity of microorganisms, most often enzymatic activity or the concentration of a molecule (coenzyme or metabolite) or some change in the medium (a variation in pH, redox potential, impedance, heat production) linked to this activity. Since a single micron-sized bacterial cell having a mass of one picogram occurring somewhere in a food cannot produce a signal sufficient for direct detection by any currently imaginable means, some kind of multiplication step is necessary.

With the exception of microscopy, which uses optical amplification, the previously examined procedures all require a rather long multiplication step in order to render each cell individually observable, most often in the form of a colony visible to the naked eye, that is, a mass of about 10^9 cells resulting from 30 successive divisions, which generally takes from 1 to 2 days.

In contrast, procedures which consist of detecting and quantifying a chemical or physical signal linked to the presence of microbial activity yield a response more rapidly as a result of the summation of the signals from individual cells and as a result of the sensitivity of the detectors available for physicochemical analyses.

Herein lies one great advantage to using these procedures. Whereas for counting on agar-based medium, the incubation time necessary for detection is independent of the initial number of cells, each cell being treated as an independent signal, in the case of physicochemical detection procedures, the incubation time necessary varies as an inverse function of the initial number of cells as a result of a summation of signals produced by the various cells. The greater the initial number of cells, the lesser the number of cell generations required in order to reach the instrumentational detection threshold. For homogeneous populations, there is actually a linear relationship between the logarithm of the initial concentration and the incubation time necessary before detection. Given the capabilities of the instruments available, the instantaneous detection thresholds vary from 10^4 to 10^9 cells, implying that the incubation times necessary are generally lower than those required for counting techniques. These procedures may be classified in two ways: (1) by the cell target of detection, and (2) by the instrumental techniques usable.

In this chapter we shall briefly provide a classification of the usable targets and describe the possibilities for their detection followed by a more detailed presentation of the instrumental techniques used, describing for each one the most commonly practiced or the more promising application.

5.2 Detection Targets

Ideally a valid assessment of risk would necessarily involve the detection of the actual molecules resulting from the destructive action associated with the microorganism. Determination of ammonia and indole in fish thus provides a direct evaluation of the degree of bacteriological spoilage. Unfortunately, the sensitivity of the available chemical methods of detection does not allow the detection of spoilage at an early stage. Usually, in order to take advantage of the most sensitive techniques available, one must settle for the detection of chemical or physical signals which are indicative of the presence and activity of microorganisms without having any direct relationship with their possible harmfulness. Primary targets, which are normal components of cells, may be distinguished from secondary targets, which are markers introduced into the cell or attached to its surface.

5.2.1 Primary Targets

Primary targets are as follows:

1. Cellular bodies themselves which may be detected notably by their diffractory effects on light.
2. Macromolecules
   - enzymes and permeases revealed by their activities which may be measured most notably by spectrophotometry, fluorimetry, radiometry, and electrochemistry or which may be detected immunochemically;
   - other proteins and polysaccharides detectable immunochemically or by specific biological tests such as the Limulus test (lipopolysaccharides)
   - nucleic acids detectable with great selectivity by fluorimetry or by molecular hybridization
3. Small intracellular biomolecules
   - coenzymes, notably ATP measurable by a specific spectrometry (ATPmetry)
4. Excreted metabolites
   - biogenic amines, ammonia, organic acids measurable by spectrophotometry, electrochemistry (particularly by enzyme electrodes) and chromatography
5. Physical parameters of the medium
   - redox potential, pH, conductance, temperature, measurable notably by electrochemical means

5.2.2 Secondary Targets

Secondary targets are the following:

1. Indicators changing color under the influence of microbial metabolism of precursors of fluorochromes likewise becoming fluorescent, detection being by spectrophotometry or fluorimetry.
2. Antibodies which bind specifically to cell wall antigens and are subsequently detected by an immunoradiometric or immunoenzymatic process.
3. Phages which bind specifically to certain bacteria and are subsequently detected under the same conditions as antibodies.

5.3 Spectroscopy

Spectroscopy is a collection of techniques for the study and measurement of electromagnetic emissions. The activity of microorganisms is manifested as signals detectable by these process.

5.3.1 Turbidometry

The principle of turbidometry is as follows: when a beam of light passes through a microbial suspension, part of the incident light is scattered by the cells, giving an exit beam of diminished intensity and a cone of scattering. The reduction in intensity of the central beam in crossing the suspension is expressed as absorption, A, such that:

$$A = \log \frac{l_0}{l}$$

where

l_0 is the intensity of the incident light
l is the intensity of the transmitted light

This absorption, which may be measured by a photometer, is proportional to the microbial concentration between certain limits and may therefore be used to measure it.

This process is used routinely for studying the growth of microorganisms (Marcelis et al., 1980; Thomas et al., 1984) and their sensitivity to antibiotics (Lamp et al., 1983) as well as for the microbiological determination of vitamins.

In the fermentation industries, food or otherwise, turbidity measurements have a variety of uses. In brewing, for example, they may be used to evaluate the richness of starter cultures, the quantity of cells in suspension during the primary fermentation, at the decanting step and in the secondary fermentation as well as for monitoring filtration (Thouvenot and Bourgeois, 1970). Industrial photometers already allow the continuous monitoring of the quantity of cells suspended in circulating liquids or in fermenters.

In the field of microbiological testing of foods, there may be interesting possibilities for the evaluation of high levels of total flora, for example, in meat and in milk (Kouomegne, 1984). In the case of meat, two approaches have been followed. Leistner (1977) and Strasser (1979) have done direct turbidometric evaluations after simple filtration, without culture, of the flora present on the surface of meat. Strasser found that the advantage to the method resides in its potential for on-site use and in yielding results in 3–5 hours, albeit with low precision. The other approach (Jorgensen and Schulz, 1985; Schulz et al., 1988) has been to use a stomacher and centrifuge, followed by measurement of the samples in a modified Cobas-type automatic turbidometer cell (Roche Diagnostica, Denmark). This device traces the growth curve of the microbial population present, the level of the initial population being automatically deduced from the incubation time necessary to obtain a certain increase in light absorption. The results correlate well with those of plate counts, with selective medium used to evaluate specific flora. Analytical costs are reduced, making the procedure seem advantageous.

In the case of milk, application of the same procedure has been proposed by Mattila and Alivehmas (1987). A procedure imparting transparency to milk has also been proposed (Kouomegne et al., 1984), providing rapid evaluation of total microflora by a simple and direct turbidometric measurement.

5.3.2 Spectrometry

Visual examination of the color change of reagents is a commonly employed process in microbiology for the evaluation of microbial population levels and for the examination of microbial properties. This type of examination can be made more objective and quantitative by the use of a spectrophotometer allowing colorimetry, that is, measurement of the concentration of colored substances.

5.3.2.1 Measurement of Variations in Redox Potential

In the process of multiplying in a medium, microorganisms absorb oxygen and excrete reducing substances which lower the redox potential of the medium. Visual observation of the delay in the color change of redox indicators such as resazurin, methylene blue, and 2, 3, 5-triphenyltetrazolium (TTC) is a commonly utilized procedure for total floral assessment.

In milk processing, resazurin and methylene blue tests are routinely used to assess the bacteriological quality of milk arriving at the processing plant. The principle consists of adding a small quantity of indicator to the milk and measuring the time required for decolorizing to occur, the more rapid the greater the level of contamination. One variation of the procedure using strips of filter paper impregnated with resazurin (Otsuka and Nakae, 1969) also seems worthy of consideration. Assessment of the decolorizing is generally visual although this may also be done using a spectrophotometer. The resazurin procedure has been thoroughly and critically studied by Thomas and Thomas (1974), who concluded that the test is valuable and better adapted for use in milk processing than are rapid methods for counting colonies as discussed earlier (Chapter 4). The indirectness of the test is probably the principal source of doubts about its validity although comparative statistical studies show that the scaling of results for methylene blue decoloration time to log number of bacteria is justifiable (Andrew and Lock, 1974). These tests are also suspected of being affected by the possible presence of pesticide residues which inhibit bacteria, but the risk of this occurrence seems to be very limited (Quadri et al., 1974).

Similar procedures have been used to estimate total aerobic microflora or psychrotrophic flora in meats (Bradshaw et al., 1961; Baumgart et al., 1975; Emswiler et al., 1976), marine products (Webb et al., 1972), vegetables to be processed (Bartholin and Boudet, 1973; Kvasnikov et al., 1974), and of syrup to be used for maple sugar manufacturing (Kissinger, 1969, 1974).

In general, these procedures give rapid and reliable indications under simple and economical conditions, although spectrophotometric measurements of indicator optical density after a specific incubation time provide a more objective evaluation of the microbial population present.

In the case of meat, in order to avoid masking of the indicator color change by colored substances naturally present, it is necessary to use a stomacher followed by decantation of the suspension or alternatively a surface sampling technique.

In the testing of beef following sampling by the hollow cylinder sampling technique, Baumgart and co-workers (1975) compared the resazurin test with a microcolony enumeration technique (the drop technique of Sharpe et al., 1971, 1972, see Chapter 4), under the following conditions: After 30 minutes of incubation in peptone broth at 30°C, resazurin is added to a concentration of 0.003% followed by further incubation for 2–3 hours. The broth is then filtered onto a 0.45 μm membrane and the optical density of the filtrate at 570 nm is measured. A correlation coefficient of 0.85 was obtained for the two techniques with 2 hours of incubation and 0.88 with 3 hours of incubation. These authors concluded that the procedure provides valid means of classifying meats upon delivery and is used for this purpose in some countries.

Another procedure worth mentioning is that of Jedlicka and associates (1975) which consists of assessing the level of surface contamination by measuring, after incubation, the reflectance of an agar surface inoculated by contact. Variation in reflectance is essentially due to the changing of 2, 3, 3-triphenyltetrazolium in the medium from colorless to red as a result of the reducing activities of microorganisms.

5.3.2.2 Measurement of Enzymatic Activity

The procedures described above are based on the overall evaluation of oxidation-reduction enzymes. Numerous other enzymes have been used to reveal the presence of microorganisms and as a means of estimating their numbers, either visually or spectrophotometrically. Since enzyme molecules have catalytic activity and measurement of enzymatic activity is based on the determination of the products of the enzymatic reaction, this type of measurement has a considerable degree of "built-in" amplification, the practical result of which is a particularly low detection threshold. An interesting repertoire of the possibilities in this area was published many years ago by Mitz (1969).

Determination of phosphatase activity has been used in milk processing for the assessment of the quality of milk to be used in cheese making. This enzyme activity, present in all organisms tested, is a good indicator of bacterial activity in milk although it is not closely correlated with the number of viable bacteria. The enzyme is in fact quite resistant to heat, remaining active after the thermal destruction of the bacterial cells which possess it.

The sensitivity of the phosphate method may be increased by several orders of magnitude by using a fluorescent substrate (Mitz, 1969), that is, one which upon hydrolysis by the enzyme, yields a highly fluorescent compound. The detection threshold is in the order of 10^7 cells for nearly instantaneous detection.

Esterases also show some interesting properties in this sense, being equally ubiquitous but with ten times as much activity level per cell compared to phosphatase, giving a sensitivity threshold in the order of 10^6 organisms, which may also be improved by using a fluorescent substrate. Extracellular lipase activity of *Pseudomonas fluorescens* in milk has been measured this way by Stead (1983) using 4-methyl ombelliferone oleate.

Measurements of phosphatase and esterase activities have been applied to assessing the general bacteriological state of milk, meat, and fish, but only as an indirect test of bacteriological quality, for example, for monitoring the effectiveness of a thermal treatment.

The search for enzymes which are more or less characteristic also allow the evaluation of particular types of flora. The detection of glutamate decarboxylase in milk, which may be automated with the aid of an autoanalyser, thus allows the estimation of the number of *Escherichia coli* present (Moran and Witter, 1976), the detection threshold being in the order of 50,000 cells/ml. Starting with samples contaminated with a few units per ml, an incubation time of around 10 hours in trypticase soy broth is necessary. This technique has been applied, with various modifications, to the detection of fecal coliforms in water by more or less automated processes in several laboratories (Trinel and Leclerc, 1977; Warren et al., 1978; Trinel et al., 1980).

Revealing the presence of β-glycuronidase using 4-methyl umbelliferone-β-D-glycuronide (MUG), a nonfluorescent substrate which liberates its fluorescent moiety upon hydrolysis, has been applied lately to the evaluation of *Escherichia coli* populations. In the Enterobacteriaceae and Vibrionaceae families, this enzyme occurs only in *E. coli*, 50% of members of the genus *Shigella*, some *Salmonella* and in *Yersinia*. The reaction is used to rapidly detect *E. coli* in water by the MPN method (Peterson et al., 1987) and by filtration (Freier and Hartman, 1987) as well as in various foods by liquid culture techniques (Robinson, 1984) and on agar-based media (Weiss and Humber, 1988).

5.3.2.2 Coenzyme Determinations: The ATP-Luciferin-Luciferase System Bioluminescence Reaction

The detection of coenzymes allows an even lower threshold than detection of enzymes themselves (Mitz, 1969).

The nicotinamide adenine dinucleotide (NAD) method uses an enzyme-substrate system consisting of lactate dehydrogenase, lactic acid, and a tetrazolium salt. Throughout the reaction the NAD serves as a catalyst alternating between the NAD and NADH forms, while lactic acid is oxidized to pyruvic acid and the tetrazolium salt is reduced to a colored formazan. The detection threshold is from 10^5 to 10^6 organisms.

The determination of adenosine triphosphate (ATP) by the bioluminescent ATP-luciferin-luciferase reaction is the most frequently utilized. The compound ATP is present in all living cells, particularly in all bacterial cells, being the universal form in which energy is transferred from energy-producing systems (e.g., the respiratory chain) to energy-consuming systems responsible for biosynthesis, muscular contraction, bioluminescence, and so on. ATP is therefore an indicator of the presence of living organisms. The bioluminescent reaction of the firefly, which employs a pigment, luciferin, and an enzyme, luciferase, thereby determines ATP with high sensitivity.

Injecting ATP into a mixture of luciferase, luciferin and Mg^{++} results in an immediate emission of light which reaches a maximum within a few seconds followed by a drop back to zero after about 1 minute. The kinetics of this response are strongly influenced by the composition of the reaction medium (Strehler, 1965). Either the peak emission intensity or the quantity of light emitted over a portion of the response may be measured. In both cases the results obtained are proportional to the quantity of ATP injected, within certain limits, providing the basis for the evaluation of the quantity of microorganisms present in a given product. This technique is characterized by numerous points that warrant discussion and continue to be objects of study.

First, whether or not the quantity of ATP is an accurate measurement of the quantity of microorganisms is certainly a valid question, given that variations in ATP content as a function of physiological state and as a function of species are quite significant, possibly in the order of 1–10 for each of these factors. Consequently, the ATP concentration provides only an order of magnitude for a composite population (average coefficient: 5×10^{-16} g ATP/cell). The procedure therefore appears to be well suited to the measurement of monospecific populations under uniform growth conditions, for example, monitoring the progress of fermentations (Jacubczak and Leclerc, 1979).

The second difficulty is with the extraction of the microbial ATP. Various reagents have been proposed, dimethyl sulfoxide (0.9 ml/ml of sample) serving as a reference, while various other solutions with patents pending are also usable.

The third problem occurs during the actual measurement of the light emission. Fluorimeters allow a detection limit of 1 μg/ml. Quantum counters with very sensitive photomultipliers are capable of counting individual photons, thereby multiplying sensitivity by 100–1,000. The speed of the reaction requires, however, that the ATP be injected directly into the reaction cell inside the machine without introducing interfering light. Liquid scintillation spectrometers may be adapted for this purpose (Hammerstedt, 1973) or a luminometer may be used. Several very advanced devices are now available on the market. Some of these are capable of measuring 10^{-13} g of ATP, which corresponds theoretically to about 200 bacteria, but given the dilution of the sampling during preparation and the volume in which the measurement is made, the limit is actually in the order of 10^5 cells/g, which is not entirely satisfactory (Lacheretz et al., 1979). Many efforts are therefore under way in order to reach an ideal limit in the order of 10^3 cells/g.

For food product analysis, sample preparation poses definite problems. When the microorganisms being sought are the only living material in the product, they contain all of the ATP and preparation is fairly straightforward. The cells are recovered on a filter and the ATP is extracted using a suitable agent. The procedure is therefore usable for the evaluation of the overall contamination of water, the concentration of a starter culture, or the effectiveness of an antibiotic. On the other hand, when microorganisms are to be detected in a product containing cellular material (biological liquid, food), the product also contains "somatic" ATP which must be distinguished from the microbial ATP (Sharpe et al., 1970). Appropriate extraction processes may deal with these difficulties in theory, although the literature is replete with mentionings of problems encountered in the case of products such as milk and meat.

Having no incubation step, the procedure does not benefit from the specificity offered by selective media and must therefore be limited to the evaluation of total microflora. A specific process for staphylococci has nevertheless been proposed by Tuncan and Martin (1987), based on the use of lyostaphin, a lytic agent relatively specific for staphylococci.

Some success has been achieved with carbonated drinks (La Rocco et al., 1985; Littel and La Rocco, 1985 and 1986), for which the detection limit for yeasts is in the order of 2 CFU/ml, based on the analysis of filter residues obtained from 100 ml of sample. The correlation coefficient with standard analysis is satisfactory.

In the dairy context, the situation is more variable, some authors reporting good results (Frankinet and Pernelle, 1983; Pernelle, 1983) while others report difficulties associated with the presence of residual somatic ATP which increases the background noise level, increasing the practical detection limit to between 10^5 and 10^6 cells/ml (Bossuyt, 1981; Theron et al., 1986a, b; Botha et al., 1986).

Interesting results have been obtained with meats. Stannard and Wood (1983) obtained good results in the 10^5–10^6 range/g, albeit using a rather onerous preparation technique. Bülte and Reuter (1985) did not surpass the detection limit of 5×10^6/g, while for Kennedy and Oblinger (1985), the limit remained at 10^7/g. Littel and co-workers (1986), using a double filtration technique, succeeded in eliminating most of the somatic ATP and lowering the detection limit to 5×10^4 CFU/g of meat. Satisfactory results were also obtained for fish by Ward and co-workers (1986).

5.3.3 Radiometric Methods

The intensity of microbial activity may also be evaluated by radioactivity measurements, some of which are extremely sensitive (e.g., liquid scintillation spectrometry). Two types of method have been proposed, based on different types of biological activity. One of these is CO_2 production resulting from energy pathways, whether respiratory or fermentative, while the other includes the assimilation of metabolically significant substances and the specific adsorption of labeled antibodies.

5.3.3.1 The $^{14}CO_2$ Method

The $^{14}CO_2$ method essentially involves placing the analyzed food or cells collected on a filter in the presence of a sugar or other metabolizable substance labeled with ^{14}C. This substance most often being glucose, the $^{14}CO_2$, which is released as a result of respiration or fermentation, is trapped in an alkaline solution for radioactivity measurement, thereby yielding an indication of the overall level of microbial activity in the product. The degrading of glucose with CO_2 production being a very widespread activity, this method permits non-selective detection of the vast majority of microorganisms, although some selectivity is possible by varying the composition of the medium and the nature of the labeled substrate.

Originally proposed in 1956 by Levin and colleagues for the detection of coliforms on a ^{14}C lactose medium, the $^{14}CO_2$ was detected in these early trials by means of barite-impregnated paper and flow-counter mesurement, requiring a radioactive dose in the 25 μCi range. The procedure gradually improved, some workers using a liquid scintillation spectrometer while others retained the flow counter, receiving the $^{14}CO_2$ directly and giving rise to the development of the Bactec device. This device works as follows: The sample is injected into a flask holding a culture medium containing 1.5 μCi as ^{14}C glucose, sealed by a rubber cap. After sufficient incubation time, around 6–8 hours for food products, the flask is flushed with a gas stream which carries the $^{14}CO_2$ to an ionization chamber in which the radioactivity is measured. Detection time for a given inoculum depends on the species, 6 hours being sufficient to detect a mixed contamination of 10^5 cells and 8 hours being sufficient for 10^4 cells. Several successive measurements of $^{14}CO_2$ are possible on the same culture fed with labeled substrate, at intervals of a few hours or even a few days. This nondestructive method provides very early detection of heavy contamination with the possibility of later detection of low-level contaminations. The process has aroused interest in medical bacteriology, notably for the rapid detection of bacteremia, as well as in food-related environments. Its application to the latter, however, after generating a few published works (Rolley et al., 1976; Stewart et al., 1980), currently seems to be dormant.

5.3.3.2 Labeled Compound Absorption or Adsorption Methods

The absorption method is based on the property possessed by many microorganisms enabling them to concentrate certain metabolically important substances intracellularly. This is an active transport phenomenon which often results in intracellular concentrations 1,000–10,000 times higher than the extracellular concentration. In fact, the adsorption of labeled substances by a given number of living cells is widely used to study active transport phenomena. Conversely, information about the number of cells present may be derived from measurements of absorbed radioactivity once all conditions are standardized.

Bourgeois and associates proposed a procedure in 1973 featuring the following details:

- The labeled compound is a radioactive amino acid, selected from among those which heterotrophic microorganisms concentrate most actively (i.e., lysine in general).
- The microorganisms are first collected on a membrane which is then applied to an absorbent pad soaked with culture medium, as in the membrane culture technique; following a brief incubation period, the membrane is recovered, rinsed on a filtration apparatus and dried.
- A liquid scintillation spectrometer is used for detection, the cell-carrying membrane being introduced into the scintillation liquid.

The performance of this method is quite remarkable, due to the degree of amino acid absorption and also to the sensitivity of liquid scintillation spectrometry. One cell on a membrane can be demonstrated in 24 hours or as little as 8 hours depending on the species. In the case of mixed contaminations (total mesophilic flora), testing based on standards of 10,000/g and 100,000/g may be effected within 3 and 2 hours, respectively.

The procedure has been the subject of experimentation for applications to brewing, milk processing, and meat processing (Mafart et al., 1976, 1978, 1981, 1985; Bourgeois et al., 1985). The major drawback associated with this very effective method is radiometry itself, even though the levels of radioactivity involved are very low (0.2 μCi/trial).

5.4 Electrochemistry

Various microbial activity associated chemical characteristics of the medium may be electrically measured and thereby form the basis of signals which indicate the presence of microorganisms.

5.4.1 Potentiometry

5.4.1.1 Redox Potential

The drop in redox potential accompanying microbial development—which, as previously mentioned, may be monitored spectrophotometrically—may also be registered as the potentiometric measurement of hydrogen evolution or of oxygen absorption.

The procedure proposed by Wilkins and co-workers (1974, 1978) consists of monitoring hydrogen production during culture by means of a system composed essentially of two reference electrodes. The liberation of hydrogen is manifested as an increase in cathode potential which may be visualized using a scrolling printer. Such variations in potential are perceptible only after a latent period whose duration is an inverse function of the inoculum size. The linear relationship existing between logarithmic number of bacteria and the latent period enables evaluation of the number of bacteria from the measured latent period. In the case of *E. coli*, this period is 1 hour for 10^6 cells and 7 hours for 1 cell.

Junter (1979, 1980) and Selegny (1980) and their co-workers proposed measuring O_2 absorption by adding lipoic acid, a redox intermediate, to the medium to regularize and amplify the signal. These works are somewhat theoretical, but other authors have attempted to apply these processes to the analysis of meat (Brown and Childers, 1978) and milk (Rowe and Gilmore, 1986). The latter authors studied the evolution of oxygen partial pressure in stirred and agitated milk in relation to the number and behavior of psychrotrophes. This evolution seems to be a good indicator of spoilage by protease and lipase production by pseudomonads, providing more valuable information than the number of bacteria, which is poorly correlated with spoilage enzyme activity.

The approach of Kroll and co-workers (1989) was to use an oxygen to assess catalase activity, which is an enzyme activity related to the level of contamination. In the case of meats, the presence of endogenous catalase gives rise to difficulties which are overcome by filtering onto large pored positively charged "Zeta-plus" filters. A linear response has been obtained above 10^5 bacteria/g.

5.4.1.2 pH

Visual observation of the color change of a pH indicator is a widely used process for demonstrating microbial activity and hence the presence of the microorganism responsible, although potentiometry may provide useful possibilities.

Harrison and associates (1974) proposed applying an electrochemical process described by Mitz (1969) to testing beer for contaminations. Microbes concentrated onto a filtering membrane are introduced into a cell containing a weak buffer having an initial pH of 7. Microbial multiplication brings about a drop in pH which may be measured and recorded using appropriate equipment. The ability to measure pH variations of 0.1 unit with certainty allows the detection of heavy contaminations in 0.5–6 hours and contaminations of a few units per ml in 15–20 hours. The value of this system appears to be limited by the difficulty in measuring and recording pH variations with sufficient precision and safety. Since pH is a logarithic function of hydrogen ion concentration, large changes in cell populations result in only small changes in pH. The procedure does not appear to have been developed to a practical level.

From this point of view it would appear to be useful to apply titrimetry after a standard incubation time. A linear relation does exist between the initial population and the variation in total acidity expressed as quantity of acid. The acidity of milk is therefore considered to be an indicator of its bacteriological quality. This approach has been examined by Shelef and Jay (1970) for the assessment of the bacteriological status of refrigerated meat, in which case the flora brings about an alkalization which is measured by titrating back to a final pH of 5.0 using dilute HCl.

5.4.1.3 Ion Concentrations

Various ions excreted into the medium by microorganisms may indicate the type and degree of contamination, and possibly the potential storage time. The latter has been exploited in the case of milk and fish, determinations being performed either by conventional chemical methods or by electrochemical methods using ion-specific electrodes or, better yet, biosensors (i.e., enzyme or immobilized microorganism electrodes) (Schmid, 1988; Schaertel and Firstenberg-Eden, 1988).

In milk processing, variations in pyruvate and ammonium ion concentration have received attention for some time, the significance of which is discussed elsewhere (Grappin et al., 1980). The determination of lactate using a lactic dehydrogenase or lactic oxidase electrode may be used for the same purpose (Matsunaga, 1982).

With respect to seafoods and sometimes meats, the most frequently used indicator metabolites are biogenic amines (histamine), trimethylamine, hypoxanthine, indole, and ammonium analyzed by various methods. Among these substances, ammonium and hypoxanthine may be measured using specific electrodes.

The technique of Ward and co-workers (1979) for ammonium determination in the case of fresh shrimp provides results well correlated with other spoilage criteria. Assessment is more difficult in the case of frozen shrimp.

The technique of Watanabe and colleagues (1983) uses an enzyme electrode, an oxygen electrode in fact, equipped with a xanthine oxidase membrane.

5.4.2 Amperometry

Amperometry is based on the fact that microorganisms themselves are negatively charged particles, which in the presence of an electrode consisting of a platinum anode and a silver peroxide cathode, yield their electrodes to the anode and regenerate their charge using the components of medium. This process creates a measurable current. A reference electrode surrounded by a membrane impermeable to bacteria can differentiate the current produced by electro-active substances from that due to microorganisms (Matsunaga et al., 1979; Nishikawa et al., 1982). The technique seems intended for monitoring cell populations in fermentors than for the bacteriological control of food products.

Patchett and co-workers (1989) have nevertheless recently attempted to apply it to testing meat and milk, meeting with some success. Prior incubation for a few hours is necessary in order to reach the threshold of 10^6/ml. The Biocheck developed by the Cranfield Institute of Technology is based on this principle and seems to harbor some interesting possibilities.

5.4.3 Conductometry

In the course of the metabolic reactions which allow the growth of microorganisms, various electrically inert molecules (e.g., sugars) are converted to ionized molecules such as acids. This occurrence obviously results in an increase in the conductance, that is, a drop in the resistance (impedance for alternating current) of the medium.

This phenomenon may be exploited by the following scheme: The sample to be analyzed is introduced into a measuring cell containing a suitable culture medium into which two electrodes are inserted. The periodic passage of an alternating current permits the measurement of impedance and the recording of its variations in comparison with a control cell containing sterile medium. A drop in the impedance in the measuring cell indicates the presence and growth of microorganisms, the time required to obtain a detectable signal being related to the number of microorganisms present.

The devices currently available appear to combine good performance characteristics (instantaneous detection of 10^7 cells/ml) with good adaptation to the needs of food product testing, that is, the ability to accept samples of variable size, simultaneous handling of numerous samples, wide temperature range, ease of use, and ability to work with turbid media.

Numerous publications cite this technique. The essential elements are presented in an article by Firstenberg-Eden and Eden (1984). More recent articles discuss the development of buffers and media adapted to the technique (Owens and Wacher-Viveros, 1986) and the detection of particular groups or the performance of particular analyses, such as for coliforms (Martins and Selby, 1980; Firstenberg-Eden and Klein, 1983), salmonella (Gibson, 1987; Ogden and Cann, 1987; Ogden, 1988; Bullock and Frodsham, 1989), histamine-producing bacteria (Klausen and Huss, 1987), and yeasts (Zindulis, 1984), as well as measurement of the lag phase (Mackey and Derrick, 1984), determination of niacin (Einarsson and Snygg, 1986), detection of antibiotics in milk (Okigbo and Richardson, 1985), and prediction of storage time (Philips and Griffiths, 1985). Other articles deal with analysis of different categories of products, such as water (Silverman and Munoz, 1979), milk and dairy products (Cady, 1977; Cady et al., 1978; Firstenberg-Eden, 1984; Griffiths and Phillips, 1984; Khayat and Richardson, 1986; Khayat et al., 1988), meat (Strasser, 1979; Firstenberg-Eden, 1983; Bülte and Reuter, 1984), fish (Gibson et al., 1984; Ogden, 1986), frozen vegetables (Hardy et al., 1977), beer (Evans, 1982 and 1985; Adams, 1989), and wine (Henschke et al., 1988).

Overall, the process appears very interesting, although the high cost of the equipment required has so far limited its development in many countries.

5.5 Other Processes

5.5.1 Chromatography

Given that the essential requirement of a process suitable for rapid microbiological detection is sensitivity, chromatographic techniques such as gaseous phase chromatography (GC) and high-pressure liquid chromatography (HPLC) are likely future candidates in this field, although developments have thus far been rare and limited to GC. Described by Mitruka and Alexander (1968), this process utilizes an electron capture detector to detect small numbers of cells following a brief incubation.

The feature of greatest interest of GC is its capacity for the analysis of not just one substance but of groups of substances, thereby providing a spectral characterization of the microorganism (e.g., the spectrum of secreted fatty acids) allowing the detection of microbial presence and a step toward identification at the same time. Strictly speaking, the characterization of a microorganism by its secreted fatty acid spectrum requires previous isolation and growth on the appropriate standard cultured medium. In this case, the analysis of the medium constitutes an identification technique.

The direct analysis of a food which is organoleptically spoiled or hygienically suspect may, in some cases, lead to the discovery of the microorganism responsible without prior incubation. This possibility obviously refers to heavy contaminations due to the exclusive development of a single microbial species. The usefulness of GC analysis for such detections stems from both its speed and its aptness for the presumptive identification of the dominating microorganism.

Gas chromatography may first be used to determine the microbial origin of a spoilage. In brewing, for example, it is well known that the fatty acid spectrum of beer is the result of the fermenting action of the yeast on the wort, that it depends on the strain of yeast used, and that it is altered by heavy contaminants in a way that allows their presumptive identification (Welhoener et al., 1971; Arkima, 1973).

Jackson and Kempton (1972) provided examples of cases in which GC analysis determined the source of a spoilage in 3 hours. This was not done using GC alone, although it contributed, along with other data (pH, usual product contaminants) to the presumptive identification of the microorganism responsible. In pineapple sauce, for example, low pH and an abnormal ethanol level detected by GC suggested a yeast was involved. In stale salami, an alcohol and acid spectrum which had remained normal suggested the exclusion of a bacteriological cause. In spoiled canned goods, GC determines rapidly whether contamination is the result of a leak or of incomplete sterilization, since *Clostridium* species responsible for spoilage after incomplete sterilization have a characteristic spectrum. Besides its speed and capacity for discrimination, GC analysis also has two other advantages over conventional methods, namely, that it addresses only the microorganisms which are truly active in spoiling the product and that it detects the agent responsible even if the latter has been destroyed by a thermal treatment or by a self-sterilizing process subsequent to the actual spoilage. Schafer and co-workers (1982) as well as Eyles and Adams (1986) have followed this procedure in order to identify the source of spoilage of canned goods.

With respect to hygiene, Mayhew and Gorbach (1975) suggested applying GC analysis to the presumptive detection of *Clostridium botulinum* in foods. They have shown that the short-chain fatty acid spectra obtained by analysis of the culture medium used to grow the microorganism as well as by analysis of juice from a can spiked with the same *C. botulinum* type B strain are very similar. Tests carried out with various species of *Clostridium* and other microorganisms led them to conclude that the procedure is a good presumptive test for the development of *C. botulinum*. Snygg and associates (1979) applied head space analysis by GC to the detection of *C. botulinum* positive samples in trout vacuum-packed under polyethylene film. By analyzing not only fatty acids but also alcohols and amines, they obtained complex chromatograms that could discriminate between toxic and nontoxic samples using a computer. A more recent publication deals with the estimation of coliforms in milk (Guerzoni et al., 1987).

In summary, GC analysis is a good method for the rapid determination of the cause of spoilage, notably for the discovery of development of the genus *Clostridium*, although it is a relatively complex technique, at least for this particular use.

5.5.2 Microcalorimetry

Microcalorimetry is based on the well-known fact that microbial activity liberates caloric energy, thereby making it possible to use a very sensitive calorimeter to detect the presence of small numbers of microorganisms in a contaminated food product or nutrient medium by comparing its thermogenesis to that of a sterile control.

Lampi and Meiselman (1974) and Lampi and co-workers (1974) thus determined that the minimum detection time for a minimum thermal flux of 0.03 cal/hr., that is, for a number of microorganisms in the range of 10^4 cells/ml, varies from 2 to 13 hours depending on the inoculum size. Caloric production passes through a maximum at the end of a time interval which is a linear function of the log number of bacteria. These authors consider this relationship usable for estimating the level of contamination of a food product. Kallings (1976) further asserted that thermograms are species characteristics and as such may be used for identification purposes. The procedure nevertheless appears to be difficult to apply to food testing. The work of Gram and Sogaard (1985), dealing with ground meat analysis, may be cited, however, which concludes that estimates in the range of 10^5–10^8 CFU/g in less than 24 hours are possible.

5.6 Conclusion

Various techniques may therefore be involved in the necessary evolution of food microbiology and its application in the food industry. Furthermore, the overview in this chapter provides only a sample of the possibilities, since the correct rapid technique must be custom designed for each situation in order to fully exploit the particular conditions prevalent in a given company (i.e., the level of contamination tolerable, the nature of the usual contaminants, etc.).

With respect to reliability, rapid techniques of analysis often have little in the way of advantages over the conventional methods which serve as references during their evaluation. They are also accompanied by specific disadvantages, of which the most often voiced are lack of specificity, technological overload, and costliness.

Speed is often achieved partly at the expense of specificity, that is, presumptive identification will be less precise and less certain than the conventional method in the case of any particular microorganism. The time saving must therefore be weighed against the loss of specificity before deciding whether or not the innovation is justifiable.

With respect to technological load, it must be borne in mind in any event that bacteriological analysis, whether conventional or otherwise, requires specific skills. Rapid techniques, which are generally biochemical, chemical, or physical in nature do not require more skill and usually lend themselves readily at least to semiautomation.

Finally, with respect to cost, rapid techniques generally require significant capital investment for equipment, although on an operational level they undoubtedly compare quite favorably since they are often miniaturized and partly automated.

The changeover to rapid techniques is therefore worthy of serious consideration.

References

Adams M. R., 1989. A medium designed for monitoring pitching yeast contamination in beer using a conductimetric technique. *Letters Appl. Microbiol.*, **8**, 55–58.

Andrew J. and Luck H., 1974. Sample size determination for estimating the mean of the logarithm of the bacterial count of milk by using the methylene blue reduction time. *Milchwissenschaft*, **29**, no. 3, 137–141.

Arkima V., 1973. Die Bildung der Fettsäuren C4–C10 bei durch wilde Hefen Hervorgerufenen Garungen. *In: Proceedings of the Congress of the European Brewery Convention*, **14**, Salzburg, Elsevier Scientific Publishing Company, Amsterdam, 1974, 309–315.

Bartholin G. and Bouder B., 1973. Recherche d'une méthode rapide de contrôle de la qualité bactériologique en usine de conserves de légumes. *Bulletin d'Informations Techniques de la Station Expérimentale de Dury*, **8**, 57.

Baumgart J., Portner A., Lassak G., 1975. Schnellmethode zur Bestimmung der aeroben Gesamtkeimzahl (Lebendkeimzahl-auf Frischfleisch mit Hilfe photometrisch messbarer Extinktionsänderungen beim Resazurintest). *Fleischwirtschaft*, **55**, no. 7, 969–973.

Bossuyt R., 1981. Determination of bacteriological quality of raw milk by an ATP assay technique. *Milchwissenschaft*, **36**, no. 5, 257–260.

Botha W. C., Lücke H., Jooste P. J., 1986. Determination of bacterial ATP in milk. The influence of adenosine triphosphate-hydrolyzing enzymes from somatic cells and Pseudomonas fluorescens. *J. Food Prot.*, **49**, no. 10, 822–825.

Bourgeois C. M., Mafart P., Thouvenot D., 1973. Méthode rapide de détection des contaminants dans la bière par marquage radioactif. *In: Proceedings of the Congress of the European Brewery Convention*, **14**. Salzburg, Elsevier Scientific Publishing Company, Amsterdam, 1974, 219–230.

Bourgeois C. M., Mafart P., Cleret J. J., 1985. Appréciation rapide de la qualité microbiologique des carcasses de volailles par une méthode radiométrique par absorption. *Sci. Aliments*, **5**, 61–72.

Bradshaw N. J., Dyett E. J., Herschdoerfer S. M., 1961. Rapid bacteriological testing of cooked or cured meats, using a tetrazolium compound. *J. Sci. Food Agric.*, **12**, April, 341–344.

Brown L. R. and Childers G. W., 1978. A rapid method of estimating total bacterial counts in ground beef. *In:* Sharpe E. N., Clark D. S. (eds). *Mechanizing microbiology*, 87–103. Charles C. Thomas.

Bullock R. D. and Frodsham D., 1989. Rapid impedance detection of salmonellas in confectionery using modified LICNR broth. *J. Appl. Bacteriol.*, **66**, 385–391.

Bülte M. and Reuter G., 1984. Impedance measurement as a rapid method for the determination of the microbial contamination of meat surfaces, testing two different instruments. *Int. J. Food Microbiol.*, **1**, 113–125.

Bülte M. and Reuter G., 1985. The bioluminescence technique as a rapid method for the determination of the microflora of meat. *Int. J. Food Microbiol.*, **2**, 371–381.

Cady P., 1977. Instrumentation in food microbiology. *Food Product Dev.*, April.

Cady P., Hardy D., Martins S., Dufour S. W., Kraeger S. J., 1978. Automated impedance measurements for rapid screening of milk microbial content. *J. Food. Prot.*, **41**, no. 4, 277–283.

Einarsson H. and Snygg B. G., 1986. Niacin assay by monitoring changes in electrical conductance caused by microbial growth. *J. Appl. Bacteriol.*, **60**, 15–19.

Emswiler B., Kotula A., Chesnut C., Young E., 1976. Dye reduction method for estimating bacterial counts in ground beef. *Appl. Environ. Microbiol.*, **31**, no. 4, 618–620.

Evans H. A. V., 1982. A note on two uses for impedimetry in brewing microbiology. *J. Appl. Bacteriol.*, **53**, 423–426.

Evans H. A. V., 1985. A note on the use of conductimetry in brewery microbiological control. *Food Microbiol.*, **2**, 19–22.

Eyles M. J. and Adams R. F., 1986. Detection of microbial metabolites by gas chromatography in the examination of spoiled canned foods and related products. *Int. J. Food Microbiol.*, **3**, 321–330.

Firstenberg-Eden R., 1983. Rapid estimation of the number of microorganisms in raw meat by impedance measurement. *Food Technol.*, January, 64–70.

Firstenberg-Eden R., 1984. Collaborative study of the impedance method for examining raw milk samples. *J. Food Prot.*, **47**, no. 9, 707–712.

Firstenberg-Eden R. and Klein C. S., 1983. Evaluation of a rapid impedimetric procedure for the quantitative estimation of coliforms. *J. Food Sci.*, **48**, 1307–1311.

Firstenberg-Eden R. and Eden G., 1984. Impedance microbiology. *Innovation in Microbiology. Series no. 3,* Letchworth, Research Studies Press, 170 p.

Frankinet J. and Pernelle M., 1983. Détermination rapide de la flore totale d'un lait cru par dosage d'ATP par bioluminescence. *Bio-Sciences*, **8**, 127–128.

Freier T. A. and Hartman P. A., 1987. Improved membrane filtration media for enumeration of total coliforms and *Escherichia coli* from sewage and surface waters. *Appl. Environ. Microbiol.*, **53**, no. 6, 1246–1250.

Gibson D. M., 1987. Some modification to the media: rapid automated detection of salmonellas by conductance measurement. *J. Appl. Bacteriol.*, **53**, 299–304.

Gibson D. M., Ogden I. D., Hobbs G., 1984. Estimation of the bacteriological quality of fish by automated conductance measurements. *Int. J. Food Microbiol.*, **1**, 127–134.

Gram L. and Sogaard H., 1985. Microcalorimetry as a rapid method for estimation of bacterial levels in ground meat. *J. Food Prot.*, **48**, no. 4, 341–345.

Grappin R., Dromard, T., Dasen A., 1980. Relation entre la qualité bactériologique du lait cru et sa teneur en pyruvate et en ammoniaque. Communication au Colloque International de Microbiologie Alimentaire de la Société Française de Microbiologie, Lille, June.

Griffiths M. W. and Phillips J. D., 1984. Detection of post-pasteurization contamination of cream by impedimetric methods. *J. Appl. Bacteriol.*, **57**, 107–114.

Guerzoni M. E., Gardini F., Cavazza A., Piva M., 1987. Gas-liquid chromatographic method for the estimation of coliforms in milk. *Int. J. Food Microbiol.*, **4**, 73–78.

Hammerstedt R., 1973. An automated method for ATP analysis utilizing the luciferin-luciferase reaction. *Analytical Biochem.*, **52**, no. 2, 449–455.

Hardy D., Kraeger S., Dufour S., Cady P., 1977. Rapid detection of microbial contamination in frozen vegetables by automated impedance measurements. *Appl. Environ. Microbiol.*, **34**, no. 1, 14–17.

Harrison J., Webb T., Martin P., 1974. The rapid detection of brewery spoilage microorganisms. *J. Instit. Brew.*, **80**, no. 4, 390–398.

Henschke P. A. and Thomas D. S., 1988. Detection of wine-spoiling yeasts by electronic methods. *J. Appl. Bacteriol.*, **64**, 123–133.

Jackson E. and Kempton A., 1972. The application of gas chromatography to selected problems in the food industry. *Dev. Ind. Microbiol.*, **14**, 285–301.

Jakubczak E. and Leclerc H., 1979. Contrôle de la fermentation lactique par mesure de l'ATP en fabrication de yaourts. Communication aux Journées de Microbiologie Alimentaire de la Société Française de Microbiologie, Lille, May.

Jedlicka G., Hill W., Heck J., 1975. A research note. A reflectance method for the enumeration of surface bacteria. *J. Food. Sci.*, **40**, no. 3, 647–648.

Jorgensen H. L. and Schultz E., 1985. Turbidimetric measurement as a rapid method for the determination of the bacteriological quality of minced meat. *J. Food Microbiol.*, **2**, 177–183.

Junter G. A., Selegny E., Lemeland J. F., 1979. Analyse théoriques des variations de potentiel dans les cultures de *Escherichia coli* K12 en présence d'un transporteur d'électrons. *Ann. Microbiol.* (Institut Pasteur), **130A**, 295–313.

Junter G. A., Lemeland J. F., Selegny E., 1980. Electrochemical detection and counting of *Escherichia coli* in the presence of a reducible coenzyme, lipoic acid. *Appl. Environ. Microbiol.*, **39**, no. 2, 307–316.

Kallings L., 1976. Application of microcalorimetry. *In: International Symposium on "Rapid methods and automation in microbiology,"* vol. 2, 140–143. Learned Information (Europe) Ltd, Cambridge, Oxford.

Kennedy J. E. and Oblinger J. R., 1985. Application of bioluminescence to rapid determination of microbial levels in ground beef. *J. Food Prot.*, **48**, no. 4, 334–340.

Khayat F. A. and Richardson G. H., 1986. Detection of abnormal milk with impedance microbiology instrumentation. *J. Food Prot.*, **49**, no. 7, 519–522.

Khayat F. A., Bruhn J. C., Richardson G. H., 1988. A survey of coliforms and *Staphylococcus aureus* in cheese using impedimetric and plate count methods. *J. Food. Prot.*, **51**, no. 1, 53–55.

Kissinger J., 1969. Modified resazurin test for estimating bacterial population in maple sap. *J. Assoc. Off. Anal. Chem.*, **52**, no. 4, 714–716.

Kissinger J., 1974. Collaborative study of a modified resazurin test for estimating bacterial count in raw sample sap. *J. Assoc. Off. Anal. Chem.*, **57**, no. 3, 544–547.

Klausen N. L. and Huss H. H., 1987. A rapid method for detection of histamine-producing bacteria. *Int. J. Food Microbiol.*, **5**, 137–146.

Kouomegne R., Bracquart P., Linden G., 1984. Application d'un réactif de transparisation du lait au dénombrement de bactéries. *Le Lait*, **64**, 418–435.

Kroll R. G., Frears E. R., Bayliss A., 1989. An oxygen electrode-based assay of catalase activity as a rapid method for estimating the bacterial content of foods. *J. Appl. Bacteriol.*, **65**, 209–217.

Kvasnikov E., Gerasimenko L., Tabarovskaja Z., 1974. Utilization de chlorure de 2,3,5-triphényl-tétrazole pour la détection rapide des bactéries anaérobies mésophiles dans l'industrie des conserves (in Russian). *Voprosi Pitanijia*, **6**, 62–65.

Lacheretz A., Jakubczak E., Leclerc H., Catsaras M., 1979. Essais de détermination de la contamination bactérienne de la viande par mesure de l'ATP. Communication aux Journées de microbiologie alimentaire de la Société Française de Microbiologie, Lille, May.

Lamp R. P. M., Moulton R. P., Mulders S. L. T., 1983. Performance of a semiautomated antibiotic susceptibility testing system (ABAC). *J. Appl. Bacteriol.*, **55**, 209–214.

Lampi R. and Meiselman H., 1974. *Calorimetric Detection of Food Contamination.* Activity report, Research Development Association, Military Packaging System Inc., **26**, 103–112.

Lampi R., Mikelson D., Rowley D., Previte J., Wells R., 1974. Radiometry and microcalorimetry: Techniques for the rapid detection of foodborne microorganisms. *Food Technol.*, **28**, no. 10, 52–58.

La Rocca K. A., Galligan P., Little K. J., Spurgash A., 1985. A rapid bioluminescent ATP method for determining yeast contamination in a carbonated beverage. *Food Technol.*, July, 49–53.

Leistner L., Bem Z., Dresel J., 1977. Schnellmethoden zur keimzallbestimmung bei Fleisch. Mitteilungsbl. der Bundesanstalt für Fleischforschung. *Kulmbach*, **56**, 3070–3072.

Levin G., Harrison V., Hess W., Gurney H., 1956. A radioisotope technique for the rapid detection of coliform organisms. *Am. J. Public Health*, **46**, 1405–1414.

Littel K. J. and La Rocco K. A., 1985. Bioluminescent standard curves for quantitative determination of yeast contaminants in carbonated beverages. *J. Food Prot.*, **48**, no. 12, 1022–1024.

Littel K. J. and La Rocco K. A., 1986. ATP screening method for presumptive detection of microbiology contaminated carbonated beverages. *J. Food Sci.*, **51**, no. 2, 474–476.

Littel K. J., Pilelis S., Spurgash A., 1986. Bioluminescent ATP assay for rapid estimation of microbial numbers in fresh meat. *J. Food Prot.*, **49**, no. 1, 18–22.

Liuzzo J., Lagarde S., Grodner R., Novak A., 1975. A total reducing substance test for ascertaining oyster quality. *J. Food. Sci.*, **40**, no. 1, 125–128.

Mackey B. M., Derrick C. M., 1984. Conductance measurements of the lag phase of injured *Salmonella* typhimurium. *J. Appl. Bacteriol.*, **57**, 299–308.

Mafart P., Bourgeois C. M., Duteurtre B., Moll M., 1976. Radiometric method for control of filtration and pasteurization. Technical quarterly. *Master Brewers Association of America*, **13**, no. 3, 157–160.

Mafart P., Bourgeois C. M., Duteurtre B., Moll M., 1978. Use for ^{14}C lysine to detect microbial contamination in liquid foods. *Appl. Environ. Microbiol.*, **35**, no. 6, 1211–1212.

Mafart P., Cleret J. J., Bourgeois C., 1981. Détection et évaluation des contaminations microbiennes dans les produits alimentaires. *Analusis*, **9**, no. 1, 32–34.

Mafart P. and Bourgeois C., 1985. Les méthodes radiométriques dans l'analyse des produits alimentaires. *Bios*, **16**, no. 11, 21–25.

Marcelis J. H., Versteeg H., Mansvelt Beck H. J., Vinke D., 1980. Semielectronic turbidimeter for automated monitoring of bacterial growth in test tubes. *Appl. Environ. Microbiol.*, **39**, no. 2, 281–284.

Martins S. B. and Selby M. J., 1980. Evaluation of a rapid method for the quantitive estimation of coliforms in meat by impedimetric procedures. *Appl. Environ. Microbiol.*, **39**, no. 3, 518–524.

Matsunaga T., Karube I., Teraoka N., Suzuki S., 1982. Determination of cell numbers of lactic acid producing bacteria by lactate sensor. *Europ. J. Appl. Microbiol. Biotechnol.*, **16**, 157–160.

Mayhew J. and Gorbach S., 1975. Rapid gas chromatographic technique for presumptive detection of *Clostridium botulinum* in contaminated food. *Appl. Microbiol.*, **29**, no. 2, 297–299.

Mitruka B. and Alexander M., 1968. Rapid and sensitive detection of bacteria by gas chromatography. *Appl. Microbiol.*, **16**, no. 4, 636–640.

Mitz M., 1969. The detection of bacteria and viruses in liquids. *Ann. NY Acad. Sci.*, **158**, 651–664.

Moran J. and Witter L., 1976. An automated rapid test for *Escherichia coli* in milk. *J. Food Sci.*, **41**, no. 1, 165–167.

Nishikawa S., Sakai S., Karube I., Matsunaga T., Suzuki S., 1982. Dye-coupled electrode system for the rapid determination of cell populations in polluted water. *Appl. Environ. Microbiol.*, **43**, no. 4, 814–818.

Ogden I. D., 1986. Use of conductance methods to predict bacterial counts in fish. *J. Appl. Bacteriol.*, **61**, 263–268.

Ogden I. D., 1988. A conductance medium to distinguish between *Salmonella* and *Citrobacter* spp. *Int. J. Food Microbiol.*, **7**, 287–297.

Ogden I. D. and Cann D. C., 1987. A modified conductance medium for the detection of *Salmonella* spp. *J. Appl. Bacteriol.*, **63**, 459–464.

Okigbo O. N. and Richardson G. H., 1985. Detection of penicillin and streptomycin in milk by impedance microbiology. *J. Food Prot.*, **48**, no. 11, 979–981.

Otsuka G. and Nakae T., 1969. Resazurin test paper method for determining the sanitary quality of raw milk. *J. Dairy Sci.*, **52**, no. 12, 2041–2044.

Owens J. D. and Wacher-Viveros M. C., 1986. Selection of pH buffers for use in conductimetric microbiological assays. *J. Appl. Bacteriol.*, **60**, 395–400.

Patchett R. A., Kelly A. F., Kroll R. G., 1989. Investigation of a simple amperometric electrode system to rapidly quantify and detect bacteria in foods. *J. Appl. Bacteriol.*, **66**, 49–55.

Paul S. and Nitsch B., 1974. Die Anwandung der elektronischen Mikrokoloniezählung, EMZ, zur Qualitätskontrolle von Rohmilch. *Deutsche Molkerei Zeit.*, **95**, no. 33, 1201–1208.

Pernelle M., 1983. Utilisations de la bioluminescence en industrie laitière. *Technique Laitière*, **980**, 33–36.

Peterson E. H., Nierman M. L., Rude R. A., Peeler J. T., 1987. Comparison of AOAC method and fluorogenic (MUG). Assay for enumerating *Escherichia coli* in foods. *J. Food Sci.*, **52**, no. 2, 409–410.

Philips J. D. and Griffiths M. W., 1985. Bioluminescence and impedimetric methods for assessing shelf-life of pasteurized milk and cream. *Food Microbiol.*, **2**, 39–51.

Quadri R., Buckle K., Edwards R., 1974. Rapid methods for the determination of faecal contamination in oysters. *J. Appl. Bacteriol.*, **37**, no. 1, 7–14.

Robinson B. J., 1984. Evaluation of a fluorogenic assay for detection of *Escherichia coli* in foods. *Appl. Environ. Microbiol.*, **48**, no. 2, 285–288.

Rowe M. T. and Gilmour A., 1986. Oxygen tension measurement as a mean of detecting incipient spoilage of raw milk by psychrotrophic bacteria? *Int. J. Food Microbiol.*, **3**, 43–49.

Rowley D., Previte J., Srinivasa H., 1976. A radiometric screening method for estimating the level of aerobic mesophilic bacteria in foods. *In: International Symposium on "Rapid methods and automation in microbiology,"* vol. 2, 3. Learned information (Europe) Ltd, Cambridge, Oxford.

Schaertel B. J. and Firstenberg-Eden R., 1988. Biosensors in the food industry: Present and future. *J. Food Prot.*, **51**, no. 10, 811–820.

Schafer M. L., Peeler J. T., Bradshaw J. G., Hamilton C. H., Carver R. B., 1982. A rapid gas chromatographic method for the identification of sporeformers and nonsporeformers in swollen cans of low-acid foods. *J. Food Sci.*, **47**, 2033–2037.

Schmid R. D., 1988. Trends in biosensors. *Biofutur*, March 37–41.

Schulze E., Jensen B., Celerynova E., 1988. Automated turbidimetry for rapid determination of the bacteriological quality of raw meat and processed meat products. *Int. J. Food Microbicol.*, **6**, 219–227.

Selegny E., Junter G. A., Charriere G., Lemeland J. F., 1980. Détection électrochimique d'*Escherichia coli* par potentiométrie en présence d'acide lipoïque. Communications au Colloque International de Microbiologie Alimentaire de la Société Française de Microbiologie, Lille, June.

Sharpe A., Woodrow M., Jackson A., 1970. Adenosine triphosphate levels in foods contaminated by bacteria. *J. Appl. Bacteriol.*, **33**, no. 4, 758–767.

Sharpe A. and Kilsby D., 1971. A rapid inexpensive bacterial count technique using agar droplets. *J. Appl. Bacteriol.*, **34**, no. 2, 425–440.

Sharpe A., Dyett E., Jackson A., Kilsby D., 1972. Technique and apparatus for rapid and inexpensive enumeration of bacteria. *Appl. Microbiol.*, **24**, no. 1, 4–7.

Shelef L. and Jay J., 1970. Use of titrimetric method to assess the bacterial spoilage of fresh beef. *Appl. Microbiol.*, **19**, no. 6, 902–905.

Silverman M. P. and Munoz E. F., 1979. Automated electrical impedance technique for rapid enumeration of fecal coliforms in effluents from sewage treatment plants. *Appl. Environ. Microbiol.*, **37**, no. 3, 521–526.

Snygg B. G., Andersson J. E., Krall C. A., Stollman U. M., Akesson C. A., 1979. Separation of botulinum-positive and -negative fish samples by means of a pattern recognition method applied to headspace gas chromatograms. *Appl. Environ. Microbiol.*, **38**, no. 6, 1081–1085.

Stannard C. J. and Wood J. M., 1983. The rapid estimation of microbial contamination of raw meat by measurement of adenosine triphosphate (ATP). *J. Appl. Bacteriol.*, **55**, 429–438.

Stead D., 1983. A fluorimetric method for the determination of *Pseudomonas fluorescens*-AR11 lipase in milk. *J. Dairy Res.*, **50**, 491–502.

Stewart B. J., Eyles M. J., Murrell W. G., 1980. Rapid radiometric method for detection of salmonella in foods. *Appl. Environ. Microbiol.*, **40**, no. 2, 223–230.

Strasser L., 1979. Prüfung ausgewählter Schnellverfahren zur Bestimmung das Oberflächenkeimgehaltes von Rinderschlacht-Tierkörpern und Wild-zugleich Angaben uber die Höke und die Zuzammensetzung dieser Microflora. Thesis, veterinary medicine, Université Libre de Berlin.

Strehler B., 1965. Adenosine-5′ triphosphate and creatine phosphate. Determination with luciferase. *In:* Bergmeyer H., *Methods of enzymatic analysis*, 559–572. Academic Press Ltd., New York.

Theron D. P., Prior B. A., Lategan P. M., 1986. Determination of bacterial ATP levels in raw milk: Selectivity of nonbacterial ATP hydrolysis. *J. Food Prot.*, **49**, no. 1, 4–7.

Theron D. P., Bernard A. P., Lategan P. M., 1986. Sensitivity and precision of bioluminescent techniques of enumeration of bacteria in skim milk. *J. Food Prot.*, **49**, no. 1, 8–11.

Thomas D., Henschke P. A., Garland B., Tucknott O. G., 1985. A microprocesssor-controlled photometer for monitoring microbial growth in multi-welled plates. *J. Appl. Bacteriol.*, **59**, 337–346.

Thomas S. B., 1974. The development of dye reduction tests for the bacteriological grading of raw milk. Part 2: Resazurin test. Dairy Industries, January–February, 31–34.

Thouvenot D. and Bourgeois C. M., 1970. Détermination photométrique de la densité cellulaire au cours du traversage. *Bios*, no. 2, 13–21.

Trinel P. and Leclerc H., 1977. Automatisation de l'analyse bactériologique de l'eau. Description d'une nouvelle méthode de colorimétrie. *Annales de Microbiologie*, **128A**, 419–432.

Tuncan E. U. and Martin S. E., 1987. Lysostaphin lysis procedure for detection of *Staphylococcus aureus* by the firefly bioluminescent ATP method. *Appl. Environ. Microbiol.* **53**, no. 1, 88–91.

Ward D. R., Finne G., Nickelson R. II, 1979. Use of a specific ion electrode (ammonia) in determining the quality of shrimp. *J. Food Sci.*, **44**, 1052–1054.

Ward D. R., La Rocco K. A., Hopson D. J., 1986. Adenosine triphosphate bioluminescent assay to enumerate bacterial numbers on fresh fish. *J. Food Prot.*, **49**, no. 8, 647–650.

Warren L. S., Benoit R. E., Jessee J. A., 1978. Rapid enumeration of fecal coliforms in water by a colorimetric β-galactosidase assay. *Appl. Environ. Microbiol.*, **35**, no. 1, 136–141.

Watanabe E., Ando K., Karube I., Matsuoka H., Suzuki S., 1983. Determination of hypoxanthine in fish meat with an enzyme sensor. *J. Food Sci.*, **48**, 496–500.

Webb N., Thomas F., Busta F., Kerr L., 1972. Evaluation of scallop meat quality by the resazurin reduction technique. *J. Milk Food Technol.*, **35**, no. 11, 664–668.

Weiss L. H. and Humber J., 1988. Evaluation of a 24-hour fluorogenic assay for the enumeration of *Escherichia coli* from foods. *J. Food Prot.*, **51**, no. 10, 766–769.

Welhoener H., Barwald G., Klein-Steuber Tims B., Scheible E., 1971. Schnelles Erkennen von Infektionen durch chemische Analyse. *In:* Congress of the European Brewery Convention, 287–296. Estoril. Elsevier Scientific Publishing Company, Amsterdam.

Wilkins J., Stoner G., Boykin E., 1974. Microbial detection method based on sensing molecular hydrogen. *Appl. Microbiol.*, **27**, no. 5, 949–952.

Wilkins J. R., Young R. N., Boykin E. H., 1978. Multichannel electrochemical microbial detection unit. *Appl. Environ. Microbiol.*, **35**, no. 1, 214–215.

Zindulis J., 1984. A medium for the impedimetric detection of yeasts in foods. *Food Microbiol.*, 159–167.

CHAPTER

6

Identification

Marielle Bouix, J.-Y. Leveau

In order to properly identify microorganisms, it is necessary to work with pure strains. The essential first step of any identification is therefore isolation followed by purification. The actual identification consists of a certain number of steps to be carried out in a specific order and which vary from one microorganism to another. These generally include examinations of the following:

- Cultural traits
- Morphological and structural traits
- Sexual traits in the case of fungi
- Biochemical and physiological traits
- General traits
- Immunological traits
- Pathogenic capabilities

Not all of these traits are sought for all microorganisms, but findings at each step narrow down the diagnosis and direct the next step towards specific tests. This process must be determined as a function of the type of microorganism with the aid of the appropriate identification key.

6.1 The Examination of Cultural Traits

Examination of cultural traits is a matter of noting the appearance of the culture in liquid medium (formation of pellicles, precipitates, gas, etc.) and on solid medium. Traits such as colonial morphology (top view and cross-section), size, color, and appearance (shiny, dull, mucilaginous, etc.) of the microorganism grown on agar-based medium in Petri plates are also noted. The appearance of a culture grown on agar slants also provides useful details (i.e., rectilinear or invasive, more mucilaginous or less so, etc.).

6.2 The Examination of Cell Morphological and Structural Traits

Examination of morphological and structural traits is done by microscopic examination, in which cell morphology, size, type of association, and often the mode of vegetative reproduction as well as structural characteristics can be observed. Microscopic examination includes the examination of a wet mount, that is, of the live microorganism between a slide and cover-slip, and the examination of a staining of a fixed smear. Various stainings may be done, the most common being the simple methylene blue stain which facilitates observation of cell morphology and the Gram stain, which aside from revealing the morphology also distinguishes the bacteria into two main groups based on the final staining characteristic of the cells. The staining differences are the result of structural differences in the bacterial cell wall, gram-positive bacterial cell walls consisting essentially of mucocomplex and retaining the violet stain of crystal violet after the decolorizing step while gram-positive bacteria with a thinner and more chemically complex cell wall lose the stain during decolorizing with alcohol and are subsequently colored pink by the safranin counter stain. This criterion is one of the most fundamental bases of bacterial classification.

More specific stains may be done in order to demonstrate spores, flagella, bacterial capsules, and various cellular inclusion bodies. And the composition and structure of the cell wall molecular constituents may be determined for more fundamental taxonomic studies.

Among yeasts, the ultrastructure of cell walls of ascomycetes and basidiomycetes differs (Kreger Van Rij and Veenhuis, 1971) as well as the composition of the polysaccharides, glucans, and mannans, of which the chain length and degree of branching vary from one species to another (Ballou, 1976).

For bacteria, the main cell wall compound is specifically peptidoglycan. Among gram-negative bacteria this peptidoglycan has a relatively homogeneous structure. Among gram-positive bacteria, on the other hand, significant variations occur as much in the amino acid composition as in the sugar composition.

This species variability makes cell wall composition a very useful taxonomic criterion for gram-positive bacteria, although the techniques required are not practicable on a routine basis (Schleifer and Kandler, 1972; Kandler and Schleifer, 1980).

6.3 The Examination of Sexual Traits

Applied mainly to yeasts, examination of sexual traits involves the examination of spore formation by culturing the microorganism on appropriate sporulation media.

6.4 The Examination of Biochemical and Physiological Traits

At this level, various aspects of the metabolism of microorganisms are examined by means of tests which essentially amount to a search for various enzymes.

6.4.1 Energetic and Nutritional Type

Some microorganisms are photosynthetic (phototrophs) but these are few in number, especially in the food and agricultural industries. The majority of microorganisms draw their energy from the degradation of chemical compounds, that is, they are *chemotrophs*. Among these, various microorganisms found in soil and water have the ability to use mineral substances as their energy source and are known as *chemolithotrophs*, while others use organic substances and are called *chemoorganotrophs*. The latter also use carbon-containing organic substances for the synthesis of cell constituents and as such are *heterotrophs*. Most of the microorganisms involved in the food industry belong to the heterotrophic chemoorganotrophic group.

6.4.2 Respiratory Type

When the substrate is completely oxidized, the metabolism is said to be respiratory. Following degradation, substrate-derived hydrogen molecules and electrons are transferred to a chain of hydrogen and electron transporters (acceptor-donor pairs). Aerobic metabolism refers specifically to the case in which the terminal acceptor is oxygen. When the terminal acceptor is necessarily some mineral compound other than oxygen, the metabolism is said to be strictly anaerobic. If either oxygen or other compounds are able to function as terminal acceptor, the term *aero-anaerobic* (facultative anaerobic) metabolism is applicable.

The determination of the type of respiration of a microorganism is done using agar-based medium (meat-yeast or meat-liver) solidified in Prevost tubes (8×180 mm) with an agar column height of 10–12 cm. After regeneration (20–30 minutes in a boiling water bath), the medium cooled to 46°C is uniformly inoculated with a suspension of the test microorganism. The tube is then rapidly cooled and incubated. The respiratory type is determined by the portion of the agar column in which the microbial growth is situated according to the following criteria:

- At the surface and down to about 1 cm: aerobic
- At the bottom but with absence of colonies in the upper portion: anaerobic
- Throughout the entire tube: aero-anaerobic (facultative anaerobic)
- In a zone approximately 1 cm long at the middle of the column height: micro-aerophilic

This sort of examination of respiratory type must be accompanied by the proper identification of oxidation-reduction chain enzymes such as the following:

- Oxidase, the enzyme catalyzing the acceptance by molecular oxygen of hydrogen and electrons at the end of the oxidation-reduction chain. The detection of its presence is based on the oxidation of dimethylparaphenylene diamine oxalate of which the oxidized form is red and the reduced form colorless. Disks of filter paper impregnated with this reagent are commercially available.
- Catalase, which breaks down the hydrogen peroxide produced at the oxidase step. The detection of this enzyme is done by simply placing a drop of dilute hydrogen peroxide solution on a microbial colony and watching for bubbling as gaseous oxygen is formed indicating the presence of a catalase.

Peroxidase, which breaks down peroxides, may be demonstrated by the benzidine test. A bluish black color indicates the presence of peroxidase.

6.4.3 Examination of Carbohydrate Metabolism

For taxonomic purposes, it is useful to test for the utilization of various compounds other than glucose as sole carbon source.

6.4.3.1 Examination of Oxidative Metabolism

Examination of oxidative metabolism is essentially an examination of oxidizing capabilities with carbon sources other than glucose. Among those tested are various hexoses, pentoses, di- and tri-saccharides, polysaccharides, alcohols, organic acids, and heterosides. A variety of techniques are applicable:

- The auxanogram technique in which the microbial cells are incorporated into a basal carbon-free medium. After solidification of the medium, the carbon substrates are deposited onto the surface of the agar, either in the form of a few crystals or as filter paper disks impregnated with the substance (these are commercially available). Assimilation of the carbon substrate is indicated by the development of colonies in its zone of diffusion.
- Liquid cultures in tubes or vials, placed in an incubator-shaker;
- Culture on agar slants in tubes with a medium containing a pH indicator;

- For some polysaccharide compounds (e.g., starch) the test is carried out using Petri plates. A medium containing starch as sole carbon source is inoculated by a stab in the center of the plate. After incubation, a reagent (e.g., iodine solution) is poured onto the agar surface in order to reveal hydrolysis of the polysaccharide and assess the diameter of the zone of hydrolysis.

6.4.3.2 Examination of Fermentative Metabolism

The purpose of these tests is to determine the carbonaceous substrate utilizing capabilities of the microorganism in the absence of oxygen and to identify the products formed (ethanol, organic acids, gas), these products characterizing specific types of fermentation. This is carried out using either liquid media, which may contain a colored indicator, in test tubes containing small inverted gas collected tubes for indicating gas formation or using agar-based media solidified in test tubes. The basal medium is carbon-free, the various carbonaceous substrates being added just prior to inoculation. Color changes and/or gas formation indicate fermentative activity. Certain specific reactions allow the characterization of the type of fermentation.

6.4.3.3 Detection of Enzyme Activities

In some cases it is possible to search for specific intracellular enzymes of which the activities are not directly demonstrable using the microbial culture itself. In such cases the characterization of the enzyme follows its extraction from the microbial cells.

6.4.4 Metabolism of Nitrogenous Substances

6.4.4.1 Tests for Assimilative Capabilities

The purpose of testing for assimilative capabilities is to determine which forms of nitrogen (e.g., ammonium, nitrite, nitrate, amino groups, etc.) a microorganism is able to incorporate. The tests are done by the auxanogram method using a nitrogen-free medium. Each nitrogen source is deposited onto a small area of the uniformly inoculated agar-based medium. The development of colonies around these areas indicates assimilation of the nitrogen source.

6.4.4.2 Amino Acid Degradation

Decarboxylation-Deamination

Some microorganisms are able to degrade amino acids using either decarboxylases or deaminases. The detection of these enzymes is used for taxononic purposes. The decarboxylation of an amino acid results in the formation of an amine. Its deamination produces an organic acid. The demonstration of these metabolic capabilities is carried out using media containing a single amino acid and a pH indicator. The color change of the indicator reveals the presence of either acidifying or alkalizing metabolism. In some cases a specific reagent for the product formed is added after culturing. Enzymes sought in this manner are mainly lysine decarboxylase, ornithine decarboxylase, lysine deaminase, tryptophan deaminase, and phenylalanine deaminase.

Other Reactions

Some bacteria hydrolyze tryptophan and liberate indole. The demonstration of this reaction is done by detecting indole after culturing a microorganism in a tryptophan-containing medium. The degrading of sulfur-containing amino acids is indicated by the liberation of H_2S. This metabolism is demonstrated with a medium containing sulfur-containing amino acids and ferrous sulfate. Production of H_2S manifests itself as the formation of a black iron sulfide precipitate.

6.4.4.3 Protein Degradation

For purposes of microbial identification, two proteins are used, namely, gelatin and casein.

Gelatin Hydrolysis

Several media have been proposed, such as the following:

- *Gelatin-based nutrient medium in test tubes.* This medium is inoculated by stabbing. The formation of a cone of liquefaction indicates hydrolysis of the gelatin.
- *Frazier's medium.* This is an agar-based medium containing gelatin, used with Petri plates and inoculated by stabbing in the center of the plate. After culturing, mercuric chloride is used to reveal hydrolysis which appears as a colorless zone whereas the absence of hydrolysis appears as opaqueness.

Casein Hydrolysis

The casein hydrolysis test is done using an agar-based medium containing lactose-free milk. The medium is poured into Petri plates and inoculated at the center of the plate. Casein hydrolysis appears as a clarification of the medium.

6.4.4.4 Hydrolysis of Urea

Some microorganisms possess a urease. This enzyme is demonstrated on a medium containing urea and a colored indicator. The hydrolysis of urea liberates ammonia and thereby reveals itself as a change in the indicator color.

6.4.5 Lipid Metabolism

Lipid metabolism testing amounts to a search for various enzymes capable of degrading lipid-like substances. The reactions of lipid metabolism begin with the hydrolysis of the macromolecules. It is generally the enzymes catalyzing this step which are sought using substrates such as Tween® 80 (polyoxyethylene sorbitan monooleate), lecithin (egg yolk medium), and various triglycerides. The hydrolysis of these compounds is indicated generally by a zone of clearing around colonies and/or a zone of opaqueness due to the precipitation of the salts of the fatty acids formed.

6.4.6 Various Physiological Properties

When dealing with bacteria it is very useful to know whether or not the microbial cells are motile. For this purpose, culturing in semisolid media is done.

Other tests with a more technological orientation may also be performed for the following:

- Osmophilic or osmotolerant traits
- Halophilic traits
- Optimal growth temperature
- Minimum and maximum growth temperatures
- Thermoresistance
- Antibiotic or antiseptic sensitivity or resistance
- Production of special compounds, aromatic compounds, enzymes, polysaccharides, etc.

6.4.7 Rapid Tests

The examination of the biochemical properties of a microorganism for identification purposes thus requires a large number of tests and as a result is rather lengthy to implement. Quick identification kits have been available for some time now, however, which allow the examination of large numbers of traits to be carried out very rapidly. These are essentially the API systems which provide galleries of 10, 20, and 50 traits for different microorganisms (e.g., enterobacteria, staphylococci, streptococci, lactobacilli, yeasts, etc.) and the "Roche tubes" (e.g., Enterotube®, Oxyfermtube®, etc.). These systems of combined tests save considerable time at the preparation and inoculation steps. Incubation times, however, remain at 24 hours. API galleries are also available for the detection of enzymatic activities, to be read after 4 hours of incubation (e.g., the "rapid 20 E" systems for enterobacteria, "staph ident" for staphylococci, and "An ident" for anaerobes). Finally, the ATB 32 systems can also identify bacteria in 4 hours by means of automation or visual reading.

6.5 The Examination of Genetic Traits

6.5.1 DNA Base Pair Composition: G + C% Content

The earliest techniques for the determination of G + C% consisted essentially of DNA base analysis. This required grinding of the microbial cells in order to extract the nucleic acids. The specific recovery of DNA was made possible by hydrolysis of RNA using a ribonuclease. Determination of the bases was done after chromatographic separation. The entire operation was time consuming and fastidious.

Sueoka and co-workers (1959) developed a G + C% determination method based on the examination of the sedimentation properties of DNA centrifuged in a cesium chloride gradient. At equilibrium, the density of the cesium chloride in the DNA sedimentation zone is proportional to the G + C%. A detailed account of this technique may be consulted in a work by Schildkraut and colleagues (1962).

Another method consists of following the thermal denaturing of the DNA molecule using ultraviolet spectrophotometry (hyperchromicity). As the increasing temperature causes the hydrogen bonds of the double-stranded molecule to break, light absorbance increases due to increased exposure of the absorbing nucleotide bases. The denaturation or melting temperature, Tm, characterizing the DNA may thus be determined in order to derive the G + C content by virtue of the linear correlation which exists between the two (Marmur and Doty, 1962).

Two identical bacteria obviously have identical G + C contents. On the other hand, the observation of identical G + C% is an insufficient basis for concluding that two strains under examination are identical.

6.5.2 DNA/DNA and DNA/RNA Homology

Two identical DNA molecular structures means two totally identical bacterial strains. For this reason, the degree of similarity between two strains can be established by determining the percentage of hybridization between single strands of DNA obtained from each strain (DNA/DNA hybridization). It may also be practical to assess the degree of homology between fragments of DNA and RNA. In this case, transfer RNAs (tRNA) and ribosomal RNAs (rRNA) are used. The degree of homology between a DNA fragment and a messenger RNA is of little interest compared to DNA/DNA homology given the relationship that exists between these two types of molecules. Moreover, since it is difficult to obtain labeled RNA messengers, this type of homology is not studied.

Interest in studying the homology between DNA and transfer and ribosomal RNA is intense, particularly for the purpose of identifying similarities between bacteria thought to be relatively unrelated. Some of the techniques used are based on the differential adsorption of the test molecules onto nitrocellulose membrane. Denatured DNA in particular has the property of adsorbing onto nitrocellulose and retaining its hybridization capability while RNA molecules are not retained.

The actual hybridization technique consists of the following steps: extraction, purification, denaturation, and adsorption of the test DNA onto nitrocellulose membrane; incubation of the membrane in a known and radioactively labeled reference DNA solution under conditions allowing hybridization; elimination of unhybridized molecules followed by measurement of the radioactivity level which indicates the level of hybridization and hence the homology of the bacteria which provided the DNA. Details of the implementation of these techniques are described by Gillespie and Spiegelman (1965).

It should be noted also that liquid phase hybridization techniques have been developed in which the reaction is followed using an ultraviolet spectrophotometer, as well as methods using hot (radioactive) or cold DNA probes.

6.5.3 16S RNA Nucleotide Sequences

Nucleotide sequencing is a very useful method for studying the taxonomic relationships between bacteria. The 16S fraction of the ribosomal RNA is the principal focus of this method. Its principle consists of first isolating the 16S fraction and then hydrolyzing it with ribonuclease T1 which selectively hydrolyzes the 3′-5′ bonds associated with guanylic acid residues. The oligonucleotides thus obtained are separated by two-dimensional electrophoresis, leading to the creation of an oligonucleotide "fingerprint" characterizing the test strain. Fingerprints obtained for different strains may be compared in order to produce similarity indexes, S_{AB} having values between 0 and 1 for very unrelated and very closely related strains, respectively.

6.6 The Examination of Immunological Traits

The specificity of the antigen-antibody reaction is exploited as a means of completing and refining the examination of a microorganism. In general, morphological, physiological, and biochemical studies are sufficient to allow species identification of microorganisms. Within a species, however, a very fine level of differentiation exists which defines types of individuals distinguishable by serological methods. In practice, a microorganism may be considered to be a mosaic of antigenic determinants, some of which are specific distinguishing characteristics.

The specificity of the antigen-antibody reaction is routinely exploited for the detection of specific microorganisms in a food product. In such cases, antisera specifically recognizing the microorganism may be used in conjunction with techniques as simple as agglutination on a slide, or with more sophisticated methods such as immunofluorescence and the enzyme-linked immunosorbent assay known as the ELISA. More recently, flow cytometry techniques have provided a coupling between morphological and immunochemical studies. Systems also exist in which culturing of microbial cells is coupled to an immunological reaction such as the AES system for salmonella.

In some cases, it is possible to extract antigens from microbial cells and then proceed with the immunological reaction using purified antigens. Such a reaction manifests itself as a precipitation at the antigen-antibody interface.

In the case of toxin-producing microorganisms, the toxin itself may be the serological target, readily detectable using immunodiffusion methods.

6.7 The Examination of Pathogenic Capabilities

Pathogenic capability may be demonstrated either by searching for specific enzymes directly or indirectly associated with the pathogenicity of the microorganism or by experiments using laboratory animals injected with the suspect sample. The detection of toxins in products by means of laboratory animals is practiced in food microbiology only in the case of searches for botulin toxins.

6.8 Use of Results

The results from the various tests are used to fit the microorganism into a classification scheme and thereby determine its species. Two theories collide at this stage with respect to the methodology for using such results.

According to classical taxonomic methodology, tests are carried out in a precise order and the results of the earlier tests determine the choice of subsequent tests to be performed, which implies that some traits, that is, the first ones tested, carry more significance than the others.

In order to avoid being led into what some consider to be an arbitrary hierarchy of traits, all of the results may be processed simultaneously using computerized schemes. A positive trait is assigned a score of 1 and a negative trait a score of 0. Each microorganism can therefore be characterized as a series of figures of the form 1101001.... and so on. By numerical taxonomy, these characteristic numbers can be used to calculate a coefficient of similarity which expresses the degree of relatedness between two microbial strains. The taxonomical distance separating two strains can thus be established and gradually the relationships between all strains can be defined in these terms.

References

Ballou C. E., 1976. Structure and biosynthesis of the mannan component of the yeast cell envelopes. *Adv. Microbiol. Physiol.*, **14**, 93–157.

Gillespie D. and Spiegelman S., 1965. A quantitative assay for DNA-RNA hybrids with DNA immobilized on a membrane filter. *J. Mol. Biol.*, **12**, 829–842.

Kandler O. and Schleifer K. H., 1980. Taxonomy I: Systematic of bacteria. *In:* Ellenberg, Esser, Kubitzki, Schepf, Ziegler (eds). *Progress in Botany (Fortschritte des botanik)*, **42**, 234–252, Springer Verlag, Berlin, Heidelberg.

Kreger Van Rij N. J. W. and Veenhuis M., 1971. A comparative study of the cell wall structure of basidiomycetous and related yeasts. *J. Gen. Microbiol.*, **68**, 87–95.

Marmur J. and Doty P., 1962. Determination of the base composition of deoxyribonucleic acid from its thermal denaturation temperature. *J. Mol. Biol.*, **5**, 109–118.

Schleifer K. H. and Kandler O., 1972. Peptidoglycan types of bacterial cell-walls and their taxonomic implications. *Bacteriol. Rev.*, **36**, 407–477.

Schildkraut C. L., Marmur J., Doty P., 1962. Determination of the base composition of deoxyribonucleic acid from its buoyant density in CsCl. *J. Mol. Biol.*, **4**, 430–443.

Sueoka N., Marmur J., Doty P., 1959. *Nature*, **183**, 1 427.

Suggested Readings, Chapters 2–6

Baker J. M., Collins-Thomson D. L., Griffiths M. W., 1991. Bacterial bioluminescence: Applications in food microbiology. *J. Food Protect.*, **55**, no. 1, 62–70; 41 ref.

Blackburn C., 1991. Detection of *Salmonella* in foods using impedance. *Europ. Food Drink Rev.*, Winter, 35, 37, 39–40; 21 ref.

Clark D. C., Pinder A. C., Poulter S. A. G., Purdy P. W., 1990. Validation of flow cytometry for rapid enumeration of bacterial concentrations in pure cultures. *J. Appl. Bacteriol.*, **69**, no. 1, 92–100; 12 ref.

Coombs P., Ligugnana R., Rovere E., 1991. Rapid determination of salmonellae using the Malthus technique. *Latte*, **16**, no. 1, 40–43; 11 ref.

Eid N., Daeschel, M. A., 1990. Advances in food diagnostics. *Dev. Ind. Microbiol.*, **331** (Suppl. 5), 151–155; 62 ref.

Lemieux L., Puchades R., Simard R. E., 1991. Overview of rapid methods and automation in food microbiology with emphasis on flow injection analysis. *Lebensmittel Wissenschaft und Technologie*, **24**, no. 3, 189–197; 110 ref.

Pafumi-Rizzo J., 1992. Quality control for microbiological media. *Food Australia*, **44**, no. 6, 272–273; 5 ref.

Powell S., 1990. Use of conductance techniques for rapid detection of food-borne pathogens. *Europ. Food Drink Rev.*, Winter, 67, 69, 71, 73; 23 ref.

Stewart G. S. A. B., 1990. In vivo bioluminescence: New potentials for microbiology. *Letters Appl. Microbiol.*, **10**, no. 1, 1–8; 18 ref.

CHAPTER

7

Microbiological Applications of Immunology

Florence Humbert, Cécile Lahellec

7.1 Introduction

Immunology or the science of immunity is a multidisciplinary study. In medicine, from which most of our knowledge about it originates, immunity represents the refractory state of an individual to disease. It is induced in an individual by contact with a substance called an *antigen*, which is foreign to the body and results in the formation of a complex substance called *antibody*. Like antigens, antibodies are protein molecules made of long polypeptide chains. When antigen and antibody have complementarily charged groups, they spontaneously undergo structural adjustments which allow them to fit together as a specific solid combination. This in vivo phenomenon may be visualized in vitro. Such reactions, although employed less frequently in the food microbiology field than in the field of infectious pathology, have undergone widespread proliferation due to the development by many laboratories of monoclonal antibodies directed against specific antigenic fractions of microorganisms or their toxins.

Immunochemistry refers to the detection of various constituents of food by the application of immunological reactions and is the subject of chapters in other volumes of this series.

All of the techniques described herein are based on the same principle of antigen–antibody specificity. Based on the nature of the antigen, however, as well as on the antigen- or antibody-marking technique, the following classification may be distinguished (Table 7.1):

Table 7.1 Some Immunological Reactions

Type of antigen	Labeling	Reaction type
Soluble	→	Precipitation
Particulate	→	Agglutination
	Fluorescent compound	Immunofluorescence
	Enzyme	ELISA
	Radioactive element	Radioimmunological assay

- Precipitation in the case of *soluble* antigens, that is, completely dissolved in the medium, which may be either liquid or solid, giving rise to the terms *liquid phase precipitation* or *solid phase precipitation.*
- Agglutination in the case of particulate antigens, that is, 1 micron or larger. Smaller antigen particles can, however, be fixed onto blood cells, latex beads, or carbon particles using appropriate techniques. This method is called *indirect agglutination*, conditioned or passive, while the specific use of blood cells is called *hemagglutination.*
- Immunofluorescence, which consists of coupling the antigen or more often the antibody to a fluorescent substance, that is, fluorescein isothiocyanate (green) or rhodamine (red), in order to detect the presence of antibody or of its corresponding antigen in the sample.
- The ELISA or enzyme-linked immunosorbent assay, which can visualize immunological reactions by means of an enzyme. The specific substrate of this enzyme liberates a colored compound in its presence. The appearance of color thus indicates a positive reaction, the color intensity being a function of the quantity of antigen or antibody which has reacted.
- Radioimmunology, in which the antigen is radioactively labeled, that is, generally with I^{125}. The widespread acceptance of this method is limited largely by its requirement for special equipment.

7.2 Precipitation

7.2.1 Characteristics

Precipitation occurs when a soluble and hence invisible antigen forms with specific antibodies an insoluble antigen–antibody complex which appears in the form of a visible precipate. This is a very simple reaction to carry out, although not very sensitive, and may give rise to zone phenomena, that is, the resolubilizing of the precipitate making the reaction appear falsely negative at certain concentrations of antigen and antibody. Precipitation reactions may be carried out in liquid or solid media.

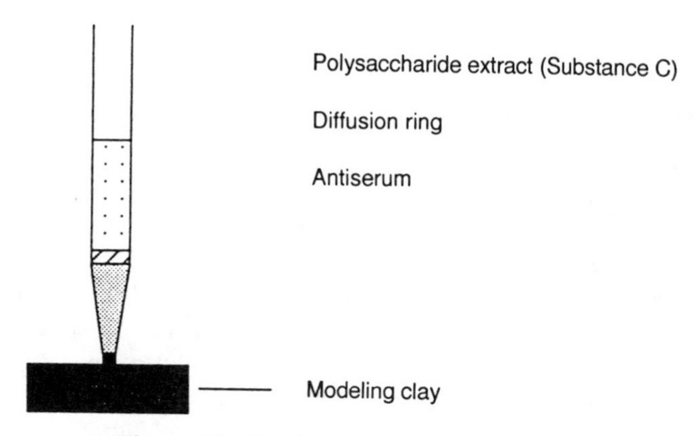

Figure 7.1 Precipitation in liquid medium.

7.2.2 Precipitation in Liquid Medium

Precipitation in liquid medium consists of placing the antiserum in a small diameter tube and carefully layering the sample solution on top of the serum. A ring forms at the interface if the sample contains sufficient antigens. This type of reaction is used for determining Lancefield group streptococci (Figure 7.1).

7.2.3 Precipitation in Solid Medium

With precipitation in solid medium, agar acts as a mesh through which antigen or antibody may pass but not the much larger antigen–antibody complex. When the complex forms in the zone where solvent fronts meet, it is unable to diffuse and the accumulation of complex in this zone brings about the formation of a line of precipitation visible to the naked eye.

Simple diffusion in tubes (the Oudin reaction) and double diffusion on slides or Petri plates (the Ouchterlony reaction) are practiced. These are suitable for the demonstration of the presence of staphylococcal enterotoxins as well as those of *Clostridium perfringens* and the toxins of *Clostridium botulinum*. Although other methods (i.e., radioimmunological, passive hemagglutination, etc.) are faster and more sensitive, these agar diffusion techniques remain widely used. The Ouchterlony reaction is the "official" technique for the detection of staphylococcal enterotoxins.

7.2.4 Immunoelectrophoresis

Immunoelectrophoresis is a refinement of the precipitation technique and is used for the same range of applications. This method begins with separation of the constituents of the antigenic mixture by means of an electric field and is completed by immunological detection of the antigenic constituents using antiserum directed against them.

7.3 Agglutination

7.3.1 Characteristics

Agglutination occurs when particulate antigens form a lattice between themselves with antibodies specific to them functioning as bridges. The resulting agglomeration or agglutinate is visible to the naked eye (Figure 7.2).

The antigens involved must have multiple antigenic sites (e.g., bacteria, parasites, proteins, etc.) or must be fixed to red blood cells, latex beads, or carbon particles in order to be able to bind with several antibodies (Figure 7.3). The antibodies themselves must also be of the "agglutinating" type (at least bivalent) in order to participate in the formation of the agglutination lattice.

7.3.2 Applications

Agglutination-based techniques, including hemagglutination inhibition and passive reversed hemagglutination, may be used for the detection of toxins of the genera *Staphylococcus* and *Clostridium* (Johnson et al., 1967). These techniques, more rapid and sensitive than those based on precipitation in agar, are also of much more delicate handling.

Other better known food microbiological applications, such as the determination of some bacterial serotypes, especially those of *Salmonella*, are carried out by means of agglutination on slides or in tubes.

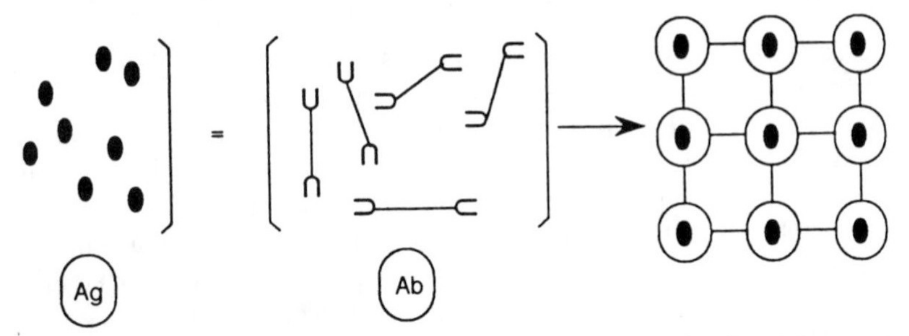

Figure 7.2 Schematic representation of an agglutination reaction involving naturally agglutinating antigens (bacteria, parasites, etc.).

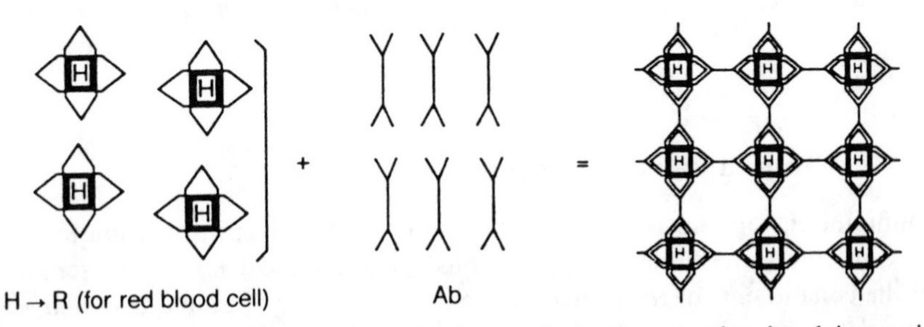

Figure 7.3 Schematic representation of a hemagglutination reaction involving antigens fixed onto red blood cells.

7.3.2.1 Agglutination on slides

Using a platinum wire to remove bacteria from an agar-grown colony and suspend them in a drop of antiserum placed on a slide, aggregates will form and increase in size, indicating a positive reaction if the serum antibodies are specific for the bacterial antigenic determinants whereas the suspension will otherwise remain smooth and opalescent.

7.3.2.2 Agglutination in Tubes

A series of dilutions of antiserum is prepared in a row of small test tubes on a rack and mixed with the bacterial suspension. After an incubation period at 37° or 40°C, the results are read generally by observing the pellet of cells gathered at the bottom of each tube. A smooth-edged pellet indicates a negative reaction while a rough-edged irregular pellet indicates a positive reaction. This procedure can also be carried out using microtitration plates instead of tubes.

7.3.2.3 Detection of Salmonella *by the "Sero-Enrichment" Technique*

One variation of the use of these agglutination reactions is the so-called sero-enrichment technique for *Salmonella* detection developed by Sperber and Deibel in 1969 and followed up by Boothroyd and Baird-Parker (1973) and Lahellec and Colin (1977).

Principle

The principle is based on the direct detection of *Salmonella* present in an enrichment medium by means of antiflagellar antisera and is the basis of a disposable test commercialized by an American firm.

Description

The technique consists of the following steps:

Preenrichment and enrichment are carried out using conventional techniques. The choice of medium (e.g., Tetrathionate, Rappaport-Vassiliadis, Selenite-Cystine, etc.) and the incubation temperature may, of course, vary according to the product being tested.

Following a 24-hour incubation of the enrichment medium, a drop of this medium is placed in a tube containing 10 ml of M broth (Difco). Following incubation of this in a water bath at 41°C for 18 hours, 0.85 ml is placed in a hemolysis tube containing 50 μl of a formol solution (formol 3 ml, NaCl 4.2 g, distilled water 100 ml). A mixture of antisera (100 μl) is added to this, prepared with 0.5 ml each of Spicer Edwards 1, 2, 3, 4, L complex, EN complex, I complex, Z6, poly D, Arizona polymonophasic sera plus 11.5 ml of an 0.85%

NaCl solution. The tube is incubated at 41°C for 2 hours with readings at 30 minute intervals.

Reading and Interpretation

The appearance of a flaky precipitate or of a clouding that may be transient indicates the presence of *Salmonella* in the tested product.

The specificity and sensitivity of this technique are considered good in the case of meat and poultry products (Lahellec and Colin, 1977). It is an excellent screening method having the advantage of detecting lactose-fermenting *Salmonella*. If, however, examination of the *Salmonella* present is desired, parallel isolation on selective media is necessary.

7.4 Immunofluorescence

7.4.1 Characteristics

The antibodies are coupled to fluorescent compounds using appropriate techniques. When these antibodies react with antigens to which they are specific, the fluorescent antigen–antibody complex may be visualized with the aid of a fluorescent microscope. This technique may be direct or indirect.

7.4.1.1 Direct Immunofluorescence

In the direct immunofluorescence technique, the specific antibody is labeled and deposited onto the test sample preparation. Fluorescent emission after washing indicates the presence of antigen.

7.4.1.2 Indirect Immunofluorescence

Indirect immunofluorescence involves two successive reactions using the same antibody, the first of which uses the antibody to recognize the antigen and the second in which the antibody itself is detected as an antigen. It is not this antibody which carries the fluorescent marker but the anti-antibody which is added for the second reaction (Figure 7.4).

7.4.2 Applications

Immunofluorescence has found numerous applications in the field of food hygiene, especially for the detection of *Salmonella* (Thomason, 1981). The details of its implementation may be obtained from the 1992 edition of the *Bacteriological Analytical Manual.*

Sharing of common antigenic domains among *Salmonella* and other enterobacteria may account for 5–7% of all false-positive results observed. The

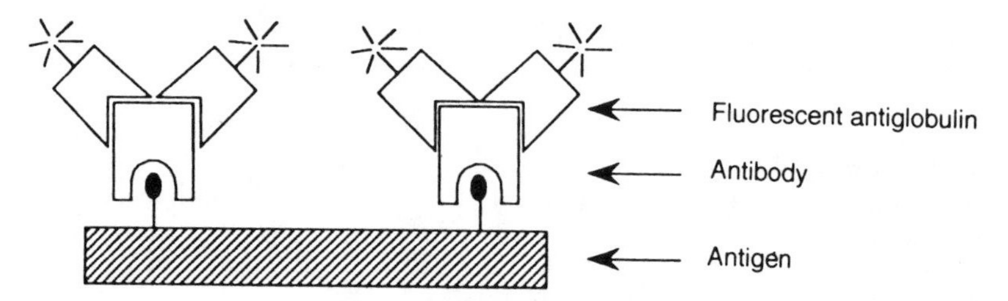

Figure 7.4 Schematic representation of an indirect immunofluorescent reaction.

low number of false-negative results (1–2%) make the technique suitable for preliminary screenings. The conditions for its use may therefore be considered to be identical to those of the "sero-enrichment" technique. The uncertainty associated with the reading of results (which may be automated for large numbers of samples) does, however, limit its routine use for the testing of foodstuffs.

7.5 ELISA

7.5.1 Principle

The acronym ELISA refers to a large number of techniques, although only those reactions said to be in the heterogenous phase (in which the antigen or antibody is attached onto a solid phase such as plastic, cupules, metal beads, tubes, etc.) have found applications in food hygiene.

Among this group of reactions, the "sandwich" technique (so-called because the antigen ends up trapped between two antibody molecules) is the most widely used. Its principle is shown in Figure 7.5.

7.5.2 Applications

ELISA has been applied by numerous laboratories to the detection or quantification of aflatoxins, botulin toxins, *Salmonella*, staphylococcal en-

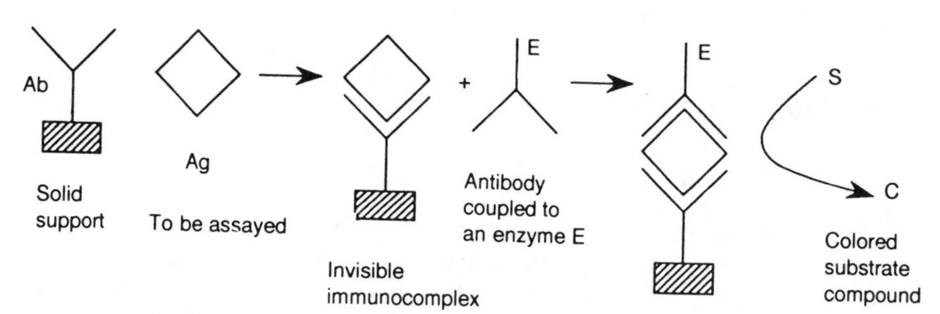

Figure 7.5 Principle of the "sandwich"-type ELISA technique.

Table 7.2 ELISA methods used in food hygiene and their relative sensitivities

Microorganism or product	Sensitivity	Time (hr.)	Reference
Staphylococcal enterotoxin A	0.4 ng/ml	20	Saunders and Bartlett, 1977
Staphylococcal enterotoxins A, B, C, D, and E	1 ng/ml	8	Freed et al., 1982
Salmonella	10^5/ml	2*	Swaminathan et al., 1985
Aflatoxins B_1, B_{2a}	1–100 pg/ml	15	Lawellin et al., 1977
Clostridium botulinum toxins types A and E	50–100 LD_{50} i.p. in mice	16–18	Notermans et al., 1978, 1979
Vibrio cholerae endotoxin	1.3 μg/ml		Holmgren and Svennerholm, 1973

Source: Data from Jay, 1984.
*Not including the 48 hours necessary for enrichment.

terotoxins, various viruses, and gram-negative bacterial lipopolysaccharides (Table 7.2).

This technique has the following advantages:

- great sensitivity sometimes approaching that of radioimmunological assays;
- automation obtainable by optical density readings (color intensity) using programmable equipment;
- relative simplicity;
- generally good stability of the reagents used.

All of this explains the tremendous rise in the popularity of the technique during recent years, although it should be noted that its specificity depends on that of the antibody or antibodies used (which may have cross-reactions with bacteria or substances other than those of interest) and that any of these detection techniques come into play only after several steps of extraction (aflatoxins) or culture (*Salmonella*, *Listeria*, etc.) which lengthens the overall time required for the analysis.

7.6 Radioimmunology

7.6.1 Characteristics

In radioimmunology, the radioactive element (generally I^{125}) is coupled to the antigen. The quantity of antigen bound to specific antibodies (which are themselves usually bound electrostatically to the walls of polystyrene tubes) is measured using a Geiger-Müller counter. Prior washings eliminate unbound antigens. Radioimmunology appears to be the most sensitive immunological technique available and there is general agreement among its users that substances may be detected down to a limit of 1 ng/ml; however, the majority of these techniques require a minimum of 24 hours.

7.6.2 Applications

Radioimmunological techniques have been applied to testing for staphylococcal endotoxins in different foods (Bergdoll and Reiser, 1980), but remain limited to specially equipped specialized laboratories.

The broad descriptions given in this chapter provide an overview of the current possibilities for the use of immunological techniques in food microbiology. In the years to come, it is very likely that the use of some of these techniques will be phased out as newer techniques from genetic engineering as well as tests using highly specific DNA probes derived from the microorganisms being tested become fully developed and commercialized.

References

Boothroyd M. and Baird-Parker A. C., 1973. The use of enrichment serology for *Salmonella* detection in human foods and animal feeds. *J. Appl. Bacteriol.*, **36**, 165–172.

Freed R. C., Evenson M. L., Reiser R. F., Bergdoll M. S., 1982. Enzyme linked immunosorbent assay for detection of staphylococcal enterotoxins in foods. *Appl. Environ. Microbial.*, **44**, 1349–1355.

Holmgren J. and Svennerholm A. M., 1973. Enzyme linked immunosorbent assays for cholera serology. *Infect. Imm.*, **7**, 759–763.

Jay J. M., 1984. Analysis of food products for microorganisms of their products by non-culture methods. *In:* Gruenwedel D. W., Whitaker J. R. (eds.) *Food analysis, principles and techniques*, vol. 3, 87–126. Marcel Dekker Inc., New York and Basel.

Johnson H. M., Hall H. E., Simon M., 1967. Enterotoxin B: Serological assay in cultures by passive hemagglutination. *Appl. Microbiol.*, **15**, 815–818.

Lahellec C. and Colin P., 1977. Application de la technique dite de "séro-enrichissement" à la mise en évidence de *Salmonella* à partir de carcasses de volailles et de certains produits transformés. *Bull. Inf. Station Expérimentale d'Aviculture de Ploufragan*, **17**, no. 2, 76–78.

Lawellin D. W., Grant D. W., Joyce B. K., 1977. Enzyme linked immunosorbent analysis for aflatoxin B_1. *Appl. Environ. Microbial.*, **34**, 94–96.

Notermans S., Dufrenne J., Van Schothorst M., 1978. Enzyme-linked immunosorbent assay for detection of *Clostridium botulinum* toxin type A. *Jap. J. Med. Sci. Biol.*, **31**, 81–85.

Notermans S., Dufrenne J., Kozkaki S., 1979. Enzyme-linked immunsorbent assay for detection of *Clostridium botulinum* type E toxin. *Appl. Environ. Microbiol.*, **37**, 1173–1175.

Saunders G. C. and Bartlett M. L., 1977. Double-antibody solid-phase enzyme immunoassay for the detection of Staphylococcal enterotoxin A. *Appl. Environ. Microbiol.*, **34**, 518–522.

Sperber W. H. and Deibel R. M., 1969. Accelerated procedure of *Salmonella* detection in dried foods and feeds involving only broth cultures and serological reactions. *Appl. Microbiol.*, **17**, no. 4, 533–539.

Swaminathan B., Aleixo J. A. G., Minnich S. A., 1985. Enzyme immunoassays for *Salmonella:* One-day testing is now a reality. *Food Technol.*, **39**, no. 3, 83–89.

Thomason B. M., 1981. Current status of immuno-fluorescent methodology for *Salmonellae*. *J. Food Prot.*, **44**, 381–384.

Note: For additional reference sources, see p. 107.

CHAPTER

8

Identification of Microorganisms by Nucleic Acid Probe Hybridization

Aline Lonvaud-Funel

8.1 Introduction

The identification of a microorganism by conventional means is based on an assemblage of morphological, physiological, and biochemical tests. Alternative approaches focus on the total cellular protein or membrane lipid composition. But in all cases, such characteristics only reflect the portion of the genome expressed under a particular set of culture conditions. Taxonomy has already made use of DNA analysis by comparing the base composition (G + C%) or by testing for the percentage of homology of nucleotide sequences. The information thus obtained often leaves room for ambiguities given the difficulty of making a proper assessment within the tolerated variations. The molecular hybridization method deals with the problems associated with the identification of microorganisms in a better and simpler way by using the more recent techniques of molecular biology which simplify the manipulation of nucleic acids and access newer tools. Identification is based on the constitutive composition of nucleic acids (DNA or RNA) rather than on the products of their expression by the cell and thereby focuses on the genome itself. The risk of errors or of uncertainty due to the disappearance or the acquisition of a physiological trait which may or may not be expressed by the reference strain is thus eliminated.

8.2 The Principle of Molecular Hybridization

The principle of molecular hybridization is based on the property which single-stranded nucleic acid chains possess of hydridizing only with fragments of complementary sequence. The specificity of the technique resides herein. The capacity of DNA to adhere to nitrocellulose or nylon membranes is also utilized.

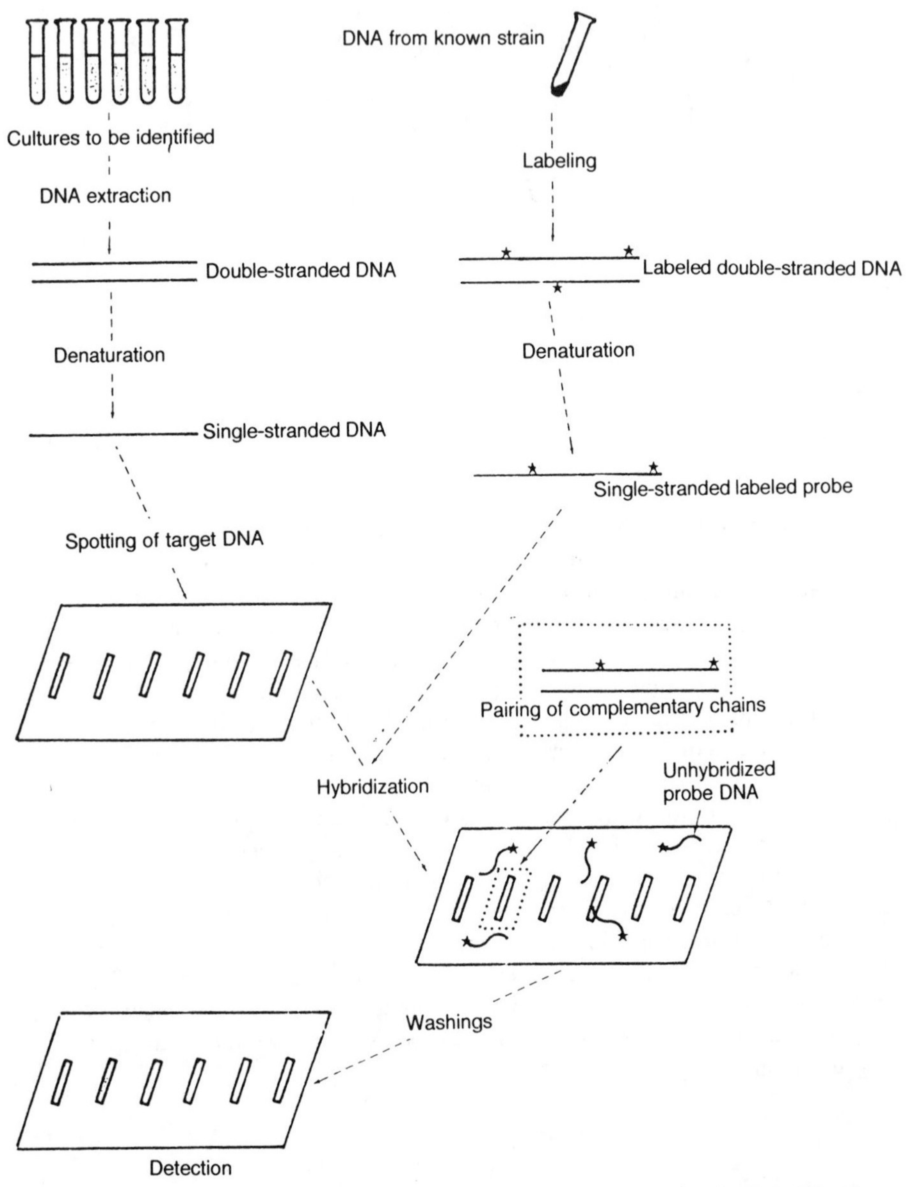

Figure 8.1 Principle of identification of microorganisms by molecular hybridization.

In order to carry out an identification, DNA from the unknown strain and DNA from a reference strain are required. The latter constitutes a probe and is labeled to allow the detection of the hybrid formed with the target DNA to be identified.

Various types of markers are possible, radioactive and otherwise. Detection is achieved by autoradiography or in the case of "cold" probes by a color-producing enzymatic reaction.

The different steps of manipulation are as follows (Figure 8.1):

- Probe preparation: extraction and marking of DNA from the known strain
- Target DNA preparation: extraction and deposition onto a membrane
- Hybridization: contacting the probe with the membrane in conditions allowing the formation of a stable hybrid
- Membrane development: repeated washings to eliminate excess probe and nonspecific hybrids, detection of the probe DNA–unknown DNA hybrid

Based on this principle, several strategies may be envisaged depending on the circumstances:

1. A specific species is to be identified among several microorganisms. In such a case the probe is from this species and the DNA of the various microorganisms is deposited onto the membrane.
2. The microorganism isolated from a medium must be identified among several most probable species encountered in the same environment. The membrane receives the DNA of these different known species and the probe is prepared from the DNA of the microorganism to be identified.
3. Finally, responding more or less to both situations, in the most general case, the DNA of the unknown strains is deposited onto the membrane and successively submitted to hybridization with different probes through hybridization–dehybridization cycles.

8.3 Radioactive Probes and Cold Probes: Detection Principles

8.3.1 Radioactive Probes

The original probes used were of the radioactive type. DNA is labeled by incorporation of a radioactive nucleotide, made so usually using ^{32}P. Hybrids are detected directly by autoradiography. After hybridization, the membrane is put in a plastic envelope and inserted into a cassette placed against a photographic film (Kodak XO Mat RP) with amplifying screens. The unit is kept at $-80°C$ for the required time which varies from a few hours to a few days depending on the activity of the probe and the quantity of DNA hybridized. After exposure, the film is developed and the spots indicating unknown DNA hybridized with the probe are identified. This procedure is still widely used, especially in laboratories where the handling of radioactive elements is not a problem.

8.3.2 Cold Probes

Identification by nucleic acid probes is unlikely to become a very widespread routine practice unless an alternative to the use of radioactive isotopes is found. For this reason, research aimed at the development of "cold" labeling techniques has been intense for several years.

Detection may be achieved indirectly by using ligand-protein and hapten-antibody affinity systems that result in color-producing reactions. Direct revealing is possible when the enzyme that catalyzes the final reaction is covalently linked to the nucleic acid. Indirect detection is, for the time being, the most widespread. The DNA probe is marked either by the incorporation of nucleotide analogs in a synthetic reaction or by a direct chemical or photochemical reaction.

8.3.2.1 Labeling by Incorporation of Labeled Nucleotide

The mechanisms of incorporation are the same as those used for radioactive probes and are examined in the following chapter. The most well-known analogs incorporated are labeled with biotin, bromine, and digoxigenin. They are revealed after hybridization using antibody systems.

Biotinylated Probes

Biotin linked to an incorporated nucleotide (biotin-11-dUTP) is recognized by streptavidine for which it has a very high affinity ($K = 10^{-15}M$). An alkaline phosphatase–biotin complex then binds onto the streptavidine (Gibco/BRL DNA detection system kit) or alternatively the biotin is recognized by an alkaline phosphatase–streptavidine conjugate (Figure 8.2). Revealing is based on the color-generating alkaline phosphatase catalyzed reaction using most

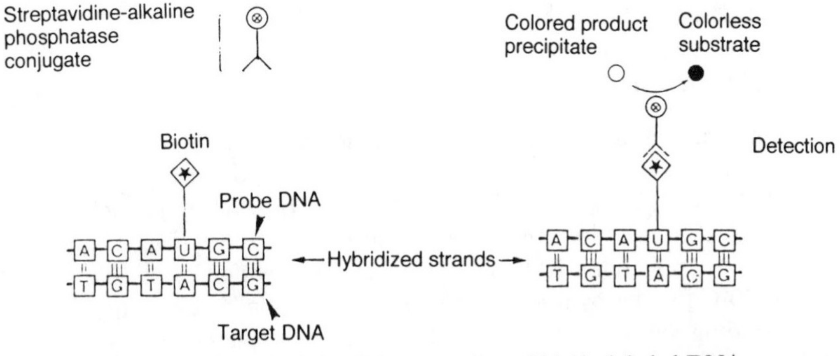

Figure 8.2 Principle of the reporting of biotin-labeled DNA.

often either BCIP (5-bromo-4-chloro-3-indolylphosphate) or NBT (nitro-blue tetrazolium, Gibco/BRL blue gene kit). This is a coupled oxidation-reduction reaction, BCIP being oxidized and dephosphorylated by phosphatase with simultaneous reduction of NBT to formazan.

Incorporation of a Uridine Analog

In this system developed by the Pasteur Institute, 5-bromodeoxyuridine incorporated into the probe DNA is detected following hybridization by an immunoenzymatic system. A mouse anti-5-BdUr monoclonal antibody binds to the marked nucleotide which is then recognized by an anti-lg mouse antibody coupled to peroxidase which in the presence of a chromogenic substrate and H_2O_2 yields a colored compound.

Incorporation of a Digoxigenin Marked UTP Derivative

For this type of labeling, Boehringer Mannheim has proposed incorporation by "random priming." Uridine is linked to digoxigenin by a bridge of 11 carbon bond lengths which improves the subsequent binding of the antidigoxigenin antibody which is itself linked to alkaline phosphatase. The latter catalyzes the color-producing reaction between BCIP and NBT.

8.3.2.2 Direct DNA Labeling

DNA itself is labeled by means of chemical or photochemical reactions. Unlike in the previous example, there is no in vitro synthesis of nucleic acid chains.

Sulfonation

The most well-known system currently consists of sulfonation of the cytidyl nucleotides of the DNA chain (the Chemiprobe process). Heat-denatured DNA is treated at ambient temperature with sodium bisulfite and methylhydroxylamine which stabilizes the sulfonated complex. The mixture may be used directly or after precipitation with ethanol and dissolution of the DNA (Lebacq et al., 1982). Approximately 10–15% of the cytosyl residues of the chain are labeled by this procedure. Detection of the hybrid involves immunoenzymatic reactions similar to those already described, with sulfonated cytidine being recognized by a specific monoclonal antibody. The antibody is detected in turn by an alkaline phosphatase conjugated polyclonal antibody with visualization using the BCIP/NBT couple.

Acetylaminofluorene (AAF) Labeling

AAF labeling, also developed at the Pasteur Institute, is based on a chemical reaction in which guanyl residues become labeled by attachment of AAF to the C8 position. This may be performed on single- or double-stranded DNA and provides very stable probes (Tchen et al., 1984). The product obtained constitutes an excellent hapten that subsequently allows specific recognition for detection purposes.

Photochemical Labeling

Reagents such as photobiotin acetate or furocumarin derivatives form stable covalent bonds with nucleic acids after irradiation. In the former case (Gibco/BRL Photobiotin Labeling system), biotin is attached to DNA by a simple procedure. When either single- or double-stranded highly purified linear DNA (other cellular constituents may react with photobiotin) is strongly irradiated in the presence of photobiotin, biotin binds to the molecule. Detection is achieved by means of the alkaline phosphatase–streptavidine system in the presence of BCIP/NBT.

Another example of photochemical labeling is given by Dattagupta and co-workers (1989) which uses a biotinylated derivative of angelicine, a furocumarin. The bond between the two active portions of the molecule is made by a polyethylene oxide bridge. The DNA mixture and the biotin-PEG-angelicine reagent are irradiated for 1 hour with ultraviolet at a distance of 10 cm.

The authors have also proposed detection by chemiluminescence or by colorizing with colloidal gold. In the former case, antibiotin antibodies bind peroxidase which uses a luminescent molecule, isoluminol, in the presence of iodophenol as activator of the reaction along with H_2O_2. In the latter case, streptavidine binds stably to colloidal gold particles and this complex recognizes the probe DNA-bound biotin. Reporting is enhanced by the addition of silver.

8.3.2.3 Direct Labeling of DNA by the Reporting Enzymes

The ultimate simplification of the system consists of binding alkaline phosphatase or peroxidase directly to the DNA. The chemical bridge is composed of polyethyleneimine in the presence of glutaraldehyde. The application of this procedure has not yet been described.

8.3.3 Comparison of the Different Systems

The use of radioactive probes has been routine in research laboratories for many years. They are extremely sensitive and the reporting process is specific and with no background response to obscure the result. However, they obviously require all of the precautions associated with the handling of radioelements as well as proper facilities that meet safety standards. One of their major inconveniences is due to their rapid loss of activity. The most commonly used isotope, ^{32}P, lasts for 14.3 days, making it necessary for all practical purposes to label the probe for each hybridization.

Cold probes have several advantages compared to radioactive ones, not the least of which is handling by normal methods for biochemical or chemical reagents. One of these reagents, AAF, is however, carcinogenic. Nevertheless, the labeled DNAs are stable and may be stored frozen for several months. Probe stocks can therefore be created for identifying the most commonly tested microorganisms, making hybridizations a simple tool for routine use.

The earliest cold probes were of the biotinylated type. These were markedly less sensitive than radioactive probes. A high level of background interference caused especially by nonspecific binding thus limited their use.

Labeling and detection systems, as well as the quality of the membranes, have undergone much evolution and continue to be refined. Most of the progress has been in the simplification of the systems. Some labelings do not require absolute purification of the DNA while the sensitivity has been increased by reducing the number of steps preceding the color reaction. The first system described used two hapten-antibody affinity pairs prior to the enzymatic reaction, but it has already been made possible to attach the enzyme directly to the DNA.

The detection limit for cold probes is on the order of picograms, although this value depends very much on the probe used and on technical proficiency. It has been well established in rigorous comparison studies that cold probes are 20–30 times less sensitive than radioactive probes. Cold probe technology will develop, however, in association with the polymerase chain reaction (PCR) which allows amplification factors of 10^6–10^7.

8.4 Principles of Labeled Nucleotide Incorporation

Nucleotides labeled radioactively or otherwise are incorporated during in vitro synthesis of DNA fragments using a known DNA template. The polymerization enzymes (polymerase I or polymerase I Klenow fragments) use the four nucleotides present in the reaction mixture, one of which is labeled. The synthesized molecule is absolutely identical to the reference DNA chosen for the creation of the probe.

8.4.1 The "Nick and Patch" Technique

The probe DNA is subjected to the action of DNAse which randomly introduces nicks into both strands of the molecule, thereby providing sites for DNA polymerase I. The exonuclease activity of the latter eliminates nucleotides one by one by hydrolysis in the $5' \rightarrow 3'$ direction while the polymerase activity incorporates added nucleotides, of which one type is labeled either radioactively or chemically, beginning at the free $3'$ ends of the nicks and using the opposite strand as a template. The result is a double-stranded marked DNA molecule having a sequence identical to the starting molecule (Figure 8.3).

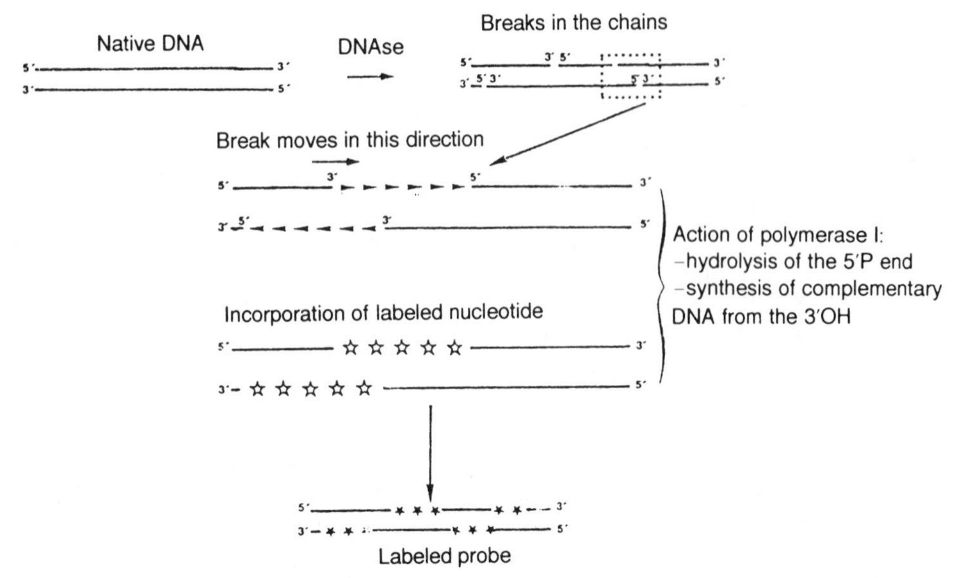

Figure 8.3 Principle of labeling by nicking and patching.

This technique is often used for radioactive labeling. It is also the principle of the Gibco/BRL biotin labeling kit of which the medium contains the following:

- a buffer at pH 8 containing 50 ml Tris-HCl, 50 mM $MgCl_2$, 500 μg/ml bovine serum albumin, and 100 mM β-mercaptoethanol
- the dNTP nucleotide mixture at 0.5 mM; in the case of radioactive probes, the labeled nucleotide may be ^{32}P-dCTP and for cold probes biotin 11-dUTP
- DNAse and DNA polymerase I at 20 pg/ml
- one μl of a DNA extract of a reference strain representing from 100 to 500 ng

The labeling reaction lasts from 1 to 2 hours at 19°C and is stopped by the addition of 500 μl of sterile water. Prior to use in hybridization, the labeled DNA is denatured by heating for 5 minutes in a water bath at 100°C.

8.4.2 Labeling by "Random Priming"

Incorporation of the labeled nucleotide takes place during the synthesis of the complementary sequence of one DNA strand by $5' \rightarrow 3'$ polymerase activity of the DNA polymerase I Klenow fragment. The enzyme requires a primer provided in the reaction mixture in the form of random sequence hexanucleotides. The template is the single-stranded DNA obtained after denaturation of the reference DNA molecule (Figure 8.4). For example, the protocol used for

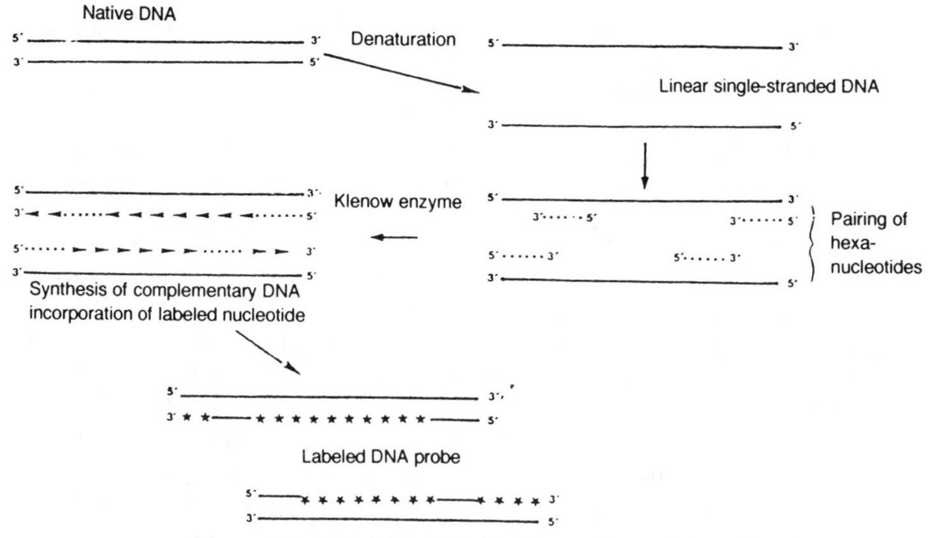

Figure 8.4 Principle of labeling by random priming.

the labeling of digoxigenin by random priming recommended by Boehringer Mannheim suggests adding the following to obtain a final volume of 20 μl:

- 10 ng to 3 μg of purified denatured DNA
- 2 μl of the hexanucleotide mixture
- 2 μl of the dideoxynucleotide mixture
- 1 μl of the Klenow enzyme (2 U)

After 1 hour of incubation at 37°C, the reaction is stopped by the addition of 0.2 M EDTA.

8.5 Hybridization Protocols

Most methods currently describe DNA hybridizations after immobilizing on a solid support. The most common way, known as the "dot-blot," consists of depositing onto membranes DNA extracts of the strains to be identified. Direct hybridization on colonies is also possible. Widely used in research for the screening of recombinant strains, this technique is of particular interest for the identification of contaminating strains in samples of media being monitored.

8.5.1 Hybridization Parameters

The parameters that control the speed of hybridization and its intensity determine the quality of the results obtained. The handling conditions are a function of this.

The double helical structure of DNA may be reversibly disrupted and restored. For a given salt concentration in the medium, a certain temperature increase leads to the separation of the strands. The melting temperature, Tm, is defined as the temperature at which 50% of the molecules are dissociated into single strands. This is a function of the ionic strength of the medium, the base composition, the length of double helix, and the possible presence of adjuncts such as formamide. For example, Wahl and colleagues (1987) reported that when the salt concentration is increased by tenfold, the melting temperature is increased by 16°C. Moreover, the stability of the hybrids is lowered when the bases are not correctly paired, that is, when the sequences are not entirely homologous. Their Tm decreases by 1° for each 1% of noncomplementary bases in the case of hybrids of more than 150 base pairs.

Medium conditions must therefore be determined which destabilize hybrids such as may form between a probe and a relatively nonhomologous DNA while not reducing the stability of hybrids between strands having a higher degree of complementarity. The parameters that determine the stringency of the medium are temperature, ionic strength, and the presence or absence of formamide. These also control the rate of strand pairing.

Since the response to the test must be more specific the more the probe DNA and target DNA are homologous, the stringency of the medium must be set higher by increasing the temperature and decreasing the salt concentration. Conversely, if the range of complementarity sought is less narrow, the stringency must be decreased.

Medium conditions during the posthybridization washings also have an influence on the test result since the purpose of these steps is to eliminate the less specifically bound probe DNA.

The rate of hybrid formation is optimal for temperatures 25°C below the melting temperature. When the probe DNA is present in excess, however, the rate depends mostly on the probe concentration and on the complexity of its sequence. It may be increased by adding dextran sulfate, an anionic polymer which by an exclusion effect increases the relative concentration of the available DNA.

8.5.2 The Various Steps of the "Dot-Blot" Hybridization Test

8.5.2.1 DNA Binding

Native or single-stranded DNA, following heat denaturation in alkaline medium or ultraviolet exposure, is immobilized by vacuum filtration on membranes with the aid of devices such as the "Hybri- slot" (BRL). Deposition of the DNA, contained in a volume of 1–5 μl, is often bound to the membrane by means of UV treatment or by baking at 80°C.

The earliest filters were made of nitrocellulose, although a variety of membranes is currently available (from Schleicher et Schull, Pall, DuPont de Nemours, Biorad, Genofit, Millipore, and others) including nitrocellulose derivatives and nylons charged differently depending on the groups attached. The performance of these is quite variable from one membrane to the next and depends on the type of probe (Durand, 1988). Generally, nitrocellulose is suitable for radioactive probes.

The immobilizing protocol must be adapted according to the chosen support. Nitrocellulose membranes bind single-stranded DNA while nylon immobilizes native DNA. Denaturation in the latter case takes place after the deposition of the sample. A medium of low ionic strength is thus more suitable for nylon membranes.

For many applications it is useful to be able to reuse a membrane holding samples to be identified. Such a membrane may be successively hybridized with a variety of probes. It is therefore important to select a durable support that can be treated to eliminate hybridized probe. Nylon membranes are recommended for this purpose.

Other strategies may be envisaged for the identification of a strain isolated from a known medium in routine testing for certain microorganisms such as contaminants. The DNAs from the known strains are fixed onto the filter while DNA from the strain to be identified is used as the probe in a conventional hybridization. A single procedure thus allows the identification of the microorganism among a collection having a particular significance. This method is described for various medical applications (Ezaki et al., 1988; Dattagupta et al., 1989).

8.5.2.2 Hybridization

Target DNAs attached to a support are contacted with probe DNA. The latter must hybridize only that target DNA with which it has the required sequence complementarity and then remain immobilized on the membrane. In order to avoid nonspecific binding of the probe to any site, the membrane is treated beforehand in a prehybridization phase, that is, adsorbing onto the membrane surface macromolecules which have no affinity for the probe. Mixtures of varying complexity are used for this purpose, containing 0.2% albumin, 5% skim milk, 10% polyethylene glycol, 0.02% polyvinyl pyrrolidine 40, and 0.02% Ficoll 400.

The ionic strength is increased to promote maximum pairing of single strands during both the prehybridization and the hybridization. It is set using a 4 to 6 × SSC buffer (20 × SSC = 3 M NaCl + 0.3 M sodium citrate). The medium also always contains 0.1% sodium dodecyl sulfate (SDS), which prevents adsorption of single-stranded DNA, hence of the probe, onto the membrane.

Dextran sulfate (10%), by exclusion of the DNA from the volume which it occupies, increases the apparent concentration of fragments greater than 250 nucleotides. It increases by 100-fold the rate of hybridization of a probe with immobilized DNA without increasing the background response (Wahl et al., 1979).

The addition of deionized 50% formamide decreases the melting temperature of DNA. Its presence of prehybridization and hybridization media consequently allows the temperature of the reaction to be decreased. For example, a hybridization under very stringent conditions takes place in a 5 SSC buffer at 68°C but at 42°C in the presence of 50% formamide. Prehybridizations generally take about 2–3 hours.

The hybridization itself takes place under the same physicochemical conditions, after addition of the probe at a concentration of about 50 ng/ml. The probe DNA must be previously denatured by boiling for 10 minutes if such a step is not part of its preparation protocol. This is the case notably for probes obtained by nicking-patching or by random priming, whereas cold probes that may be sulfonated for example are already in single-strand form.

During hybridization, the addition of unlabeled, totally heterologous DNA (such as denatured salmon sperm DNA) increases the specificity of the test by saturating membrane binding sites other than those of the target DNA.

At the end of hybridization, the probe DNA strands are paired, not only with the complementary target DNA but also to a lesser degree with other deposited DNAs corresponding to other microbial species. Detection must not be done until after nonspecifically hybridized labeled strands have been eliminated.

8.5.2.3 Washing of Filters and Stabilizing of Hybrids

The specificity and sensitivity of the test also depends largely on a final step in which the membranes are washed several times under conditions allowing only those probe DNA molecules which have correctly paired to remain immobilized.

As during hybridization, the temperature and ionic force determine the associations. As a general rule, filters undergo several successive washes in buffers of decreasing ionic strength thereby increasing the stringency of the medium. Washings at ambient temperature are usually in 2 or 3 SCC and 1% SDS followed by washings at the hybridization temperature in 1 SCC and 1% SDS and finally at ambient temperature again in 0.1% SCC. Protocols using cold probes usually involve a stabilization of the hybrids by treating the filters with a "blocking" solution containing albumin. Well practiced and proven in the case of sulfonated probes, this solution consists of albumin, skimmed milk, Tween 20, and heparin and increases the color signal response (Lebacq et al., 1987). The final step is the detection per se of the hybrids following the various principles already described according to the type of probe used.

The hybridization yield depends on several physicochemical factors such as temperature, ionic strength, and pH but also the probe concentration and the length of the homologous strands. No standardized conditions have really been established. In absolute terms, the protocol must be adapted to the particular problem to be solved. In practice, the instructions suggested by the suppliers of the kits and reagents or described by articles in their bibliographies allow satisfactory results to be obtained rapidly. Minor modifications may be required in order to improve responses.

8.5.3 Hybridization on Colonies

Hybridization on colonies depart somewhat from the methods developed to test for the presence of cloned DNA on recombinant bacteria. Three possibilities exist. One of these consists of transferring the colonies, after surface culture on solid media, by simple application of the membrane for a few minutes. The bacteria to be identified may also be grown directly on the filter placed on the surface of the solid nutrient medium. Finally, in other applications, pure cultures of the strains to be identified are deposited onto the filter which may be cultured in a Petri plate if the deposits are at an insufficient concentration. In all cases the filter is then treated, subjected to the steps of cell lysis, hybridization, washing, and detection. The various solutions required are poured directly onto the colonies and then eliminated by slow filtration or they may be contacted with the cells by capillary action. The first such method described by Grunstein and Hogness in 1975 was for *E. coli*. In this particular case, lysis is achieved easily in alkaline solution with 0.5 M NaOH and 1.5 M NaCl.

This treatment is not sufficient for other microorganisms. Generally, cell lysis requires treatment with lysozyme and SDS. For lactic acid bacteria, both bacilli and cocci, our laboratory has treated cell walls for 45 minutes at 37°C with lysozyme (10 mg/ml) in Tris-glucose to form protoplasts. Lysis follows in alkaline medium as per the protocols usually described elsewhere.

The efficiency of the lysis may thus differ depending on the microorganisms present. This is the principal difficulty encountered in the practical case of mixtures of bacteria of which the cell walls have variable sensitivity to lysozyme action. Ford and Olson (1988) suggested evaluating the effectiveness of this step by a methyl green coloration test described by Grunstein (1983). After denaturation and binding of the DNA liberated in alkaline medium, the filters are washed to eliminate unbound material. Prehybridization, hybridization, and detection all follow as for dot-blot hybridizations.

Hybridization on colonies is routinely practised with radioactive probes. It has also been adapted to cold probes, in particular to sulfonated probes (Chemiprobe, Lebacq et al., 1988) and to digoxigenin probes (Boehringer Mannheim process). Biotin or photobiotin labeled probes, however, are not suited to this procedure. These require that the hybridization be performed on pure DNA completely free of proteins and other cellular constituents (Zwadujk, 1986).

The identification of phages may be carried out according to the same principle as for colonies using a filter applied to a solid medium surface for hybridization. The probe, corresponding to fields of lysis, hybridizes with the DNA of homologous phages.

8.5.4 Dehybridization

After hybridization with a first probe and detection thereof, filters may be reused and rehybridized with a new probe. This is obviously highly advantageous, since several target DNAs can be tested with probes obtained from different species.

Dehybridization of radioactive probes requires washings under medium and temperature conditions that differ slightly depending on the author. The filter is placed in a buffer containing 0.1 SSC, 0.1% SDS, 0.2 M Tris HCl pH 7.5 and heated to 65°C for 30 minutes or simply in a boiling water bath in 0.1% SDS. Dehybridization may be verified by exposure of the film to the membranes.

In the case of cold probes, decolorizing of the filter is the first step. This is possible in the case of nylon membranes which are the only ones able to withstand the decolorizing treatment with dimethylformamide (Chemiprobe Bioprobe Systems SA and Boehringer Mannheim processes). The filters are then dehybridized in the usual way and are then ready for exposure to the new probe, via the original protocol at the prehybridization step with virtually no changes. In general, the more resistant nylon membranes are chosen when the performance of multiple hybridizations is anticipated.

8.6 Technical Developments

Identification using nucleic acid probes is still a relatively new procedure, with intense research efforts constantly aimed at numerous refinements. All of these are intended to increase the specificity and sensitivity of the system while simplifying the procedures.

8.6.1 Specific Probes

Most of the time probes are prepared from "total" DNA without any distinction between chromosomes and plasmids. It is possible, however, to target more specific sequences which code for particular properties such as pathogenicity. Probes for such purposes may be prepared from cloned whole plasmids or portions thereof or even from synthesized nucleotide sequences. The latter may be deduced from the sequence of a protein or some peptide using the genetic code. Since many nucleotide sequences may correspond to the same polypeptide, the probes are used in mixtures. The stringency conditions during the procedure ensure the specificity of the hybridization.

These considerations suggest a future for probes as diagnostic kits in industry. Probes for pathogenic microorganisms that contaminate foodstuffs may in theory be easily produced. Equipment for oligonucleotide synthesis is becoming increasingly reliable and efficient, making the procedure more and more routine in laboratories, while the production of DNA by recombinant bacteria is steadily improving and making yields on the order of 5 mg of DNA/l of culture commonplace (Lebacq, 1987).

8.6.2 Application of PCR

Great progress is expected to accrue from the amplification of DNA by the polymerase chain reaction commonly known as PCR. This process is essentially the repetitive synthesis of a portion of DNA to be detected. Schematically, the principle is as follows: double-stranded DNA is heat denatured and two oligonucleotide primers hybridize to each single strand delimiting the portion of the DNA to be amplified. A polymerase synthesizes complementary DNA from the 3′-OH end of the primers. From one DNA sequence, two are obtained, absolutely identical. This double-stranded DNA is then denatured and cycled through the same process which may be repeated up to 40 times. Since at each synthesis cycle the quantity of DNA determined by the primers is doubled, increasing at an exponential rate, the amount of DNA theoretically obtainable from n cycles is 2^n. In practice the yield factor must be considered, which reduces this value substantially. After 20 amplification cycles the quantity of DNA is closer to 1.85^{20} rather than to 2^{20} (Larzul, 1989).

The chain amplification is possible because of a special polymerase, Taq polymerase isolated from an archebacterium (*Thermus aquaticus*) capable of resisting high temperatures and having an optimal temperature in the 70–80°C range. This property is indispensable for the cycling of all of the steps involved, included heat denaturation, hybridizing of the primers, and synthesis. Automation of amplification is now possible by means of specially designed equipment that carefully controls heating and cooling. A programming example is given by Schowalter and Sommer (1989) with denaturing for 1 minute at 94°C, hybridization of the primers for 2 minutes at 50°C, and chain elongation for 3 minutes at 72°C.

Labeling of probes during amplification is done by incorporation of labeled nucleotides as in the random priming procedure. The great progress implicit in this method is largely due to more efficient labeling which makes the reliable detection of short probes possible.

Starting with extremely small quantities of DNA, down to a single molecule, amplification of specific probes allows sensitive detection even by cold labeling. This technique established itself in laboratories at a time when much effort had already been devoted to increasing the sensitivity of cold probe signal detection sensitivity. There is no longer any doubt that with the support of this technique, the detection of microorganisms by DNA hybridization will become simplified and automated and thus enter the realm of routine diagnostics.

8.7 Food Applications

Up to the present, molecular hybridization has not had any real applications to the identification of microorganisms in the food and agricultural industries. Species recognition is essential nonetheless, both from a technological point of view and in testing for undesirable contaminations. At the present time, however, only a few examples exist which describe at the research stage various possibilities for the technique and the refinement of specific probes.

With respect to microorganisms involved in food processes, DNA hybridization has been applied to lactic acid bacteria. Using this method, Jimeno and co-workers (1989) have differentiated the species but not the subspecies of bacteria used in cheese making, *Streptococcus thermophilus* and *Lactobacillus lactis*.

For wine making, the identification of *Leuconostoc oenos* is now practiced using this method in our laboratory. It provides a considerable time saving and a high reliability compared to conventional methods, which are always limited by the difficulties in growing the species and by the ambiguous responses to the tests (Lonvaud-Funel et al., 1989). The technique has also been adapted to other must and wine species, particularly lactobacilli, pediococci, and other leuconostocs.

Strains isolated from liqueur wines that are extremely difficult to grow in culture and that have been next to impossible to identify until now have been classified among *Lactobacillus homohiochii* and *L. hilgardii.* In the case of some lactobacilli, hybridizations between strains belonging to species considered distinct by biochemical criteria, and, on the other hand, the absence of hybridization in other cases, have led to the incubation of the phylogenetic relationships between them.

In the previous applications using hybridization of probes prepared from total DNA, discrimination is necessarily limited, even though already effective. It is possible to go further using more specific probes that are focused on much more limited portions of the genome. An excellent demonstration of this is provided by the identification and reclassification of lactic acid bacteria isolated from meats. A distinction shows up between *Lactobacillus curvatus* and *L. sake*, of which the conventionally used sugar fermentation profiles leave

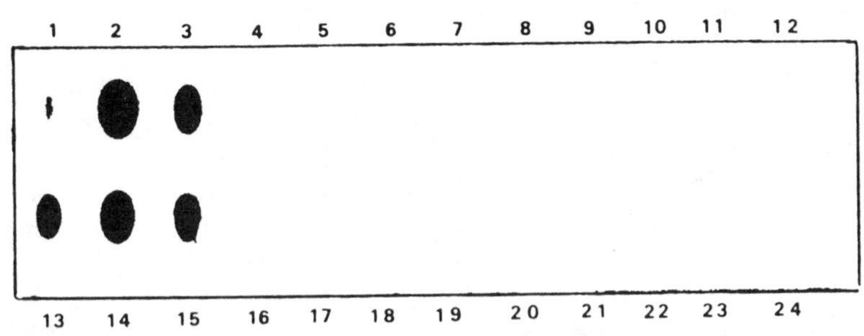

Figure 8.5. Result of the DNA hybridization of various strains of lactic acid bacteria from wine with a *Leuconostoc oenos* probe. *Leuconostoc oenos* (1, 2, 3, 13, 14, 15), *Lactobacillus hilgardii* (4, 5), *Lactobacillus brevis* (6, 7, 16, 17), *Lactobacillus plantarum* (9, 10, 20, 21, 22), *Lactobacillus casei* (11, 12), *Pediococcus damnosus* (18, 19, 23), *Leuconostoc mesenteroides* (24).

considerable doubt about their respective taxonomic positions and of which the percentage of DNA homology is high.

From an *L. curvatus* gene bank, Petrick and associates (1988) isolated a specific 1.2 kb probe which did not hybridize with *L. sake* nor with other strains of *L. curvatus*, but which hybridized with strains previously classified as *L. sake*.

For the detection of contaminating microorganisms, a few examples describe the use of specific probes. Cloned fragments of plasmid or chromosomal DNA or of synthetic oligonucleotides serve as a template for the preparation of these probes.

The detection of *E. coli* enterotoxin in particular has been studied with different protocols. Among these is hybridization on colonies done with a probe representing an 850 bp cloned fragment coding for a toxin (Hill et al., 1983b) or a 22 bp oligonucleotide (Hill et al., 1985) homologous to a portion of the gene deduced from the amino acid sequence of the toxic protein. The advantage of using the oligonucleotide is the ease with which the probe is obtained. Its limitation, however, is that any single point mutation occurring over this length may diminish hybridization with the probe without affecting synthesis of active toxin by the bacteria. The simultaneous use of two such probes corresponding to nonoverlapping regions of the same gene eliminates this problem.

The detection of the same sort of contaminant has also been done using biotinylated probes, hybridized on purified extracts of total DNA. The author has also demonstrated the impossibility of hybridizing such probes on colonies due to the presence of cellular biotin (Bialkowska-Hobrzanska, 1987).

Similar work has addressed *Yersinia enterocolitica.* The probes correspond to a plasmid fragment coding for virulence (Hill et al., 1983a), to a group of plasmid fragments or to a portion of a gene coding for Ca^{2+} dependence, a trait retained by all strains having the virulence-associated plasmid (Jagow et al., 1985). Oligonucleotide probes appear to be equally reliable (Miliotis et al., 1989; Kapperud et al., 1990).

Similarly, detection of salmonella by nucleic acid hybridization is also possible. In the initial work (Fitts et al., 1983), the test was done after enrichment of the sample, including with nonselective media; however, culturing may cause the loss of the plasmids which code for virulence. Under these conditions it is therefore preferable to use chromosomal DNA probes.

In most cases, molecular hybridization for the enumeration or detection of food-contaminating microorganisms avoids sample enrichment. The test is positive by virtue of addressing the DNA of the cell even if the latter, for any of a variety of reasons, no longer expresses the toxicity gene. The genus *Yersinia*, for example, which is very difficult to isolate from foods, does not always retain its pathogenicity. The risk of false-negatives is also eliminated, an omnipresent risk with conventional methods since virulent strains may be inhibited in conventional media.

8.8 Conclusion

For the moment at least, molecular hybridization is one of the most promising technical contributions of molecular biology to food microbiology. It is henceforth possible to identify microorganisms with absolute certainty and to classify them to at least the species level using simple procedures. Membrane supports for DNA as well as labeling and detection kits currently available are very effective and are constantly being improved. Cold labeling techniques now allow the introduction of these procedures into quality control laboratories without any requirement for special equipment. Furthermore, the use of synthetic oligonucleotides and the rapid development of DNA amplification by PCR will undoubtedly lead to the production of specific probes with obvious advantages for the detection of food product contaminations. Moreover, great simplifications can now be expected to occur at the detection system level. Being a rapid and sensitive method, the identification of microorganisms by nucleic acid probes is certain to lighten the workload associated with routine diagnosis and guarantee, among other things, exceptional reliability.

References

Bialkowska-Hobrzanska H., 1987. Detection of enterotoxigenic *Escherichia coli* by dot-blot hybridization with biotinylated DNA probes. *J. Clin. Microb.*, **25**, no. 2, 338–343.

Boehringer Mannheim GMBh Biochemica, 1989. DNA labeling and detection nonradioactive. *Applications Manual*, 61 p.

Dattagupta N., Rae P. M. M., Huguenel E. D., Carlson E., Lyga A., Shapiro J. A., Abbarella J. P., 1989. Rapid identification of microorganisms by nucleic acid hybridization after labeling the test sample. *Anal. Biochem.*, **177**, 85–89.

Durand B., 1988. Le transfert des acides nucléiques sur membranes. *Technoscope Biofutur*, **18**, 3–13.

Ezaki T., Dejsirilert S., Yamamoto H., Takeuchi N., Liu S., Yabuuchi E., 1988. Simple and rapid genetic identification of *Legionella* species with photobiotin-labeled DNA. *J. Gen. Appl. Microbiol.*, **34**, no. 2, 191–199.

Fitts R., Diamond M., Hamilton C., Neri M., 1983. DNA/DNA hybridization assay for detection of *Salmonella* spp. in foods. *Appl. Environ. Microbiol.*, **46**, no. 5, 1146–1151.

Ford S. and Olson B., 1988. Methods for detecting genetically engineered microorganisms in the environment. *Adv. Microbiol. Ecol.*, **10**, 45–79.

Grunstein M., 1983. *Methods for screening colony hybridization, plaque hybridization.* Schleicher and Schuell, Keene, New Hampshire.

Grunstein M. and Hogness D. S., 1975. Colony hybridization: A method for the isolation of cloned DNAs that contain a specific gene. *Proc. Natl. Acad. Sci. USA*, **72**, no. 10, 3961–3965.

Hill W. E., Payne W. L., Aulisio C. C. G., 1983a. Detection and enumeration of virulent *Yersinia enterocolitica* in food by DNA hybridization. *Appl. Environ. Microbiol.*, **46**, no. 3, 636–641.

Hill W. E., Madden J. M., McCardell B. A., Shah D. B., Jagow, J. A., Payne W. L., Boutin B. K., 1983b. Foodborne enterotoxigenic *E. coli* detection and enumeration by DNA colony hybridization. *Appl. Environ. Microbiol.*, **45**, no. 4, 1324–1330.

Hill W. E., Payne W. L., Zon G., Moseley S. L., 1985. Synthetic oligodeoxyribonucleotide probes for detecting heat-stable enterotoxin-producing *Escherichia coli* by DNA colony hybridization. *Appl. Environ. Microbiol.*, **50**, no. 5, 1187–1191.

Jagow J. and Hill W. E., 1986. Enumeration by DNA colony hybridization of virulent *Yersinia enterocolitica* colonies in artificially contaminated food. *Appl. Environ. Microbiol.*, **51**, no. 2, 441–443.

Jimeno J., Casey M., Gruskovnjak J., Furst H., 1989. Identifizierung von Milchsäurebakterien. *Swhweiz. Milchw. Forschung*, **18**, no. 2, 19–23.

Kapperud G., Dommarsnes K., Skurnik M., Hornes E., 1990. A synthetic oligonucleotide probe and a cloned polynucleotide probe based on the yop A gene for detection and enumeration of virulent *Yersinia enterocolitica*. *Appl. Environ. Microbiol.*, **56**, no. 1, 17–23.

Larzul D., 1989. Une révolution dans le diagnostic: l'Amplification enzymatique *in vitro*. *Biofutur*, **75**, 36–49.

Lebacq P., 1987. Spécifiques, sensibles et inoffensives: Voici les sondes nucléiques non-radioactives. *Technoscope Biofutur*, **11**, 12–18.

Lonvaud-Funel A., Biteau N., Fremaux C., 1989. Identification de *Leuconostoc oenos* par utilisation de sondes d'ADN. *Sci. Aliments*, **9**, 533–541.

Miliotis M. D., Galen J. E., Kaper J. B., Moris J. C., 1989. Development and testing a synthetic oligonucleotide probe for the detection of pathogenic *Yersinia* strains. *J. Clin. Microbiol.*, **27**, 1667–1670.

Petrick H. A. R., Ambrosio R. E., Holzapfel W. H., 1988. Isolation of a DNA probe for *Lactobacillus curvatus*. *Appl. Environ. Microbiol.*, **54**, no. 2, 405–408.

Schowalter D. B. and Sommer S. S., 1989. The generation of radiolabeled DNA and RNA probes with polymerase chain reaction. *Anal. Biochem.*, **177**, 90–94.

Tchen P., Fuchs R. P. P., Sage E., Leng M., 1984. Chemically modified nucleic acids as immunodetectable probes in hybridization experiments. *Proc. Natl. Acad. Sci. USA*, **81**, 3466–3470.

Wahl G. M., Stern M., Stark G. R., 1979. Efficient transfer of large DNA fragments from agarose gels to diazobenzyloxymethyl-paper and rapid hybridization by using dextran sulfate. *Proc. Natl. Acad. USA*, **76**, no. 8, 3683–3687.

Wahl G. M., Berger S. L., Kimmel A. R., 1987. Molecular hybridization of immobilized nucleic acids: Theoretical concepts and practical considerations *In: Methods in enzymology*, vol. 102, 399–407.

Zwadyk P. Jr., Cooksey R. C., Thornsberry C., 1986. Commercial detection methods for biotinylated gene probes: Comparison with ^{32}P-labeled DNA probes. *Curr. Microbiol.*, **14**, 95–100.

CHAPTER

9

Mechanization and Automation of Techniques

R. Grappin, Christine Piton

9.1 Introduction

As emphasized by Sharpe (1974), the word *automation* is often used in the analytical field to designate techniques or apparatuses which in fact represent only the mechanization of simple manual operations. Even though the boundary between mechanization and automation is sometimes blurred, the term *mechanization* is used for techniques done by machines which more or less accurately reproduce repetitive operations usually done by hand, such as the suspension of solid samples, distribution of media, or the counting of colonies. When an analysis consists of several distinct phases, each of these may be mechanized, the human operator intervening only at points between. This simple level of mechanization may be increased in sophistication by adding microprocessors and sensors which replace the human to some extent by a decision-making process. At this point we have automation.

In general, an automatic method is fixed in a specific configuration for the purpose of executing, often at very long intervals, only a single type of analysis. At the beginning of the 1980s, however, robotics began to make their way into laboratories, which may be considered an intermediate step between mechanization and automation. By means of suitable tools or "hands," a programmed robot is able to automatically carry out a series of unit operations such as sampling, dilution, agitation, weighing, and so on. Robotics offer great flexibility of use and operational versatility but at a lower frequency of analysis than truly automated equipment.

During the past several years, conventional microbial enumeration techniques or food products have begun to be replaced either by mechanized convenional methods or by rapid methods, some of these entirely automatic, which ensure better product quality management. Very often, total automation is possible only by switching to physical, chemical, or biochemical measurements such as spectroscopy, electrochemistry, chromatography, or immunology which reduce or eliminate the incubation step required for conventional colony counts. In such cases the microbial population is evaluated by means of signals that signify either the presence or the activity of microorganisms, for example, assays of enzyme activity, of a cell constituent or metabolite concentration, or measurement of a medium modification linked to such activity. By extension, these indirect techniques are often considered automatic techniques even if the automation involves only one of the steps of the microbiological analysis. For example, the ATP assay by bioluminescence is not really an automatic technique unless all of the sample preparation and reagent adding operations are performed by the machine.

The choices to be implemented for mechanization and automation purposes obviously vary as a function of the type of food product and the analysis to be performed, that is, whether for total flora, spoilage flora, fecal contamination indicators, or pathogens. The following text provides an outline of the basic elements of these choices, by way of presenting the objectives, limits, and constraints of the automation of microbial enumeration techniques.

9.2 The Aim of Automation

Regardless of the intended field of application, the following advantages are to be expected from the mechanization and automation of laboratory analyses:

1. A decrease in the unit cost per analysis. This decrease may be obtained by introducing simple laboratory devices such as medium distributors, automatic diluters, sample suspending machines, or colony readers which increase the number of analyses performed without increasing labor costs. Automation often also results in decreased consumption of products (i.e., reagents, pipettes, Petri plates, etc.) and in decreased utilities costs (water, power). For example, the use of the spiral plating technique described by Gilchrist and co-workers (1973) reduces culture medium consumption by 61% and Petri plate use for dilutions of 10^{-1}–10^{-6} by 83% (Briner et al., 1978). When very elaborate and expensive equipment is involved, reductions in the cost per analysis are realized only when the system is constantly in operation.

2. Improvements in working conditions for laboratory personnel by freeing them from tedious and tiring repetitive tasks such as counting colonies on Petri plates or counting bacteria by epifluorescent microscopy. This is

especially important for highly specialized laboratories which use only one or two methods to carry out very large numbers of analyses.

The introduction of automated techniques has often been presented as offering the possibility of employing less qualified personnel, the machine supposedly being able to perform all operations by itself. In fact, experience shows that monitoring of this type of equipment is not always a simple matter, especially in the case of highly perfected instruments, since the technician must be capable of detecting equipment malfunctions, verifying calibrations, and ensuring proper maintenance.

3. Measurement with greater precision by significantly decreasing the risk of human error, whether at the dilution step, during colony counting, or while logging results. In addition, the use of machinery designed around perfectly defined and controllable technical characteristics may improve the fidelity of the measurements. For example, the DEFT technique (direct epifluorescent filter technique) used for raw milk total microfloral counts has coefficients of repeatability varying from 15 to 20% (Pettipher et al., 1983; Suhren et al., 1984; Dasen et al., 1987); whereas the entirely automated Bacto-Scan machine, also based on epifluorescent microscopy, may reach coefficients as low as approximately 6% (Grappin et al., 1985).
4. Faster response. Methods based on the measurements of physical or chemical parameters associated with microorganisms generally obtain results more rapidly than any technique based on the growth of colonies on solid medium. In this case speed is not a result of automation but due rather to the method employed. Nevertheless, for some automated techniques, the growth of the bacteria for several hours is required for the analysis. This is the case for measurements of the sensitivity of microorganisms to antibiotics using an automatic turbidometer or for microbial measurements by impedance, radiometry, or microrespirometry.

In practice, the various objectives associated with the automation of microbiological techniques are rarely achieved simultaneously since numerous economic, regulatory, and microbiological constraints come into play.

9.3 The Limitations of Automation

9.3.1 Economic Limitations

The development and refinement of an automatic device are determined by economic considerations on the part of the manufacturer who requires that the potential market be considerable as well as on the part of the end users in their laboratories. In the field of food and agriculture, bacteriological analyses are done either to test starting materials or products during processing or to verify the bacteriological quality of products in terms of national or international standards. In either case, automatic techniques are used in only two situations:

either the laboratory has a sufficient volume of analyses to justify the purchase of what is likely to be very expensive equipment or automation is the only way to achieve the necessary speed of response, the time factor being in this case the fundamental technoeconomic element. This is especially the case for the testing of perishable starting materials prior to processing or testing of products during processing.

Essentially two types of food testing laboratories get involved with automation. First, there are those multidisciplinary laboratories whose diversity is a reflection of the type of product being analyzed or of the type of bacteriological analysis. For this category, methods and devices must be adapted either to performing analyses on a great variety of products and over a large range of concentrations or for purposes other than enumeration or to mechanizing steps that are common to various analyses (dilution, colony counting). An automated technique such as impedance measurement appears to meet the first two criteria. It is usable, in conjunction with judicious choice of culture media and incubation temperatures, for the enumeration of different microbial groups (total flora or total psychroptrophic flora, coliforms) in products as diverse as milk and frozen vegetables (Cady, 1978; Cady et al., 1978; Firstenburg-Eden and Tricarico, 1983; Firstenberg-Eden et al., 1984). It also has other applications such as the detection of antibiotics in milk (Okigbo and Richardson, 1985), measurement of lactic ferment activity (Okigbo et al., 1985), or the detection of postpasteurization contamination of milk (Bossuyt and Waes, 1983). In addition, the spiral plating technique may be employed with numerous food products (Campbell and Gilchrist, 1973; Jarvis et al., 1977; Kramer et al., 1979) which allows selective counting of several microbial groups on the same Petri plate with satisfactory precision as a function of colonial diameter by using electronic laser scanning (Richard, 1982). Devices described by Sharpe and colleagues (1972) or Engelbrecht (1978), which allow sample dilutions and successive additions of reagents and media at long intervals, may be used for a variety of analyses. And finally, still in this category of equipment, methods using laboratory robots may be mentioned, such as the technique described by Richardson and associates (1988). With the aid of a Zymark-type robot used for the preparation of microtitration plates and of a second robot for reading, a whole series of milk analyses, such as total microfloral and total coliform counts, leucocyte counts, and antibiotic assays, may be performed simultaneously or consecutively.

Highly specialized laboratories, of which the best example is probably provided by European milk analysis laboratories, make up the second group of laboratories for which automation is a going concern. In these installations, the bacteriological quality of thousands of milk herd samples per month is tested using a single technique, generally a simplified method of enumerating mesophilic bacteria on Petri plates or in rolling tubes. There currently are 37 such laboratories in France analyzing approximately 700,000 samples per month. Under these conditions it is possible to switch over to practically total automation of the techniques. Devices such as the Petri-Foss (Grappin and

Jeunet, 1974) automatically carry out at rates of 300 samples/hour all of the operations associated with the preparation of Petri plates: sampling of the milk with a calibrated loop, inoculation, addition of agar, mixing, cooling, and labeling of the plate. Similar devices are used in Dutch laboratories for the preparation of rolling tubes using either a calibrated loop (Posthumus et al., 1974) or the dilution method (Jaartsveld and Swinkels, 1974). For colony counting, all laboratories are currently using automatic counters which provide readings with good precision in a few seconds (Grappin and Jeunet, 1974; Grappin, 1975).

Aside from the colony counters, these systems are simply mechanized versions of manual techniques and as such still have the drawback of requiring 3 days of incubation in order to produce an answer. Since 1980, a newer and entirely automatic device specially designed for total microflora in raw milk, the Bacto-Scan (Foss Electric, Denmark), has been commercialized and has been the subject of numerous studies (Kielwein and Daun, 1981; Grappin et al., 1985; Nieuwenhof and Hoolwerf, 1988; Suhren et al., 1988; Heiss et al., 1989). It is based on the automatic counting of milk bacteria by epifluorescence after chemical treatment and centrifugation of the sample. The response time is low (7 minutes) and the more recent versions (the Bacto-Scan 800) allow an analysis of 80 samples/hour.

Outside of the milk analysis business, few food analysis laboratories can contemplate the use of fully automated techniques because the number of different analyses which these very specialized systems perform is often too low. It has been noted, for example, by Read and colleagues (1978) that out of 800 American laboratories performing regulatory testing of milk quality, and in particular microbial counts, only 30 were equipped with automatic systems. In fact, only concentration of the analysis in a few centralized locations would allow laboratories to process sufficient numbers of samples, but this practice is seriously limited by the transportation requirements for samples intended for bacteriological analysis.

9.3.2 Regulatory Constraints

Two major types of methods exist for the purpose of determining numbers of microorganisms: there are reference methods which by definition or by convention give the "true values" of these numbers and there are the indirect techniques which do not measure numbers of bacteria but measure instead a signal associated with their presence. In either case, these methods may be employed as an official method for cases of conflict requiring a judgment or as a routine method for the monitoring and continuous testing of food products or as a rapid screening method.

When a reference method is mechanized, its use in automated form does not generally pose any major problem whether in routine applications or for official purposes as long as the principle of measurement is not changed. Devices for Petri plate preparation which mechanize the simplified count method using a calibrated loop (Thompson et al., 1960) or the spiral plater are examples of this. The usefulness of such devices is essentially due to their technoeconomical values, that is, their reliability, flexibility and ease of use, cost per analysis, and so on.

Alternatively, the situation is much more delicate when the automatic device is based on a principle different from that of the reference method based on growth of microorganisms on nutritional media. The precision of the estimation by the rapid method then becomes a key element in the decision. Caution must be exercised when evaluating an analytical method by comparison to another method chosen as a reference, especially when the latter has its imperfections. In the dairy field, this is the case for total microfloral enumeration in milk which is influenced by the type of agitation, diluent, culture medium, and the incubation time-temperature combination (Wilson, 1935; Huhtanen et al., 1970, 1972, 1975, 1976; Richard, 1980; Fryer, 1982). Nevertheless, even if the reference technique does not necessarily give an exact picture of the true number of microorganisms present in milk, it is generally recognized that the number adequately reflects hygienic conditions during production and storage. In the context of milk pricing, it is of utmost importance to ensure that a sufficient correlation has been obtained between the automatic method and the reference method. Use of the Bacto-Scan has been authorized for milk-pricing purposes in France and Germany.

Moreover, the interpretation of the results of microbiological analyses is often made in terms of thresholds, which limit the importance of the accuracy of the methods. In the regulatory context, the presence or absence of pathogenic microorganisms may be expressed in terms of 25 or 100 g of product. Similarly, a food product becomes unfit for consumption when it does not meet certain criteria, for example, in the case of fresh fish fillets, 10^5 aerobic mesophiles/g, 10 fecal coliforms/g, 10^2 *S. aureus/g*, 10 sulfite-reducing anaerobes/g, and absence of *Salmonella* in 25 g. At this point it becomes necessary to know the probabilities of obtaining false-positive or false-negative results with the new technique. An example provided by Munson and co-workers (1978) for the determination of *Salmonella* in food products by fluorescence shows that by choosing the fluorescent threshold, only positive samples need be confirmed by the reference technique.

9.3.3 Analytical Aspects

As mentioned earlier, one of the benefits to be expected from automatic devices is an improvement in the precision of measurements. It is important to bear in mind the numerous aspects that determine the precision of an analytical technique (Grappin, 1976; Anonymous, 1985).

- Fidelity, which represents the closeness of the agreement between the results obtained, either under identical experimental conditions (repeatability) or by different operators in different laboratories (reproducibility).

- Accuracy, which is defined as the closeness of the agreement between the true value of the parameter to be measured (as given by the reference method) and the result obtained with the experimental technique.
- The detection limit or analytical end-point which is defined as the lowest value of the measured parameter which can be detected by the device with a known level of certainty. This criterion is particularly important in microbiology where it may be necessary to determine whether or not a single microbial cell is present in 10 or 25 g of product. The detection threshold of a device, determined by the signal to noise ratio of the method, is often one of the limiting factors governing the precision of indirect enumeration techniques. In fact, physicochemical methods are often not directly usable mainly because of the weak signal produced by microorganisms compared to interfering signals (e.g., leucocytes in raw milk). It is therefore necessary to either eliminate interference by chemical or physical treatments, as is done for raw milk analysis with the Bacto-Scan or by ATP-metry, or diminish its relative significance by amplifying the bacterial signal with a microbial growth step, as is the case of impedance measurement techniques.
- Sensitivity, which represents the smallest variation in the parameter being measured (e.g., number of bacteria) that the method is capable of detecting.

In fact, the only criterion routinely influenced by automation is fidelity since a machine is capable of reproducing hand-performed operations with greater precision. The other precision criteria, although sometimes influenced by the use of an automatic device, have little relationship to automation itself but are more directly associated with the principle of measurement on which the system is based.

The repeatability of automated techniques is generally at least as good as that of conventional methods and often better because the use of machines reduces the main source of random error input which is human involvement. It is therefore not too surprising to note that reference methods that provide the "true" value and that are often entirely manual, are also sometimes of lower fidelity than automated techniques. For example, Grappin and associates (1985) observed that for total microfloral counts in raw milk, the coefficients of variation of repeatability are around 12% for the reference method and only 6% for the Bacto-Scan.

In the case of automatic devices for the preparation of Petri plates used for counts of total aerobic mesophiles in raw milk, it has been noted that the repeatability is identical to that obtained by manual preparation (Grappin and Jeunet, 1974). Comparisons between manual and automated techniques give an excellent correlation with a variance standard error of 14% or identical to the standard error of repeatability characterizing each method.

The problem is more complex for automatic colony counters which are becoming more and more widely used in laboratories. In practice, results may be identical to those obtained by visual counting (Fruin and Clark, 1977), provided that the following conditions are met:

- The resolving power of the device is identical to that of the naked eye, i.e., 0.1–0.15 mm for the enumeration of mesophile flora, and the instrument is properly calibrated.
- The colonies are evenly distributed, well contrasted, and not surface penetrating, spreading, in beadlike strings or surrounded by debris. Since this condition is rarely met, Petri plates must be sorted accordingly in order to eliminate those which will provide faulty readings (Grappin and Jeunet, 1974).

In spite of these precautions, a residual imprecision always persists in colony counts. In France, the standard for the use of automatic counters in milk analysis laboratories specifies spreads between visual and instrument counts of less than ± 30 colonies for plates with 30–100 colonies, less than ± 50 for 100–300 and ± 70 for 300–700 at least 95% of the time. For counts of specific microorganisms such as *E. coli*, *S. faecalis*, or *C. perfringens*, Guthertz and Fruin (1979) showed that visual counts were superior to automatic counts. In any event, it is certain that the use of automatic counters leads to a theoretical drop in the precision, even though visual counts are not error free, with coefficients of variation between automatic counts ranging from 3.6% for a single operator to 4.7% for different operators (Peeler et al., 1982).

9.4 Conclusion

The concepts guiding the mechanization and automation of microbiological techniques are very similar to those found in analytical chemistry. Moreover, the most thoroughly explored route to the automation of these techniques is marked by the use of physical and chemical measurement techniques that readily lend themselves to automation. The use of these indirect techniques, however, poses numerous analytical and regulatory problems. In addition, the results obtained with them are strongly influenced by the nature and the physiological state of the microorganisms present in the food sample. Automated techniques are thus often limited to use in rapid methods of testing and screening. Yet, partial or total mechanization of proven techniques often constitutes the simplest and most sure method for the automation of microbiological counting techniques. This is the case for automation of which the principle very closely resembles that of the corresponding manual technique, such as the spiral plating technique or culture on membrane with counting of microcolonies by an automatic device (Sharpe and Michaud, 1978).

Total automation as defined in the introduction is possible only for a limited number of cases. In fact, as soon as it becomes necessary to resort to isolation of bacteria on a medium for identification purposes or for signal amplification by means of microbial growth, automation is limited to addressing only certain steps of the analysis. Finally, an automatic technique refined for one type of product is not generally transferable to other fields of application without a thorough study of the technical characteristics of the method and of the impact of the microbial flora on its precision.

References

Anonymous, 1985. Lait. Définition et évaluation de la précision globale des méthodes indirectes d'analyse de lait. Application au calibrage et au contrôle de qualité dans les laboratoires laitiers. Norme provisoire FIL 128.

Bossuyt R. G. and Waes G. M., 1983. Impedance measurements to detect post-pasteurization contamination of pasteurized milk. *J. Food Prot.*, **46**, 622–624.

Briner W. W., Wunder J. A., Blair D. W., Parran J. J., Blanney T. L., Jordan W. E., 1978. Use of spiral bacterial plating and laser colony counting techniques in studies of the microbial ecology of man *In:* Sharpe A. N., Clark D. S. (eds). *Mechanizing miocrobiology*, 154–169. Charles C. Thomas Publisher, Springfield, IL.

Cady P., 1978. Progress in impedance measurements in microbiology. *In:* Sharpe A. N., Clark D. S. (eds). *Mechanizing microbiology*, 199–239. Charles C. Thomas Publisher, Springfield, IL.

Cady P., Hardy D., Martins S., Dufour S. W., Kraeger S. J., 1978. Automated impedance measurements for rapid screening of milk microbial content. *J. Food Prot.*, **41**, 277–283.

Campbell J. E. and Gilchrist J. E., 1973. Spiral plating technique for counting bacteria in milk and other foods. *Dev. Ind. Microbiol.*, **14**, 95–102.

Dasen A., Piton C., Grappin R., Guerry P., 1987. Evaluation de la technique DEFT associée à un comptage visuel ou à un comptage par analyse d'images pour la numération de la flore totale du lait cru. *Le Lait*, **67**, 77–95.

Engelbrecht E., 1978. The biodilutor and bioreactor for automated tests on serial dilutions of samples. *In:* Sharpe A. N., Clark D. S. (eds). *Mechanizing microbiology*, 187–198. Charles C. Thomas Publisher, Springfield, IL.

Firstenberg-Eden R. and Tricharico M., 1983. Impedimetric determination of total mesophilic and psychrotrophic counts in raw milk. *J. Food Sci.*, **48**, 1750–1754.

Firstenberg-Eden R., Van Sise M. L., Zindulis J., Kahn P., 1984. Impedimentric estimation of coliforms in dairy products. *J. Food Sci.*, **49**, 1449–1452.

Fruin J. T. and Clark Jr. W. S., 1977. Plate count accuracy: analysis and automatic colony counter versus a true count. *J. Food Prot.*, **40**, 552–554.

Fryer T. F., 1982. Factors affecting the enumeration of microorganisms from raw milk. *Kiel. Milchwirtsch. Forschungsber.*, **34**, 96–101.

Gilchrist J. E., Campbell J. E., Donnelly C. B., Peeler J. T., Delaney J. M., 1973. Spiral plate method for bacterial determination. *Appl. Microbiol.*, **25**, 244–252.

Grappin R., 1975. Mise au point sur les appareils automatiques utilisés pour la numération des germes totaux du lait: Préparation des boites de Petri et comptage des colonies. *Rev. Lait. Fr.*, **335**, 629–641.

Grappin R., 1976. Guide pour l'évaluation des méthodes d'analyse de routine. *Le Lait*, **56**, 608–621.

Grappin R. and Jeunet R., 1974. Progrès récents dans la mécanisation d'une méthode simplifiée de numération des germes totaux du lait. XIXème Congrès International de Laiterie, New Dehli, India, B8, 576–577.

Grappin R., Dasen A., Favennec P., 1985. Numération automatique et rapide des bactéries du lait cru à l'aide du Bacto-Scan. *Le Lait*, **65**, 123–147.

Guthertz L. S. and Fruin J. T., 1979. Colony count accuracy using selective media: Analysis versus automatic colony counters. *J. Food Prot.*, **42**, 420–423.

Heiss E., Ellner R., Kreuzer K., 1989. Versuche zur Keimzahlbestimmung mit dem Bactoscan 8000. *Dtsch. Molk.-Ztg.*, **110**, 848–851.

Huhtanen C. N., Brazis A. R., Arledge W. L., Cook E. W., Donnelly C. B., Ginn R. E., Murphy J. N., Randolph H. E., Sing E. L., Thompson D. I., 1970. Effect of dilution bottle methods on plate counts of raw milk. *J. Milk Food Technol.*, **30**, 259–273.

Huhtanen C. N., Brazis A. R., Arledge W. L., Cook E. W., Donnelly C. B., Ginn R. E., Jezeski J. J., Pusch D., Randolph H. E., Sing. E. L., 1972. A comparison of two and three days incubation for enumerating raw milk bacteria. *J. Milk Food Technol.*, **35**, 136–140.

Huhtanen C. N., Brazis A. R., Arledge W. L., Donnelly C. B., Ginn R. E., Randolph H. E., Koch E. J., 1975. A comparison of phosphate buffered and distilled water dilution blanks for the standard plate count of raw milk. *J. Milk Food Technol.*, **38**, 264–268.

Huhtanen C. N., Brazis A. R., Anderson H. J., Arledge W. L., Donnelly C. B., Ginn R. E., Koch E. J., Nelson P. E., La Grange W. S., Peterson D. E., Randolph H. E., Sing E. L., Thompson D. I., Hehr H. M., 1976. Comparison of incubation at 20 and 32°C for 48 and 72 hours for enumeration of raw milk bacteria. *J. Milk Food Technol.*, **39**, 417–420.

Jaartsveld F. H. J. and Swinkels R., 1974. A mechanized roll tube method for the estimation of the bacterial count of milk. *Neth. Milk Dairy J.*, **28**, 93–101.

Jarvis B., Lach V. H., Wood J. M., 1977. Evaluation of the spiral plate maker for the enumeration of micro-organisms in foods. *J. Appl. Bacteriol.*, **43**, 149–157.

Kielwein G., Daun U., 1981. Das Bacto-Scan Verfahren zur automatischen Bakterien Zählung in der Anlieferungsmilch. *Dtsch. Molk.-Ztg*, **102**, 874–878.

Kramer J. M. and Kendall M., Gilbert R. J., 1979. Evaluation of the spiral plate and laser colony counting techniques for the enumeration of bacteria in foods. *Europ. J. Appl. Microbiol. Biotechnol.*, **6**, 298–299.

Munson T. E., Schrade J. P., Bisciello N. B., 1978. Automated fluorescent antibody test for *Salmonella*. *In:* Sharpe A. N., Clark D. S. (eds). *Mechanizing microbiology*, 104–119. Charles C. Thomas Publisher, Springfield, IL.

Nieuwenhof F. F. J. and Hoolwerf J. D., 1988. Suitability of Bactoscan for the estimation of the bacteriological quality of raw milk. *Milchwissenschaft*, **43**, 577–586.

Okigbo L. M., Oberg C. J., Richardson G. H., 1985. Lactic culture activity tests using pH and impedance instrumentation. *J. Dairy Sci.*, **68**, 2521–2528.

Okigbo O. N. and Richardson G. H., 1985. Detection of penicillin and streptomycin in milk by impedance microbiology. *J. Food Prot.*, **48**, 979–981.

Peeler J. T., Leslie J. E., Danielson J. W., Messer J. W., 1982. Replicate counting errors by analysts and bacterial colony counters. *J. Food Prot.*, **45**, 238–240.

Pettipher G. L., Fulford R. J., Mabbit L. A., 1983. Collaborative trial of the direct epifluorescent filter technique (DEFT), a rapid method for counting bacteria in milk. *J. Appl. Bacteriol.*, **54**, 177–182.

Posthumus G., Klijn C. J., Giesen T. J. J., 1974. A mechanized loop method for the total count of bacteria in refrigerated supplier's milk. *Neth. Milk Dairy J.*, **28**, 79–92.

Read R. B., Campbell J. E., Winbush J. S., 1978. Status of mechanized microbiological standard methods for food. *In:* Sharpe A. N., Clark D. S. (eds). *Mechanizing microbiology*, 67–73. Charles C. Thomas Publisher, Springfield, IL.

Richard J., 1980. Influence de l'agitation du lait cru sur les résultats de dénombrement de sa flore à l'aide d'une anse calibrée. *Le Lait*, **60**, 211–235.

Richard J., 1982. Dénombrement sélectif des bactéries du lait en fonction de la taille des colonies totale à l'aide d'un compteur électronique à laser. *Sci. Aliments*, **2**, 31–40.

Richardson G. H., Grappin R., Yuan T. C., 1988. A reflectance colorimeter instrument for measurement of microbial and enzymatic activities in milk and dairy products. *J. Food Prot.*, **51**, 778–785.

Sharpe A. N., 1974. The need for automation in microbiology labs. Sixth Latin American Congress in Microbiology, 25 November–7 December 1974, Caracas, Venezuela.

Sharpe A. N., Biggs D. R., Oliver R. J., 1972. Machine of automatic bacteriological pour plate preparation. *Appl. Microbiol.*, **24**, 70–76.

Sharpe A. N. and Michaud G. L., 1978. Evaluation of bacteria using hydrophobic grid-membrane filters *In:* Sharpe A. N., Clark D. S. (eds). *Mechanizing microbiology*, 140–153. Charles C. Thomas Publisher, Springfield, IL.

Suhren G. and Heeschen W., 1984. Untersuchungen zur Keimzahlbestimmung in Rohmilch mit der direkten Epifluoreszenz-Filter-Technik DEFT. *Kiel. Milchwirtsch. Forschungsber.*, **36**, 87–136.

Suhren G., Heeschen W., Reichmuth J., 1988. Zur Messung der bakteriologischen Beschaffenheit von Rohmilch mit dern Bactoscan-Gerät. *Dtsch. Molk.-Ztg.*, **109**, 380–387.

Thompson D. I., Donnelly C. B., Black L. A., 1960. A plate loop method for determining viable counts of raw milk. *J. Milk Food Technol.*, **23**, 167–171.

Wilson G. S., 1935. The bacteriological grading of milk: a critical study. Medical Research Council, Special Report Series no. 206, London.

CHAPTER

10

Prediction of Product Life Span

C. M. Bourgeois

10.1 The Value and Importance of Such Predictions

The quality and safety characteristics required of a product must be ensured not just on leaving the factory but all the way to the moment of consumption. This implies that there must be some method for predicting the behavior of a product after it leaves the factory, based on knowledge of its composition, the treatments it has undergone during its production, and the environmental conditions under which it will be stored. In microbiological terms, there must be a method for predicting the qualitative and quantitative evolution of the microflora.

In actual fact, methodological approaches to this challenge are relatively recent and not yet widespread. The protection of the consumer by public agencies still relies essentially on inspection and on analyses of finished products, which results in the numerous inconveniences generally acknowledged by all parties concerned. In order to be able to work within regulatory guidelines, manufacturers have implemented internal testing systems that monitor starting material and finished product quality as well as the manufacturing process.

The current trend towards reducing heat treatments and minimizing use of preservatives while lengthening storage life spans all without lowering microbiological safety has greatly increased the demands being made on microbiologists.

The HACP system (hazard analysis critical control point; ICMSF, 1988), which consists of systematically localizing the critical quality and safety points in a process and devising methods for monitoring these, has brought about considerable improvement, limited nevertheless by the delay in response inherent in microbiological analyses (Chapter 2).

"Predictive microbiology," based on the modelization of microfloral evolution, is complementary to this system (Roberts, 1989) to the extent that it reduces these delays in response and allows earlier prediction. The thermal destruction of *Clostridium botulinum* is a striking example of the effectiveness of this approach. Knowledge of the exponential kinetics of this process and of the specific thermosensitivity of the microorganism in question has allowed the modelization of its thermal destruction, correlative specification of the type of treatment to apply, and ascertainment of product safety by a simple examination of the heat treatment record printout. Applied to microbial growth in food products, the goal of modelization is to obtain by similar means earlier and more useful information about product storage, safety, and sensorial quality than is obtainable from the enumeration of significant microflora.

10.2 Predictive Microbiology

Predictive microbiology requires that the product life span limiting parameter be identified and that the principle of its evolution be analyzed and understood sufficiently for the prediction of its shelf life.

10.2.1 Identification of the Product Life Span Limiting Factor

First, a definition of what is meant by product life span or storage life span is required. This is the length of time between the moment a finished product leaves the production line and the consumption deadline or "best before" date, beyond which date there is a risk of the product becoming dangerous or developing perceptible organoleptic changes. The length of time is therefore slightly longer than the product "shelf life."

Information about the underlying principle of the limitations of the storage life span of a food product may be drawn from the reference section to this chapter or from examination of products returned to the manufacturer due to storage defects. This limit may have to do with safety or with sensorial quality. Objective methods of measurement are indispensable and may be of the following types depending on the case:

- analysis of a chemical substance (nutrient, levels of peroxide or free fatty acids, etc.);
- microbiological analysis;
- sensorial analysis.

This chapter deals with cases in which the onset of risk or of deterioration is of microbiological origin. In such cases, the agent responsible for the storage limitations is the microbial population directly responsible for the deleterious effect. It is therefore this population that must be characterized. Where organoleptic spoilage is involved, it is obviously useful to envisage the use of sensorial analyses as well. It is also necessary to determine from the reference section, from experience acquired in the business, or by means of experiments, the level at which the agent limiting the storage life begins to manifest its undesirable effects, that is, the effect threshold. This determination is complicated by the fact that the level of intensity of the undesirable agent (e.g., the concentration of a malodorous molecule) is not always directly correlated with the number of microorganisms of the type responsible for the effect.

10.2.2 Prediction of Product Life Span

The notion of predicting product life span in fact includes two distinct ideas:

- prediction of the total product life span for an entire lot coming off the production line or of the residual life span of a product on the shelf, in either case of finished products or
- prediction of the product life span from the design or formulation stage.

Various possibilities exist in either case, although modelization provides the most satisfactory route. Two types of models may be envisaged (Roberts, 1989):

- Probabilistic models which predict the probability of a microbial response under certain conditions, for example, the probability of toxin production by *C. botulinum*, this being the specific subject of the works which led to the development of this type of model
- Kinetic models which predict the concentration reached by a given strain or species and which are therefore applied to predicting the onset of the risk of intoxication or infection, for example, by *Salmonella*, and also to predicting storage life span when the limiting factor is a microbiological spoilage process

10.3 Modelizing the Effect of Temperature

The combined knowledge of the behavior of a microorganism at a given temperature, of the relationship between growth and temperature, the initial number of microorganisms, and the threshold of effect determines the prediction of the time for which a product may be stored at a given temperature. This factor is of such importance that it has been the subject of essential works and of a procession of proposed models.

10.3.1 The Arrhenius Model

The Arrhenius model is expressed by the following relation:

$$r = Ae^{-\mu/RT}$$

in which

r is the growth rate
μ is the activation energy in cal/mol
R is the universal gas constant
A is another constant
T is the storage temperature in degrees Kelvin

Very often, especially in the case of processed foods in which the microorganisms have undergone a "stress," the latter do not begin to grow until after a latent phase of variable length. Prediction of storage life span requires determination of this length in addition to the growth rate or generation time (Figure 10.1).

Separate evaluation of these two factors is important since the specific growth rate depends uniquely on the characteristics of the immediate environment of the microorganism (i.e., composition of the medium and atmosphere temperature) while the length of the latent phase depends on the initial state of the cells resulting from all of the treatments undergone. Heat, cold, and radiation all subject microorganisms to stress, and a latent phase of variable length is necessary for them to recover their ability to grow.

Using data obtained at a given temperature, the Arrhenius equation may be applied to determining the two kinetic parameters at other temperatures and thereby predict the levels of bacteria which will be reached after a given time of storage. This prediction may take into consideration the slowing as the

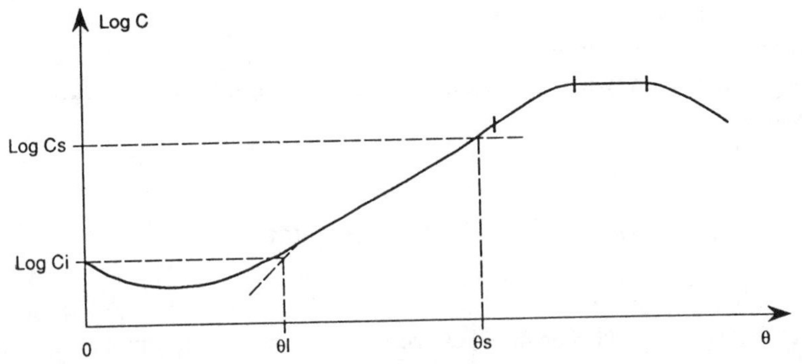

Figure 10.1 Prediction of storage life span. C = microbial concentration; Ci = initial concentration; Cs = threshold concentration of effect; θl = latent period; θs = storage life span. The latent period θl is read off the figure. The generation time G is calculated from the slope of the curve.

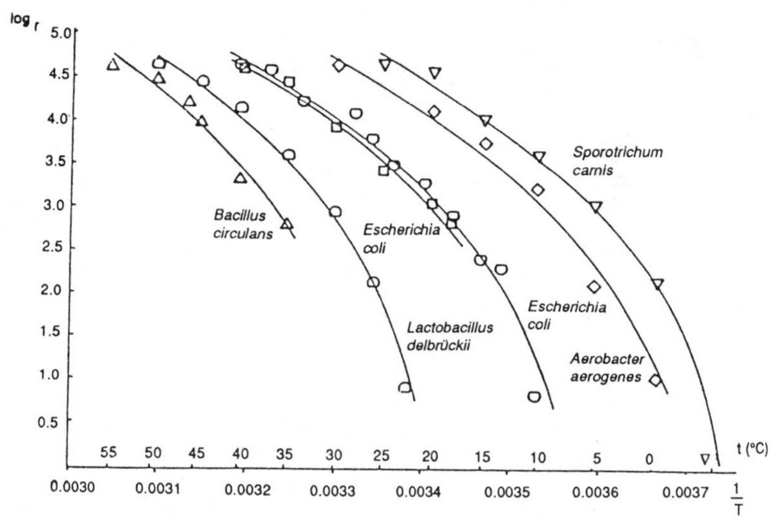

Figure 10.2 Arrhenius plots of data by Johnson and co-workers for six bacteria (adapted from Ratkowsky et al., 1982).

maximum is approached as well as the series of temperatures to which the product will be subjected during its storage.

In any event, it has been shown that the Arrhenius model, although perfectly suitable for the modelization of chemical spoilage processes, does not appear to be entirely satisfactory for the modelization of microbiological processes (Ratkowsky et al., 1982). In practice, the relation $r = Ae^{-\mu/RT}$, which may be expressed in the following form:

$$\text{Log } r = \text{Log } A - \frac{\mu}{RT} \quad \text{or even Log } r = K_1 - K_2 \times \frac{1}{T}$$

is represented graphically by a line of slope $-K_2$. Figure 10.2, from the work of Ratkowsky and co-workers (1982), shows, however, that in the case of microbial growth, this relation is not linear.

10.3.2 The Schoolfield and Ratkowsky Models

Schoolfield and colleagues proposed in 1981 a nonlinear regression derived from the Arrhenius equation describing the effect of temperature on biological systems. For the details of this very complex expression, the reader is referred to either the original work or to that of Adair et al. (1989). This model has been successfully applied to processing of microbiological data by Broughall et al. (1983) and by Broughall and Brown (1984).

Broughall and Brown (1984) provide a hypothetical applied example to pasteurized milk of which the factor limiting storage life span is considered to be multiplication of staphylococci. Figure 10.3 shows the times neces-

Temperature	pH	Water activity								
		0,9	0,91	0,92	0,93	0,94	0,95	0,96	0,97	0,98
10°C	4,0	-	-	-	-	-	-	-	-	-
	4,5	-	-	-	-	-	-	-	-	-
	5,0	-	-	-	-	-	-	-	-	-
	5,5	-	-	-	-	-	-	-	-	-
	6,0	-	-	-	-	-	-	-	-	-
	6,5	-	-	-	-	-	-	-	-	-
	7,0	-	-	-	-	-	-	-	-	-
20°C	4,0	-	-	-	-	-	-	-	-	-
	4,5	-	-	-	-	-	-	-	-	1,5/2,0
	5,0	-	-	-	-	-	-	-	2,0	1,5
	5,5	-	-	-	-	-	-	-	1,5	1,0/1,5
	6,0	-	-	-	-	-	-	1,5	1,0/1,5	0,5/1,0
	6,5	-	-	-	-	2,0	1,0/1,5	1,0/1,5	0,5/1,0	0,5/1,0
	7,0	-	-	-	2,0	1,0/1,5	1,0/1,5	0,5/1,0	0,5/1,0	0,5/1,0
30°C	4,0	-	-	-	-	-	1,5/2,0	1,0/1,5	1,0/1,5	0,5/1,0
	4,5	-	-	-	-	2,0	1,0/1,5	0,5/1,0	0,5/1,0	< 0,5
	5,0	-	-	-	-	1,0/1,5	0,5/1,0	0,5/1,0	0,5	< 0,5
	5,5	-	-	-	1,0/1,5	1,0	0,5/1,0	0,5/1,0	< 0,5	< 0,5
	6,0	-	-	1,5/2,0	1,0/1,5	0,5/1,0	0,5/1,0	< 0,5	< 0,5	< 0,5
	6,5	-	1,5/2,0	1,0/1,5	0,5/1,0	0,5/1,0	0,5/1,0	< 0,5	< 0,5	< 0,5
	7,0	-	1,0/1,5	1,0/1,5	0,5/1,0	0,5/1,0	0,5/1,0	< 0,5	< 0,5	< 0,5

Figure 10.3 Predicted time (days) for a 1,000-fold increase in number of *S. aureus* (adapted from Broughall and Brown, 1984). A hyphen means more than 2 days.

sary for a 1,000-fold multiplication at 10°, 20°, and 30°C for different a_w/pH combinations.

In 1982, Ratkowsky and co-workers proposed another model, based on purely empirical concepts and expressed by the following equation:

$$\sqrt{r} = b(T - T_0)$$

in which

r is the growth rate
b is the regression coefficient
T is the storage temperature
T_0 is a hypothetical temperature, without any special metabolic significance but characterizing the microorganism under consideration

Figure 10.4 shows that for populations of *Pseudomonas*, the relation between $\sqrt{r}$ and T is for all practical purposes linear, at least for the range between the minimum and optimum growth temperatures. Prior to this, in 1983, Ratkowsky and his co-workers proposed an equation derived from the displayed expression and which described growth over the entire range of temperatures including those beyond the optimum. This model has been used successfully by some authors, for example, to predict the life span of milk (Chandler and

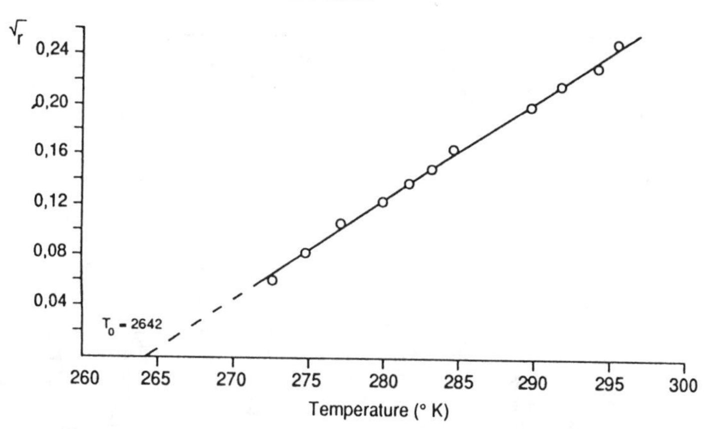

Figure 10.4 Typical linear relationship for *Pseudomonas* Group 1 strain 16L16 between square root of growth rate and temperature (adapted from Ratkowsky et al., 1982). The growth rate was measured as the reciprocal of the time required to reach a turbidity equal to 25% of saturation.

McMeekin, 1985a, b). Some time-temperature integrators also function on the basis of this model (see Section 10.5).

Pooni and Mead (1984), however, have observed marked deviations between predicted and actual storage life spans for chicken at higher shelf temperatures when applying this model. They considered this anomaly to be partly due to the specific behavior of *Pseudomonas* and related organisms, of which the growth rate and therefore the spoilage effect increase more rapidly as a function of temperature than the Ratkowsky relation predicts, but also partly due to other microorganisms not taken into consideration as well as differential temperature effects on the metabolic activities of the microorganisms.

More recently, a few publications have compared the predictions of the Schoolfield and Ratkowsky models to real data. According to Baird-Parker and Kilsby (1987), the predictions of both models are satisfactory, with those of Schoolfield being closer to reality at lower temperatures. Adair and associates (1989) have also found the Schoolfield predictions to be more reliable over the entire temperature range, especially at low temperatures.

10.4 Multifactorial Models

Predicting the evolution of a microbial population requires analysis of the contribution of essential factors to successful management of the microbiological quality of products. Pioneering works by Roberts and Ingram (1973) have provided multidimensional surface representations of the results of multifactorial experiments which show the effect of some of the essential factors on growth. Other authors have developed bi-dimensional models (Broughall et al.,

Temperature (°C)	Growth	Water Activity								
		0.88	0.89	0.90	0.91	0.92	0.93	0.94	0.95	0.96–0.98
8	L	*	*	*	*	*	*	*	33.1	4.3
	G								2.4	1.1
10	L	*	*	*	*	*	*	21.6	4.9	1.1
	G							2.2	1.1	0.60
12	L	*	*	*	*	31.2	15.8	5.2	1.4	0.72
	G					3.0	1.7	0.96	0.57	0.34
14	L	*	*	33.0	22.7	12.8	5.4	1.9	0.83	0.55
	G			2.6	1.7	1.1	0.73	0.48	0.32	0.21
16	L	35.8	26.8	18.7	11.4	5.7	2.3	1.0	0.61	0.42
	G	1.9	1.4	1.0	0.73	0.53	0.38	0.27	0.19	0.14
18	L	24.5	17.1	11.0	6.1	2.9	1.3	0.72	0.47	0.33
	G	1.2	0.84	0.61	0.44	0.32	0.24	0.17	0.13	0.09
20	L	17.1	11.2	6.7	3.5	1.7	0.88	0.55	0.37	0.25
	G	0.82	0.59	0.42	0.31	0.22	0.16	0.12	0.09	0.07
22	L	12.1	7.5	4.3	2.2	1.1	0.66	0.43	0.29	0.20
	G	0.58	0.42	0.31	0.22	0.16	0.12	0.09	0.07	0.05
24	L	8.7	5.2	2.9	1.5	0.83	0.51	0.34	0.23	0.15
	G	0.41	0.31	0.23	0.17	0.13	0.09	0.07	0.05	0.04
26	L	6.3	3.7	2.0	1.1	0.63	0.40	0.27	0.18	0.12
	G	0.3	0.22	0.17	0.13	0.10	0.07	0.05	0.04	0.03
28	L	4.7	2.7	1.5	0.82	0.50	0.32	0.21	0.14	0.10
	G	0.21	0.16	0.13	0.10	0.07	0.06	0.04	0.03	0.03
30	L	3.6	2.0	1.1	0.64	0.40	0.26	0.17	0.11	0.08
	G	0.15	0.12	0.09	0.07	0.06	0.05	0.04	0.03	0.02

Figure 10.5 Latent periods and generation times (in days) predicted for *Staphylococcus aureus* (adapted from Broughall et al., 1983). *predicted generation time >35 days; L = latent period; G = generation time.

1983, Figure 10.5) and tri-dimensional models (Broughall and Brown, 1984, Figures 10.6 and 10.7) showing the effects on latent period or on generation time of factors such as temperature, water activity, and pH. After these initial attempts, the time came to broach the subject by more methodical routes (as proposed by Roberts in 1989).

10.4.1 The Parameters Under Consideration and the Principles of Their Modelization

Since the most significant parameter associated with the life span of a product is usually the concentration of a particular species of microorganism in the product, the first action to consider is the direct modelization of this concentration as a function of the significant factors such as temperature, pH, a_W, atmospheric composition, the concentrations of the principal substrates, or inhibitors.

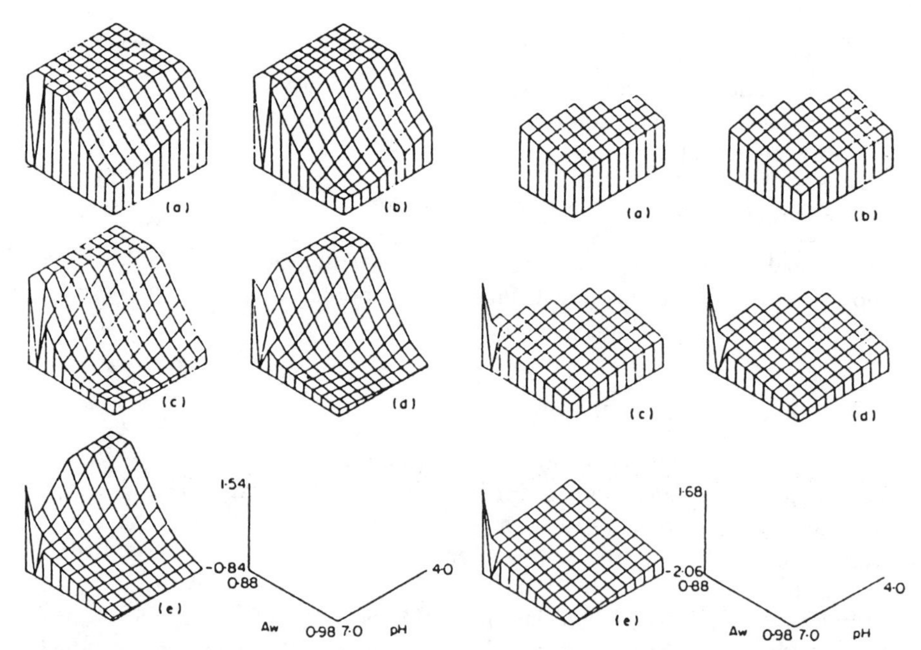

Figure 10.6 Modelistic predictions for the kinetics of the latent phase of *S. Aureus.*

Figure 10.7 Modelistic predictions for the kinetics of the generation time of *S. aureus.*

Surface (a) 10°C, surface (b) 15°C, surface (c) 20°C, surface (d) 25°C, surface (e) 30°C. The scales used for the five figures are given on the sixth figure. The vertical axis represents the $\log_{10}$ of the kinetic parameter (adapted from Broughall and Brown, 1984).

Given the complex nature of the growth process, however, it has been generally recognized as preferable to modelize the generation time and the growth rate separately, that is, the two parameters that are in principle both necessary and sufficient to characterize the growth.

Precise determination of these parameters is difficult, especially given that growth in a food product frequently occurs without any true exponential growth phase. This has given rise to the approach used by Gibson and colleagues (1987 and 1988) which consists of adjusting the experimental curves obtained with various combinations of pH, temperature, and a_w using the Gompertz function which expresses the microbial concentration Lt at time t as a function of four parameters: A = the limit of Lt as t approaches 0; C = the limit of Lt as t approaches infinity; B = the growth rate of M; and M = the time elapsed when the growth rate has reached its maximum. From these four parameters, the latent phase and the generation time may be objectively deduced. The function used to modelize the four parameters A, C, B, and M is a polynomial of the form

$$y = a + b1s + b2t + b3p + b4s^2 + b5t^2 + b6p^2 + b7st + b8sp + b9tp + e$$

where

y = modelized parameter, e.g., B = maximum growth rate
s = NaCl concentration (% w/v)
t = temperature (°C)
p = pH
e = random error

This model has no theoretical basis but may be considered a Taylor series approximation of the theoretical function. It should be noted that the parameters have a clear biological significance.

10.4.2 Data Acquisition

Data acquisition consists of producing numerous growth curves for various factors under consideration, according to a suitable experimental design. Bratchell and co-workers (1989), using the method described earlier, have more recently shown the importance of gathering sufficient data using a well-defined time distribution—i.e., a minimum of 15–20 points per growth curve was considered by them to be essential for reliable estimates of the generation time. Moreover, in order for the model to effectively predict the behavior of products with varying characteristics, it is necessary to obtain data for a great variety of combinations of factors. Furthermore, in order to obtain reliable data for a microbial species, many strains must be studied. It is therefore essential to have effective examination techniques which allow the continuous and simultaneous study of numerous microbial cultures. Some of the techniques or devices which appear to lend themselves to this sort of study are as follows:

- impedance measurements;
- turbidometry, especially using the Bioscreen system which follows the evolution of large numbers of cultures in wells of microtitration plates (McClure et al., 1989);
- bi-dimensional gradient plates (McClure et al., 1989), which allow continuous variation of the intensity of the factors under consideration.

A major problem arises when designing procedures to gather such data, due to the basic characteristics of the experimental medium chosen, of which the a_W characteristics, pH, and so on must be varied. These media are, in increasing order of difficulty and realism, as follows:

- conventional bacteriological media;
- a liquid medium simulating the food product;
- the food product itself.

10.4.3 Validation of the Models

Once constructed, the model must obviously be validated by comparing the values it predicts with a sufficient number of experimental results and actual

	Packaging film			
	Polyethylene		EVA-SARAN-EVA	
Temperature (°C)	0	4	0	4
Days in storage	8,0	3,5	20,0	8,0
Predicted values for:				
- log N	5,34	5,78	5,17	5,41
- σ log N	0,09	0,09	0,06	0,07
Experimental value: log N	5,40	5,80	5,20	5,43
Log % difference	1,11	0,34	0,58	0,37

Figure 10.8 Comparison of predicted microbial populations and experimentally counted microorganisms (adapted from Zamora and Zaritsky, 1985).

field observations. This type of validation leads to preferences such as the Schoolfield model over the Ratkowsky model for lower temperatures. Figure 10.8, reproduced from Zamora and Zaritsky (1985), shows the results of a validation experiment.

Significant deviations may result from flaws inherent in the chosen mathematical model as well as from the number of factors taken into account being insufficient. For practical reasons, it is impossible to take into consideration more than three or four factors and most authors feel that this is adequate in order to establish sufficiently precise models. Nevertheless, in some cases, large spreads between predicted and observed values may suggest taking additional factors into consideration.

10.5 Prediction of the Life Span of a Manufactured Product

Given the initial bacteriological status of a product, it is possible to predict its state after a given shelving time by extrapolation. One may arrive at such predictions by any of the following three routes:

- accelerated staling tests;
- modelistic predictions based on results of a rapid analysis or an estimation of total initial microflora;
- use of time-temperature integrators.

10.5.1 Accelerated Spoilage Potential Tests

Accelerated spoilage potential tests involve drawing inferences from the correspondence which exists between the delay in the onset of spoilage at different temperatures and the delay at a test temperature allowing the prediction of the delay at the normal storage temperature.

With respect to product life span, the reference section provides some usable but sporadic data which must first be converted into kinetic values, for example, by applying the simplified Arrhenius law expression or Q_{10} model (Labuza and Schmidl, 1985). The following relations are based on chemical preservation but may apply somewhat to the storage application under consideration from a microbiological point of view.

Relation 1

$$Q_{10} = \frac{\text{Spoilage rate at temp. T} + 10}{\text{Spoilage rate at temp. T}} = \frac{V_{(T+10)}}{V_T}$$

$$Q_{10} = \frac{r_{(T+10)}}{r_T}$$

$$Q_{10} = \frac{\text{Shelf life at temp. T}}{\text{Shelf life at temp. T} + 10} = \frac{\theta_{s(T)}}{\theta_{s(T+10)}}$$

If the temperature spread contemplated between T_1 and T_2 is not 10 but Δ, the law becomes as follows:

Relation 2

$$Q_{10}^{\Delta/10} = \frac{\theta_{s(T1)}}{\theta_{s(T2)}}$$

For example, if the Q_{10} is 3 and the storage life span θ_s is 6 months at 35°C, the storage life span at 20°C is:

$$\theta_{20} = \theta_{35} \times Q_{10}^{\Delta/10} = 6 \times 3^{15/10} = 31.2 \text{ months}$$

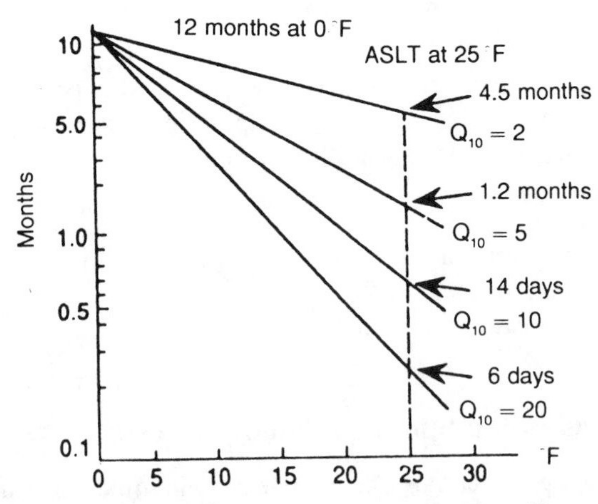

Figure 10.9 Length of storage tests at 25°F equivalent to 12 months at 0°F for different Q_{10} values. ASLT = accelerated spoilage potential life time (adapted from Labuza and Schmidl, 1985).

Frozen foods (°C)	Dry foods at intermediate humidity (°C)	Heat-treated foods (°C)
−40 (control)	0 (control)	5 (control)
−15	23 (room temperature)	23 (room temperature)
−10	30	30
−5	35	35
	40	40
	45 (if possible)	

Figure 10.10 Recommended shelf temperatures for rapid evaluation of the storage life span of foods (adapted from Labuza and Schmidl, 1985).

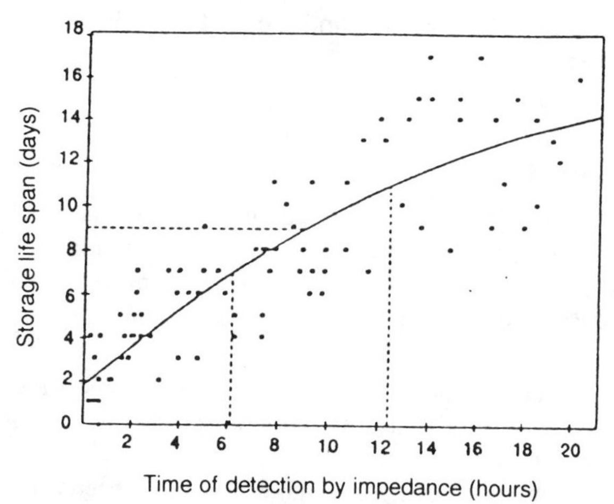

Figure 10.11 Relation between potential storage life span and detection time by impedance at 21°C (adapted from Bishop et al., 1984).

The Q_{10} is not necessarily provided in the literature but may be determined from relations 1 or 2 if the storage life span at two different temperatures is known. Accelerated spoilage potential tests may thus be contemplated for the prediction of the storage life span (Figures 10.9 and 10.10).

The measurement of impedance is considered by some to be a very rapid test for spoilage potential (Figure 10.11, Bishop et al., 1984; Gibson, 1985), albeit very approximate. This method is only applicable, however, to finished product lots sampled at the end of the production line.

10.5.2 Modelistic Prediction

Once the mathematical model describing the evolution of the concentration of a target microorganism has been established and validated, the initial concentration and possible level of activity in the lot under consideration must be determined for prediction purposes. The initial concentration may be determined by a rapid technique insofar as such techniques may exist in comparison with the product life span. Otherwise, if the factory employs an HACCP-type test method, an estimate of the range of the microbial concentration will provide an adequate basis for prediction. Rapid techniques that may be implemented for this purpose must have the following characteristics:

- They must be selective, deterioration or sanitary risk being attributable to particular groups (genera, species, or even strains charcterized by particular genes).
- They must preferably provide information about not only the number but also the state and activity of microorganisms present, if these are variable. Immunological techniques, nucleic acid probes, bioluminescence, impedance measurements, enzymatic methods, and the DEFT method have at least one of these characteristics.

10.5.3 Integrators

Another method of prediction uses time-temperature integration devices which are included with the product and which allow the cumulative temperature fluctuations which it has undergone up to a given instant to be easily observed. Various types of devices or systems are available for this purpose:

- biochemical, i.e., enzymatic or microbial preparations which gradually change the color of a reactant as a function of the thermal effects undergone (Wells and Singh, 1988);
- Electronic time-temperature integrators which have been calibrated for a certain type of microbial population (temperature effect on spoilage rate) and zeroed at the outset for the initial population (requirement for rapid analysis or for a good estimate) and which integrate the time-temperature profile undergone by the product, thereby providing an instant reading of the residual storage life span at all times.

Taoukis and Labuza (1989a and b) have systematically studied the available devices on the basis of their theoretical capacity to indicate with sufficient precision the residual storage life span, including during storage at variable temperature. They show that those devices for which the kinetic characteristics are well matched with those of the product do indeed have interesting potential.

10.6 Prediction of the Potential Life Span of a New Product

Predicting the potential life span of a new product may be done by either inoculating samples of the product with microbial cells which are capable of spoiling it or by using suitable models.

10.6.1 Inoculation of the Product with Test Strains

Inoculation of the product with test strains is commonly known as a "challenge test." This procedure is used in cases where the total absence of a pathogenic microorganism in a product is not a certainty and consumer safety therefore depends on certainty that the microorganism is incapable of multiplication in the product. An example of this is the assessment of the risk of *Clostridium botulinum* development in delicatessen products stored under conditions which a priori do not provide sufficient guarantees (pasteurized products or those subjected to ionizing treatment).

The disadvantage of this approach is that all parameters being fixed, the results obtained apply only to a specific product formulation, produced and stored under specific conditions. This becomes a costly solution when many successive formulations of a new product in development must be tested.

10.6.2 Use of Models

The modelistic approach, on the contrary, evaluates the storage potential of a product starting at the very beginning of its design or formulation.

10.7 Future Possibilities

In spite of its somewhat underdeveloped state, the predictive approach appears to be capable of meeting many of the real demands currently faced by industrial food processers. The representation of models in the form of surfaces in a tri-dimensional coordinate system has the advantage of allowing the visualization of interactions but the disadvantage of taking into consideration only a limited number of factors. This limitation may be of little consequence if, as many believe, the number of truly significant factors is small. In any event, the use of computerized methods express a microbial response as a function of greater numbers of factors as long as no graphical representation of the response is sought.

The real obstacle to the further development of this method is of an experimental nature. Before great benefits can accrue, it will be necessary to accumulate a large mass of data, which implies widespread international collaboration and aceptance of standardized working methods so that the results obtained in a given context may be useful to everybody.

References

Adair C., Kilsby D. C., Whittall P. T., 1989. Comparison of the Schoolfield (nonlinear Arrhenius) model and the square root model for predicting bacterial growth in foods. *Food Microbiol.*, **6**, 7–18.

Baird-Parker A. C. and Kilsby D. C., 1987. Principles of predictive food microbiology. *J. Appl. Bacteriol.* Symposium supplement, 435–495.

Bishop J. R. and White C. H., 1985. Estimation of potential shelf-life of pasteurized fluid milk utilizing bacterial numbers and metabolites. *J. Food Prot.*, vol. 48, August, 663–667.

Bratchell N., Gibson A. M., Truman T. M. Kelly, Roberts T. A., 1989. Predicting microbial growth: The consequences of quantity of data. *Int. J. Food Microbiol.*, **8**, 47–58.

Broughall J. M., Anslow P. A., Kilsby D. C., 1983. Hazard analysis applied to microbial growth in foods: Development of mathematical models describing the effect of water activity. *J. Appl. Bacteriol.*, **55**, 101–110.

Broughall J. M. and Brown C., 1984. Hazard analysis applied to microbial growth in foods: Development and application of three-dimensional models to predict bacterial growth. *Food Microbiol.*, **1**, 13–22.

Chandler R. E. and McMeekin T. A., 1985. Temperature function integration and the prediction of the shelf-life of milk. *Australian J. Dairy Technol.*, March, 10–13.

Chandler R. E. and McMeekin, T. A. 1985b. Temperature function integration and its relationship to the spoilage of pasteurized, homogenized milk. *Australian J. Dairy Technol.*, March, 37–39.

Chandler R. E. and McMeekin T. A., 1989. Modelling the growth response of *Staphylococcus xylosus* to changes in temperature and glycerol concentration/water activity. *J. Appl. Bacteriol.*, **66**, 543–548.

Gibson A. M., Bratchell N., Roberts T. A., 1987. The effect of sodium chloride and temperature on the rate and extent of growth of *Clostridium botulinum* type A in a pasteurized pork slurry. *J. Appl. Bacteriol.*, **62**, 479.

Gibson A. M., Bratchell N., Roberts T. A., 1988. Predicting microbial growth: Growth responses of *salmonellae* in a laboratory medium as affected by pH, sodium chloride and storage temperature. *Int. J. Food Microbiol.*, **6**, 155.

ICMSF, 1988. Microorganisms in foods. 4: Application of the hazard analysis critical control point (HACCP) system to ensure microbiological safety and quality. *Int. Commission on Microbiological Specifications for Food.* Blackwell Scientific Publications, Oxford.

Labuza T. P. and Schmidl Mary K., 1985. Accelerated shelf-life testing of foods. *Food Technol.*, September, 57–64.

McClure P. J., Roberts T. A., Otto Oguru P., 1989. Comparison of the effects of sodium chloride, pH and temperature on the growth of Listeria monocytogenes on gradient plates and in liquid medium. *Letters Appl. Microbiol.*, **9**, 95–99.

Ratkowsky D. A. Olley J., McMeekin T. A., Ball A., 1982. Relationship between temperature and growth rate of bacterial cultures. *J. Bacteriol.*, vol. 149, no. 1, January, 1–5.

Ratkowsky D. A., Lowry R. K., McMeekin T. A., Stokes A. N., Chandler R. E., 1983. Model for bacterial culture growth rate throughout the entire biokinetic temperature range. *J. Bacteriol.*, June, 1222–1226.

Roberts T. A., 1989. Combinations of antimicrobials and processing methods. *Food Technol.*, January, 156–162.

Roberts T. A. and Ingram M., 1973. Inhibition of growth of *Clostridium botulinum* at different pH values by sodium chloride and sodium nitrite. *J. Food Technol.*, **8**, 467.

Schoolfield R. M., Sharpe P. J. H., Magnuson C. E., 1981. Non-linear regression of biological temperature-dependent rate models based on absolute reaction rate theory. *J. Theor. Biol.*, **88**, 719–731.

Sharpe A. N., 1979. An alternative approach to food microbiology for the future. *Food Technol.*, March, 71–74.

Singh R. P. and Wells J. H., 1985. Use of time-temperature indicators to monitor quality of frozen hamburger. *Food Technol.*, December, 42–50.

Taoukis P. S. and Labuza T. P., 1989a. Applicability of time-temperature indicaors as shelf life monitors of food products. *J. Food Products, J. Food Sci.*, vol. 54, **4**, 783–788.

Taoukis P. S. and Labuza T. P., 1989b. Reliability of time-temperature indicators as food quality monitors under non-isothermal conditions. *J. Food Sci.*, vol. 54, **4**, 789–792.

Wells J. H. and Singh R. P., 1988. Application of time-temperature indicators in monitoring changes in quality attributes of perishable and semiperishable foods. *J. Food Sci.*, **53**, no. 1, 148–152.

Zamora M. C. and Zaritsky N. E., 1985. Modeling of microbial growth in refrigerated packaged beef. *J. Food Sci.*, **50**, 1003–1004.

Suggested Readings, Chapters 7–10

Abbaszadegan M., Huber M. S., Gerba C. P., Pepper I. L., 1993. Detection of enteroviruses in groundwater with the polymerase chain reaction. *Appl. Environ. Microbiol.*, **59**, 1318–1324.

Alvarez A. J., Toranzos G. A., 1992. Solid-phase polymerase chain reaction: Applications for direct detection of enteric pathogens in waters. *Can. J. Microbiol.*, **38**, no. 5, 365–369; 23 ref.

Anderson M. R. and Omiecinski C. J., 1992. Direct extraction of bacterial plasmids from food for polymerase chain reaction amplification. *Appl. Environ. Microbiol.*, **58**, 4080–4082.

Atmar R. L., Metcalf T. G., Neill F. H., Estes M. K., 1993. Detection of enteric viruses in oysters by using the polymerase chain reaction. *Appl. Environ. Microbiol.*, **59**, 631–635.

Bej A. K., DiCesare J. L., Haff L., Atlas R. M., 1991. Detection of *Escherichia coli* and *Shigella* spp. in water by using the polymerase chain reaction and gene probes for *uid*. *Appl. Environ. Microbiol.*, **57**, 1013–1017.

Bej A. K., Mahbubani M. H., Atlas R. M., 1991. Detection of viable *Legionella pneumophila* in water by polymerase chain reaction and gene probe methods. *Appl. Environ. Microbiol.*, **57**, 597–600.

Bej A. K., Steffan R. J., DiCesare J., Haff L., Atlas R. M., 1990. Detection of coliform bacteria in water by polymerase chain reaction and gene probes. *Appl. Environ. Microbiol.*, **56**, 307–314.

Bej A. K., Mahbubani M. H., DiCesare J. L., Atlas R. M., 1991. Polymerase chain reaction-gene probe detection of microorganisms by using filter-concentrated samples. *Appl. Environ. Microbiol.*, **57**, 3529–3534.

Bej A. K., Mahbubani M. H., Boyce M. J., Atlas, R. M., 1994. Detection of *Salmonella* spp. in oysters by PCR. *Appl. Environ. Microbiol.*, **60**, 368–373.

Bhaduri S. and Fratamico P. M., 1992. A simplified method of colony hybridization using radiolabeled probes in sealed petri dishes. *J. Rapid Methods Automation Microbiol.*, **1**, 109–115.

Blais B. W. 1994. Transcriptional enhancement of the *Listeria monocytogenes* PCR and simple immunoenzymatic assay of the product using anti-RNA:DNA antibodies. *Appl. Environ. Microbiol.*, **60**, 348–352.

Blais B. W. and Phillippe L. M., 1993. Affinity concentration of *Listeria monocytogenes* cells on concanavalin A-coated polyester cloth and subsequent detection by the polymerase chain reaction. *J. Rapid Methods Automation Microbiol.*, **2**, 235–245.

Bopp C. A., Cameron D. N., Kiehlbauch J. A., Strockbine N. A., Wachsmuth I. K., Wells J. G., 1991. The use of plasmid profiles and nucleic acid probes in epidemiologic investigations of foodborne, diarrheal diseases. *Int. J. Food-Microbiol.*, **12**, no. 1, 77–90; 34 ref.

Brauns L. A., Hudson M. C., Oliver J. D., 1991. Use of polymerase chain reaction in detection of culturable and nonculturable *Vibrio vulnificus* cells. *Appl. Environ. Microbiol.*, **57**, 2651–2655.

Breit A., Kneifel W., Manafi M., 1991. Adaptation of two commercially available DNA probes for the detection of *Escherichia coli* and *Staphylococcus aureus* to selected fields of dairy hygiene—an exemplary study. *Zentralbl. Hygiene Umweltmedizin*, **192**, no. 6, 544–553; 34 ref.

Brewer K., Schwartzentruber K., Taylor A., Use of robotics to transfer cultures on agar-containing media to fresh media. *Appl. Environ. Microbiol.*, **56**, no. 3, 819–822; 2 ref.

Buchanan R. L., Fratamico P. M., Schultz F. J., 1992. Rapid isolation of *Escherichia coli* 0157:H7 from enrichment cultures of foods using an immunomagnetic separation method. *Food Microbiol.*, **9**, no. 2, 105–113; 26 ref.

Buchanan B., Speirs J., Stavric S., 1992. Comparison of a competitive enzyme immunoassay kit and the infant mouse assay for detecting *Escherichia coli* heat-stable enterotoxin., *Letters Appl. Microbiol.*, **14**, no. 2, 47–50; 10 ref.

Butman B. T., Durham R. J., Robinson B. J., 1990. Rapid detection of food-borne pathogens. *Europ. Food Drink-Rev.*, Winter, 51 53; 3 ref.

Candlish A. A. G., 1991. Immunological methods in food microbiology. *Food-Microbiol.*, **8**, no. 1, 1–14; 58 ref.

Candrian U., Furrer B., Hoefelein C., Meyer R., Jermini M., Luethy J., 1991. Detection of *Escherichia coli* and identification of enterotoxigenic strains by primer-directed enzymatic amplification of specific DNA sequences. *Int. J. Food Microbiol.* **12**, no. 4, 339–352; 26 ref.

Candrian U., Luethy J., Meyer R., 1991. Direct detection by polymerase chain reaction (PCR) of *Escherichia coli* in water and soft cheese and identification of enterotoxigenic strains. *Letters Appl. Microbiol.*, **13**, no. 6, 268–271; 7 ref.

Cao W. W., Johnson M. G., Wang R. F., 1992. 169 rRNA-based probes and polymerase chain reaction method to detect *Listeria monocytogenes* cells added to foods. *Appl. Environ. Microbiol.*, **58**, no. 9, 2827–2831; 14 ref.

Characche P., Goodrum G. R., Jollick J. D., Kaufmann C., Scholl D. R., York C. K., 1990. Clinical application of novel sample processing technology for the identification of salmonellae by using DNA probes. *J. Clin. Microbiol.*, **28**, no. 2, 237–241; 8 ref.

Clarke R. C., de Grandis S. A., Gyles C. L., Hii J., McEwen S., Read S. C., 1992. Polymerase chain reaction for detection of verocytotoxigenic *Escherichia coli* isolated from animal and food sources. *Mol. Cell. Probes*, **6**, no. 2, 153–161; 44 ref.

Cousin M. A., 1990. Development of the enzyme-linked immunosorbent assay for detection of molds in foods. A review. *Dev. Ind. Microbiol.*, **31** (Suppl. 5), 165–173; 68 ref.

D'Aoust J. Y., Greco P., Sewell A. M., 1991. Commercial latex agglutination kits for the detection of foodborne *Salmonella*. *J. Food Prot.*, **54**, no. 9, 725–730; 21 ref.

Datta A. R., 1990. Identification of foodborne pathogens by DNA probe hybridization techniques. *Dev. Ind. Microbiol.*, **31** (Suppl. 5), 165–173; 68 ref.

Davidson B. E., Hiller A. J., Tanskanen E. I., Tulloch D. L., 1990. Pulsed-field gel electrophoresis of small intestine digests of lactococcal genomic DNA, a novel method of strain identification. *Appl. Environ. Microbiol.*, **56**, no. 10, 3105–3111; 27 ref.

de los Reyes-Gavilan C. G., Limsowtin G. K. Y., Tailliez P., Sechaud L., and Accolas J. P., 1992. A *Lactobacillus helveticus*-specific DNA probe detects restriction fragment length polymorphisms in this species. *Appl. Environ. Microbiol.*, **58**, 3429–3432.

Delfgou E., Notermans S., Soentoro P. S., Wernars K., 1991. Successful approach for detection of low numbers of enterotoxigenic *Escherichia coli* in minced meat by using the polymerase chain reaction. *Appl. Environ. Microbiol.*, **57**, no. 7, 1914–1919; 20 ref.

Delley M., Mollet B., Hottinger H., 1990. DNA probe for *Lactobacillus delbrueckii. Appl. Environ. Microbiol.*, **56**, 1967–1970.

Desnier I. and Labadie J., 1992. Selection of cell wall antigens for the rapid detection of bacteria by imunological methods. *J. Appl. Bacteriol.*, **72**, no. 3, 220–226; 9 ref.

Doyle M. P. and Kim M. S., 1992. Dipstick immunoassay to select enterohemorrhagic *Escherichia coli* 0157:H7 in retail ground beef. *Appl. Environ. Microbiol.*, **58**, no. 5, 1764–1767; 24 ref.

Eley A., 1990. New rapid detection methods for *Salmonella* and *Listeria. Br. Food-J.*, **92**, no. 4, 28–31; 17 ref.

Emond E., Fliss I., Pandian S., 1993. A ribosomal DNA fragment of *Listeria monocytogenes* and its use as a genus-specific probe in an aqueous-phase hybridization assay. *Appl. Environ. Microbiol.*, **59**, 2690–2697.

Fenwick S. G. and Murray A., 1991. Detection of pathogenic *Yersinia enterocolotica* by polymerase chain reaction. *Lancet*, **337**, no. 8739, 496–497; 1 ref.

Ferraro K., Foster K., Garramone S., Groody E. P., 1992. Modified colorimetric DNA hybridization method and conventional culture method for detection of *Salmonella* in foods: Comparison of methods. *J. AOAC-Int.*, **75**, no. 4, 685–692; 7 ref.

Ferreira J. L., Hamdy M. K., McCay S. G., Baumstark B. R., 1992. An improved assay for identification of type A *Clostridium botulinum* using the polymerase chain reaction. *J. Rapid Methods Automation Microbiol.*, **1**, 29–39.

Ferreira J. L., Baumstark B. R., Hamdy M. K., McCay S. G., 1993. Polymerase chain reaction for detection of type A *Clostridium botulinum* in foods. *J. Food Prot.*, **56**, 18–20.

Fluit A. C., Torensma R., Visser M. J. C., Aarsman C. J. M., Poppelier M. J. J. G., Keller B. H. I., Klapwijk P., Verhoef J., 1993. Detection of *Listeria monocytogenes* in cheese with the magnetic immuno-polymerase chain reaction assay. *Appl. Environ. Microbiol.*, **59**, 1289–1293.

Fluit A. C., Widjojoatmodjo M. N., Box A. T. A., Torensma R., Verhoef J., 1993. Rapid detection of *Salmonellae* in poultry with the magnetic immuno-polymerase chain reaction assay. *Appl. Environ. Microbiol.*, **59**, 1342–1346.

Franco B. D. G. M., Gomes T. A. T., Jakabi M., Marques L. R. M., 1991. Use of probes to detect virulence factor DNA sequences in *Escherichia coli* strains isolated from foods. *Int. J. Food Microbiol.*, **12**, no. 4, 333–338; 23 ref.

Gannon V. P. J., King R. K., Kim J. Y., Thomas, E. J. G., 1992. Rapid and sensitive method for detection of Shiga-like toxin-producing *Escherichia coli* in ground beef using the polymerase chain reaction. *Appl. Environ. Microbiol.*, **58**, 3809–3815.

Giesendorf B. A. J., Quint W. G. V., Henkens M. H. C., Stegeman H., Huf F. A., Niesters H. G. M., 1992. Rapid and sensitive detection of *Campylobacter* spp. in chicken products by using the polymerase chain reaction. *Appl. Environ. Microbiol.*, **58**, 3804–3808.

Gouvea V., Glass R. I., Woods P., Taniguchi K., Clark H. F., Forrester B., Fang Z. Y., 1990. Polymerase chain reaction amplification and typing of rotavirus nucleic acid from stool specimens. *J. Clin. Microbiol.*, **28**, no. 2, 276–282; 34 ref.

Grant K. A. and Kroll R. G., 1993. Molecular biology techniques for the rapid detection and characterization of foodborne bacteria. *Food Sci. Technol. Today*, **7**, no. 2, 80–88; 48 ref.

Hatano S. and Miyamoto T., 1993. [Rapid detection of Salmonella by ELISA using monoclonal antibodies.] *J. Antibact. Antifungal Agents, Jpn.*, **21**, no. 7, 397–404; 38 ref.

Herman L. and De Ridder H., 1992. Identification of *Brucella* spp. by using the polymerase chain reaction. *Appl. Environ. Microbiol.*, **58**, 2099–2101.

Hill W. E. and Keasler S. P., 1991. Ientification of foodborne pathogens by nucleic acid hybridization. *Int. J. Food Microbiol.*, **12**, no. 1, 67–76; 51 ref.

Hsu H. Y., Chan S. W., Sobell D. I., Halbert D. N., Groody E. P., 1991. *J. Food Prot.*, **54**, no. 4, 249–255; 24 ref.

Ibrahim A., Liesack W., Stackebrandt E., 1992. Polymerase chain reaction-gene probe detection system specific for pathogenic strains of *Yersinia enterocolitica*. *J. Clin. Microbiol.*, **30**, no. 8, 1942–1947; 29 ref.

Islam M. S., Hasan M. K., Miah M. A., Sur G. C., Felsenstein A., Venkatesan M., Sack R. B., Albert M. A., 1993. Use of the polymerase chain reaction and fluorescent-antibody methods for detecting viable but nonculturable *Shigella dysenteriae* type 1 in laboratory microcosms. *Appl. Environ. Microbiol.*, **59**, 536–540.

Jones D. P., Law R., Bej A. K., 1993. Detection of *Salmonella* spp. in oysters using polymerase chain reactions (PCR) and gene probes. *J. Food Sci.*, **58**, 1191–1197, 1202.

Josephson K. L., Gerba C. P., Pepper, I. L., 1993. Polymerase chain reaction detection of nonviable bacterial pathogens. *Appl. Environ. Microbiol.*, **59** 3513–3515.

Jothikumar N., Aparna K., Kamatchiammal S., Paulmurugan R., Saravanadevi S., Khanna P., 1993. Detection of hepatitis E virus in raw and treated wastewater with the polymerase chain reaction. *Appl. Environ. Microbiol.*, **59**, 2558–2562.

Kapperud G., Dommarsnes K., Skurnik M., Hornes E., 1990. A synthetic oligonucleotide probe and a cloned polynucleotide probe based on the yop A gene for detection and enumeration of virulent *Yersinia enterocolitica*. *Appl. Environ. Microbiol.*, **56**, no. 1, 17–23.

Keasler, S. P. and Hill, W. E., 1992. A research note: Polymerase chain reaction identification of enteroinvasive *Escherichia coli* seeded into raw milk. *J. Food Prot.*, **55**, 382–384.

Kim C., Swaminathan B., Cassaday P. K., Mayer L. W., Holloway B. P., 1991. Rapid confirmation of *Listeria monocytogenes* isolated from foods by a colony blot assay using a digoxigenin-labeled synthetic oligonucleotide probe. *Appl. Environ. Microbiol.*, **57**, 1609–1614.

Klijn N., Weerkamp A. H., de Vos W. M., 1991. Identification of mesophilic lactic acid bacteria by using polymerase chain reaction-amplified variable regions of 16S rRNA and specific DNA probes. *Appl. Environ. Microbiol.*, **57**, 3390–3393.

Knight I. T., Shults S., Kaspar C. W., Colwell R. R., 1990. Direct detection of *Salmonella* spp. in estuaries by using a DNA probe. *Appl. Environ. Microbiol.*, **56**, 1059–1066.

Koch W. H., Payne W. L., Wentz B. A., Cebula T. A., 1993. Rapid polymerase chain reaction method for detection of *Vibrio cholerae* in foods. *Appl. Environ. Microbiol.*, **59**, 556–560.

Lampel K. A., Jagow J. A., Trucksess M., Hill W. E., 1990. Polymerase chain reaction for detection of invasive *Shigella flexneri* in food. *Appl. Environ. Microbiol.*, **56**, 1536–1540.

Lee C., Chen L. H., Liu M. L., Su Y. C., 1992. Use of an oligonucleotide probe to detect *Vibrio parahaemolyticus* in artificially contaminated oysters. *Appl. Environ. Microbiol.*, **58**, 3419–3422.

McMeekin T. A., Ross T., Olley J., 1992. Application of predictive microbiology to assure the quality and safety of fish and fish products. *Int. J. Food Microbiol.*, **15**, no. 1/2, 13–32.

Morgan M. R. A., Smith C. J., Williams P. A. (eds), 1992. Food safety and quality assurance. Applications of immunoassay systems. xiii + 496p. ISBN 1 85166 747 4.

Nesbakken T., Kapperud G., Dommarsnes K., Skurnik M., Hornes E., 1991. Comparative study of a DNA hybridization method and two isolation procedures for detection of *Yersinia enterocolitica* 0:3 in naturally contaminated pork products. *Appl. Environ. Microbiol.*, **57**, no. 2, 389–394; 30 ref.

Nielsen K. H. and Tsang R. S. W., 1992. Immunoassays for *Salmonella. Genet. Eng. Biotechnologist*, **12**, no. 2, 14–18; 67 ref.

Oyofo B. A. and Rollins D. M., 1993. Efficacy of filter types for detecting *Campylobacter jejuni* and *Campylobacter coli* in environmental water samples by polymerase chain reaction. *Appl. Environ. Microbiol.*, **59**, 4090–4095.

Palmer C. J., Tsai Y. L., Paszko-Kolva C., Mayer C., Sangermano L. R., 1993. Detection of *Legionella* species in sewage and ocean water by polymerase chain reaction, direct fluorescent-antibody, and plate culture methods. *Appl. Environ. Microbiol.*, **59**, 3618–3624.

Palmer L. M. and Colwell R. R., 1991. Detection of luciferase gene sequence in nonluminescent *Vibrio cholerae* by colony hybridization and polymerase chain reaction. *Appl. Environ. Microbiol.*, **57**, 1286–1293.

Pollard D. R., Johnson W. M., Lior H., Tyler S. D., Rozee K. R., 1990. Rapid and specific detection of verotoxin genes in *Escherichia coli* by the polymerase chain reaction. *J. Clin. Microbiol.*, **28**, no. 3, 540–545; 32 ref.

Rotbart H. A., 1990. Enzymatic RNA amplification of the enteroviruses. *J. Clin. Microbiol.*, **28**, no. 3, 438–442; 25 ref.

Rotbart H. A., 1991. Nucleic acid detection systems for enteroviruses. *Clin. Microbiol. Rev.*, **4**, no. 2, 156–168; 115 ref.

Salama M., Sandine W., Giovannoni S., 1991. Development and application of oligonucleotide probes for identification of *Lactococcus lactis* subsp. *cremoris. Appl. Environ. Microbiol.*, **57**, 1313–1318.

Salama M. S., Sandine W. E., Giovannoni S. J., 1993. Isolation of *Lactococcus lactis* subsp. *cremoris* from nature by colony hybridization with rRNA probes. *Appl. Environ. Microbiol.*, **59**, 3941–3945.

Samadpour M., Liston J., Ongerth J. E., Tarr P. I., 1990. Evaluation of DNA probes for detection of Shiga-like-toxin-producing *Escherichia coli* in food and calf fecal samples. *Appl. Environ. Microbiol.*, **56**, 1212–1215.

Snyder A. P., Miller M., Shoff D. B., Eiceman G. A., Blyth D. A., Parsons J. A., 1991. Enzyme-substrate assay for the qualitative detection of microorganisms by ion mobility spectrometry. *J. Microbiol. Methods*, **14**, no. 1, 21–32; 24 ref.

Stanley P. E., McCarthy B. J., Smither R. (eds), United Kingdom, Society for Applied Bacteriology, ATP luminescence. *Rapid methods in microbiology*, xy + 302p. ISBN 0 632 02716 9.

Thomas E. J. G., King R. K., Burchak J., Gannon V. P. J., 1991. Sensitive and specific detection of *Listeria monocytogenes* in milk and ground beef with the polymerase chain reaction. *Appl. Environ. Microbiol.*, **57**, 2576–2580.

Tsai Y. L., Palmer C. J., Sangermano L. R., 1993. Detection of *Escherichia coli* in sewage and sludge by polymerase chain reaction. *Appl. Environ. Microbiol.*, **59**, 353–357.

Tsai Y. L., Sobsey M. D., Sangermano L. R., Palmer C. J., 1993. Simple method of concentrating enteroviruses and hepatitis A virus from sewage and ocean water for rapid detection by reverse transcriptase-polymerase chain reaction. *Appl. Environ. Microbiol.*, **59**, 3488–3491.

Tsen H. Y., Wang S. J., Green S. S., 1991. *Salmonella* detection in meat and fish by membrane hybridization with chromogenic/phosphatase/biotin DNA probe. *J. Food Sci.*, **56**, 1519–1523.

Wang R. F., Cao W. W., Blore P. J., Slavik M. F., 1992. Development of DNA probes specific for *Campylobacter jejuni*. *J. Rapid Methods Automation Microbiol.*, **1**, 83–92.

Wang R. F., Cao W. W., Johnson M. G., 1991. Development of a 16S rRNA-based oligomer probe specific for *Listeria monocytogenes*. *Appl. Environ. Microbiol.*, **57**, 3666–3670.

Wang R. F., Cao W. W., Johnson M. G., 1992. 16S rRNA-based probes and polymerase chain reaction method to detect *Listeria monocytogenes* cells added to foods. *Appl. Environ. Microbiol.*, **58**, 2827–2831.

Wang R. F., Slavik M. F., Cao W. W., 1992. A rapid PCR method for direct detection of low numbers of *Campylobacter jejuni*. *J. Rapid Methods Automation Microbiol.*, **1**, 101–108.

Way J. S., Josephson K. L., Pillai S. D., Abbaszadegan M., Gerba C. P., Pepper I. L., 1993. Specific detection of *Salmonella* spp. by multiplex polymerase chain reaction. *Appl. Environ. Microbiol.*, **59**, 1473–1479.

Wegmuller B., Luthy J., Candrian U., 1993. Direct polymerase chain reaction detection of *Campylobacter jejuni* and *Campylobacter coli* in raw milk and dairy products. *Appl. Environ. Microbiol.*, **59**, 2161–2165.

Wernars K., Delfgou E., Soentoro P. S., Notermans S., 1991. Successful approach for detection of low numbers of enterotoxigenic *Escherichia coli* in minced meat by using the polymerase chain reaction. *Appl. Environ. Microbiol.*, **57**, 1914–1919.

Wiedmann M., Barany F., Batt C. A., 1993. Detection of *Listeria monocytogenes* with a nonisotopic polymerase chain reaction-coupled ligase chain reaction assay. *Appl. Environ. Microbiol.*, **59**, 2743–2745.

Wiedmann M., Czajka J., Barany F., Batt C. A., 1992. Discrimination of *Listeria monocytogenes* from other *Listeria* species by ligase chain reaction. *Appl. Environ. Microbiol.*, **58**, 3443–3447.

Wirtanen G., Mattila-Sandholm T., Manninen M., Ahvenainen R., Roenner U., 1991. Application of rapid methods and ultrasound imaging in the assessment of the microbiological quality of aseptically packed starch soup. *Int. J. Food Sci. Technol.*, **26**, no. 3, 313–324; 21 ref.

Wolcott M. J., 1991. DNA-based rapid methods for the detection of foodborne pathogens. *J. Food Prot.*, **54**, 387–401; 88 ref.

PART II

METHODS OF EVALUATION OF VARIOUS TECHNOLOGICALLY SIGNIFICANT MICROFLORA

CHAPTER

11

Total Aerobic Mesophilic Microflora

C. M. Bourgeois

Total aerobic mesophilic microflora includes all microorganisms capable of multiplying in the presence of air of ambient temperatures, specifically those having an optimal growth temperature situated between 25° and 40°C, whether they be pathogens or spoilage microorganisms. It should be noted that the widespread practice of low temperature storage has diminished their imporance in the area of food spoilage in favor of psychrotrophic bacteria.

With respect to food products, this microflora is often defined in more methodological terms, being described as those microorganisms which produce visible colonies after 3 days at 30°C on a "plate count medium." This definition is obviously more restrictive than the first one, since more fastidious microorganisms such as lactobacilli grow poorly or not at all on such media. The microfloral count as evaluated under these conditions has been adopted nevertheless as an indicator of hygiene and product quality considered worthy of disscussion.

11.1. Value of the Total Aerobic Mesophilic Microfloral Count as an Indicator

In spite of some of the discussion on the subject, the meaning of this test is important. From a technological standpoint, an abundant mesophilic flora indicates that the microbial spoilage process is well under way, although there is no precise correlation between the quantitative measurement of the total flora and the time which elapses before the spoilage becomes organoleptically perceptible. Spoilage may, after all, result from the action of a specialized group of microorganisms initially representing only a small portion of the total population.

From a hygienic point of view, there is also no precise correlation between the level of total microfloral counts and the presence of pathogenic microorganisms in food products, although it has been suggested by Miskimin and co-workers (1976) that the same may be said for the results of other tests. Enumeration of total flora remains the best method for the assessment of the general microbiological quality of foods, the detection of pathogens by direct methods being the only sure approach to hygienic safety. The same opinion has been expressed by Thomas and Thomas (1974a,b,c) with respect to the dairy industry.

It should also be borne in mind that some microorganisms which are not normally considered pathogens (e.g., various species of *Proteus* as well as some bacilli, entrococci, and pseudomonads) may cause intoxications if present in foods in sufficient numbers. Consequently, any food containing an overabundant microbial flora is to be considered unfit for consumption, thereby justifying the very general application of the total mesophilic microfloral evaluation test. This test is inadequate, however, in the following cases:

- Fermented products normally contain large numbers of microorganisms responsible for fermentation. It should be noted that lactobacilli, which are the most widely used fermentation agents, grow poorly on "plate count agar."
- In the case of sterilized products, the absence of microorganisms does not determine the bacteriological qauality of the starting material used. Direct enumeration of the total flora by the Breed method does, however, provide an assessment of this.
- Cold storage destroys a portion of the microorganisms, which varies depending on the species. Results obtained from the analysis of refrigerated and especially frozen products must therefore be interpreted with caution. The practice of freezing samples intended for analysis, although often practically unavoidable, does compromise the value of the results obtained.

In conclusion, assessment of the total aerobic mesophilic microflora is an important test, particularly for monitoring purposes in industrial settings. It is important for the information it provides and for the ease with which it may be implemented, in particular by virtue of the broad range and high numbers of organisms targeted which permit the use of the rapid physicochemical detection methods.

11.2 Counting Methods

11.2.1 Direct Microscopic Examination

Direct microscopic examination, after staining on a slide (the Breed method) or on a filtering membrane (the DEFT method), allows an evaluation of the total flora present, including dead cells and living cells.

In the case of infant foods in sterilized jars, this regulatory examination can ony reveal the presence of dead cells, although the numbers of such cells does provide an indication of the bacteriological status of the starting materials used.

For foods not subjected to thermal treatments, most of the microbes counted are alive and the method thus amounts to a rapid estimation of the total aerobic mesophilic microflora. This has been applied to testing of milk and meats and of vegetables intended for freezing. The method is rapid since no incubation is required but has the disadvantages of being not very sensitive—the detection limit being in the range of 10^5–10^6 per ml of suspension—and also very tedious, unless an electronic counter is available. A short incubation period on solid medium may be used to increase the sensitivity, albeit at the expense of the speed of the analysis. A similar incubation in liquid medium has the same effect, in addition to making the technique less quantitative (see Chapter 4).

11.2.2 Solid Media Culture Methods

Enumeration by plate count is the reference method for the assessment of aerobic mesophilic microflora. A number of refinements resulting in varying degrees of automation of the inoculation and plate reading steps have contributed to the success of this method, especially in the dairy industry (see Chapter 8). The agar drop culture and spiral plating systems are among the systems which, without being automatic, simplify the analysis and reduce its cost (see Chapter 4). Research aimed at decreasing the delays in obtaining the result is ongoing.

The conventional plating analysis may be carried out at 37°C in only 24 hours as long as it is understood beforehand that there is a risk that the microflora detected will not be exactly the same as under the usual conditions. Procedures based on examination at the microcolony stage, particularly in droplets, offer the same advantage and no doubt also the same disadvantage. In the case of industrial testing, this is probably a minor inconvenience since the results are sought more for comparative purposes than for absolute values.

Finally, very simplified procedures, which consist of dipping a slide or special strips into a bacterial suspension and reading the bacterial concentration after incubation, are very approximate methods which may be of practical use in industrial settings.

11.2.3 MPN Methods

The field of application for MPN methods is the assessment of low level contaminations, which is not really the case for the generally quite numerous total mesophilic microflora. They may, however, be useful in certain cases, especially for the HGMF method.

11.3 Methods of Overall Evaluation

Methods of overall evaluation are measurements of physicochemical parameters of the culture from which information about the level of contamination of the sample may be deduced (see Chapter 5).

Assessment of the total aerobic mesophilic flora is the field of choice for these procedures, given that no specificity is required and that requirements with respect to detection limit are minimal. Given the inverse linear relationship between the logarithm of the detection limit and the minimum detection time, aiming for the highest tolerable limit will often shorten the detection time. As a result, a suitable method can be chosen which can perform the analysis within a few hours. The only requirements are that the same species be detected as by the reference technique and that the results obtained be correctly correlated with those of the reference technique. This latter point continues to be the subject of much discussion.

With respect to the impedance method, Cady and co-workers (1978) show, for example, that the results obtained are not correlated with total mesophilic microflora as well as with the delay in the onset of product spoilage. In reality, methods based on microbial activity provide responses which vary depending on the species, making the overall response of a population only weakly related to its actual numbers. On the other hand, the response being a function of both microbial activity and number does assess the spoilage capacity of the population to a certain extent.

When detected by these methods, those microorganisms that are very active in the incubation medium will certainly have more impact on the measurement than the others. It should not be forgotten, however, that even the reference method exercises some discrimination between different microorganisms present. The colonies counted on a plate are, in fact, colony-forming units or CFUs which appear because the multiplicative activity of the cells or cell clusters in the sample dilution was sufficient to give rise to a colony visible to the naked eye within 3 days. In fact, it is routinely observed that the number of countable colonies increases with incubation time. This is due to more slowly multiplying microorganisms which were not yet visible as colonies after 3 days but which are present nonetheless and which would have made a small contribution to a signal measured by the other methods of detection.

Furthermore, slow multiplication on a detection medium should not be considered synonymous with little capacity for spoiling the food under consideration. Consequently, at least some of these methods have the potential to provide a more accurate assessment of the spoiling capacity of a flora than the reference method does. This is especially the case when the parameter taken into consideration for containment level assesment purposes is directly related to the deleterious action of the population (e.g., degradation enzyme, malodorous metabolite, ammonium, amines, indole, etc.).

In any event, as a general rule, there is hesitation to give up the traditional criteria, and a method is considered applicable only when the results it provides are correctly correlated with those of the plating reference method. Among the procedures which have been previously studied, visual or spectrophotometric study of variations in redox potential is already routinely applied. In certain fields at least, turbidometry and electrical measurements of pH, ion concentration, or impedance variations also appear to be applicable at the present time, and the measurement of ATP-generated bioluminescence and of various enzymatic activities should also have some future in this field.

References

Cady P., Hardy D., Martins, S., Dufour S. W., Kraeger S. J., 1978. Automated impedance measurements for rapid screening of milk microbial content. *J. Food Prot.*, **41**, no. 4, 277–283.

Catsaras M. and Dorso Y., 1976. Méthode d'analyse microbiologique, en 24 heures, des denrées alimentaires. 1: Principes et application aux plats cuisinés. *Bull. de l'Aacadémie Vétérinaire de France*, **49**, no. 2, 237–241.

Miskimin D., Berkowitz K., Solberg M., Riha W., France W., Buchanan R., 1976. Relationships between indicator organisms and specific pathogens in potentially hazardous foods. *J. Food Sci.*, **41**, no. 5, 1001–1006.

Thomas S. and Thomas B., 1974a. The development of dye reduction tests for the bacteriological grading of raw milk. Part 2: Resazurin test–1. *Dairy Ind.*, **39**, nos. 1–2, 31–34.

Thomas S. and Thomas B., 1974b. The development of dye reduction tests for the bacteriological grading of raw milk. Part 2: Resazurin test–2. *Dairy Ind.*, **39**, no. 3, 59–62.

Thomas S. and Thomas B., 1974c. The development of dye reduction tests for the bacteriological grading of raw milk. Part 2: Resazurin test–3. *Dairy Ind.*, **39**, no. 4, 113–116.

CHAPTER
12
Psychrotrophic Microflora

Cécile Lahellec, P. Colin

12.1 Introduction

Etymologically, the term *psychrotroph* means "that which nourishes itself at low temperatures." In a broad sense, therefore, psychrotrophic microflora include all microorganisms which are able to develop at cold temperatures, that is, those which will spoil products in storage and/or which may be pathogenic to humans. The following text focuses primarily on microorganisms which are harmful to the market quality of food products stored at low temperatures.

Since the discovery by Forster in 1887 of microorganisms which develop at low temperatures, numerous difficulties with terminology have cropped up, due to the fact that some of these organisms require low temperatures for development while others do develop under these conditions but do not require them. A desire to see these organisms separated into two groups has been expressed as early as 1965 by Ingram. The definition given by Catsaras and Grebot in 1969 does clarify the situation, specifying that microorganisms which actively multiply at commonly employed refrigeration temperatures (0° ± 6°C), with a generally higher optimal growth temperature (+10°, +20°C, or sometimes higher) may be considered psychrotrophs while those which not only grow at low temperatures but require them with an optimum around 0°C may be considered psychrophiles.

Different temperatures and incubation times have been proposed for the enumeration of psychrotrophic bacteria. The 7°C/10 day combination is proposed by the International Federation for Milk and Dairy Product Testing, although works by Søgard and Lund (1980) showed that the use of a preincubation at 17°C for 17 hours followed by a 72 hour incubation period at 7°C led to identical results. Some laboratories recommend incubation at 14°C for 10 days. At a working reunion held on January 25, 1978, which focused on the psychrotrophic flora of poultry carcasses, it was decided that enumerations should be carried out after 2 weeks of incubation at temperatures between 0 and +4°C. Although there is still no precise agreement accepted by all laboratories regarding incubation conditions for psychrotrophs, the current trend is towards the use of temperatures around 0°C for relatively long incubation periods.

12.2 Techniques for the Examination of Psychrotrophic Bacteria

12.2.1 The Media Employed

The media used on a routine basis for the overall quantitative examination of the flora are the same as those used for enumeration of the total aerobic flora. For qualitative examinations, two possibilities may be contemplated:

- After incubation of the inoculated media under previously defined conditions, the largest possible number of microbial colonies (at least ten) is transferred from each plate. After purification, ordinary broth is seeded with the strains. The cultures obtained at laboratory temperature may then be frozen or lyophilized and examined when convenient. This technique allows the microorganisms to remain in as close to natural conditions as possible.
- The use of selective media may also be contemplated to the extent that rapid evaluation of certain microorganisms belonging to the psychrotrophic microflora is desired. Various media have been proposed. For pseudomonads, Mead and Adams (1977) proposed a medium which has appreciable usefulness in overall quantitative evaluation. Other selective media have been devised for lactic flora and for yeasts and molds. These are discussed in other chapters.

12.2.2 The Principal Microorganisms Encountered

Numerous psychrotrophic microorganisms may be isolated from agar media incubated at low temperature. Their relative percentages vary as a function of the type of sampling, the method of storage of the product, and so on, although qualitatively the same types of microorganisms are found. These may be summarized as follows:

- The following genera or species of gram-negative bacteria:
  - *Pseudomonas*, pigmented or otherwise
  - *Alcaligenes*
  - *Alteromonas putrefaciens*
  - *Aeromonas*
  - *Vibrio*
  - *Flavobacterium* or *Cytophaga*
  - *Xanthomonas*
  - *Serratia liquefaciens*
  - *Acinotobacter*
- The following genera or species of gram-positive bacteria:
  - *Bacillus*
  - *Corynebacterium*
  - *Brochothrix thermosphacta*
  - *Micrococcus* and *Staphylococcus*
  - *Lactobacillus* and *Streptococcus*
- A variety of yeast and molds may also be found.

12.2.3 Identification of Psychrotrophic Bacteria

The following intentionally very simplified description of the principles of psychrotrophic bacterial determination refers only to very conventional techniques used routinely in laboratories. Examination of colonies grown on ordinary nutrient agar instantly provides some useful observations with respect to appearance and pigments which may be produced. Microscopic examination of a drop of 24-hour broth placed between a slide and cover-slip provides information about the morphology and mobility of the bacteria being studied, information which may be maximized by using an oil-immersion lens with phase contrast. The principle of transforming phase variations into amplitude variations permits detection of a number of details which are not visible in stained preparations.

Morphology is a particularly important detail in the study of psychrotrophic bacteria. The Gram stain, of course, distinguishes as negative, positive, or variable on the basis of staining affinity, while motility is best assessed by observation between a slide and cover-slip. The motility results obtained from soft agar stabs are sometimes difficult to interpret for psychrotrophic bacteria since many are strict aerobes and are therefore limited in any event to surface growth. A preliminary outline for classification purposes may thus be established from a simple microscopic examination. The following may be distinguished schematically:

- Gram-negative rods
- Gram-positive rods
- Coccoid forms staining gram-positive (young forms) or gram-negative (more aged forms)
- Gram-positive cocci

12.2.3.1 Gram-Negative Rods

The test results for the following three easily observed simple reactions determine the orientation of the diagnosis.

- The presence or absence of an oxidase
- The mode of glucose utilization in Hugh and Leifson medium
- Capacity for glucose or lactose fermentation, H_2S production, and gas production in Kligler medium

The presence of oxidase may be easily demonstrated by streaking a portion of an 18 to 24-hour-old agar colony onto a disk impregnated with para-amino-dimethylamine using a hooked Pasteur pipette. The immediate appearance of a violet-red color confirms the presence of an oxidase in the enzymatic repertoire of the bacterial strain.

The mode of glucose utilization is tested using Hugh and Leifson medium in two tubes, one of which is overlaid with paraffin. If no color change occurs in either tube, the microorganism is said to be inert, that is, it does not utilize the sugar being tested. A change to yellow only in the tube which is not overlaid with oil indicates an oxidative process while the same change in both tubes is evidence of a fermentative process.

Kligler medium is used for the observation of glucose fermentation (change to yellow from the bottom of the tube), lactose fermentation (change to yellow at the surface), gas production, and production of hydrogen sulfide (blackening of the tube).

12.2.3.1.1 Gram-Negative Motile Rods (With Some Exceptions) Possessing Oxidase, Oxidative, Alkalinizing, or Inert in Hugh and Leifson Medium, Never Fermentative, no H_2S Production in Kligler Medium

These may be *Pseudomonas* or possibly *Alcaligenes* if the bacteria is alkalinizing. *Pseudomonas* and *Alcaligenes* may be distinguished by the flagellar positions, the former having polar flagella while the latter have peritrichous flagella.

Various complementary tests may be used for the classification of psychrotrophic species of *Pseudomonas*, in particular pyoverdine production in King's medium B (streak plating). Strains of *Pseudomonas* isolated from meats after some time in refrigerated storage are, however, often unpigmented.

If colonies have a very pale salmon-pink color on nutrient agar and the bacterial species otherwise fits the earlier description except for producing H_2S in Kligler medium, it is a strain of *Alteromonas putrefaciens*, which possesses motility by lateral cilia.

12.2.3.1.2 Gram-Negative Motile or Nonmotile Rods, Oxidase Positive, Fermenting Glucose and not Producing H_2S in Kligler Medium

These are either of the genus *Aeromonas* or the genus *Vibrio*. These two genera may be easily distinguished by a few tests, including the vibriostatic compound 0/129 sensitivity test (using disks impregnated with a 0.1% solution). *Vibrio* species are sensitive, *Aeromonos* species are not.

12.2.3.1.3 Gram-Negative Rods, Usually Nonmotile, Usually Oxidase-Positive, Oxidative and Weakly Fermentative After 4 or 5 Days of Incubation or Alkalinizing or Inert in Hugh and Leifson Medium

Colonies are yellow-orange pigmented on ordinary nutrient agar, or sometimes very pale yellow and growable only on enriched media. These bacteria belong either to the *Flavobacterium* or *Cytophaga* genera.

12.2.3.1.4 Gram-Negative Rods, Motile by Means of Polar Flagellae, Oxidase Negative, Oxidizing Glucose in Hugh and Leifson Medium (Sometimes with a Lagging Fermentative Tendency)

Those producing yellow-pigmented colonies on ordinary nutrient agar belong to the genus *Xanthomonas*.

12.2.3.1.5 Gram-Negative Rods, Motile by Means of Peritrichous Cilia or Nonmotile, Oxidase Negative, and Always Fermenting Glucose in Hugh and Leifson Medium

These are classified as *enterobacteria*. Psychrotrophic enterobacteria are in fact relatively rare. Those found are essentially belonging to a few strains of *Escherichia coli* or strains of *Serratia liquefaciens*, particularly in certain vacuum-packed products.

12.2.3.2 Gram-Positive Rods

12.2.3.2.1 Large, Very Rectangular Rods, Developing Abundantly and Forming Unfigmented Colonies on Ordinary Nutrient Agar and Sporulating in Broths at Ambient Temperature

These belong to the genus *Bacillus*. All are motile except for *Bacillus anthracis*.

12.2.3.2.2 Polymorphic Rods of Variable Size

At times, these are small, thin, and clustered; sometimes thick and club-shaped; sometimes long and giving rise to bacillococcal forms by fission after 24–48 hours; sometimes unable to retain the Gram stain; sometimes developing with difficulty on ordinary agar (colonies may or may not be pigmented), although may be isolated on various media, particularly in nalidixic acid medium. Such bacteria belong to the genus *Corynebacterium*.

Various complementary traits, for example, detection of β-galactosidase and urease, utilization of sodium citrate in particular, may be used for the classification of the Corynebacteriaceae.

One particularly important species in the field of food storage is *Brochothrix thermospacta* (formerly *Microbacterium thermospacta*), a small gram-positive rod often in short chains with some gram-negative elements. A selective medium is available for this microorganism, called Gardner isolation medium.

12.2.3.3 Coccoid Forms of Variable Size; Younger Forms Gram Positive, Aged Forms Gram Negative. Unpigmented Colonies on Ordinary Agar. Cells Always Nonmotile; Also Oxidative, Alkalinizing or Inert in Hugh and Leifson Medium and Never Fermentative. May or May not Possess an Oxidase

Such bacteria belong to the *Neisseriaceae* family and most often to the *Moraxella* (oxidase +) genus or to the *Acinotobacter* (oxidase −). Significant morphological variations may be observed as a function of the growth phase.

12.2.3.4 Gram-Positive Cocci

These microorganisms belong to one of two families differentiated by their response to the test for catalase: the Micrococcaceae family (catalase +) or the Lactobacillaceae family (catalase −).

Lactic flora are discussed in the next chapter. Key members of the Micrococcaceae family may be distinguished by the following two tests:

- examination of the respiratory type;
- examination of the mode of glucose utilization in Hugh and Leifson medium. Members of the genus *Staphylococcus* are facultative anaerobic and fermentative while *Micrococcus* is aerobic and may be oxidative or possibly alkalinizing or inert.

Finally, with respect to psychrotrophic bacteria, it should not be forgotten that a few remain that have not been considered in this chapter and that may represent pathogenic risks to humans. These are *Yersinia enterocolitica, Listeria monocytogenes, Clostridium botulinum* type E and others, each of which warrants detailed discussion.

12.3 Appendix: Culture Media

1. Nalidixic acid medium (for the isolation of Corynebacteriaceae):

Formula

Meat extract	3 ml
Sodium chloride	5 g
Powdered agar	15 g
Tryptose	10 g
Nalidixic acid	0.04 g
Distilled water	1000 ml

Preparation

pH is 7–7.2. Autoclave for 20 minutes at 121°C. Immediately prior to use, add aseptically 1.5 ml of horse blood to each 20 ml of medium. All Corynebacteriaceae grow on this medium.

2. Gardner's medium (for the isolation of *Brochothrix thermosphacta*):

Formula

Peptone	2 g
Yeast extract	0.2 g
Glycerol	1.5 g
KH_2PO_4	0.1 g
$MgSO_4 \cdot 7H_2O$	0.1 g
Agar	1.3 g
Distilled water	101 ml

Preparation

pH is 7. Autoclave for 15 minutes at 121°C. Sterilize and add separately:

Streptomycin	500 μg/ml
Actidione	500 μg/ml
Thallium acetate	50 μg/ml

References

Catsaras M. and Grebot D., 1969. Etude complémentaire sur les bactéries psychrotrophes des viandes. *Ann. Inst. Pasteur* in Lille, **20**, 231–238.

Lahellec C., Colin P., Bennejean G., Catsaras M., 1979. La flore psychrotrophe des carcasses de volailles. IV: Variations quantitatives et qualitatives chez les dindes en fonction de la technique de prélèvement (écouvillonnage ou broyage de peau). *Revue Méd. Vét.*, **130**, no. 3, 1613–1621.

Mead G. C. and Adams B. W., 1977. A selective medium for the rapid isolation of *Pseudomonas* associated with poultry meat spoilage. *Br. Poult. Sci.*, **18**, 661–670.

Søgard H. and Lund R., 1981. A comparison of three methods for enumeration of psychrotrophic bacteria in raw milk. IAMS XI. *International symposium on food microbiology and hygiene*, summary no. 2, 1–6.

CHAPTER

13

The Lactic Microflora

J.-Y. Leveau, Marielle Bouix, H. de Roissart

13.1 Definition, Classification, and General Properties

The lactic acid bacteria are generally associated implicitly with their various roles in the food and agricultural industries. For some technologies their activity is sought for the purpose of processing a starting material and contributing to the formulation of a fermented product. The principal species of lactic acid bacteria and their fields of application in the agrifood industries are summarized in Table 13.1. In other cases lactic acid bacteria are spoilage agents, for example, lactic souring of fruit juices. In this instance, only the market (organoleptic) quality of the product is spoiled, not its hygienic quality. Not considered as belonging to this group are certain bacterial species having some metabolic traits in common with lactic acid bacteria but which are responsible for spoilage of the hygienic quality of products which harbor them (e.g., enterobacteria).

The lactic acid bacteria which constitute this microflora are a very heterogeneous group. The principal metabolic function of a lactic acid bacterium is the excretion of lactic acid in D(−), L(+), or DL forms, which is both an imprecise and unspecific notion. Specifying the carbohydrate substrate (lactose, glucose) from which the excreted acid is produced narrows down the notion somewhat, as does measuring the quantity of acid excreted per quantity of substrate consumed, which amounts to taking into consideration the metabolic pathways leading to lactic acid production.

- the Embden-Meyerhof-Parnas pathway which leads to homofermentation in which lactic acid is the principal excreted metabolic product derived from the carbohydrate substrate;

Table 13.1 Principal Species of Bacteria Used in the Agrifood Industries

Genus	Species	Fermentative type	Lactic acid isomer	Growth temperature (°C)	Main field of use
Lactococcus	*lactic* ssp. *lactis*	Homo	L(+)	10–40	Cheese, fermented milks and creams
	lactis ssp. *cremoris*	Homo	L(+)	10–40	Cheese, fermented milks and creams
	lactis ssp. *diacetylactis*	Homo	L(+)	10–40	Cheese, fermented milks and creams
Streptococcus	*thermophilus*	Homo	L(+)	40–45	Fermented milks, cheese
Leuconostoc	*mesenteroides* ssp. *mesenteroides*	Hetero (O)	D(−)	10–30	Cheese, fermented milks and creams
	mesenteroides ssp. *cremoris*	Hetero (O)	D(−)	10–30	Cheese, fermented milks and creams
	oenos	Hetero (O)	D(−)	12–37	Wines
Pediococcus	*pentosaceus*	Homo	DL	25–35	Sausages, silage
	acidilactici	Homo	DL	35–50	Sausages, silage
Lactobacillus	*delbrueckii bulgaricus*	Homo	D(−)	45–50	Fermented milks, cheese
	dembrieckii lactis	Homo	D(−)	45–50	Cheese
	helveticus	Homo	DL	45–50	Cheese
	acidophilus	Homo	DL	35–45	Probiotics
	casei ssp. *casei*	Hetero (F)	L(+)	15–40	Cheese
	casei rhamnosus	Hetero (F)	L(+)	15–45	Probiotics
	plantarum	Hetero (F)	DL	15–40	Sausages, bread, silage, probiotics
	sake	Hetero (F)	DL	5–40	Meat, sausages
	curvatus	Hetero (F)	DL	5–40	Meat, sausages
	brevis	Hetero (F)	DL	15–40	Sausages, breads, kefir

Homo = homolactic; Hetero (F) = facultative heterolactic; Hetero (O) = obligate heterolactic.

- the pentose pathway leading to heterofermentation in which lactic acid is excreted along with a mixture of other metabolic by-products (e.g., carbon dioxide, ethanol, acetic acid, depending on the strain under consideration).

These biochemical properties are insufficient for the correct characterization of the lactic flora. Their microbiological characteristics are thus also taken into consideration.

- Lactic acid bacteria are gram-positive.
- They are not spore-formers.
- Most are nonmotile.
- They contain no cytochromes, being unable to complete the synthesis of the heme portion of porphyrins. As a result, they are incapable of respiration and are limited to fermentative metabolism.
- They are facultative anaerobes (micro-aerophiles).
- Their synthetic capabilities are weak and they are often auxotrophic for amino acids, nucleotide bases, vitamins, and some fatty acids.

Bacterial species of the lactic acid group having these general properties belong to the following five genera:

- *Lactococcus* (Lc.)
- *Streptococcus* (Sc.)
- *Leuconostoc* (Ln.)
- *Pediococcus* (Pc.)
- *Lactobacillus* (Lb.)

These five genera appear in only two sections, 12 and 14, out of the 33 sections listed in Bergey's classification (Sneath et al., 1986).

Section 12 of Bergey's classification regroups spherical nonsporulating gram-positive bacteria and includes 15 genera of aerobes, facultative anaerobes, and strictly anaerobes (Figure 13.1). Of these 15 genera, only 3 include lactic acid bacteria such as genera *Streptococcus* (including *Lactococcus*), *Leuconostoc*, and *Pediococcus*.

Section 14 regroups gram-positive nonsporulating bacilli of regular morphology and includes seven genera: *Lactobacillus*, *Erysipelothrix*, *Brochothrix*, *Listeria*, *Kurthia*, *Caryophanon*, and *Renibacterium*. This section constitutes in fact a very heterogeneous group of bacteria. The genus *Lactobacillus* has by far the largest number of species (n = 44). Among the other genera, *Listeria*, which consists of five species including *Listeria monocytogenes*, has gained in importance in recent years in the agrifood industries, particularly in cheese making. Only the genus *Lactobacillus* in this section includes lactic acid bacteria.

Although section 13 regroups the gram-positive sporulating bacteria, it warrants mention here. Among the six genera included is the genus *Sporolactobacillus* which in terms of properties is quite close to the genus *Lactobacillus* except for having the additional property of being capable of endospore formation. This genus includes only one species, *Sporolactobacillus inulinus*,

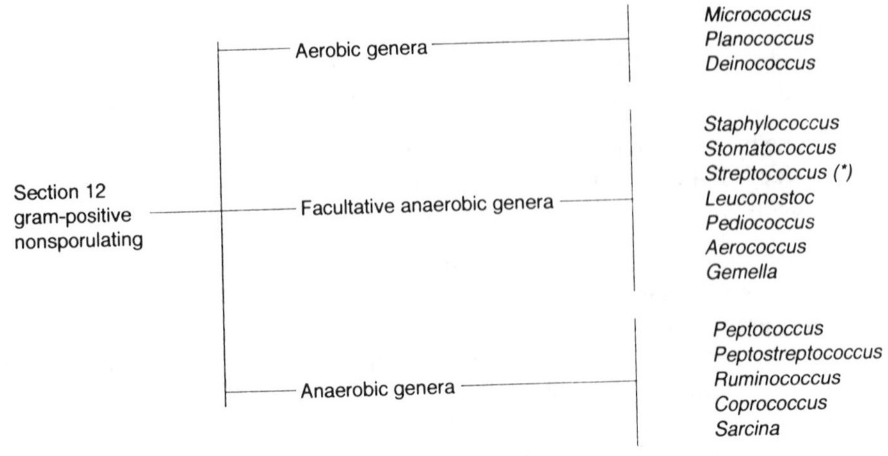

Figure 13.1 Schematic presentation of Section 12 of Bergey's classification.
*The recently defined *Lactococcus* genus is to be considered within the genus *Streptococcus.*

which is homofermentative, produces D(−) lactic acid, possesses neither catalase nor cytochromes, and does not reduce nitrates. Electrophoretic analysis of 16S RNA obtained by enzymatic hydrolysis has revealed that *Sporolactobacillus* is more closely related to *Bacillus* than to *Lactobacillus.* The spore-forming capacity of this bacteria makes it of particular concern in food-processing technologies which rely on heat treatments.

13.2 Common Techniques of Examination

The techniques to be described here are applicable to bacteria making up the lactic flora. In some cases each of the different genera (indeed certain species) require very particular conditions, for example, incubation temperatures (see Part III of this book).

13.2.1 Isolation and Enumeration

Numerous media exist for the isolation and enumeration of lactic acid bacteria. None of these gives entirely satisfactory results in all cases. Lactic acid bacteria are in fact relatively demanding nutritionally and requirements vary from one strain to another. It is therefore difficult to grow all lactic acid bacteria on a single medium. Nevertheless, Elliker's medium (1956) is often employed:

Tryptone	20 g
Yeast extract	5 g
Gelatin	2.5 g
Lactose	5 g

Sucrose	5 g
Glucose	5 g
Sodium acetate	1.5 g
Sodium chloride	4 g
Ascorbic acid	0.5 g
Distilled water	1,000 ml

Final pH 6.8. Sterilize by autoclaving for 15 minutes at 121°C.

This medium, used as broth or with agar for surface isolation, is useful for *Streptococcus* and *Lactobacillus*. Incubation is at 35°C for 18–48 hours.

A medium called MRS broth or agar (De Man, Rogosa, Sharpe, 1960) was initially proposed for the culture of lactobacilli but is also useful for *Leuconostoc* and *Pediococcus*. *Streptococcus* (including *Lactococcus*) also grows on this medium, albeit more slowly than the other lactic strains. Its composition is as follows:

Peptone	10 g
Meat extract	8 g
Yeast extract	4 g
Sodium acetate	5 g
Potassium phosphate dibasic	2 g
Ammonium citrate	2 g
Magnesium sulfate $7H_2O$	0.2 g
Manganese sulfate $4H_2O$	0.05 g
Glucose	20 g
Tween 80	1 ml
Agar	15 g
Distilled water	1,000 ml

Final pH 6.2. Sterilize by autoclaving for 15 minutes at 121°C. This medium may be used as a broth.

Lactic acid bacteria are relatively sensitive to oxygen. Cultures are therefore incubated under conditions designed to decrease the oxygen tension. Petri plates may be incubated in a CO_2 rich atmosphere. The incubation temperature is normally 30°C.

Bacteria of the lactic flora are of great significance in many food industries where they may be contaminating and potentially spoilage agents. This is especially the case for beverages. The problem here is to enumerate in finished products or in products during production, bacteria which are viable (or revivable) and capable of developing in the product. Many culture media have been proposed with the aim of meeting the challenge of providing these microorganisms with conditions analogous to those in the tested food product while obtaining sufficient growth to minimize the delay in obtaining the response to the test. Media based on tomato juice have been proposed (e.g., Bacto tomato juice agar, Bacto tomato juice agar special, and Bacto tomato juice broth, Difco). Various media have also been proposed for the detection of bacteria which contaminate beer.

- WL and WL differential media with actidione (cycloheximide) for the inhibition of certain yeasts (Bacto WL and Bacto WL Differential Difco);
- VLBS7 medium according to Emeis (1962);
- medium of Lee (1975); Lee's multidifferential agar (LMDA);
- medium of Hsu (1975), Hsu's *Lactobacillus-Pediococcus* medium (HLP medium) and Hsu's rapid medium (HRM);
- orange serum medium (Bacto orange serum Difco).

13.2.2 Identification

13.2.2.1 Morphology

It is absolutely necessary to distinguish spherical lactic acid bacteria (*Streptococcus, Lactococcus, Leuconostoc,* and *Pediococcus*) from cylindrical forms (*Lactobacillus*). This is easily done by microscopic examination of Gram stains, even though all lactic acid bacteria are gram-positive. Some strains of *Leuconostoc* and *Lactobacillus* may be morphologically difficult to differentiate.

13.2.2.2 Physiology

Respiratory Type

Lactic acid bacteria are facultative anaerobes. When placed in a long, thin tube containing a solid medium to a height of 70–80 mm, they grow throughout the agar column. This test must be accompanied by testing for catalase, an enzyme easily demonstrated by emulsifying the bacterial culture in hydrogen peroxide. The presence of catalase is indicated by the appearance of oxygen bubbles. Lactic acid bacteria are catalase-negative, although some may synthesize the protein portion of the enzyme without heme (pseudocatalase).

Peroxidase is detected by the benzidine method. The test culture is contacted with 1 ml of a mixture of equal volumes of 1% benzidine solution in 1N acetic acid and 1% hydrogen peroxide. The presence of peroxidase is indicated by the appearance of a bluish-black color. Some lactic acid bacteria are peroxidase-positive. It should be noted that lactic streptococci possess a superoxide dismutase.

Motility

The motility test may be carried out by any of the usual methods. The majority of lactic acid bacteria are nonmotile.

Growth Temperature

Growth temperature is an important test because it allows mesophilic lactic acid bacteria to be distinguished from thermophilic species. After inoculating a liquid medium (skim milk, nutrient broth, Elliker, or MRS broth) with a pure culture of the test organism, the tubes are incubated for 7–10 days at 10°C and 24–48 hours at higher temperatures (45°–50°C). Growth is then assessed by examining the media for milk coagulation, turbidity, and so on. Mesophilic lactic acid bacteria grow at 10°C while thermophilic species do not.

Test for Growth in Hostile Media

The Sherman milk test. This tests the ability of bacteria to grow in the presence of methylene blue. The 0.1% methylene blue medium is prepared by autoclaving a 1% methylene blue solution for 20 minutes at 121°C and aseptically adding 1 ml to 9 ml of sterilized milk in a tube. This test is particularly useful for differentiating *Streptococcus* and *Lactococcus*. *L. lactis* grows in the presence of 0.3% methylene blue, while fecal streptococci develop in the presence of 0.1% methylene blue. *S. thermophilus* is very sensitive to the dye.

Hypertonic broth. Testing for growth in the presence of different concentrations of sodium chloride provides valuable information for identification purposes. Media are prepared with 6.5% sodium chloride:

Glucose	5 g
Meat extract	5 g
Peptone	15 g
NaCl	65 g
Distilled water	1,000 g

Final pH 7.5. Sterilize by autoclaving for 20 minutes at 121°C.

The salt concentration may obviously be varied, from 4 up to 18%, the level tolerated by *Pediococcus halophilus*. Growth is indicated by the appearance of turbidity.

13.2.2.3 Biochemical Traits

Homo- or Heterofermentation

The fermentation test allows the assessment of the type of metabolism by which the carbon source is utilized. It is designed to show gas formation (CO_2) by use of tubes containing liquid culture in which an inverted Durham tube has been placed to collect any gas which may be produced. An alternative is the Gibson-Abdel-Malek test which consists of the following steps:

Step 1

Skim milk with 0.01% litmus	1,600 ml
Glucose	110 g

Step 2

Agar	8 g
Peptone	4 g
Meat extract	4 g
NaCl	2 g
Distilled water	400 ml

Dissolve by gentle heating.

Step 3

Tomato juice	200 ml
Yeast extract	5.6 g

Dissolve by gentle heating. Mix the resulting three liquids. Adjust the pH to 7. Distribute into 16 × 160 mm tubes. Sterilize by boiling on 3 consecutive days (Tyndallization). Solidify vertically. Carry out a sterility test by incubation. Prepare sterile plain agar in parallel (20 g/l). Inoculate melted culture medium, cool to solidify vertically, and pour an overlay (a plug) of sterile plain agar. Incubate at 30°C for several days.

The development of a homofermentative bacteria does not disturb the medium/plain agar interface whereas the gas produced by heterofermentative metabolism pushes the plain agar plug towards the top of the tube.

It is necessary to be wary of false-negative reactions which lead to labeling heterofermentative bacteria as homofermentative. Ensuring that sufficient microbial growth has in fact occurred in the culture media used is therefore important.

Lactic Acid Isomerism

Depending on their enzymatic systems, lactic acid bacteria may produce any of three forms of lactic acid: D(−) lactic acid, L(+) lactic acid, or the racemic DL lactic acid mixture. This determination is performed enzymatically with the aid of stereospecifically purified enzymes.

Carbon Substrate Metabolism

This is a matter of assessing the ability of the strains to metabolize various carbon compounds, particularly sugars. An identification series may be set up using a liquid culture medium, such as MRS broth with glucose replaced by a solution of the compound to be tested, membrane filtered and added to the solution to a final concentration of 1%. Since meat extract and citrate may give rise to false reactions, it is necessary to observe the rection in the medium without the test compound and compare the test results to the control. Growth is assessed either according to the turbidity (and gas formation should the microorganism turn out to be heterofermentative) or the color change of a pH indicator (bromocresol purple or dichlorophenol red at a concentration of 0.05%).

Rapid identification is also possible by means of the API 50CH gallery. This consists of a plastic plate which holds 50 microtubes and mini-wells. The tube portion provides anaerobic conditions and allows the examination of fermentative traits. The mini-well portion provides an aerobic zone which in the case of lactic acid bacteria is filled with paraffin oil, since the metabolism examined is fermentative. Each microtube contains a precisely determined quantity of a different substance, except for the first which serves as a negative control.

Colonies isolated on MRS medium are suspended in a suitable sterile suspension liquid or in sterile distilled water. This suspension is used to inoculate the API 50CH medium which subsequently inoculates the gallery. Incubation is at 30° or 37°C and the tests are read at 24 or 48 hours. In each of the tubes except the 25th, the change from bromocresol purple to yellow indicates acidification. In the 25th tube, metabolism of esculine brings about a change from purple to black.

Interpretation of these results may be done by consulting the table proposed by API which provides the percentages of positive reactions after 48 hours at 37°C for each test and for 26 lactobacilli strains, 5 *Lactococcus* strains, 1 strain of *Streptococcus thermophilus*, 4 strains of *Leuconostoc*, and 1 *Periococcus* strain. The company proposes the use of its APILAB software for the interpretation of the results and the calculation of the indexes which allow the identification of the bacteria examined.

Detection of Enzymes

Arginine dihydrolase. The detection of this enzyme, which liberates ammonia and citrulline from arginine, is useful for the characterization of lactic acid bacteria. The basal medium for this test is prepared as follows:

Peptone	5 g
Meat extract	5 g
Pyridoxal	0.005 g
Aqueous bromocresol purple solution, 2%	5 ml
Glucose	0.5 g
Distilled water	1,000 ml

pH adjusted to 6.4.

Arginine medium is obtained by adding 10 g of arginine. The medium is distributed into deep tubes. Sterilization is done by autoclaving for 15 minutes at 121°C. For the determination, a tube with the arginine-containing medium and a tube with the basal medium are inoculated with the test organism. Growth in the basal medium results in the color changing to yellow as glucose is metabolized while release of ammonia in the arginine-containing medium neutralizes the lactic acid produced, thereby preventing the color change.

Enzyme analysis by the API-ZYM system. The API-ZYM system is a semiquantitative micromethod for the detection of the enzymatic activities of microorganisms and is particularly useful for lactic acid bacteria. It allows the rapid detection and approximate evaluation of 19 enzymatic activities by following the evolution of a color reaction. The test involves 20 mini-wells at the bottom of which an inert fibrous mesh with various synthetic substrates is situated. The test bacterium, grown on a solid or liquid medium, is recovered and resuspended in a physiological solution, the concentration of this suspension determining the quantity of enzymes introduced. Two drops (65 μl) per well are added using a Pasteur pipette and the system is incubated at 37°C for 4 hours. Reagents are added to facilitate the reading of the results, the color intensity providing an indication of the amount of enzyme activity. Lactic acid bacteria do not possess all of the 19 activities tested by the API-ZYM system. It may be necessary to turn to conventional methods to test for other activities such as threonine aldolase which liberates acetaldehyde from threonine, an important enzyme considering that acetaldehyde contributes to the flavor of some products (e.g., yogurt).

13.2.2.4 Serology

Serological differentiation of lactic acid bacteria has been developed primarily for streptococci and the available reagents are, in fact, specific for medically significant streptococci, that is, β hemolyzing streptococci. The principle of this differentiation has, however, been extended to lactobacilli (Sharpe, 1955) and is based on a method refined by Lancefield which classified streptococci into 18 groups (A to U) by detection of group antigens (substance C) contained in the cell wall. After culture in buffered glucose broth, the cells are recovered by centrifugation and subjected to an extraction procedure. The Lancefield procedure uses 0.2N HCl in a boiling water bath for 10 minutes, followed by precipitation with ethanol if desired. The Fuller method requires more time but is more effective. Cells are treated with formamide for 15 minutes in an oil bath at 160°C. Treatment with an ethanol/HCl mixture follows in order to eliminate extraneous material by precipitation and then the specific antigen is precipitated with acetone. The extract is contacted with various sera corresponding to the Lancefield groups. The precipitation reaction, carried out in capillary tubes in order to conserve reagent, takes place when the antigen (substance C is in fact a hapten) and the corresponding serum are mutually present. The sera used for this determination may be obtained from various companies.

The Slidex-Streptokit (bioMérieux) is a latex particle-based test for β-hemolyzing streptococcal groups A, B, C, D, F, and G. Extraction of the group antigen is accomplished in 10 minutes with incubation at 37°C in the presence of an enzyme. The extract is then contacted with the sensitized latex particles. Reading of the agglutination is done after 2 minutes.

Lactococci belong to group N of the Lancefield classification. *Streptococcus thermophilus* does not possess substance C and is therefore considered not groupable.

Precipitating antigens obtainable from acetic acid extracts of *Lactobacillus* cells allow the distinguishing of seven antigenic groups. Antigenic specificity is carried most often by glycerol and ribitol teichoic acids located in the cell wall except in the case of *Lactobacillus casei* for which the antigen is a polysaccharide (Knox and Wicken, 1973).

13.2.2.5 Molecular Identification

Modern identification methods for bacteria rely on the examination of cell constituents.

DNA Base Composition: G+C%

Bacteria of the lactic flora have very different G+C% values from one strain to the next. For *Lactococcus, Streptococcus thermophilus, Leuconostoc*, and *Pediococcus*, the value is in the 34–46% range. For *Lactobacillus*, the range is from 32 to 53%, indicating a heterogeneity which is incompatible with the existence of a natural taxonomic group for the species in this genus.

DNA/DNA and DNA/RNA Homology

The homology between strains is established by means of hybridization techniques, which are now being applied to bacteria of the lactic flora. With respect to streptococci first, DNA/DNA hybridization results have provided the basis for regrouping three mesophilic species, *S. lactis, S. cremoris*, and *S. diacetylactis* into a single species with three subspecies (Garvie and Farrow, 1982). Results of DNA/rDNA led Kilpper-Balz and co-workers (1982) to propose unifying the group N mesophilic streptococci into one new genus called *Lactococcus* (Schleifer et al., 1985).

DNA from *S. thermophilus* does not hybridize with that from *Enterococcus faecalis* nor from *Lactococcus lactis* but does so enough with *S. salivarius* to suggest their closer relationship. Based on the results obtained, Farrow and Collins (1984) suggested considering *S. thermophilus* as a subspecies of *S. salivarius*, a proposition also supported by DNA/RNA hybridization studies (Garvie and Farrow, 1981).

In the case of the genus *Leuconostoc*, DNA/DNA hybridization has enabled Garvie (1983) to propose bringing together *L. mesenteroides, L. dextranicum*, and *L. cremoris* as subspecies of *L. mesenteroides*. The other species, *L. lactis* and *L. paramesenteroides* are clearly distinct. As for *L. oenos*, DNA/RNA hybridization has confirmed its distinctiveness with respect to the other species of *Leuconostoc* (Garvie, 1982). Hybridization has not yet been extensively applied to strains of *Pediococcus*.

Within the heterogeneous genus *Lactobacillus*, hybridization has also led to regrouping of some species. According to Weiss and colleagues (1983), *Lactobacillus delbreuckii*, *L. leichmanii*, *L. lactis*, and *L. bulgaricus* may all be considered as subspecies of *L. delbrueckii* on the basis of their degree of DNA homology (>80%).

16S RNA Nucleotide Sequence

This useful technique of 16S RNA identifies and compares bacteria and shows a basis for regrouping the genera *Lactococcus*, *Streptococcus*, *Leuconostoc*, *Pediococcus*, and *Lactobacillus* into one gram-positive group (Stackebrandt et al., 1983a,b) and supports the hypothesis of Orla Jensen in 1919 that these bacteria fall into one natural taxonomic group.

Cell Wall Composition

The main constituent of the cell wall of gram-positive bacteria is peptidoglycan. In the case of the lactic flora bacteria, determination of the composition and structure of the peptide chains allows their characterization to a fine degree. The methods for such determinations are difficult to use on a routine basis. The reader is invited to refer to the works of Schleifer and Kandler (1972) and Kandler and Schleifer (1980).

The peptides of *Lactococcus* (the mesophilic streptococci) are similar in composition and structure to those of *S. pyogenes*, but with D-isoasparagine. In this sense, *S. thermophilus* is similar to *S. faecalis*.

The peptide chains of *Leuconostoc* are always composed of alanine, serine, and lysine.

The peptidoglycan peptide chains of *Pediococcus* species include L-lysine, L-alanine, and D-aspartate. *P. urinae*, however, does not contain D-aspartate.

The detailed composition of the peptidoglycan peptides is known for all species of *Lactobacillus* described. For most species the peptides consist of lysine and aspartate. The peptides of some species, however, include diaminopimelic acid (e.g., *L. plantarum*). It is useful for characterization purposes to detect this compound, which is done by hydrolysis and thin layer chromatography.

Teichoic acids, which are also cell wall consituents and which have been studied in *Lactobacillus*, are made up of glycerol and ribitol.

Examination of Enzymes

The characterization of certain enzymes is a useful tool for the recognition, differentiation, and, in some cases, identification of various strains. For species of *Lactobacillus*, the electrophoretic mobility of lactic dehydrogenases (both

D-LDH and L-LDH for strains possessing both) have been determined. Esterases of 113 strains of lactic acid bacteria have been studied by Morichi and co-workers (1968) who established electrophoretic profiles for these along with electrophoregrams of soluble cellular proteins. Patterns obtained by polyacrilamide gel electrophoresis of the esterases are developed by a special protocol.

Profiles obtained for mesophilic streptococci are species specific, albeit including bands common to more than one species.

Electrophoregrams obtained for strains of *Leuconostoc* have provided a basis for regrouping some strains.

In the case of *Lactobacillus*, highly variable profiles have been observed for the various strains, some of these among the thermophilic species having weak esterase activity.

Plasmid Mapping

The cells of most lactic acid bacteria contain plasmids. These are small DNA molecules independent of the chromosome and which confer some technologically important characteristics to the cells such as lactose and citrate metabolism, protein hydrolysis, phage resistance, and others.

Mesophilic streptococci (lactococci) may carry as many as 14 different plasmids (Chopin and Langella, 1982), although this number is usually between 4 and 7. The functional properties of the plasmids of lactococci are described by McKay (1983).

Streptococcus thermophilus carries a limited number of plasmids and many strains do not carry any (Girard et al., 1987).

Studies of the plasmid content of *Leuconostoc*, *Pediococcus*, and *Lactobacillus* are still rare. These strains appear to have few or no plasmids at all. The presence of plasmids has been confirmed in *Leuconostoc* by O'Sullivan and colleagues (1982). Moreover, Smiley and Fryrer (1978) and Vescovo and co-workers (1981) have demonstrated plasmids in *Lactobacillus*.

Since cells may spontaneously lose their plasmids (Efstathiou and McKay, 1976), their taxonomic impact has been limited. Given the technological importance of some of the traits carried by plasmids, their detailed characterization has been sought. The plasmid profiles or maps which have been obtained have been judged sufficiently stable (Davies et al., 1982) to be used for strain differentiation and for the study of their traits. This observation has implications for the stability of strains used industrially.

Various techniques have been used to establish plasmid profiles (Currier and Nester, 1976; Klaenhammer et al., 1978; Orberg and Sandine, 1984; Anderson and McKay, 1983). The technique described by Birhboim and Doly (1979) forms the basis of the extraction of *E. coli* plasmids. It consists of an alkaline denaturation followed by a renaturation and precipitation of the DNA by ethanol in a high salt concentration. Electrophoresis is then carried out in

TAE buffer (40 mM tris acetate, 1 mM EDTA, pH 8). Development is done using ethidium bromide with visualization under ultraviolet light (Maniaitis et al., 1982).

13.2.2.6 Technological Properties

Acidifying Ability

This takes into consideration three criteria: (1) the quantity of lactic acid produced; (2) the nature of the acids and type of lactic acid produced, i.e., D(−), L(+), or DL lactic acids; and (3) the rate of lactic acid production.

The quantity of lactic acid produced is associated with the type of fermentative metabolism (i.e., (homo- or heterofermentation) which also determines the nature of the acids produced. In the case of homofermentation, only lactic acid is produced while heterofermentation may also yield acetic acid.

The rate of acidification is also an important technological criterion, the more rapidly acidifying strains usually being preferred. In the case of lactococci, it has been clearly established that this property is directly linked to the presence of plasmids in the cell, that is, Lac^+ (strains capable of utilizing lactose) and Prt^+ (proteolytic strains) (McKay, 1983; Kutte et al., 1979; McKay and Baldwin, 1975). In order to assess this criterion, reconstituted skim milk is inoculated at 1% with the test strain in the logarithmic growth phase and incubated at 22°C to see whether or not the strain is able to coagulate milk. If this occurs in less than 18 hours, the strain is called a *fast acid producer*. If coagulation requires 18–28 hours, the strain is considered a *slow acid producer*.

Huggins and Sandine (1984) proposed a solid medium called FSDA or fast slow differential agar for this determination containing skim milk with litmus and sodium glycerophosphate. Incubation is under anaerobic conditions at 30°C for 24–48 hours. On this medium, Lac^+ and Prt^+ colonies (fast) are red and have a diameter between 1 and 3 mm while Lac^+ and Prt^- colonies (slow) are red with a diameter of 0.2–0.5 mm. Due to the anaerobic conditions, the medium around colonies is white but turns red after 15 minutes of aerobiosis as a result of oxidation by air. The medium turns blue with Lac^- colonies.

Aromatic Compounds

Due to the organic compounds they secrete during growth on a medium, lactic acid bacteria play an important role in determining the organoleptic properties of products in which they develop. In milk, citrate has a principal role, being converted by bacteria into diacetyl, acetaldehyde, acetate, and carbon dioxide which are the primary agents responsible for the flavor of fermented dairy products.

Diacetyl may be reduced to acetoin and butanediol. Among the strains which metabolize citrate are primarily *Lactococcus lactis* ssp. *diacetylactis*, the genus *Leuconostoc*, and some strains of *Lactobacillus*. It has been established that the metabolism of citrate by *Lactococcus lactis* ssp. *diacetylactis* is linked to the presence in the cell of a small (5.5 megadalton) plasmid which codes for the synthesis of citrate permease (Kemper and McKay, 1980).

Acetaldehyde is an important flavor compound in yogurt. Its formation from threonine is catalyzed by threonine aldolase, an enzyme possessed by *L. bulgaricus* (Accolas et al., 1980).

To test the aromatizing ability of lactic acid bacteria, a gas chromatography protocol for the general determination of volatile compounds may be employed. The head-space gas from milk cultures may be analyzed directly using such a technique. Aromagrams characterizing strains grown in a given medium may thus be obtained (Degorce-Dumas et al., 1986).

Texturizing Ability

Some strains of lactic acid bacteria have the ability to synthesize exopolysaccharides such as glucans (dextrans) and fructosans (levans) which make up the capsular material surrounding the cells. These macromolecules contribute to textural modifications of products in which synthesizing strains develop.

Various routes are possible for the assessment of texturizing ability. One is to observe the production of exopolysaccharide by the strain in a liquid medium containing 5% sucrose. Such polysaccharides may be readily precipitated from the culture supernatant by adding ethanol. Immunochemical detection is also possible using antipneumococcal serum. A simpler and more routine way is to observe the appearance of colonies on agar containing 5% sucrose (Colman and Ball, 1984). Under conditions approaching the technological application, the rheological characteristics of the culture may be examined, most often using skim milk as medium.

Antagonistic Abilities

Lactic acid bacteria are well known and often used for the antagonistic influence they develop (Schaack and Marth, 1988; Daeschel, 1989) based on the excretion of lactic and other organic acids, diacetyl, hydrogen peroxide, and especially antibiotics and bacteriocines. It is the latter two which are evaluated here.

Nisin is probably the best known of these substances. Considered an antibiotic, this 3,500 dalton polypeptide is produced by *L. lactis* spp. *lactis* and imparts marked inhibitory properties to the strain (Mattick and Hirsch, 1944). Similarly, *L. lactis* spp. *cremoris* produces diplococcin which is considered more of a bacteriocin, a thermostable proteinlike substance having a quite narrow spectrum antibacterial activity and inactivated by proteases (Davey and Richardson, 1981). Numerous other compounds of this type have been found in cultures of *S. thermophilus*, *L. bulgaricus* (acidophiline), *L. lactis* spp. *diacetylactis* (lactostrepcine) (Juillard et al., 1987).

Various works have provided proof that the antagonistic ability of lactic acid bacteria is linked to the presence of plasmids in cells (McKay, 1983). Scherwitz and co-workers (1983) have shown that bacteriocin secretion by *L. lactis* spp. *diacetylactis* is linked to the presence of an 88 megadalton plasmid. They propose a technique for assessing bacteriocin production based on measurement of zones of inhibition on solid media inoculated with a sensitive test strain. Confirmation that inhibition is due to bacteriocin action is obtained by observing the disappearance of the zones after applying a proteolytic enzyme.

Bacteriophage Resistance

Bacteriophages or bacterial viruses of variable specificity may attack lactic acid bacteria causing their lysis and leading to industrial fermentation failure by lack of acidification. Bacteria possess some defense mechanisms against bacteriophage attack, notably their restriction-modification systems (Arber and Linn, 1989). The presence of these systems in lactic streptococci has been described by Lawrence and associates (1978) but not all strains possess them, which has led Limsowtin and his co-workers (1978) to attribute these technologically significant particularities to the presence of a plasmid in competent cells. Sanders and Klaenhamer (1981) and Daly and Fitzgerald (1986) have shown that the plasmids in question are of variable molecular weight.

The simplest method of testing for phage resistance is to inoculate tubes of litmus milk with a freshly coagulated milk culture at 10% with and without the bacteriophage of interest and incubating at 22°C. Sensitivity to the phage is indicated by absence of coagulation in the tube containing it (King et al., 1983). This is discussed further in Chapter 14 in this volume.

Probiotic Abilities

The probiotic virtues of lactic acid bacteria have been acknowledged for a long time (Botazzi, 1983). This characteristic is attributable to live bacteria, at least some of which are capable of implanting in the large intestine, to metabolic products such as lactic acid mainly, but also to organic acids such as hydroxymethyl glutaric acid, antibiotics and bacteriocins, and finally to enzymes which they liberate. The benefits result from the inhibition of various undesirable microorganisms, especially gram-negative bacteria. Blood cholesterol lowering and antitumoral powers have also been attributed to some strains.

Probiotic abilities of strains can only be assessed by animal studies. By definition, the bacteria must be fed live and in large numbers in the rations. The animal studies evaluate the average daily weight gain, the consumption index, the protein efficiency ratio, and the egg-laying rate for birds.

13.3 The Examination of Different Genera

13.3.1 Streptococcus

13.3.1.1 Classification

The Bergey classification (Sneath et al., 1986) lists 29 main species to which 8 species decribed as "newly decribed" have been added. These 29 species are distributed among six groups:

- pyogenic hemolytic streptococci
- buccal streptococci
- fecal streptococci or enterococci
- lactic streptococci or lactococci
- anaerobic streptococci
- other streptococci

The lactococci group corresponds to mesophilic streptococci of the lactic flora. One other species must therefore be considered, *Streptococcus thermophilus*, which is in the group of "other streptococci," but which seems to be very close to *S. Salivarius* (by DNA/DNA hybridization) which belongs to the buccal group.

In the lactococcal group, Schleifer and colleagues (1985) distinguishes two species, *L. lactis* and *L. raffinolactis* as the Bergey manual (1986) does under the old designations *S. lactis* and *S. raffinolactis*. The species *L. cremoris* and *L. diacetylactis* are considered as subspecies of *L. lactis* (Garvie and Farrow, 1982) which thus has three subspecies:

- *L. lactis* ssp. *lactis*
- *L. lactis* ssp. *cremoris*
- *L. lactis* ssp. *diacetylactis*

The latter subspecies differs from *L. lactis* ssp. *lactis* by its plasmid-linked capacity for diacetyl production from citrate, involving a reaction step between acetyl coenzyme A and hydroxyethylthiamine pyrophosphate (Cogan, 1976).

13.3.1.2 Isolation and Enumeration

In addition to the media described in Section 13.2.1, certain media particularly suited to the culture of streptococci should be noted:

M17 Medium

M17 medium (Terzaghi and Sandine, 1975) should be used for the enumeration of *S. thermophilus* in yogurt (official method: *J.O.* 4 January 1978). It is well suited for the culture of mesophilic streptococci.

Basal medium

Tryptic casein peptone	2.5 g
Peptic meat peptone	2.5 g
Papainic soy peptone	5 g
Dehydrated yeast extract	2.5 g
Meat extract	5 g
Sodium glycerophosphate	19 g
Magnesium sulfate, $7H_2O$	0.25 g
Ascorbic acid	0.5 g
Agar	9–18 g
Water	950 ml

Dissolve the ingredients in boiling water. Cool to 50°C. Adjust the pH to arrive at 7.1–7.2 after sterilization. Distribute 95 ml each to 125 ml flasks. Sterilize by autoclaving for 20 minutes at 121°C.

Lactose solution

Lactose	10 g
Water	100 ml

Dissolve the lactose and sterilize the solution by autoclaving for 20 minutes at 121°C.

Complete medium

Basal medium previously melted and brought to 48–50°C	95 ml
Lactose solution heated to 48°C	5 ml

Mix by stirring. Incubate for 2 days at 37°C. This medium may be made selective for *S. thermophilus* by adding 0.04 g/l nalidixic acid (Beerens and Luquet, 1989).

Chalmers's Medium

Chalmers's medium is convenient for the culture of lactic streptococci:

Lactose	10 g
Evans peptone	3 g
Meat extract	3 g
Yeast extract	3 g
Agar	15 g
Calcium carbonate	15 g
1% aqueous neutral red solution	5 ml
Distilled water	1,000 ml

Adjust the pH to 6.8. Autoclave for 20 minutes at 121°C. The calcium carbonate must be kept in suspension by shaking. Incubation at 25° or 30°C.

Lactic streptococci produce red surface colonies surrounded by a colorless zone due to calcium lactate formation. Mundt (1976) proposed growing lactic streptococci on MRS medium (De Man et al., 1960) modified with 0.02% sodium azide, 0.01% tetrazolium, and 40% of the glucose replaced by mannitol and containing agar at a concentration of 15 g/l.

13.3.1.3 Identification

Morphology and Structure

These bacteria are gram-positive and spherical with a diameter between 0.5 and 1 μm. Their division is parallel to a single plane, leading to groupings by pairs (diplococci), chains, and chains of pairs (streptodiplococci).

Physiology and Biochemical Traits

In addition to the general lactic acid bacteria tests, the following are worthwhile:

- *The reductase test.* This is done using sterilized skim milk at pH 7 with litmus to give a blue coloring. Lactococci reduce the litmus coloring before coagulating the milk. *S. thermophilus* acidifies and coagulates without prior reduction.
- *Citratase detection.* The medium used for this test is sterilized skim milk with previously sterilized sodium citrate in plain agar. Decomposition of citrate appears as gas formation in the medium within 3–5 days of incubation at 30°C. This enzyme is present only in *L. lactis* ssp. *diacetylactis.* This is the first reaction in the conversion of citrate into diacetyl and acetoin.
- *Acetoin production test.* Acetoin production can be tested directly by the Voges-Proskauer reaction using the milk culture. These reactions are important because they allow the distinguishing of acidifying ferments from aromatizing ferments.
- *Characterizing of hemolytic type.* Hemolysis is tested on horse blood agar:

Meat extract	3 g
Neopeptone	10 g
NaCl	5 g
Agar	15 g
Distilled water	1,000 ml

Dissolve by gentle heating. Sterilize by autoclaving for 20 minutes at 121°C. Prior to use, 25 drops of horse blood are added to each 20 ml of melted medium after cooling to 48°C. The mixture is stirred and poured into Petri plates. After 24–36 hours of incubation at 37°C, streptococcal colonies may have one of the following appearances:

- Surrounded by an obvious zone of hydrolysis. The colonies are beta-hemolyzing streptococci;
- Surrounded by a greenish coloring resulting from hydrogen peroxide production by the bacteria. This is alpha hemolysis.
- No visible change in the surrounding medium. This is called gamma hemolysis.

Lactococci are not hemolytic. *S. thermophilus* may show alpha hemolysis.

Identification may be done using API galleries. The API 50 CH gallery described earlier in this chapter may be used with a medium specifically adapted for streptococci, designated API 50 CHS. This does not allow the use of the database for interpretation of the results.

Another gallery also proposed for the rapid identification of streptococci is the API 20 STREP which includes a total of 20 enzyme activity or sugar fermentation tests. Interpretation is done by referring to the API 20 STREP

Table 13.2 Principal traits of the lactic streptococci

	L. lactis ssp.				
	lactis	*cremoris*	*diacetylactis*	*L. raffinolactis*	*S. thermophilus*
Morphology	0.5–1 μm	0.5–1 μm	0.6–1 μm	ND	0.7–0.9 μm
Growth at 10°C	+	+	+	+	−
at 40°C	+		+	−	+
at 45°C	−	−	−	−	+
Milk culture with 0.1% methylene blue	+	+	+	ND	−
Milk culture with 0.3% methylene blue	+	+	+	−	−
Growth in NaCl					
2.5%	+	+	+	+	+
4%	+	+	+	−	−
6.5%	V	+	V	−	−
Reductase	+	+	+	+	−
Citrase	−	−	+	ND	−
Acetoin	−	−	+	ND	−
Arginine dihydrolase	+	−	+	−	−
Hemolysis	Gamma	Gamma	Gamma	Gamma	Alpha
Serological group	N	N	N	N	−

ND = no data; V = variable, slow or weak depending on the strain.

catalog or by using the APILAB software. The principal traits of streptococci of the lactic flora are given in Table 13.2.

13.3.2 Leuconostoc

13.3.2.1 Classification

The Bergey classification (Sneath et al., 1986) lists four species:

- *L. mesenteroides*, which has three subspecies: *L. mesenteroides* ssp. *mesenteroides*, *L. mesenteroides* ssp. *dextranicum*, *L. mesenteroides* ssp. *cremoris*
- *L. paramesenteroides*
- *L. lactis*
- *L. oenos*

These germs correspond to the beta-coccus of the Orla Jensen classification.

13.3.2.2 Isolation and Enumeration

Two strains, *L. mesenteroides* ssp. *mesenteroides* and *L. mesenteroides* ssp. *dextranicum*, have the ability to produce a thick polysaccharide capsule from sucrose. This property may be used to detect and enumerate germs responsible for product modifications, particularly in sugar-containing products. A solid medium with extra sugar may be used for this purpose:

Meat extract	10 g
Yeast extract	3 g
Bactopeptone	2.5 g
Sucrose	150 g
K_2HPO_4	2 g
NaCl	1 g
$MgSO_4 \cdot 7H_2O$	0.2 g
Agar	15 g
Distilled water	1,000 ml

Adjust pH to 6.8–7. Sterilize by autoclaving for 20 minutes at 121°C.

Enumeration is done by plating. Dextran-producing *Leuconostoc* species produce large, very mucilaginous, and runny colonies in 24–48 hours at a maximum temperature of 30°C.

Culture of *Leuconostoc* is also possible on the medium of Mayeux, Sandine, and Elliker (1962) prepared as follows:

Tryptone	10 g
Yeast extract	5 g
Sucrose	100 g
Sodium citrate	1 g
Glucose	5 g
Gelatin	2.5 g
Agar	15 g
Distilled water	1,000 ml

The glucose is added after the other ingredients are completely dissolved. The agar is melted beforehand separately in a small volume of water in order to prevent the formation of heat-decomposition products in the medium during use of the hotplate/stirrer. The medium is sterilized by autoclaving for 15 minutes at 121°C.

This medium may be made selective by the addition of sodium azide which may be prepared as a 1% solution in distilled water sterilized by autoclaving for 15 minutes at 120°C. A final concentration of 0.075% in the medium is required. The medium must be surface inoculated after oven drying. Incubation is at 25°C and the cultures are examined daily. Dextran-producing strains form mucilaginous colonies while others form small colonies. The sodium azide inhibits the growth of gram-negative organisms and lactococci. This is not, however, a rigorously selective medium for *Leuconostoc*. Proper identification methods are therefore needed in order to confirm the strain.

13.3.2.3 Identification and Properties

Morphology

The cells are spherical and often lens-shaped, which may make it difficult to distinguish them from heterofermentative *Lactobacillus* strains. Dimensions are 0.5–0.7 μm by 0.7–1.2 μm and the cells are most often grouped in pairs rather than in chains, which distinguishes them from lactococci and streptococci. The capsule, when present, may be demonstrated by means of a negative stain using India ink between slide and cover-slip. These bacteria are gram-positive and nonsporulating.

Physiology

All species of *Leuconostoc* are mesophilic. The optimal growth temperature is around 25°C. No species grows at 45°C. Neither *Leuconostoc mesenteroides* ssp. *cremoris* nor *L. oenos* grow at 37°C. The minimum temperature for the development of these strains is 10°C.

Biochemical Traits and Properties

All species of *Leuconostoc* are heterofermentative and do not possess fructose 1,6 diphosphate aldolase. Carbon dioxide, D(−) lactic acid, and ethanol are produced from glucose. Some strains possess an oxidative mechanism and excrete acetic acid instead of ethanol. This metabolism does not allow *Leuconostoc* to produce enough acid to coagulate milk, although *L. lactis* is able to do so.

Leuconostoc species occur mostly on vegetables, although *L. mesenteroides* ssp. *cremoris* and *L. lactis* make up part of the microflora of milk products and may be introduced with the starter cultures for some cheeses and butters. These two species are able to metabolize citrate-producing acetoin, diacetyl, and CO_2 although *L. mesenteroides* ssp. *cremoris* is more effective than *L. lactis*. These strains contribute to flavor development in fresh dairy products and to the openness of cheeses in which they develop.

Leuconostoc species are nutritionally very fastidious with respect to vitamins and amino acids. The dextran-producing strains occur frequently on sugar beet and sugarcane and are often responsible for spoilage and technological failures. The cell walls of these organisms contain dextran sucrase, an enzyme which may be used for the industrial production of dextran.

Leuconostoc oenos occurs specifically in grape must in which it is the agent responsible for malolactic fermentation which converts malic acid into L(+) lactic acid and CO_2. Proper control of this step in wine production is very important and has justified numerous works concerning this bacterial species, including the development of DNA probes for its rapid and reliable identification (Lonvaud-Funel et al., 1989). The principal traits of the genus *Leuconostoc* are summarized in Table 13.3.

13.3.3 Pediococcus

13.3.3.1 Classification

The Bergey classification (Sneath et al., 1986) lists eight species for this genus:

- *P. damnosus*
- *P. parvulus*
- *P. inopinatus*
- *P. dextrinicus*
- *P. pentosaceus*
- *P. acidilactici*
- *P. halophilus*
- *P. urinae-equi*

The species *P. cerevisiae* listed in the old 1974 classification has been discontinued. *P. damnosus* is recognized as the type species.

Table 13.3 Principal traits of the genus *Leuconostoc*

	L. mesenteroides ssp.					
	mesenteroides	*dextranicum*	*cremoris*	*L. paramesenteroides*	*L. lactis*	*L. oenos*
Growth at 10°C	+	+	+	+	+	+
at 37°C	+	+	−	+	+	−
at 39°C	−	−	−	−	+	−
at 45°C	−	−	−	−	−	−
Heteroformation	+	+	+	+	+	+
Citratase	−	−	+	−	+	+
Dextran formation	+	+	−	−	−	−
Arginine dihydrolase	+	+	+	+	+	+

These microorganisms are widespread on natural products of vegetable origin (livestock feed silage). *P. damnosus* is a spoilage organism in brewing where it is known as "sarcine," a type of spoilage characterized by the appearance of clouding and especially of a defect associated with the production of diacetyl and acetoin. *P. parvulus*, *P. inopinatus*, and *P. dextrinicus* may occur in beer as well as in wine and cider. *P. pentosaceus* and *P. acidilactici* may occur in milk and dairy products, which are generally not their preferred media and they develop poorly in them; however, they may be used in brining for their acidifying power. *P. halophilus* is present in brines. These bacteria are not pathogenic. They correspond to *Tetracoccus* in the Orla Jensen classification.

13.3.3.2 Isolation and Enumeration

There does not appear to be any culture medium for the specific isolation and enumeration of these bacteria. They are fastidious with respect to growth factors and amino acids and grow on the complex media already listed with the fundamental examination techniques. Given the great importance of these organisms in brewing and in the beverage industry, various media have been suggested in the trade literature. Extract of hopped malt may be used for the culture of *P. damnosus*, which tolerates the antiseptic effect of the active ingredients (humulones, lupulones) in hops. A high concentration of NaCl (15%) in culture media allows the selective growth of *P. halophilus*, the only species able to tolerate this level of salinity. Addition of thallium acetate and crystal violet brings about the inhibition of gram-negative and spore-forming bacteria which are associated with pediococci. Actidione added to the culture medium effectively inhibits yeast growth.

The media WL and WL Differential mentioned earlier are suitable for the culture of pediococci. In any case, when the products to be examined are only slightly contaminated, which is often the case of beers, it is preferable to proceed with an enrichment culture, for example, in Elliker liquid medium prior to isolation on WL, WLD, or other isolation media.

13.3.3.3 Identification

Morphology and Structure

The cells are spherical with a diameter between 0.6 and 1 μm. They have the peculiarity of reproducing alternatingly on either of two perpendicular planes, giving rise to pairs and the characteristic tetrads easily observable under the microscope. The cells are gram-positive.

Physiology and Biochemical Traits

The genus *Pediococcus* is homofermentative, producing DL or L(+) lactic acid. They are differentiated generally by examining their tolerance to different temperatures, pH, and sodium chloride concentrations. Such tests are useful for evaluating the technical potential of strains used especially for brining. Table 13.4 summarizes the essential traits of these bacteria.

13.3.4 Lactobacillus

13.3.4.1 Classification

The new Bergey classification (Sneath et al., 1986), lists 44 species for this genus. This is a heterogeneous genus in which very marked differences exist between some species.

The genus *Lactobacillus* occurs in a very wide variety of natural environments on natural products of animal and vegetable origin, and consequently are of great importance to most food technologies either as spoilage microbes or as agents of fermentations which are controlled to some degree in order to bring about some industrial benefit such as acidification, aromatization, textural modification, and antagonistic or probiotic effects. For a given product, the development of certain *Lactobacillus* species may be sought while that of related species naturally present at the beginning of a fermentation may represent total process failure. The exact nature of this specificity depends on the particular product being made (e.g., cheese versus dry sausages or sauerkraut).

The 44 species of *Lactobacillus* (Kandler and Weiss, 1986) are distributed among three groups corresponding to genera defined by Orla Jensen (1919). Group 1 consists of the obligate homofermentative and thermophilic species corresponding to the genus *Thermobacterium* which includes 15 species. This group is divided into two complexes of species. The first of these is made up of

Table 13.4 Principal traits of the genus *Pediococcus*

	Pediococcus							
	damnosus	*parvulus*	*inopinatus*	*dextrinicus*	*pentosaceux*	*acidilactici*	*halophilus*	*urinae-equi*
Lactic acid produced	DL	DL	DL	L(+)	DL	DL	L(+)	L(+)
Growth at 35°C	−	+	+	+	+	+	+	+
at 50°C	−	−	−	−	−	+	−	−
Growth at pH 4.2	+	+	−	−	+	+	−	−
at pH 8.5	−	−	−	−	±	±	+	+
Growth in NaCl								
at 4%	−	+	+	+	+	+	±	+
at 6.5%	−	+	±	−	+	+	+	+
at 18%	−	−	−	−	−	−	+	+

three subspecies of *L. delbrueckii:*

- *L. delbrueckii* ssp. *delbrueckii*
- *L. delbrueckii* ssp. *bulgaricus*
- *L. delbrueckii* ssp. *lactis*

The other, unlike the first, is a heterogeneous complex based on *L. acidophilus:*

- *L. acidophilus*
- *L. gasseri*
- *L. crispatus*
- *L. helveticus*

Group II corresponds to the Orla Jensen classification genus *Streptobacterium* and consists of 11 mesophilic species incapable of growth at 45°C but able to grow at 15°C, homofermentative on glucose but heterofermentative on pentoses or gluconate. This group includes three species complexes, the first of which corresponds to *L. plantarum* and the second consisting of *L. casei* and its subspecies:

- *L. casei* ssp. *casei*
- *L. casei* ssp. *pseudoplantarum*
- *L. casei* ssp. *tolerans*
- *L. casei* ssp. *rhamnosus*

The third complex regroups the following species:

- *L. sake*
- *L. curvatus*
- *L. bavaricus*

Group III includes the obligate heterofermentative lactobacilli of which certain species make up the genus *Betabacterium* of the Orla Jensen classification. The 18 species of this group ferment hexoses and produce lactic acid, acetic acid (or ethanol), and carbon dioxide in 1:1:1 proportions. They are thus not to be confused with species of the genus *Bifidobacterium* which are also heterofermentative but which do not produce gas and produce lactic and acetic acids in proportions of 2:3. Included in this group are the following:

- *L. bifermentans*
- *L. brevis*
- *L. buchneri*
- *L. fermentum*
- *L. kefir*
- *L. confusus*
- *L. viridescens*

13.3.4.2 Isolation and Enumeration

The composition of media used for the culture of the genus *Lactobacillus* takes into equal consideration its essential acidogenic, acidophilic, and nutritionally demanding traits. MRS medium (cited in Section 13.2.1) is suitable for the isolation and enumeration of lactobacilli in media in which they make up the majority of the germs present. Acidified MRS medium contains all the same ingredients plus concentrated acetic acid to obtain a poststerilization (autoclaving for 20 minutes at 121°C) pH of 5.4 at 25°C. The official analytical method for the specific flora of yogurt (*Journal Officiel*, 4 January 1978) suggests this medium for the enumeration of *L. delbrueckii* ssp. *bulgaricus*.

For the enumeration of bacteria of malolactic fermentations, MRS medium modified in the following manner may be used:

Tryptone		10 g
Meat extract		10 g
Yeast extract		5 g
K_2HPO_4	(MW = 174.18)	2 g
Sodium acetate	(MW = 82.03)	5 g
Ammonium sulfate	(MW = 132.14)	2 g
$MgSO_4 \cdot 7H_2O$		0.2 g
$MnSO_4 \cdot H_2O$		0.05 g
Citric acid $\cdot H_2O$	(MW = 210.14)	2 g
DL malic acid	(MW = 134.09)	5 g
Tween 80		1 ml
Distilled water		950 ml

Adjust the pH to 4.8–4.9 with 35% NaOH solution. Add 17 g of Difco agar. Autoclave for 15 minutes at 121°C. When ready for use, add a filter sterilized solution (50 ml) containing the following:

Glucose		2 g
Fructose		2 g
Sucrose	(MW = 342.30)	2 g
L(+) arabinose	(MW = 150.13)	2 g
D(+) xylose	(MW = 150.13)	2 g

Inoculate by deep stab or with agar overlay and incubate anaerobically at 25°C for 7 days.

Rogosa medium (Rogosa, Mitchell, and Wiseman, 1951) is a selective medium used for the enumeration of lactobacilli of food origin (dairy products, meats) and intestinal origin. It has the following composition:

Tryptone	10 g
Yeast extract	5 g
Glucose	20 g
Sodium acetate	25 g
Ammonium citrate	2 g
Potassium phosphate monobasic	6 g
Tween 80	1 g
Saline solution	5 ml
Agar	15 g
Distilled water	1 l

pH is adjusted to 5.4 with acetic acid. The saline solution contains the following:

$MgSO_4 \cdot 7H_2O$	11.5 g
$MnSO_4 \cdot 2H_2O$	2.4 g

or

$MnSO_4 \cdot 4H_2O$	2.8 g
$FeSO_4 \cdot 7H_2O$	0.68 g
H_2O	100 ml

The growth of undesirable contaminating bacteria is inhibited by the high sodium acetate content and the low pH. Growth of lactobacilli is favored by the manganese, magnesium, and iron salts and by the presence of Tween 80. Inoculated plates are incubated in jars with a CO_2-enriched atmosphere at 30°C.

Sharpe (1981) gives the following composition for this medium:

Trypticase BBL	10 g
Yeast extract	5 g
KH_2PO_4	6 g
Ammonium citrate	2 g
$MgSO_4 \cdot 7H_2O$	0.58 g
$MnSO_4 \cdot 4H_2O$	0.28 g
Glucose	10 g
Arabinose	5 g
Saccharose	5 g
Tween 80	1 g
Sodium acetate, $3H_2O$	2.5 g
Acetic acid qsp pH 5.4	
Agar	15 g

It should be noted first that in addition to glucose, the medium contains arabinose and sucrose which favor the growth of some heterofermentative lactobacilli. Moreover, the sodium acetate content is ten times lower than in the previous formulation. These media are commercially available with different amounts of sodium acetate and are used industrially for the enumeration of lactobacilli in meat. They are somewhat inhibitory for streptococci but pediococci, leuconostocs, and yeasts may grow.

APT medium (Evans and Niven, 1951) is a nonselective medium described for the isolation of lactobacilli and having the following composition:

Tryptone	10 g
Yeast extract	5 g
KH_2PO_4	5 g
Sodium citrate	5 g
Sodium chloride	5 g
Glucose	10 g
Tween 80	1 g
$MgSO_4 \cdot 7H_2O$	0.8 g
$MnCl_2 \cdot 4H_2O$	0.14 g
$FeSO_4 \cdot 7H_2O$	0.04 g
Water	1 l

pH is adjusted to 6.7–7. The medium is sterilized at 121°C for 15 minutes.

In the case of fermented milk containing strains of *Bifidobacterium*, it may be necessary to enumerate these organisms specifically, for which purpose MRS medium may be made selective by preparing the following solution:

Lithium chloride	6 g
Nalidixic acid	0.03 g
Neomycin sulfate	0.2 g
Water	100 ml

pH is adjusted to 7.2–7.5 with 0.1 N sodium hydroxide solution. Sterilization is done by filtering through a 0.22 μ membrane. Five ml of this solution is added to each 100 ml of MRS medium just prior to use. Incubation for the detection and enumeration of *Bifidobacterium* is done anaerobically at 37°C for 72 hours.

13.3.4.3 Identification and Properties

Morphology and Structure

Members of the genus *Lactobacillus* are gram-positive nonsporulating bacilli. They may become gram-negative towards the end of a culture cycle and in the presence of high levels of acidity. Cell dimensions vary considerably from one species to another, some being long and thin while others may be short and thick with a coccobacillary appearance. Towards the end of a culture especially, the cells may be slightly bent or appear in chains of varying length. A few species have motility by means of peritrichous flagellae.

Some species are capable of synthesizing exopolysaccharides. This is especially the case for *L. confusus* and other heterofermentative species grown on sucrose. Some species (*L. delbrueckii* ssp. *bulgaricus*) which are able to produce this type of compound may be used in dairy technology to impart certain textures to products.

Physiology

The majority of *Lactobacillus* species are facultative anaerobes. In fact, they are microaerophiles. Surface growth on solid media is improved by anaerobiosis or by reduced oxygen tensions and CO_2 enrichment of the atmosphere (5–10%). They possess neither cytochromes nor catalase. They may, however, decompose hydrogen peroxide by means of pseudocatalase. No reaction with benzidine occurs since peroxidase is absent. The degree of oxygen requirement varies with species and strain.

Being fermentative microorganisms, *Lactobacillus* species break down sugars by lactic homo- or heterofermentation. Lactic acid is always present among the excreted metabolic products and is not utilized. Other metabolic products include carbon dioxide, ethanol as well as acetic, formic and succinic acids. They may grow at any temperature between 2° and 53°C. The optimal growth temperature is generally in the 30°–40°C range. It is often useful to test the behavior of strains with respect to temperature in order to distinguish *Thermobacterium* from the other species of *Lactobacillus*. This is examined by incubating tubes of MRS broth at 15°C for 2 weeks and at 45°C for 24–48 hours. Higher temperatures (e.g., 50°C) may also be tested. The genus *Lactobacillus* is not capable of growth at temperatures higher than 55°C. This genus is slightly acidophilic, the optimal growth pH being between 5.5 and 6.2. Growth continues until pH 5 and often lower. The growth rate is slower when the culture begins at neutral pH or higher.

Unable to liquify gelatin or break down casein, these microorganisms nevertheless have some proteolytic capabilities due to proteases and peptidases. Similarly, some species possess lipolytic ability while others are amylolytic. The principal traits of the genus *Lactobacillus* are summarized in Table 13.5.

Properties

The genus *Lactobacillus* is of enormous technological importance. These microorganisms may be found in abundance in any medium which provides conditions favoring their development—i.e., presence of sugars and proteins or breakdown products thereof (peptides and amino acids), nucleic acids, growth factors, reduced oxygen tension, and favorable temperature. In some cases, they may be bioprocessing agents whose activity is to be favored while

Table 13.5 Principal traits of the genus *Lactobacillus*

	Growth							
	at 15°C	at 45°C	Lactose	Sucrose	Gluconate	Ribose	Xylose	Arginine dihydrolase
L. delbrueckii ssp. *delbrueckii*	−	+	−	+	−	−	−	±
L. delbrueckii ssp. *bulgaricus*	−	+	+	−	−	−	−	−
L. delbrueckii ssp. *lactis*	−	+	+	+	−	−	−	±
L. acidophilus	−	+	+	+	−	−	−	−
L. gasseri	−	+	±	+	−	−	−	−
L. crispatus	−	+	+	+	−	−	−	−
L. helveticus	−	+	+	−	−	−	−	−
L. plantarum	+	−	+	+	+	+	±	−
L. casei ssp. *casei*	+	−	±	+	+	+	−	−
L. casei ssp. *pseudoplantarum*	+	−	+	+	+	+	−	−
L. casei ssp. *tolerans*	+	−	+	−	−	−	−	−
L. casei ssp. *rhamnosus*	+	−	+	+	+	+	−	−
L. sake	+	−	+	+	+	+	−	−
L. curvatus	+	−	±	−	+	+	−	−
L. bavaricus	+	−	+	+	+	+	−	−
L. bifermentans	+	−	−	−	−	+	−	−
L. brevis	+	−	±	±	+	+	±	+
L. buchneri	+	−	±	±	+	+	±	+
L. fermentum	+	−	±	±	+	+	±	+
L. kefir	+	−	+	−	+	+	−	+
L. confusus	+	−	−	+	+	+	+	+
L. viridescens	+	−	−	±	−	−	−	−

in other cases they may be spoilage agents whose development is detrimental. They modify the organoleptic properties of the products in which they develop by secreting, in addition to acids, organic compounds such as diacetyl by the conversion of citrate, acetaldehyde from threonine, and a variety of metabolites resulting from amino acid conversions.

It is generally agreed that under clean cònditions, milk freshly drawn from the udder does not contain *Lactobacillus*. Contamination of milk by lactic acid bacteria in the immediate environment occurs very early. The streptococci reproduce the fastest, while the concentration of lactobacilli remains low. They become the majority later on, especially if the milk is incubated, due to their greater tolerance to the acidity produced. *L. helveticus* and *L. delbreuckii* spp. *bulgaricus* are the bacteria most commonly used as lactic starter cultures.

L. plantarum, *L. brevis*, and *L. casei* are always present in large numbers in ripened cheeses. Strains such as *L. maltoromicus* and *L. bifermentans* are occasionally responsible for technological failures in cheese making.

The genus *Lactobacillus* plays an essential acidification role in the process of aging of pickled products. The principal species encountered are *L. plantarum*, *L. sake*, and *L. curvatus* along with other lactic acid bacteria belonging to the genera *Lactococcus*, *Streptococcus*, and *Pediococcus*. Some species (*L. viridescens*) are responsible for spoilage (greening).

The genus *Lactobacillus* is present in low numbers on vegetable products. The beginnings of decomposition of vegetable biomass favor their growth considerably. For this reason, *Lactobacillus* is involved especially in the production of silage and of sauerkraut. Among the species responsible, *L. plantarum* and *L. brevis* are principally encountered, but also *L. curvatus*, *L. casei*, and *L. fermentum*. In beers, the genus *Lactobacillus* is responsible for spoilage in most cases. In wine, some species bring about the conversion of malic acid into lactic acid and carbon dioxide by means of the malolactic enzyme. Lactic souring may also be found in these products.

Humans also harbor various species of *Lactobacillus*. Aside from dental caries, lactobacilli are generally considered to have no pathogenic capabilities. The most common intestinal species is *L. acidophilus*, which provides the basis for its use in the preparation of fermented milks and pharmaceutical preparations used for reestablishing the intestinal flora. There seems to be a great degree of heterogeneity within this species, the exact nature of which remains the subject of controversy.

References

Accolas J. P., Hemme D., Desmazeaud M. J., Vassal L., Bouillanne C., Veaux M., 1980. Les levains lactiques thermophiles: propriétés et comportement en technologie latière. *Le Lait*, **60**, 487–524.

Anderson D. G. and McKay L. L., 1983. Simple and rapid method for isolating large plasmid DNA from lactic *Streptococci*. *Appl. Environ. Microbiol.*, **46**, 549–552.

Arber W. and Linn S. S., 1989. DNA modification and restriction. *Ann. Rev. Biochem.*, **38**, 467–500.

Beerens H. and Luquet F. M., 1989. Guide pratique d'analyse microbiologique des laits et produits laitiers. Lavoisier, Technique et Documentation.

Birnboim H. C. and Doly J., 1979. A rapid alkaline extraction DNA procedure for screening recombinant plasmid DNA. *Nucleic Acids Res.*, **7**, 1513–1523.

Botazzi V., 1983. Other fermented products. *In:* Rehm H. J., Reed G. (eds). *Biotechnology*, vol. 5, Verlag Chemie, 315–363.

Chopin A. and Langella P., 1982. Analogie de profils plasmidiques chez les streptocoques du groupe N. *Le Lait*, **62**, 705.

Cogan T. M., 1976. The utilisation of citrate by lactic acid bacteria in milk and cheese. *Dairy Ind. Int.*, **41**, 12–16.

Colman G. and Ball L. C., 1984. Identification of *Streptococci* in a medical laboratory. *J. Appl. Bacteriol.*, **57**, 1–14.

Currier T. C. and Nester E. W., 1976. Isolation of covalently closed circular DNA of high molecular weight from bacteria. *Anal. Biochem.*, **76**, 431–441.

Daeschel M. A., 1989. Antimicrobial substances from lactic acid bacteria for use as food preservatives. *Food Technol.* **1**, 164–166.

Daly C. and Fitzgerald G., 1986. Mechanisms of bacteriophage insensitivity in the lactic streptococci. *In:* Ferretti J. J., Curtis R. (eds). Streptococial genetics, American Society for Microbiology, Washington D.C.

Davey G. P. and Richardson B. C., 1981. Purification and some properties of diplococcin from *Streptococcus cremoris* 346. *Appl. Environ. Microbiol.*, **41**, 84–89.

Davies F. L., Underwood H. M., Gasson M. J., 1981. The value of plasmid profiles for strain identification in lactic *Streptococci* and the relationship between *Streptococcus lactis* 712, ML3 and C2. *J. Appl. Bacteriol.*, **51**, 325–337.

DeGorce-Dumas J. R., Goursaud J., Leveau J. Y., 1986. Analyse des composés volatils du yaourt par chromatographie en phase gazeuse-espace de tête (head-space). *Ind. Alim. Agric.*, **8**, 805–808.

De Man J. C., Rogosa M., Sharpe M. E., 1960. A medium for cultivation of *Lactobacilli*. *J. Appl. Bacteriol.*, **23**, 130.

Efstathiou J. D. and McKay L. L., 1976. Plasmids in *Streptococcus lactis*: Evidence that lactose metabolism and proteinase activity are plasmid linked. *Appl. Environ. Microbiol.*, **32**, 38–44.

Elliker P. R., Anderson A. W., Hannesson G., 1956. An agar culture medium for lactic acid *Streptococci* and *Lactobacilli*. *J. Dairy Sci.*, **39**, 1611.

Emeis C. C., 1962. A study of methods for the detection of *Sarcinae* in biological brewery control. *Brew. Digest*, **37**, no. 12, 32.

Evans J. B. and Niven C., 1951. Nutrition of the heterofermentative *Lactobacilli* that cause greening of cured meat products. *J. Bacteriol.*, **62**, 599–603.

Farrow J. A. E. and Collins M. D., 1984. DNA base composition, DNA/DNA homology and long chain fatty acid studies of *Streptococcus thermophilus* and *Streptococcus salivarius*. *J. Gen Microbiol.*, **130**, 357–362.

Garvie E. I., 1981. Sub-divisions within the genus *Leuconostoc* as shown by RNA/DNA hybridization. *J. Gen. Microbiol.*, **127**, 209–212.

Garvie E. I., 1983. *Leuconostoc mesenteroïdes ssp cremoris* (Knudsen et Sorensen). Comb. nov. and *Leuconostoc mesenteroïdes ssp dextranicum* (Beijerinck)(Comb. nov. *Int. J. System Bacteriol.*, **33**, 118–119.

Garvie E. I. and Farrow J. A. E., 1981. Subdivisions within the genus *Streptococcus* using deoxyribonucleic acid-ribosomal ribonucleic acid hybridization. *Zentralbl. Bakteriol. Mikrobiol. Hyg. Abt. I. Orig. C.*, **2**, 299–310.

Garvie E. I. and Farrow J. A. E., 1982. *Streptococcus lactis ssp cremoris* (Orla Jensen) comb. nov. and *Streptococcus lactis ssp diacetylactis* (*Matuszewski et al.*) nom. rev., comb. nov. *Int. J. System Bacteriol.*, **32**, 453–455.

Girard F., Lautier M., Novel G., 1987. DNA-DNA homology between plasmids from *Streptococcus thermophilus. Le Lait*, **67**, no. 4, 537–544.

Hsu W., Taparowsky J., Brener M., 1975. Schnellzuchtung von Brauerei-milch saurebakterien. *Brauwissenschaft (Nurnberg)*, **28**, no. 6, 157–160.

Huggins A. R. and Sandine W. E., 1984. Differentiation of fast and slow differential coagulating isolates in strains of lactic streptococci. *J. Dairy Sci.*, **67**, 1674–1679.

Juillard V., Spinner H. E., Desmazeaud M. J., Boquien C. Y., 1987. Phénomènes de coopération et d'inhibition entre les bactéries lactiques utilisées en industrie laitière. *Le Lait*, **67**, no. 2, 149–172.

Kandler O. and Schleifer K. H., 1980. Taxonomy I: systematic of bacteria *In:* Ellenberg, Esser, Kubitzki, Schepf, Ziegler (eds). *Progress in botany* (*Fortschritte des botanik*), **42**, 234–252, Springer Verlag, Berlin, Heidelberg.

Kandler O. and Weiss N. Genus *Lactobacillus*, Beijerinck, 1901, 212 *In:* Sneath P. H. A. et al. (eds). 1986 *Bergey's manual of systematic bacteriology*, vol. 2, 1209.

Kemper G. M. P. and McKay L. L., 1980. Characterization of plasmid DNA in *Streptococcus lactis ssp diacetylactis*: Evidence for plasmid linked citrate utilization. *Appl. Environ. Microbiol.*, **37**, no. 2, 316–323.

Kilpper-Balz R., Fischer G., Schleifer K., 1982. Nucleic acid hybridization of group N and group D *Streptococci. Ann. Microbiol.*, **7**, 245–250.

King W. R., Collins E. B., Barret E. L., 1983. Frequencies of bacteriophages resistant and slow acid producing variants of *Streptococcus cremoris. Appl. Environ. Microbiol.*, **45**, no. 5, 1481–1485.

Klaenhammer T. R., McKay L. L., Baldwin K. A., 1978. Improved lysis of group N *Streptococci* for isolation and rapid characterization of plasmid deoxyribonucleic acid. *Appl. Environ. Microbial.*, **35**, 592–600.

Knox K. W. and Wicken A. J., 1973. Immunological properties of teichoic acids. *Bacteriol. Rev.*, **37**, 215–257.

Lawrence R. C., Thomas T. D. Terzachi B., 1976. Reviews of the progress of dairy science: Cheese starters. *J. Dairy Res.*, **43**, 141–193.

Lee S., Jangaard N., Coors J., Hsu W., Fuchs C., Brenner M., 1975. Lee's multi-differentical agar-LMDA; a culture medium for enumeration and identification of brewery bacteria. *Am. Soc. Brew. Chem. Proc.* (St. Paul), **33**, no. 1, 18–25.

Limsowtin G. K. Y., Heap H. A., Lawrence R. C., 1978. Heterogenicity among strains of lactic *Streptococci. New Zealand J. Dairy Sci. Technol.*, **13**, 1–8.

Lonvaud-Funel A., Biteau N., Fremaux C., 1989. Identification de Leuconostoc oenos par utilisation de sondes d'ADN. *Sci. des Aliments*, **9**, 533–541.

McKay L. L., 1983. Functional properties of plasmids in lactic *Streptococci. Antonie Van Leewenhoek*, **49**, 259–274.

McKay L. L. and Baldwin K. A., 1975. Plasmid distribution and evidence for a proteinase plasmid in Streptococcus lactis C2. *Appl. Microbiol.*, **29**, 546.

Maniatis T., Fritsch E. F., Sambrook J., 1982. Molecular cloning: A laboratory manual. Cold Spring Harbor Laboratory, Cold Spring Harbor, N.Y.

Mattick A. T. R. and Hirsch A., 1944. A powerful inhibitory substance produced by group N *Streptococci. Nature*, **154**, 551.

Mayeux J. V., Sandine W. E., Elliker P. R., 1962. A selective medium for detecting *Leuconostoc* in mixed strain starter cultures. *J. Dairy Sci.*, **45**, 655.

Morichi T., Sharpe M. E., Reiter B., 1968. Esterases and other soluble proteins of some lactic acid bacteria. *J. Gen. Microbiol.*, **53**, 405–414.

Mundt J. O., 1976. *Streptococci* in dried and frozen foods. *J. Milk Food Technol.*, **39**, 413–416.

Orberg P. K. and Sandine W. E., 1984. A microscale method for rapid isolation of covalently closed circular plasmid DNA from group N *Streptococci. Appl. Environ. Microbiol.*, **47**, 677–680.

Orla Jensen, 1919. The lactic acid bacteria. Hostet Son, Copenhagen.

O'Sullivan T. and Daly C., 1982. Plasmid DNA in leuconostoc species. *J. Food Sci. Technol.*, **6**, 206.

Rogosa M., Mitchell J. A., Wiseman R. F., 1951. A selective medium for isolation and enumeration of oral and fecal *Lactobacilli. J. Bacteriol.*, **62**, 132–133.

Sanders M. E. and Klaenhammer T. R., 1981. Evidence for plasmid linkage of restriction and modification in *Streptococcus cremoris* KH. *Appl. Environ. Microbiol.*, **42**, 944–950.

Schaack M. M. and Marth E. H., 1988. Interaction between lactic acid bacteria and some food borne pathogens: A review. *Cultured Dairy Prod. J.*, **23**, no. 4, 14–20.

Scherwitz K. M., Baldwin K. A., McKay L. L., 1983. Plasmid linkage of a bacteriocin-like substance in *St. lactis ssp diacetylactis* strain WM4: Transferability to *St. lactis. Appl. Environ. Microbiol.*, **45**, no. 5, 1506–1512.

Schleifer K. H. and Kandler O., 1972. Peptidoglycan types of bacterial cell-walls and their taxonomic implications. *Bacteriol. Rev.*, **36**, 407–477.

Schleifer K. H., Kraus J., Dvorak C., Kilpper-Balz R., Collins M. D., Fisher W., 1985. Transfer of *Streptococcus lactis* and related *Streptococci* to the genus *Lactococcus. Gen. Nov. System Appl. Microbiol.*, **6**, 183–195.

Sharpe M. E., 1955. A serological classification of *Lactobacilli. J. Gen. Microbiol.*, **12**, 107–122.

Sharpe M. E., 1981. The genus *Lactobacillus. In:* Starr M. P., Stolp H., Truper H. G., Balons A., Schlegel H. G. (eds). The proaryotes, vol. 2, chap. 131 1653–1679.

Smiley M. B. and Fryder V., 1978. Plasmids lactic acid production and N acetyl D glucosamine fermentation in *Lactobacillus helveticus ssp Jugurti. Appl. Environ. Microbiol.*, **35**, 717–781.

Sneath P. H. A., Mair N. S., Sharpe M. E., Holt J. G., 1986. *Bergey's manual of systematic bacteriology*, vol. 2, Williams & Wilkins.

Stackebrandt E., Fowler V. J., Woese C. R., 1983a. A phylogenetic analysis of *Lactobacilli, Pediococcus pentosaceus* and *Leuconostoc mesenteroïdes. System Appl. Microbiol.*, **4**, 326–337.

Stackebrandt E., Ludwig W., Seewaldt E., Schleifer K. H., 1983b. Phylogeny of non spore forming members of the order Actinomycetales. *Int. J. Syst. Bacteriol.*, **33**, 173–180.

Terzaghi B. E. and Sandine W. E., 1975. Improved medium for lactic *Streptococci* and their bacteriophage. *Appl. Microbiol.*, **29**, 807–813.

Vescovo M., Botazzi V., Sarra P. G., Dellaglio F., 1981. Evidence of plasmid deoxyribonucleic acid in *Lactobacillus. Microbiologia*, **4**, 413–419.

Weiss N., Schillinger U., Kandler O., 1983. *Lactobacillus lactis. Lactobacillus leichmanii* and *Lactobacillus bulgaricus* subjective synonyms of *Lactobacillus delbrueckii ssp lactis* comb. nov. and *Lactobacillus delbruckii ssp bulgaricus* comb. nov. *System Appl. Microbiol.*, **4**, 552–557.

CHAPTER

14

Bacteriophages of Lactic Starters: Detection and Enumeration

J.-P. Accolas, Marie-Christine Chopin,** G. K. Y. Limsowtin†*

14.1 Review of the Characteristics of Bacteriophages

Bacteriophages (or phages) are viruses capable of multiplying inside bacterial cells and causing cell lysis at the end of their multiplicative cycle. In the case of lactic acid bacteria (lactococci, streptococci, and thermophilic lactobacilli) used as starter cultures in dairy technology, phage attacks are the most frequent cause (although not always recognized as such by all professionals) of serious and occasionally catastrophic problems encountered in food production. Phages disturb and even completely stop lactic fermentations and in doing so are a cause of defective products.

Lactic acid bacterial phages currently constitute a major problem for the dairy industry. The concentration of dairy operations in huge installations processing several hundred thousands of liters of milk per day, the frequent use for certain types of cheese making from prewarmed milk having undergone a prolonged incubation (15–18 hours) with a preproduction starter, and the use of lactic starters selected by empirical means have all contributed in varying degrees to creating the present phage problem which is one of the biggest to be solved in the dairy industry.

Knowledge of lactic acid bacterial phages has grown considerably during the 1980s thanks to the techniques of molecular biology (see the review of

*Station de Recherches Laitières and
**Laboratoire de Génétique Microbienne, INRA, 78350 Jouy-en-Josas
†New Zealand Dairy Research Institute, Palmerston North, New Zealand

Jarvis, 1989). These techniques have made possible comparisons of phage proteins or genomes giving detailed information at the molecular level serving as a very useful complement to previously obtained conventional data on the morphology, host specificity, and serology of these phages. This has been the case as much for lactococci (Braun et al., 1989) as for *Streptococcus salivarius* ssp. *thermophilus* (Mercenier and Lemoine, 1989; Neve et al., 1989; Prevots et al., 1989; Larbi et al., 1990; Mercenier, 1990) or for the lactocobacilli (Sechaud et al., 1988). Some of these phages have been very well studied and are considered reference phages. This is the case, for example, of *Lactococcus lactis* phage P008 isolated in Germany (Teuber and Loof, 1987), of *Lactobacillus casei* phages PL-1 and ØFSW (see mainly Japanese works cited by Sechaud et al., 1988) or of *Lactobacillus delbrueckii* ssp. *lactis* phage LL-H isolated in Finland (Alatossava, 1987).

Accumulated data suggest that the numbers of different phages that attack strains of a given species (or of related species) is generally low. This applies to lactococci, for which it is often observed that bacterial and phage variation is practically continuous for a given phage type (Chopin et al., 1976; Lawrence, 1978). The same applies to *Lactobacillus helveticus* phages, for which recent studies have shown evidence of two phage types (Sechaud, unpublished data) and to *Lactobacillus delbrueckii* ssp. *lactis* and ssp. *bulgaricus*, even though the obvious morphological variety of the phages of these latter two lactobacilli suggested at first glance a greater phage diversity than that which was ultimately demonstrated (Sechaud et al., 1988).

With respect to the origin of phages, lysogenic bacteria (those which possess one or several phage DNAs integrated into their own genome) constitute one of the main natural reservoirs of phages. It has been demonstrated that lysogeny occurs among lactococci (Reyrolle et al., 1982; Relano et al., 1988; Braun et al., 1989), in *S. salivarius* ssp. *thermophilus* (Smaczny and Krämer, 1984; Mercenier, 1990), among leuconostocs (Shin and Sato, 1980) and among lactobacilli (Sechaud et al., 1988 and 1989). Strains which are sensitive (so-called indicator strains) to attenuated phages liberated by lysogenic strains, have been found in numerous cases. Furthermore, it has also been demonstrated that virulent phages capable of hyperinfecting a lysogenic strain may have come from the attenuated phage harbored by such a strain of *Lactobacillus casei* (Shimizu-Kadota et al., 1983 and 1985).

Other interactions between lactic acid bacteria and their specific phages must also be taken into consideration in order to explain the behavior of certain lactococcal strains. For example, in pseudolysogeny, a lytic phage maintains a relatively stable equilibrium in a heterogeneous bacterial population consisting of a phage-resistant strain giving rise to sensitive mutants at a certain rate (Limsowtin and Terzaghi, 1977; Daly and Fitzgerald, 1987). Also to be considered is the more general case of the mixed starters used in northern European countries which consist of a complex lactic flora and which always contain phages (Stadhouders et al., 1976). Phages may also be propagated by lactic acid (or related) bacteria in raw milk which constitutes a constantly renewed phage input. This is apparently the major source of phages attacking lactic starter lactococci within the ecosystem of large-scale cheese production in New Zealand (Jarvis, 1987). These sources of phage, varied and poorly monitored, and still all too often the archaic use of starters of different origins both explain how phages are able to arise and establish themselves in French dairies. They may undergo mutations, modifications, or recombinations and give rise to new phages. Some of these are able to attack the constituent strains of lactic starters and decimate them and in doing so complete and feed the phage infection cycle in a given factory (Lawrence and Heap, 1986; Heap and Lawrence, 1988).

In order to prevent phage infections, it is necessary to implement all of the presently well-defined preventive antiphage measures (Cogan, 1980; Shaw, 1983). It is also important to give special attention to the choice and the implementation of the starters used. Different starter systems have proven themselves in the daily practice of cheese production. Schematically, these are two major types: the mixed starter system expecially well studied in the Netherlands (Stadhouders and Leenders, 1984; Stadhouders, 1986) and the selected pure strain system developed in New Zealand (Lawrence and Heap, 1986; Heap and Lawrence, 1988). These systems are conventional and characterized by a pragmatic approach to the starter problem, particularly with respect to the consideration of phage risk.

During the 1980s, progress in lactic acid bacterial genetics led to the development of strains possessing increased phage resistance. Most of the work done up to the present has dealt with lactococci. A certain number of plasmid coded mechanisms imparting a generalized phage resistance to carrying strains have been described. These are the classical restriction-modification systems as well as other mechanisms which affect the adsorption of phages onto the bacterial cell or injection of phage DNA or the intracellular multiplication of the phage (abortive infection). The corresponding plasmids have been isolated and the resistance genes carried by them have been cloned in order to construct new plasmids carrying several mechanisms and thus obtaining a cumulative and generalized phage resistance. It is hoped that this approach will bring about great improvements in the resistance of dairy starter strains to phages and that phage attack will thereby be adequately controlled (see the review by Sanders, 1988).

Thus, dairy-processing plants must be equipped with testing laboratories which allow the daily monitoring of the evolution of phages which may be present in production areas and careful planning of the composition of the lactic starter used in production. Routine methods exist in order to do this but require experienced personnel, that is, capable of extracting from test results all of the information useful for managing the daily use of lactic starters. When a dairy-processing plant receives its starters from a starter culture supplier, a spirit of cooperation and trust must exist in this matter between the two parties. In view of these concepts, the methods for the detection and enumeration of lactic acid bacterial phages are described in this chapter.

14.2 Review of Methods: Meaning and Significance of Results

Any unexpected variation in the behavior of a starter culture, whether during its preparation or during production (e.g., a long lag or a sudden interruption and delay of the acidification process) must a priori be attributed to a phage attack. Various methods allow the confirmation of this presumption, some having been specially designed for finding phages in mixed starters while others are well adapted to the search for phages infecting selected strains of multi-strain starters.

14.2.1 The Double Agar Layer Method

The conventional method and to date the most precise, is the so-called double agar layer method which detects and enumerates phages. This method, however, is usable only for starters composed of selected strains. When a bacterial suspension of suitable concentration containing a small amount of agar is spread evenly as a thin film over an ordinary agar surface, a confluent, uniform and opaque lawn of bacterial growth develops after a few hours of incubation. In the presence of high concentrations of a lytic phage specific for the bacterial strain used, total lysis of the bacterial lawn is observed, that is, total clearing of the Petri plate. When the incorporated phage suspension is more dilute, zones of lysis more or less confluent are observed throughout the surface of the plate, separated by strips of bacterial lawn. When the phage suspension is sufficiently diluted, small, circular, clear, and well-isolated fields of lysis (plaques) appear having a diameter reaching 2–3 mm (Figure 14.1). It can be demonstrated that each field of lysis (plaque) results from a single phage particle initially present in the test suspension. These plaques may be counted and multiplied by the dilution factor to express the number of phage particles present in one ml of analyzed sample.

14.2.2 The pH or Redox Indicator Method

A second method takes advantage of the behavior of lactic acid bacteria grown in milk with an added pH or redox indicator. For example, a tube of litmus milk (initially a lavender color) inoculated with a strain of lactococci, allows the successive observations of reduction (white color developing from the bottom of the tube upwards), acidification (color change towards pink near the air interface of the medium), and finally coagulation of the milk, giving it its final appearance, that is, white and coagulated with a thin pink-colored layer at the surface. In the presence of a sufficient number of specific lytic phages, the bacterial strain is infected and destroyed at some variable rate. In such a case, the colored indicator remains unchanged or the characteristic reactions appear later and usually incompletely in comparison with the control culture

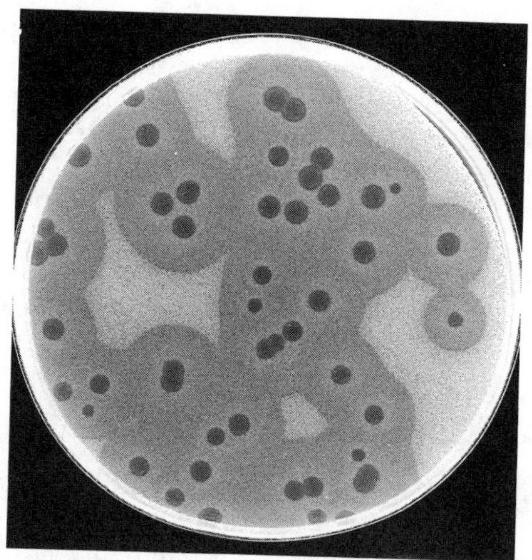

Figure 14.1 Plaques obtained by the double agar layer technique on M17 agar after suitable dilution of cheese whey containing lytic phages specific for the starter strain *Lactococcus lactis* ssp. *cremoris*.

After overnight incubation at 30°C, round plaques are formed, each resulting from the development of a single phage particle initially present in the dilution. Plaques appear as dark gray areas in contrast with the surrounding bacterial lawn. Each plaque is surrounded by a concentric halo of large diameter running into the halos of nearby plaques. These are the result of incomplete lysis of the bacterial lawn brought about by the slow diffusion of phage lysin from the plaque while the plate sat at ambient temperature for 3–4 days. A few strips of intact bacterial lawn remain which appear as light gray regions beyond the halos of lysis (INRA photograph).

(e.g., absence of coagulation). The differences between the control culture and the phage-containing culture persist and intensify with repeated transfers. In the presence of a small number of phages, the first culture may show no difference between control and test and several transfers may be required in order to observe a difference.

14.2.3 Rapid Lysis Method

A third method takes advantage of the nondevelopment (sudden lysis) or rapid clearing of a culture grown in a clear broth in the presence of a sufficient number of phages while the control culture becomes turbid during the same incubation period. If the initial number of lytic phages present is low, clearing of the culture due to bacterial lysis may not appear until after one or several transfers.

Both of the previously described methods require a minimum of work and are able to detect the presence of a lytic phage specific for a lactic acid bacterial strain with reasonable certainty. As such, they constitute rapid methods useful in practice for routine testing. They may also be used for performing quantitative assays of phages. The same technique is employed, but with decimal dilutions of the sample in order to determine the last dilution in which phage is still present and active. The requirement for transferring the cultures in which the phage, initially present in low numbers because of the dilution, will not appear until after several transfers; the requirement for three to five tubes at each dilution in order to determine the most probable number of phage particles in the sample; and finally the lack of precision associated with liquid media counting methods, all weigh against the use of these methods on a routine basis.

14.2.4 Milk Acidification Test

A fourth method in its simplest form is a milk acidification test applicable to all types of starters. Under well-standardized conditions (reconstituted and warmed powdered milk; inoculum, temperature and incubation time determined), the acidification is compared between a control culture and a similar culture with the suspected phage-containing sample. Thus, in the case of lactococci, for example, phages are likely present when the difference in acid production is at least 10% after 6 hours of incubation at 30°C. This acidification test may be made to approximate the characteristics of the processing plant production cycle (e.g., temperature variations). At the end of the incubation period, the pH of the control is compared to that of the tested sample. If the difference exceeds some threshold based on previous experience (e.g., 0.3 pH units), the presence of phages is strongly suspected.

A slightly more elaborate acidification test has been developed in the Netherlands for the enumeration of "disturbing phages" which cause acidification problems in cheese factories using mixed lactococcal starters. It should be kept in mind that mixed starters which normally contain a complex lactic flora with its own resident phages do not really lend themselves to the detection and enumeration of target phages by the previously described methods in Sections 14.2.1 and 14.2.3. The Dutch test of Stadhouders and Leenders consists of two successive steps, a first step for enrichment and a second step, which is the actual acidification test. Schematically, the first step consists of preparing decimal dilutions (10^{-3}–10^{-9}) of the sample to be analyzed (which may be samples of milk, cheese whey, cheese, or starter used by the factory) and transferring an aliquot of a given dilution to a flask of autoclaved milk inoculated with an uncontaminated culture or the mixed starter under study. A series of seven flasks corresponding to seven dilutions is thus prepared. An eighth flask, inoculated with the mixed starter only, serves as a control. These are incubated for 24 hours at 20°C to allow the development of problematic phages present in a given dilution. Sterile filtrates of each dilution and of the control are then prepared and the acidification test itself is performed (6 hours at 30°C) on a series of vials of milk inoculated with a fresh uncontaminated culture of the mixed starter plus an aliquot of filtrate. After incubation, the acidity (or pH) of each culture is measured. Disturbing phages are said to be present in a given dilution when the differences in acidity between the corresponding culture and the control is at least 5°N (value corresponding to 4.5°D or 0.045% lactic acid) as indicated by the example in Table 14.1.

This sort of acidification test performed on a routine basis takes 2 days which quickly becomes fastidious and tedious. It does, however, allow a reliable presumptive diagnosis when problematic phages appear in a factory and attack the mixed starter used there.

14.2.5 Rapid Methods

The conventional methods for the detection and enumeration of phages generally require time, for example, 6 hours at 30°C for a simple acidification test or between 6 hours (minimum) and 15 hours in an incubator to obtain well-formed plaques on a solid medium. Under these conditions, the results obtained have only a retrospective value and are therefore usable only a posteriori for implementing measures designed to guard against renewed phage attacks (changing of the starter used in production; reducing the overall average level of phage contamination by rigorous general disinfection). The development of new indirect methods for estimating the levels of bacterial flora in milk harbor possibilities for the rapid presumptive diagnosis of phage infections and earlier responses at the production level.

14.2.5.1 Impedance Measurement

Impedance measurement takes advantage of the increase in the conductance of a culture medium such as milk resulting from the growth and metabolism of the starter lactic flora. Given that a decreasing linear relationship exists

Table 14.1 Enumeration of disturbing phages by the acidification test of Stadhouders and Leenders, in cheese factories using mixed starters

	Dilution of the analyzed sample							
	−3	−4	−5	−6	−7	−8	−9	Control
Factory no. 1	59	58	58	59	59	59	57	60–60
Factory no. 2	49	48	48	48	48	45	60	60–62
Factory no. 3	47	48	47	57	63	64	64	62–62

Acidity expressed as °N units used in the Netherlands. One °N is the number of ml of 0.1M NaOH required to neutralize 100 ml of milk to the phenolphtalein end-point. A difference ⩾5°N between control and test indicates the presence of problematic phages.

Factory no. 1: sample does not contain any problematic phages; Factory no. 2: sample contains 10^8 problematic phages/ml; Factory no. 3: sample contains 10^6 problematic phages/ml.

Source: Adapted from Cox, 1980; reproduced with the permission of the FIL-IDF.

between the time required to observe a significant increase in conductance (the detection time) and the initial level of microorganisms present in the sample, a lengthening of the detection time may be expected when a phage infection slows down or stops starter growth and metabolism.

Such disturbances have been experimentally verified and a rapid method has been proposed which uses the comparative measurement of the evolution of impedance in a test sample and in a control in milk supplemented with yeast extract (which decreases the detection time by stimulating starter growth). Significant inhibition of the drop in impedance ($\geqslant 50\%$) has thus been observed after only 2 hours when 10^5 problematic phages were initially present in the yeast extract enriched milk inoculated with a mixed cheese starter. This method is supported by a commercially available device which lends itself to repetitive analyses.

14.2.5.2 ATPmetry

The bioluminescent measurement of ATP, that is, by the ATP-dependent reaction of the luciferin luciferase complex, is an excellent indicator of microbial activity. ATPmetry has recently been applied to the detection of lactococcal phages, the first results obtained suggesting that the comparison of the ATP evolution (total or free) in control and phage-containing milk cultures reveals significant differences after 1–3 hours of incubation when the phage contamination is between 10^6 and 10 phages/ml. This is therefore a very sensitive method. In any event, its implementation is delicate and requires qualified personnel. The commercially available equipment lends itself to repetitive analyses.

14.2.5.3 Immunoenzymatic Phage Detection

For the sake of completeness, the immunoenzymatic detection of lactic bacterial phages must be mentioned, which is expected to be marketed by a specialized laboratory in the very near future. This test would employ a mixture of antiphage antibodies obtained from aminals immunized using serotypes representative of different phages infecting lactic acid bacteria used in dairy technology (i.e., lactococci, *S. salivarius* ssp. *thermophilus*, lactobacilli, and leuconostocs). The existence of an antibody-phage antigen complex, if any phage is present in the analyzed sample, would be detected by means of an ELISA-type enzymatic reaction. This test could be offered in two forms, one allowing rapid searches during production (a qualitative response may be obtained in 15 minutes with a phage contamination threshold fixed in advance), and the other being a quantitative analysis carried out in the processing plant laboratory (determining the level of phage contamination using microtitration plates and requiring 1 hour). Such a test would obviously have major advantages in terms of specificity, sensitivity, and speed. This raises the question of whether or not such high sensitivity would result in false-positive results in processing plants using mixed lactic starters which always contain "resident" phages.

14.2.6 General Recommendations and Related Remarks Concerning Mode of Operation and Interpretation of Results

14.2.6.1 Sampling

The sample must be as representative as possible of the phage population present in the processing plant. In cheese production, for example, this will be preferably a weighted sampling of the whey drawn from the various production vats or, better yet, a sample coming from the whey tank withdrawn after skimming and as late as possible in the work day when the same starter is used throughout production. The purpose of the search is in effect to detect new phage(s), these being perhaps mutants of phages found earlier and which infect the starter used. Such phages may appear at any moment and the objective set will not be fully achieved unless they are detected early enough, that is, before their numbers (which may incrase very rapidly throughout the processing plant) become such that the acidifying activity of the starter is almost certainly going to be diminished or stopped during production or, even worse, the starter vat itself becomes contaminated with phage due to insufficient protection. The sampling frequency may be increased in order to detect a phage contamination site which is suspected of being particularly loaded and resistant to disinfection; however, this practice must remain the exception rather than the rule.

14.2.6.2 Sample Preparation

Following withdrawal, samples must be sterilized by filtration as quickly as possible, either with 0.45 μm membranes or with Seitz-type filtering plates in order to eliminate the microflora and prevent the rapid drop in phage titer due to adsorption to bacteria present in the sample. The use of disposable filters mounted on syringes or accelerating the sterilizing filtration by centrifugation as suggested by Hull (1977) (Figure 14.2) are recommended for carrying out repetitive analyses. Once filtered, the samples must be cooled and stored in a refrigerator at below 5°C.

In order to speed up the filter sterilizing and avoid early blockage of the filter, it is usually necessary to clarify the samples beforehand. In the case of cheese whey, centrifugation or filtering on filter paper may be done in order to eliminate suspended curd particles and fat. For samples of milk which have been acidified to some degree (e.g., in fermented milk production), acidifying all samples to pH 4.6 with lactic acid, centrifuging to remove coagulated casein, carefully decanting the whey supernatants, and adjusting the pH of these to 6.5

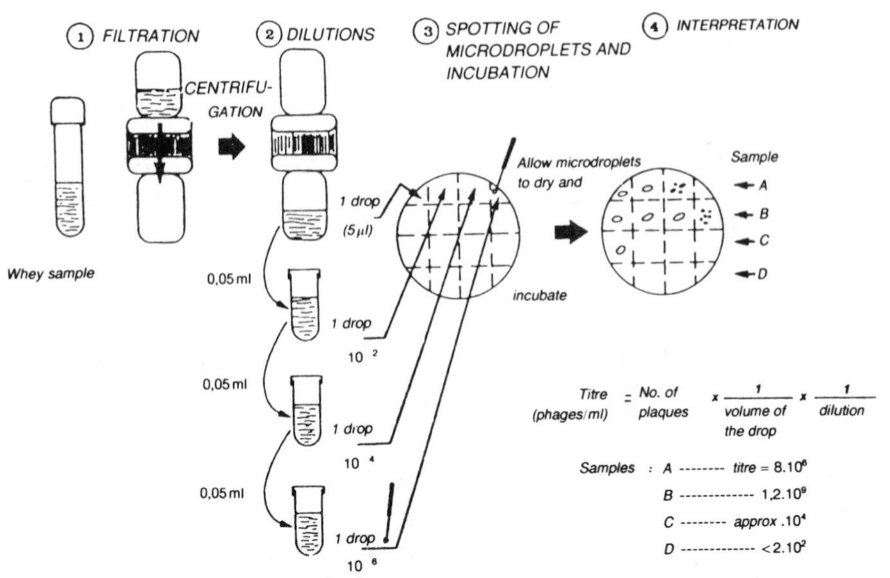

Figure 14.2 Detection and enumeration of bacteriophages by the double agar layer technique (adapted from Hull, 1977).

1. *Filtration:* The sample is placed in a filtration device fitted with a Seitz filtering plate (the sample may be centrifuged beforehand in order to facilitate filter sterilizing). Whey samples are generally difficult to filter, but the volume required for the analysis is small (1 to a few ml).

2. *Dilutions:* Autoclaved reconstituted skim milk (0.1% w/v) is used as a diluent, 0.05 ml of sample or previous dilution being added to 4.95 ml giving dilutions of 10^{-2}. Each dilution is carefully shaken before moving on to the next dilution.

3. *Deposing of microdroplets and incubation:* A Petri plate containing a suitable medium (M17 or MRS) seeded with the proven pure lactic strain (streptococcal or lactobacillary) is placed on squared paper (*dashed lines*). Using an inoculating loop, a microdroplet is placed in each of the zones defined by the gridlines, flaming the loop between each transfer. A whole Petri plate should be used for each pure strain of lactic acid bacteria routinely used in the processing plant lactic starter rotation scheme. Four different samples including all of their dilutions may be deposited as microdroplets on the same Petri plate. The plates are left on the counter for 30 minutes in order for the droplets to be absorbed by the agar. Incubation is aerobic at 30°C for lactococci and anaerobically at 37°C for thermophilic lactic acid bacteria for at least 6–8 hours to allow well-formed, visible plaques to develop.

4. *Interpretation of results:* Plaques, that is, small round holes "punched" in the bacterial lawn, are counted, each one resulting from the development of a single phage particle.

with sodium hydroxide is recommended. The electrode used for the pH measurements must be thoroughly rinsed with distilled water, kept in a bleach solution with a free chlorine concentration of 400 ppm for 2 minutes, and rinsed again between each adjustment.

14.2.6.3 Indicator Cultures Used

Regardless of the method used, a phage specific for a bacterial strain will be detected with certainty only in the presence of the corresponding pure strain, to the exclusion of all other strains not sensitive to the phage. It is necessary to be fully aware of this requirement, since dairies routinely use commercial starter cultures provided by various suppliers, or they prepare their own complex mixtures of lactic flora. When the commercial starter contains a mixture of different strains (a multistrain starter), its inappropriate use as a test culture may be the source of chronic problems and errors of interpretation. The dairy-testing laboratory may overcome this problem if the producer of lactic starter is willing to provide pure cultures of each of the strains making up the multistrain starter. Otherwise it is necessary to isolate beforehand the strains representing the dominant flora of the commercial starter used. This may be done notably in the case of empirically mixed starters (e.g., the mesophilic commercial lactic starters from northern Europe widely used in France, or the "natural" thermophilic lactic starters used by artisanal cheese makers in eastern France for "cooked curd" cheeses) to obtain indicator strains and ultimately to check for phages. The other possibility is to carry out the two-step acidification test as described in Section 14.2.4, but this is time consuming and tedious.

The cultures used must be active and incubated at the same temperature as the starter (e.g., freshly coagulated milk cultures). A suitable broth may also be used. Either way, any difference in treatment between the test culture and the starter used for production (culture medium, temperature, incubation time) may cause strain modifications which may give rise to different responses to phages.

14.2.6.4 Choice of a Method and Interpretation of Results

Not all phages are detected to equal degrees by any of the methods described thus far. In otherwords, variable results are to be expected depending on the method used. For example, some phages will easily form plaques on a bacterial lawn on agar without inhibiting in comparable proportions the growth and acid production of the same bacterial strain cultured in milk. Conversely, other phages will readily lyse a bacterial culture in liquid media such as milk or suitable broth but will not easily form plaques on agar. It is therefore difficult a priori to recommend one method over another. It is certainly preferable, insofar as the technical burden can be kept tolerable, to use two methods in parallel in order to decrease the probability of phages escaping detection. Titers obtained by a plaque on agar method and by a liquid medium enumeration method (such as the most probable number determined by means of dilution limits for observing inhibition of bacterial activity) may have to be compared, that is, by calculating the ratio of the titers obtained and deducing the efficiency of one method versus the other. In any event, due to the amount of work involved, this sort of comparison can be done only occasionally.

14.2.6.5 Precautions to be Taken During Technical Manipulations

It is absolutely imperative that all handling of phages be done in a laboratory completely separate from the areas where lactic acid bacteria are maintained and the other cultures are prepared. Different persons must be assigned, if possible, to culture maintenance and phage detection. In addition, the equipment used and the laboratory must be systematically disinfected. This means prolonged soaking of glassware (any flasks, pipettes, tubes, etc.) in bleach at a free chlorine concentration of 400 ppm immediately after use and/or autoclaving of any supplies in which phage has been propagated (culture or dilution tubes, Petri plates, etc.) before cleaning or trashing. All sinks and drain panels must also be systematically disinfected with bleach at the end of each working day.

14.3 Recommended Methods

Readers requiring in-depth knowledge relative to these methodologies are strongly advised to consult bulletin FIL/IDF 129, which gives a complete description of the acidification test used in the Netherlands for enumerating phages in mixed starters (Cox, 1980) and the article by Stadhouders and Leenders (1984), which provides a detailed explanatory scheme of the implementation of this same method. Articles by Anderson and Meanwell (1942), describing a presumptive acidification test; by Valles (1955), describing the development of a phage detection method using milk containing a redox indicator; by Terzaghi and Sandine (1975), giving the original description of M17 medium and of litmus milk buffered with sodium β-glycerophosphate; by Mullan (1979) and Keogh (1980), devoted to factors affecting plaque formation by lactococcal phages; and two articles by Accolas and Spillman (1979a,b), describing the methodology used to detect and isolate phages of *S. salivarius* ssp. *thermophilus* and those of various thermophilic lactobacilli, are all also worth consulting.

In the field of rapid methods, review articles by Quesneau (1983a,b) and by Piton and Dasen (1988) are beneficial as are articles on phage detection by Waes and Bossuyt (1984) using impedence measurements and by Seven and co-workers (1988) using ATP measurements. Finally, the development of an immunoenzymatic test has been undertaken by Société Transia (8 bis rue Saint-Jean-de-Dieu, 69007 Lyon, France).

This chapter deals specifically with the detailed description of the two methods which appear to be among the most recommendable for the detection and counting of phages in dairies. These are the double agar layer technique as adapted by Hull (1977) and the revealing of phages in litmus milk, buffered or otherwise.

14.3.1 Enumeration of Phages on Solid Medium: The Hull Microdroplet Test (1977)

14.3.1.1 Phages of Lactococci and of S. Salivarius *ssp.* Thermophilus

Recommended Medium: M17 Medium of Terzaghi and Sandine (1975)

The ready-to-use M17 media in dehydrated form are now commercially available in France from various suppliers. Although these are entirely suitable, it may be desirable to turn to the original description of the medium, for example, should strain growth be insufficient. The original medium uses ingredients manufactured by the American company BBL and prepared under license for sale under various labels by the French firm bioMérieux.

The composition of the basal medium, valid for M17 agar, is as follows (g or ml/l): polytone (bioMérieux) 5 g; sodium β-glycerophosphate 19 g; ascorbic acid 0.5 g; $MgSO_4 \cdot 7H_2O$ (1M solution) 1 ml; distilled water 940 ml; pH 7.15 ± 0.05 at ambient temperature, adjusted if necessary with NaOH or HCl. Autoclave at 121°C for 15 minutes. Solutions sterilized separately (121°C for 15 minutes): 10% lactose (w/v) and 1M $CaCl_2 \cdot 2H_2O$.

Hard M17 agar (bottom layer). Basal M17 medium with sufficient agar added to obtain a hard gel (9–18 g of agar/l depending on manufacturer's specifications) is autoclaved at 121°C for 15 minutes. Preparation of the solid medium may be done in a variety of ways. One of the simplest and quickest is to carefully weigh directly into 250 ml flasks, using a precise top-loading balance, the quantity of agar corresponding to 200 ml of medium (between 1.8 and 3.6 depending on the agar used). The corresponding volume of basal M17 liquid medium is then added (188 ml) and all flasks are autoclaved at 121°C for 15 minutes. Overheating of the medium is thus avoided and time is saved. A stock of flasks representing several liters of basal agar medium may be prepared in this manner and used as required. The M17 hard agar is used as follows: After melting in a boiling water bath or autoclave, the medium is cooled and held at 48°C. Lactose solution (prewarmed to 48°C) is added (10 ml to each flask or 50 ml to 940 ml of basal medium) after adding the sterile molar $CaCl_2$ solution (2 ml to each flask or 10–940 ml of basal medium). The medium becomes slightly cloudy due to the addition of the calcium salt. These steps should be performed slowly and with gentle stirring in order to minimize precipitation of the medium and bubble formation. The complete medium thus prepared is immediately poured into sterile Petri plates (20 ml per plate) which are left undisturbed overnight at ambient temperature in order to obtain a sufficiently dry surface, at which point the plates are usable. They may be kept in a refrigerator for use as required. All plates should be examined prior to use in order to eliminate those which have become contaminated.

Soft M17 agar (top layer). The formulation and preparation are the same as for the complete hard M17 agar except that the quantity of agar is reduced by half. The soft agar may also be prepared by mixing equal volumes of melted agar (bottom) medium and M17 broth at 48°C in a sterile flask.

Mode of Operation. Preparation of Petri Plates

Aliquots of a fresh culture of the strain under study in M17 broth (0.1 ml) or in milk (0.2 ml of a 1/10 dilution of a freshly coagulated culture in autoclaved, skim milk) are distributed to sterile tubes (10 × 80 mm). The tubes are then prewarmed to 48°C in a water bath and 2.5 to 3 ml of complete M17 top agar melted and held at 48°C is added to each. The contents of each tube is rapidly mixed by rolling the tube between the palms of both hands in order to avoid bubble formation and then poured onto the surface of a dry layer of M17 bottom agar in a Petri plate. The top agar is evenly distributed by tilting the plate back and forth in different directions and left to solidify. The plates thus prepared may be used the same day after sufficient drying or stored for up to 1 week in a refrigerator (temperature <5°C) and used as required. Storage for more than 1 week may lead to uneven and irregular bacterial lawns as well as the development of contaminating germs.

Suspected phage-containing samples and successive decimal dilutions thereof are all deposited as microdroplets onto the seeded M17 agar surfaces by means of a calibrated loop (e.g., 5 μl) as indicated in Figure 14.2. The plates are left undisturbed on the counter in order for the agar to completely absorb the microdroplets (about 30 minutes) and then incubated at the normal starter incubation temperature, aerobically for lactococci or preferably anaerobically for *S. salivarius* ssp. *thermophilus* which does not form a sufficiently dense and uniform lawn under aerobic conditions. Anaerobic jars may be used for this purpose.

14.3.1.2 Phages of the Thermophilic Lactobacilli (Lactobacillus delbrueckii *ssp.* bulgaricus *and ssp.* lactis; L. helveticus)

Recommended Medium: MRS Medium of De Man and Co-workers (1960)

Ready-to-use MRS media in dehydrated form, with or without agar, are widely available in a variety of commercial forms, but the experience of some people (i.e., the authors) suggests that these do not always give entirely satisfactory results. Careful choice of a complete instant medium is therefore important.

Mode of Operation

The procedure for pouring plates, preparing top agar, and deposing microdroplets is identical to that described for M17 medium. Both MRS bottom and top agars are thus prepared. Calcium chloride is added to sterile 1M solution, 10 ml/l of basal agar medium (which already contains glucose as carbohydrate substrate) just prior to use. Incubation is at the temperature normally used for the starters, preferably anaerobically in order to obtain reproducible results as in the case of *S. salivarius* spp. *thermophilus*.

14.3.1.3 Interpretation of Results

When the samples and their successive dilutions contain lytic phages specific to the bacterial strain under study, zones of confluent lysis and plaques of lysis will be successively observed. The plaques are counted and the number of phages present in the sample is calculated using the dilution factor (see Figure 14.2, in which four examples are given: A, B, C, and D).

Zones of nonspecific lysis may be observed at low dilutions. These are due to the action of a virolysin which disrupts the cell wall of the bacteria thereby provoking a nonspecific lysis of the bacterial strain. Unlike the phage, lysin is not infectious. A strain of lactic acid bacteria may be resistant to a phage but sensitive to the lysin produced by the phage during multiplication of the latter in its host strain. When a culture contains two bacterial strains, both may be lysed by one phage, one by specific phage infection and the other by nonspecific enzymatic attack. In the case represented in Figure 14.2, the lysis produced by sample C and not accompanied by plaque formation in subsequent dilutions may be the result of nonspecific action of a virolysin.

Various methods distinguish the nonspecific action of a lysin from the specific action of a lytic phage. A phage may be multiplied by transfers to milk, broth, or agar whereas the action of lysin disappears. Phage lysin can lyse dead cells while the phage cannot. A bacterial lawn obtained by the double agar layer technique can be killed by 30 minutes of exposure to chloroform vapors (by inverting the plate with a solvent-soaked paper filter placed in the lid). By placing microdroplets of sample on the chloroform-treated bacterial lawn, nonspecific lysin activity is seen as zones of lysis appearing within a few hours. Finally, it should be remembered that as phage lysin is diluted, the zones of lysis fade while a phage suspension brings about the formation of plaques when diluted. In practice, phage lysin activity is rarely visible beyond dilutions of 1/100 or 1/1,000 while bacteriophage titers may reach numbers as high as 10^{10}/ml or more.

14.3.2 Detection of Lactic Acid Bacterial Phages Using Buffered or Unbuffered Litmus Milk

Litmus milk is simultaneously a redox indicator (turning from lavender to white) and a pH indicator (turning from lavender to pink due to acidification).

14.3.2.1 Recommended Media for Lactococci and for S. salivarius *ssp.* thermophilus

Litmus milk buffered with sodium β-glycerophosphate, using Terzaghi and Sandine formulation (1975) prolongs the growth of lactococci and thermophilic streptococci by preventing the pH from falling rapidly and in doing so also favors the multiplication of specific lytic phages which may be present. Under these conditions, the difference in behavior between a control culture and a phage-containing culture appear more rapidly and decisively than when using unbuffered milk.

Medium Composition and Preparation

Litmus milk: powdered skim milk free of antibiotics 100 g; litmus (sterile concentrated RAL solution) 10 ml or 1 g dry litmus; distilled water 1,000 ml. Distribute into test tubes (10 ml per tube) and autoclave at 115°C for 15 minutes. Solution of β-GP, autoclaved separately (121°C, 15 minutes): 95 g of salt added to 100 ml distilled water. Complete solubilization of the salt takes place during autoclaving. The concentrated β-GP solution has a pale yellow color. To be made ready for use, 0.2 ml of sterile β-GP solution is added to each tube containing 10 ml of litmus milk, giving buffered medium with a β-GP content of 1.9% as for M17 medium.

14.3.2.2 Recommended Media for Thermophilic Lactobacilli

Ordinary litmus milk or litmus milk weakly buffered with β-GP (0.04 ml of sterile concentrated solution per 10 ml of milk, i.e., a concentration five times less than that used to grow lactococci and thermophilic streptococci) are preferred over strongly buffered milk since the salt has a marked inhibitory effect on the development of lactobacilli.

14.3.2.3 Mode of Operation

A series of tubes of litmus milk or buffered litmus milk, depending on the strain, are inoculated with aliquots (0.1 or 0.2 ml) of a fresh litmus milk culture of the microorganism under study. The filter-sterilized samples to be tested (and dilutions thereof) are added to the tubes (1 ml to each), leaving one as a control. Incubation then follows at the normal culture temperature for the starter and the color of the tubes is periodically observed. It is advisable to shake the tubes periodically (a single inversion is sufficient if the tubes are of the screw-cap sealing type) to homogenize the contents and color until the onset of coagulation. This shaking has no lasting effect on the evolution of the indicator.

In the case of lactococci, the sequence of the colored reactions is as follows: A reduction is observed (changing to white at the bottom of the tube), then an acidification (the surface on contact with air turning from lavender to pink, without being reduced), and finally coagulation of the milk (noted as RAC).

In the case of *S. salivarius* ssp. *thermophilus*, the sequence is acidification (changing to pink throughout the tube), then coagulation, and finally late and often incomplete reduction of the litmus (noted as Acr). In the case of thermophilic lactobacilli, acidification followed by coagulation of the milk is observed and finally a late but complete reduction of the litmus except near the surface which remains pink as in the other cases (noted as ACR). As soon as a delay is observed in a tube with respect to the control tube, each culture is transferred at 1 or 2% and the incubation and periodical observation of the transferred cultures is continued. If a lytic phage is present in the test tube, the delay on the phage-containing culture with respect to the control culture must increase. When the test samples and the control progress in the same manner, they are all transferred as soon as the indicator begins to change (reduction or acidification as the case may be) and the incubation is continued. When a lytic phage is present in small numbers or has low activity under the test conditions used, three or four transfers may be necessary in order to demonstrate its presence.

Under the best conditions, with rapidly growing strains such as *S. salivarius* spp. *thermophilus*, it is possible to carry out four transfers in 1 working day. In all other cases it is acceptable, should no confirmation be forthcoming or complete by the end of a working day, to cool the initial cultures or the transferred cultures in an ice bath, refrigerate them overnight, and resume the incubation the next morning. Cultures may also be transferred at the end of a day and the newly transferred cultures cooled and refrigerated during the night for incubation beginning in the morning.

Prolonged incubation of the cultures (24 hours or more, nights included) without frequent periodic observation or transfers must be avoided at all costs. Otherwise test samples identical to the control culture will almost certainly be observed. The development of phage-resistant secondary cultures will, in the long run, set the course for comparable progression (RAC, ACr, ACR, and so on, depending on the strains) in the sample and control cultures.

14.3.2.4 Interpretation of Results

Any inhibition observed during the incubation of the initial culture or of transferred cultures which is confirmed during the incubation of subsequent transfers indicates the likely presence of a phage. On the other hand, an inhibition observed during the initial culture but which disappears with subsequent transfers is likely due to some other cause. In either case, a confirming test may be carried out by means of the double agar layer technique.

Acknowledgments

We thank *The Australian Journal of Dairy Technology* and the *International Dairy Federation* (IDF) for kindly authorizing the reproductions of the illustration by R. R. Hull (1977) and the table by W. A. Cox (1980), respectively.

References

Accolas J.-P. and Spillman H., 1979a. Morphology of bacteriophages of *Lactobacillus bulgaricus, L. lactis* and *L. helveticus. J. Applied Bacteriology,* **47**, 309–319.

Accolas J.-P. and Spillmann H., 1979b. The morphology of six bacteriophages of *Streptococcus thermophilus. J. Applied Bacteriology*, **47**, 135–144.

Alatossava T., 1987. Molecular biology of *Lactobacillus lactis* bacteriophage LL-H. *Acta Universitatis Oulouensis,* series A 191, University of Oulu, Finland, 65 + 30 p.

Anderson E. B. and Meanwell L. J., 1942. The problem of bacteriophage in cheese-making. I: Observations and investigations on slow acid production. *J. Dairy Res.*, **13**, 58–72.

Braun V., Hertwig S., Neve H., Geis A., Teuber M., 1989. Taxonomic differentiation of bacteriophages of *Lactococcus lactis* by electron microscopy, DNA-DNA hybridization and protein profiles. *J. Gen. Microbiol.*, **135**, 2551–2560.

Chopin M.-C., Chopin A., Roux C., 1976. Definition of bacteriophage groups according to their lytic action on mesophilic lactic streptococci. *Appl. Environ. Microbiol.*, **32**, 741–746.

Cogan T. M., 1980. Les levains lactiques mésophiles. Une revue. *Le Lait*, **60**, 397–425.

Cox W. A., 1980. Mise en évidence et dénombrement des bactériophages des streptocoques lactiques mésophiles. *In:* Les levains utilisés dans la fabrication du fromage. *Bull. FIL/IDF* **129**, chap. 6, 29–36.

Daly C. and Fitzgerald G., 1987. Mechanisms of bacteriophage insensitivity in the lactic streptococci. *In:* Ferretti J. J., Curtiss III R. (eds). *Streptococcal Genetics*, 259–268, ASM, Washington.

De Man J. C., Rogosa M., Sharpe M. E., 1960. A medium for the cultivation of lactobacilli. *J. Appl. Bacteriol.*, **23**, 130–135.

Heap H. A. and Lawrence R. C., 1988. Culture systems for the dairy industry. *In:* Robinson R. K. (ed). *Developments in food microbiology*, vol. 4, 149–185. Elsevier Applied Science.

Hull R. R., 1977. Methods for monitoring bacteriophage in cheese factories. *Aust. J. Dairy Technol.*, **32**, 63–64.

Jarvis A. W., 1987. Sources of lactic streptococcal phages in cheese plants. *New Zealand J. Dairy Sci. Technol.*, **22**, 93–103.

Jarvis A. W., 1989. Bacteriophages of lactic acid bacteria. *J. Dairy Sci.*, **72**, 3406–3428.

Keogh B. P., 1980. Appraisal of media and methods for assay of bacteriophages of lactic streptococci. *Appl. Environ. Microbiol.*, **40**, 798–802.

Larbi D., Colmin C., Rousselle L., Decaris B., Simonet J. M., 1990. Genetic and biological characterization of nine *Streptococcus salivarius* subsp. *thermophilus* bacteriophages. *Le Lait*, **70**, 107–116.

Lawrence R. C. A lecture: Action of bacteriophages on lactic acid bacteria. Consequences and protection. 20ème Congrès International de Laiterie. Sessions Scientifiques et Techniques (2 ST), 15 p.

Lawrence R. C. and Heap H. A., 1986. The New Zealand starter system. *Bull. FIL/IDF*, **199**, 14–20.

Limsowtin G. K. Y. and Terzaghi B. E., 1977. Characterization of bacterial isolates from a phage-carrying culture of *Streptococcus cremoris. New Zealand J. Dairy Sci. Technol.*, **12**, 22–28.

Mercenier A., 1990. Molecular genetics of *Streptococcus thermophilus. FEMS Microbiol. Rev.*, in press.

Mercenier A. and Lemoine Y., 1989. Genetics of *Streptococcus thermophilus:* A review. *J. Dairy Sci.*, **72**, 3444–3454.

Mullan W. M. A., 1979. Lactic streptococcal bacteriophage enumeration. A review of factors affecting plaque formation. *Dairy Ind. Int.*, **44**, 11–15.

Neve H., Krusch U., Teuber M., 1989. Classification of virulent bacteriophages of *Streptococcus salivarius* subsp. *thermophilus* isolated from yoghurt and Swiss-type cheese. *Appl. Microbiol. Biotechnol.*, **30**, 624–629.

Piton C. and Dasen A., 1988. Evaluation de la mesure d'impédance comme technique rapide d'appréciation de la qualité bactériologique du lait cru. *Le Lait*, **68**, 467–484.

Prevots F., Relano P., Mata M., Ritzenthaler P., 1989. Close relationship of virulent bacteriophages of *Streptococcus salivarius* subsp. *thermophilus* at both the protein and the DNA level. *J. Gen. Microbiol.*, **135**, 3337–3344.

Quesneau R., 1983a. La luminescence. *Technique Laitière*, **974**, 49–59.

Quesneau R., 1983b. La mesure de l'impédance. *Technique Laitière*, **974**, 62–64.

Relano P., Mata M., Bonneau M., Ritzenthaler P., 1987. Molecular characterization and comparison of 38 virulent and temperate bacteriophages of *Streptococcus lactis. J. Gen. Microbiol.*, **133**, 3053–3063.

Reyrolle J., Chopin M.-C., Letellier F., Novel G., 1982. Lysogenic strains of lactic acid streptococci and lytic spectra of their temperate bacteriophages. *Appl. Environ. Microbiol.*, **43**, 349–356.

Sanders M. E., 1988. Phage resistance in lactic acid bacteria. *Biochimie*, **70**, 411–422.

Sechaud L., Cluzel P.-J., Rousseau M., Baumgartner A., Accolas J.-P., 1988. Bacteriophages of lactobacilli. *Biochimie*, **70**, 401–410.

Sechaud L., Callegari M.-L., Rousseau M., Muller M.-C.., Accolas J.-P., 1989. Relationship between temperate bacteriophage 0241 and virulent bacteriophage 832-B1 of *Lactobacillus helveticus. Netherlands Milk Dairy J.*, **4**, 261–277.

Seven V., Onno B., Pinet X., 1988. Application du dosage de l'ATP à la détection des phages de bactéries lactiques. *Science des Alim.*, **8**, n.s., 3 –45.

Shaw M., 1983. La lutte contre les bactériophages dans les fromageries. *Technique Laitière*, **976**, 51–59.

Shimizu-Kadota M., Sakurai T., Tsuchida N., 1983. Prophage origin of a virulent phage appearing in fermentations of *Lactobacillus casei* S-1. *Appl. Environ. Microbiol.*, **45**, 669–674.

Shimizu-Kadota M., Kiwaki M., Hirokawa H., Tsuchida N., 1985. ISL1: A new transposable element in *Lactobacillus casei. Mol. Gen. Genetics*, **200**, 193–198.

Shin C. and Sato Y., 1980. Lysogeny in leuconostocs. *Japanese Journal of Zootechnical Science*, **51**, 478–484.

Smaczny T. and Kramer J., 1984. Acidification disturbance in manufacture of yogurt, Bioghurt and Biogarde caused by bacteriocins and bacteriophages of *Streptococcus thermophilus. Deutsche Molkerei-Zeit.*, **105**, 614–618.

Stadhouders J., 1986. The control of cheese starter activity. *Netherlands Milk Dairy J.*, **40**, 155–173.

Stadhouders J., Bangma A., Driessen F. M., 1976. Control of starter activity and the use of starter concentrates. *Nordeuropaeisk Mejeri-Tidsskrift*, **42**, 191–198.

Stadhouders J., Leenders G. J. M., 1984. Spontaneously developed mixed-strain cheese starters. Their behaviour towards phages and their use in the Dutch cheese industry. *Netherlands Milk Dairy J.*, **38**, 157–181.

Teuber M. and Loof M., 1987. Genetic characterization of lactic streptococcal bacteriophages. *In:* Ferretti J. J., Curtiss III R. (eds). *Streptococcal genetics*, 250–258, ASM, Washington.

Terzaghi B. E. and Sandine W. E., 1975. Improved medium for lactic streptococci and their bacteriophages. *Appl. Microbiol.*, **29**, 104–126.

Valles E., 1955. Sur l'emploi de différents indicateurs d'oxydoréduction pour l'étude des bactériophages des streptocoques lactiques. *Le Lait*, **35**, 241–258.

Waes G and Bossuyt R., 1984. Impedance measurements to detect bacteriophage problems in cheddar cheesemaking. *J. Food Prot.*, **47**, 349–351.

Suggested Readings, Chapters 11–14

Bottazzi V., 1988. An introduction to rod-shaped lactic-acid bacteria. *Biochimie*, **70**, no. 3, 303–315; 68 ref.

Braun V. Jr, Hertwig S., Neve H., Geis A., Teuber M., 1989. Taxonomic differentiation of bacteriophages of *Lactococcus lactis* by electron microscopy, DNA-DNA hybridization, and protein profiles. *J. Gen. Microbiol.* Reading: Society for General Microbiology. Sept. 1989. **135** (pt. 9), 2551–2560. ill.

Champagne C. P., Girard F., Morin N., 1988. Bacteriophage development in an immobilized lactic acid bacteria system. *Biotechnol. Lett.* Kew: Science and Technology Letters. July 1988. **10** no. 7, 463–468.

Davidson B. E., Powell I. B., Hillier A. J., 1990. Temperate bacteriophages and lysogeny in lactic acid bacteria. *FEMS Microbiol. Lett. Fed. Europ. Microbiol. Soc.* Amsterdam: Elsevier Science Publishers. Sept. 1990. **87**, no. 1/2, 79–90.

Eckner K. F., 1992. Bacteriocins and food applications. *Dairy Food Environ. Sanitation*, **12**, no. 3, 204–209; 65 ref.

Jarvis A. W., 1989. Bacteriophages of lactic acid bacteria. *J. Dairy Sci.* Champaign, IL: American Dairy Science Association. Dec. 1989. **72**, no 12, 3406–3428. ill.

Lautrou Y., Rainard P., Poutrel B., Zygmunt M. S., Venien A., Dufrenoy J., 1991. Purification of the protein X of *Streptococcus agalactiae* with a monoclonal antibody. *FEMS Microbiol. Lett.*, **80**, no. 2/3, 141–146; 23 ref.

Lodics T. A. and Steenson L. R., 1990. Characterization of bacteriophages and bacteria indigenous to a mixed-strain cheese starter. *J. Dairy Sci.* Champaign, IL: American Dairy Science Association. Oct. 1990. **73**, no. 10, 2685–2696. ill.

Miller R. and Galston G. 1989. Rapid methods for the detection of yeast and lactobacillus by ATP bioluminescence. *J. Inst. Brew.* London: The Institute. Sept./Oct. 1989. **95**, no. 5, 317–319.

Mistry V. V. and Kosikowski F. V., 1986. Influence of milk ultrafiltration on bacteriophages of lactic acid bacteria. *J. Dairy Sci.* Champaign, IL: American Dairy Science Association. Oct. 1986. **69**, no. 10, 2577–2582.

Schaak M. M. and Marth E. H. 1988. Interaction between lactic acid bacteria and some foodborne pathogens: A review. *Cultured Dairy Products J.*, **23**, no. 4, 17–18, 20; 8 ref.

Shearman C., Underwood H., Jury K., Gasson M., 1989. Cloning and DNA sequence analysis of a *Lactococcus* bacteriophage lysin gene. *MGG Mol. Gen. Genst.* Berlin: Springer International. Aug. 1989. **218**, no. 2, 214–211.

Tanskanen E. I., Tullock D. L., Hillier A. J., Davidson B. E., 1990. Pulsed-field gel electrophoresis of small intestine digests of lactococcal genomic DNA, a novel method of strain identification. *Appl. Environ. Microbiol.;* **56**, no. 10, 3105–3111; 27 ref.

Toba T., Kotani T., Adachi S., 1991. Capsular polysaccharide of a slime-forming *Lactococcus lactis* ssp. *cremoris* LAPT 3001 isolated from Swedish fermented milk "langfil." *Int. J. Food Microbiol.;* **12**, no. 2/3, 167–172; 15 ref.

Valyasevi R., Sandine W. E., Geller B. L., 1990. The bacteriophage kh receptor of *Lactococcus lactis* subsp. *cremoris* KH is the rhamnose of the extracellular wall polysaccharide. *Appl. Environ. Microbiol.* Washington, D.C.: American Society for Microbiology. June 1990. **56**, no. 6, 1882–1889.

CHAPTER

15

The Yeasts

Marielle Bouix, J.-Y. Leveau

15.1 Introduction

Yeasts have been employed unknowingly by humans for thousands of years, particularly for making bread and alcoholic beverages. It took the work of Pasteur (1866–1876) to demonstrate the role of yeast in alcoholic fermentation. Since then, the ease with which they are cultured and the innocuity of many species have made them the most widely used microorganisms for the production of alcoholic beverages and bakery products and for protein or vitamin supplements in human and animal diets. Improved knowledge of yeast molecular biology combined with genetic engineering techniques has made feasible the use of yeasts for the production of human and animal proteins such as rennin, growth hormone, and hepatitis B vaccine.

Although yeasts are very valuable microorganisms of great industrial interest both for their intrinsic capabilities and as expression vectors for new capabilities, they also may become food product spoilage agents if their development is not properly monitored.

The purpose of this chapter is to examine the general characteristics of yeasts as well as the techniques by which yeasts are detected, enumerated, and identified in food and agricultural products and in fermentation worts.

GENERAL CHARACTERISTICS OF YEASTS

15.2 Definition and Classification of Yeasts

A yeast may be defined as a unicellular fungus reproducing by budding or fission. Taxononically, both morphological and physiological traits distinguish yeasts from other fungi. Fungi which go through a yeastlike phase under certain conditions (e.g., *Aureobasidium, Geotrichum, Endomyces*) are not included with the yeasts (Kreger Van Rij, 1984).

The current reference classification is that of Kreger Van Rij (1984), which has some significant changes with respect to the previous classification by Lodder (1971). New taxonomic criteria as DNA base composition, cell wall structure, and type of coenzyme Q are taken into consideration in the new version, allowing more rigorous examination.

In 1970, the Lodder classification included 39 genera and 349 species. The current classification consists of 60 genera and 500 species. A species is defined as an assemblage of clones or strains. The description of a species is therefore based on several strains. The species criterion is based on infertility versus the formation of viable spores in the case of heterothallic strains while for homothallic and asporogenous strains, the criterion is DNA-DNA hybridization.

Yeasts belong to three classes of fungi: the Ascomycetes, the Basidiomycetes, and a third class called the Deuteromycetes which regroups the imperfect forms of yeasts having affinities with Ascomycetes or Basidiomycetes (Table 15.1). Within these three classes, the yeasts are grouped in orders, families, subfamilies, and ultimately genera and species.

Table 15.2 details the class Ascomycetes. These yeasts reproduce by means of a sexual process whereby a single cell undergoes meiosis to produce an ascus. Two families are distinguished, the Spermophthoraceae with three genera and the Saccharomycetaceae with four subfamilies and 30 genera. The latter family includes many yeasts which are involved in the fermentation or spoilage of food and agricultural products.

Table 15.1 The classification of yeasts in the Eumycetes

Ascomycetes	Hemiascomycetes	Endomycetales	Spermophthoraceae Saccharomycetaceae
Basidiomycetes		Ustilagenales	Filobasidiaceae Teliospore-forming yeasts
		Tremellales	Sirobasidiaceae Tremellaceae
Deuteromycetes		Blastomycetes Sporobolomycetaceae	Cryptococcaceae

Table 15.2 Classification of ascosporogenous yeasts

Family	Subfamily	Genus
Spermophthoraceae		*Coccidiascus*
		Metschnikowia
		Nematospora
Saccharomycetaceae	Schizosaccharomycetoideae	*Schizosaccharomyces*
	Nadsonioideae	*Hanseniaspora*
		Nadsonia
		Saccharomycodes
		Wickerhamia
	Lipomycetoideae	*Lipomyces*
	Saccharomycetoideae	*Ambrosiozyma*
		Arthroascus
		Citeromyces
		Clavispora
		Cyniclomyces
		Debaryomyces
		Dekkera
		Guilliermondella
		Hansenula
		Issatchenkia
		Kluyveromyces
		Lodderomyces
		Pachysolen
		Pachyticospora
		Pichia
		Saccharomyces
		Saccharomycopsis
		Schwaniomyces
		Sporopachydermia
		Stephanoascus
		Torulaspora
		Wickerhamiella
		Wingia
		Zygosaccharomyces

Table 15.3 shows the four families and the various genera of yeasts belonging to the class Basidiomycetes. These are characterized by sexual reproduction with the formation of basidiospores supported on a basidium. Few yeasts in this class are capable of fermentation and they are rarely implicated in food product spoilage.

Table 15.4 lists the imperfect yeasts within the Cryptococcaceae family which includes 15 genera and the Sporobolomycetaceae family with 2 genera. The former includes many yeasts which may be found in food products. Moreover, the genera *Cryptococcus* and *Candida* include species which are pathogenic to humans.

Table 15.3 Classification of basidiosporogenous yeasts

Filobasidiaceae	*Chionosphaera*
	Filobasidiella
	Filobasidium
Teliospore-forming yeasts	*Leucosporidium*
	Rhodosporidium
	Sporidiobolus
Sirobasidiaceae	*Fibulobasidium*
	Sirobasidium
Tremellaceae	*Holtermannia*
	Tremella

Table 15.4 Classification of imperfect yeasts

Cryptococcaceae	*Aciculoconidium*
	Brettanomyces
	Candida
	Cryptococcus
	Kloerckera
	Malassezia
	Oosporidium
	Phaffia
	Rhodotorula
	Sarcinosporon
	Schizoblastosporion
	Sterigmatomyces
	Sympodiomyces
	Trichosporon
	Trigonopsis
Sporobolomycetaceae	*Bullera*
	Sporobolomyces

15.3 Biological Characteristics

Cells tend to be ovoid and vary in size from a few microns up to 25 or 30 microns. Some may form cellular associations or appear in filamentous form at certain stages of their life cycle.

Yeasts distinguish themselves from bacteria primarily by their eukaryotic cell structure. The cytoplasm of yeast cells contains the usual organelles of higher nonphotosynthetic vegetative cells (i.e., mitochondria, Golgi apparatus, endoplasmic reticulum, ribosomes, vacuoles, storage bodies). The nucleus consists of a nuclear membrane and contains chromosomes. The cells are protected by a rigid cell wall, consisting essentially of polysaccharide (i.e., glucans or glucans and mannans). All yeasts are nonmotile.

Yeasts multiply by an asexual process, that is, mitotically. Some also manifest a sexual process with alternating diploid and haploid phases. In order to grow, yeasts require organic carbon compounds (which are their carbon and energy sources at the same time), reduced nitrogen in the form of ammonium ion (a few yeasts are able to use oxidized nitrogen [nitrate] or organic nitrogen) for protein and nucleic acid synthesis, and various mineral elements. Vitamin requirements vary depending on the species.

All yeasts are capable of breaking down glucose, fructose, and mannose in the presence of oxygen by oxidative metabolism leading to the formation of carbon dioxide and water:

$$C_6H_{12}O_6 + 6O_2 \rightarrow 6CO_2 + 6H_2O$$

The energy yield of this reaction is very high and allows great cellular multiplication. In addition to simple sugars, carbon compounds such as other carbohydrates (i.e., mono-, di-, and trisaccharides) are utilized by some species, polysaccharides such as inulin and starch are utilized by a few species, as are alcohols, acids, and alkanes. In general, however, yeasts have much less macromolecular hydrolytic capability than molds.

Aside from oxidative metabolism, some yeasts are able to break down carbohydrates by a fermentative metabolism without the requirement for oxygen, leading to the formation of carbon dioxide and primarily ethanol:

$$C_6H_{12}O_6 \rightarrow 2C_2H_5OH + 2CO_2$$

In addition to these primary compounds, higher alcohols as well as aldehydes, esters, acids, and so on are formed in small quantities, which all make important qualitative contributions to the bouquet of fermented beverages. The cellular multiplication resulting from this type of metabolism is less copious than for oxidative metabolism.

15.4 Physiological Characteristics

15.4.1 Temperature

Most yeasts grow well at temperatures between 25° and 30°C. The optimal growth temperatures to which yeasts are exposed in their natural habitats are not necessarily in this range. In fact, like other microorganisms, yeasts may be classed as psychrophiles, mesophiles, or thermophiles, depending on their growth temperature limits:

- Psychrophilic yeasts are those whose maximal growth temperature is 20°C or lower. The minimum growth temperature depends on the medium but may be around 5°C. These yeasts are limited in number but nine have been found in water or soils in the Antarctic (Table 15.5).

Table 15.5 Obligate psychrophilic yeasts according to Watson (1987)

Cryptococcus vishniacii
Leucosporidium antarcticum
Leucosporidium trigidum
Leucosporidium gelidum
Leucosporidium nivalis
Leucosporidium scottii
Leucosporidium stokesii
Torulopsis austramarina
Torulopsis psychrophila

Table 15.6 Obligate thermophilic yeasts according to Watson (1987)

Candida slooffii
Cyniclomyces guttulata
Saccharomyces telluris
Torulopsis bovina
Torulopsis pintolopesii

- Thermophilic yeasts are those having a minimum growth temperature of 20°C and the ability to develop at a maximum temperature of 48°–50°C. Only five yeasts are listed in this category (Table 15.6). They usually live in the digestive tracts of animals.
- Mesophilic yeasts include all other species which are capable of developing at temperatures between 0° and 48°C.

In general, yeasts are not very heat-resistant, their destruction beginning at 52°C. Cells in the exponential growth phase are more sensitive than those in the stationary phase (Parry et al., 1976). Thermal resistance also depends on growth temperature. Cells which have grown at 36°C throughout their exponential phase are more resistant than those grown at 23°C (Walton and Pringle, 1980) due to increased resistance to thermal denaturation of the cytoplasmic membrane. Yeasts are sensitive to freezing and lyophilization, with large variations between genera and species. Lyophilization may be used to preserve yeasts (Russell and Stewart, 1980) as long as temperature, freezing rate, and medium composition are carefully considered. Culture age is also important, exponential phase cells being less resistant than stationary phase cells.

15.4.2 Osmotic Pressure and Water Activity

The effect of osmotic pressure varies from one strain to another. Most strains are unable to develop at water activities lower than 0.90 while some may tolerate higher osmotic pressures corresponding to water activities around 0.60, albeit at a reduced metabolic rate (Leveau and Bouix, 1979). This is the case for yeasts of the genus *Zygosaccharomyces* (formerly *Saccharomyces* in the Lodder classification) as well as *Debaryomyces hansenii, Hansenula anomala, Pichia ohmeri, Schizosaccharomyces pombe, Torulopsis candida,* and *T. lactis condensi.* These yeasts are called *xerotolerant.*

In any case, the influence of osmotic pressure is related to the nature of the components of the medium, in particular the salts which have an inhibitory effect over and above that of the osmotic pressure. For example, *Zygosaccharomyces rouxii,* the most osmotolerant yeast known, develops in sugar-containing media at a water activity of 0.60 but can only tolerate water activities of 0.85 and higher in salty media.

The mechanism of low water activity resistance in yeasts is the intracellular accumulation of polyols which offsets the difference in osmotic pressure between the cell and the medium (Rose, 1987). *Debaryomyces hansenii* is thus able to develop in a medium containing 3M sodium chloride by accumulating glycerol within its cytoplasm to the exclusion of the salt. For an increase in the NaCl concentration of the medium from 0.004 to 2.7 M, the intracellular glycerol concentration increases from 0.2 to 2.6 M (Gustafsson and Norkraus, 1976).

In alcoholic fermentation, alcohol concentration and osmotic pressure are related, the alcohol excretion rate being decreased by high osmotic pressures which causes intracellular ethanol to accumulate and reach toxic levels (Leveau and Bouix, 1986).

15.4.3 Oxygen

All yeasts are capable of developing in the presence of oxygen. There are no strictly anaerobic yeasts. Some are strictly aerobic, particularly those of the genera *Rhodotorula, Rhodosporidium, Lipomyces, Saccharomycopsis, Cryptococcus,* and *Sporobolomyces* as well as a few species of the genera *Hansenula, Pichia, Torulopsis,* and *Debaromyces.* All other yeasts are facultative anaerobic, these including the following types:

- yeasts which prefer fermentative metabolism even in the presence of oxygen, i.e., *Saccharomyces, Schizosaccharomyces,* and *Brettanomyces* plus a few species of *Torulopsis;*
- yeasts which prefer respiratory metabolism if oxygen is present, i.e., *Candida, Kluyveromyces,* most *Pichia* and *Hansenula* and a few *Torulopsis.*

Oxygen is also required for the synthesis of sterols and of nicotinic acid. If these are supplied by the medium, the yeasts can develop under strictly anaerobic conditions. Otherwise 1.5 μg of oxygen/g of biomass are required in order to carry out sterol synthesis for survival.

15.4.4 Chemical Agents

15.4.4.1 Acids

The cellular envelope is impermeable to H^+ and OH^- ions. Yeasts therefore tolerate very large pH ranges, in theory from 2.4 to 8.6. Between these values, the intracellular pH varies only from 5.8 to 6.8.

Organic acids in their undissociated form have an inhibitory effect, being able to penetrate into the cell even though H^+ ions alone do not. The sensitivity of a yeast to an organic acid therefore depends on the pH, regardless of whether or not the acid is dissociated, and on the ability of the yeast to metabolize or eliminate the acid should it penetrate into the cell. Thus, sorbic and propionic acids are more inhibitory than acetic, lactic, and citric acids.

The longer chain acids, C_6, C_8, C_{10}, have an inhibitory effect on *Saccharomyces* (Geneix, 1984). Benzoic acid and its derivatives have an inhibitory effect similar to that of the short chain aliphatic acids.

The precise action of the acids is poorly understood. Cell multiplication appears to stop while the other activities continue. It has been shown, for example, that *Saccharomyces cerevisiae* in the presence of 50–100 μg/ml of sorbic acid (pH = 4) stops growing but continues to respire and ferment sugars (Reinhard and Radler, 1981).

15.4.4.2 Ethanol

Yeasts do not all have the same sensitivity to alcohol. The most resistant is the genus *Saccharomyces* used in alcoholic fermentation processes for the production of beverages or of industrial ethanol. Ethanol tolerance depends on cytoplasmic membrane composition, particularly on the lipid composition, this in turn depending on the culture medium and temperature (Leveau and Bouix, 1986).

15.4.4.3 Sulfite

Sulfur dioxide has a more pronounced inhibitory effect on bacteria than on yeast, although sensitivity does vary among the yeasts. For example, *Zygosaccharomyces bailli* and *Brettanomyces* are more resistant than *Kloerckera apiculata* and *Pichia membranaefaciens* (Hammond and Carr cited in Rose, 1987). The action of sulfate depends on the pH. The lower the pH, the more the $SO_2 \leftrightarrow HSO_3^- \leftrightarrow SO_3^{2-}$ equilibrium is displaced towards SO_2, which is the most antimicrobially active of the three chemical species.

Schimz and Holzer (1979) have shown that a concentration of 1 mM at a pH < 4 brings about a decrease in cellular ATP in *Saccharomyces cerevisiae* leading to cell death if the exposure time is greater than 1 hour.

15.4.4.4 Antibiotics

Cycloheximide (actidione)

The sensitivity of yeasts to cycloheximide is variable and three groups may be distinguished on this basis (Rose, 1987):

- yeasts inhibited by 1 μg/ml; *Saccharomyces* is in this group;
- yeasts inhibited by 25 μg/ml; this is the case for *Schizosaccharomyces*;
- yeasts which tolerate 1 mg/ml; these are *Zygosaccharomyces*, *Hanseniaspora*, and *Kloeckera* as well as *Kluyveromyces marxianus.*

Nystatin and amphotericin are used to treat *Candida* infections. Chloramphenicol inhibits the synthesis of mitochondrial proteins without affecting the synthesis of cytoplasmic proteins. In the presence of chloramphenicol, only those yeasts capable of fermenting the fermentable substrates that may be present are able to develop.

15.5 Genetic Characteristics

15.5.1 Genetic Material

15.5.1.1 Chromosomes

Yeasts are eukaryotic organisms and as such possess a true nucleus with linear chromosomes. The genus *Saccharomyces* has 17 chromosomes or 17 chromosomal pairs depending on whether the cell is in the haploid or diploid form.

From a structural point of view, the chromosomes possess autonomously replicating or ARS sequences occurring approximately every 20,000 base pairs, a centromer ensuring equal distribution of the chromosomes during mitosis by hooking onto the fibers of the achromatic spindle and telomers which are specific structures at each end of the chromosome. All of these sequences have been studied and are used to construct artificial chromosomes for the genetic transformation of yeasts. Yeasts have continuous structural genes like those of bacteria and discontinuous genes made up of introns and exons like those of higher organisms. Moreover, the regulating genes are specific to yeasts.

15.5.1.2 Plasmids

Aside from chromosomes, the nucleus also contains small circular DNA molecules of about 6,000 base pairs, the 2 μ plasmids, present in quantities of 50–100 copies/cell. These plasmids autoreplicate independently of the chromosomes and are autotransferable without affecting cell viability. They carry the genetic information for a few nonessential traits and are of considerable importance in genetic engineering manipulations.

15.5.1.3 Mitochondrial DNA

Each mitochrondrion contains several circular DNA molecules in the 75,000 base pair size range. This DNA carries the information for the respiratory chain enzymes and therefore has an essential role in the energetic metabolism of the cells. Poor distribution of the mitochondria between mother and daughter cell occasionally arises during mitosis giving rise to cells without mitochondria. Mutations can also change the DNA causing the functional loss of respiratory capability. These cells can only ferment (as is the case for the fermenting yeasts). Morphologically, colonies appear smaller on solid media and are called *small* mutants.

15.5.1.4 Killer Yeasts

Some strains of *Saccharomyces* have the ability to secrete a protein which is toxic to other yeasts. These killer yeasts contain viral RNA particles in their cytoplasm, which code for an exotoxin and for a resistance protein which lodges in the cell wall.

15.5.2 Reproductive Cycle

The haploid strain of *S. cerevisiae* belong to two sexual groups called *a* and *alpha* (α): They are capable of vegetative multiplication by budding as well as fusion under certain conditions, resulting in a diploid cell which may also multiply vegetatively. These cells may sporulate by undergoing meiosis yielding four haploid spores.

Industrial strains are often polyploid (3, 4, 5n chromosomes) and therefore possess several genes for a given trait. They are genetically more stable because of the difficulty for mutations to be expressed but, on the other hand, make crossings impossible since they are usually incapable of sporulating. Genetic improvement of these strains involves protoplast fusion or cytoduction techniques or transformation by genetic engineering.

15.6 Ecological Characteristics and Consequences of Yeast Development

In nature, yeasts occur principally on fruits and vegetables rich in readily assimilated sugars, developing especially around lesions and scars exuding sugar-containing juice. Fresh, unblemished fruits have a much lower level of yeast contamination than damaged fruits.

Generally speaking, plant-based food products (fruits, fruit juices, vegetables, dried fruits, syrups, jams, etc.) as well as sweetened products (candies, biscuits, honey, etc.) are particularly vulnerable to spoilage by yeasts; however, it is quite possible for yeasts to spoil other food products as well.

Since the yeasts of food products are nonpathogenic, they will not cause food poisoning but their development may spoil the market qualities of products by formation of turbidity (yeast cells themselves), of abnormal odors or tastes (ethanol, pH changes), or by swelling of products and/or of their packaging (CO_2). Table 15.7 gives a general indication of the sorts of yeasts which may be found in some products.

In certain types of manufacturing, yeast development is desired in order to process a starting material. This is the case in bread making, in which yeast is used to raise dough; in the production of fermented beverages (beer, wine, cider, and others depending on the starting material and the country); in the

Table 15.7 The major spoilage yeasts in food products

Products with high sugar concentration: 40–70% (honey, jam, syrup, candy, etc.)
Zygosaccharomyces rouxii
Zygosaccharomyces bisporus
Zygosaccharomyces bailli
Sweetened drinks: sugar concentration around 15%
Torulopsis stellata
Torulopsis cantarelli
Pichia kluyveri
Pichia fermentans
Saccharomyces ssp.
Candida (a few species)
Brettanomyces (a few species)
Wine
Saccharomyces cerevisiae (ex *bayanus*)
Zygosaccharomyces bisporus
Brettanomyces intermedius
Pichia membranaefaciens
Beer
Brettanomyces
Saccharomyces cerevisiae (ex *diastaticus*)
Dairy products
Kluyveromyces fragilis
Candida
Torulopsis
Rhodootorula (cream)
Pickling, brine solutions
Debaryomyces hansenii

production of industrial alcohol from the juice of sugar beets or sugarcane or from molasses or grain hydrolysates; or in the production of yeast itself as a source of protein and B vitamins from various industrial by-products such as whey, ruined batches of wines, and so on. Yeasts also participate in the ripening of some cheeses. In addition, as mentioned earlier, certain yeasts are now used to produce enzymes or vaccines.

In industries which utilize the fermenting capabilities of yeasts, the strains employed are being increasingly selected as a function of the technical objectives sought: rapid fermentation in bread making, aromatic characteristics in beer and wine making, ethanol tolerance and good kinetics in industrial alcohol production, and so on. Control of these processes must constantly improve to meet demands, and in-depth knowledge of the selected strains is indispensable. Furthermore, from a hygienic point of view, yeast analysis is not required for food products. From a technological point of view, however, they are capable of causing process modifications significant enough to render the products uncommercializable. For the manufacturer, producing goods which are susceptible to spoilage due to yeast development during storage, quantitative and even qualitative analyses of the yeasts in the finished products may be necessary. However, particular attention must be given to starting materials in order to be able to adjust the treatment parameters as a function of the initial numbers of yeasts and the potential danger they represent. Analysis of the products during the manufacturing process is also required to monitor the effectiveness of any adjustments implemented. Familiarity with the techniques for the examination of yeasts is thus necessary whether the yeasts are being used as a processing agent or whether it is a matter of controlling product microbiological quality in anticipation of prolonged storage.

YEAST EXAMINATION TECHNIQUES

15.7 Detection and Culture Techniques

15.7.1 Enumeration Techniques

Enumerations can be done using the various techniques already described, namely

- direct enumeration by cell counter;
- tedious enumeration using Petri plates, with either pour plate or surface plating;
- approximate enumeration in liquid media (with inverted gas-collecting tubes);
- filtering onto membranes in the case of liquid or soluble products with low level contaminations (less than one yeast cell per ml).

Products containing yeasts may also contain molds and bacteria. Media used for counting may be made selective against bacteria by acidifying or by adding antibiotics. Selecting against molds is more difficult but mold colonies are unlikely to be mistaken for yeast colonies. A single procedure thus allows the enumeration of both yeasts and molds.

Any glucose-containing medium is suitable for the culture and counting of yeasts. The most commonly used (i.e., in France) are the following:

Sabouraud medium

Peptone Chapoteaut	10 g
Glucose monohydrate	20 g
Agar	15 g
Distilled water	1,000 ml

pH is 6–6.3. Sterilize by autoclaving for 20 minutes at 121°C.

This medium may be made selective by adding chloramphenicol (0.5 mg/ml) or gentamycin (0.04 mg/ml). Chloramphenicol is not very effective against gram-negative bacilli. Gentamycin, while much more selective, still allows the growth of some enterobacteria.

MYGP medium

Malt extract	3 g
Yeast extract	3 g
Glucose	10 g
Peptone	5 g
Agar	15 g
Distilled water	1,000 ml

pH is 5–6. Sterilization by autoclaving for 15 minutes at 121°C.

Malt extract medium

Malt extract	20 g
Agar	15 g
Distilled water	1,000 ml

Sterilization by autoclaving for 20 minutes at 115°C.

Like the Sabouraud medium, these two are suitable for the culture of yeasts but are not selective.

Wort media: beer wort or grape must

These media are used principally in brewing and enology. They are worts that have been deproteinized by heating to 120°C followed by filtration, mixed with agar, and poured.

Submerged inoculation (pour plate) is employed with all of these media when using Petri plates. They may also be used in broth form. Incubation is generally at 28°C for 24–72 hours.

The most commonly used media for the enumeration of yeasts in food products are oxytetracycline glucose agar (OGA) and potato dextrose agar (PDA) acidified to pH 4.

OGA medium

Yeast extract	5 g
Glucose	20 g
Agar	16 g
Distilled water	1,000 ml

pH is 6.8–7. Sterilize by autoclaving for 20 minutes at 115°C. Add to the cooled medium (50°C) an oxytetracycline (terramycin) solution such that the antibiotic concentration in the medium is 0.10 mg/ml. This medium is surface inoculated and must be incubated for 2–5 days at 20°–25°C. It supports the growth of all yeasts (and molds) found in food products while totally inhibiting the development of bacteria. Growth is usually rapid and reading can be done after 36–48 hours of incubation.

PDA medium

Potato extract	4 g
Glucose	20 g
Distilled water	1,000 ml

Sterilize by autoclaving for 15 minutes at 121°C. After cooling to 50°C, acidify to pH 3.5 or 4 by adding dilute hydrochloric acid or phosphoric acid (determine the amount to add by testing with a small volume of medium). This medium is for submerged inoculation and should be incubated for 3–5 days at 20°–25°C. Growth is slower than on OGA. For enumeration by filtration, the membranes may be incubated either on one of these agar media or on a pad soaked with a yeast-mold liquid medium.

Enumeration of Xerotolerant Yeasts

The media just described are for total yeast counts. In some cases it may be necessary to analyze specifically for xerotolerant yeasts. The following medium is recommended for this purpose:

Whalley's medium

Yeast extract	5 g
Peptone	2 g
Soluble starch	2 g
Glycerol	1 g
Ammonium chloride	2 g
Glucose	20 g
Sucrose	400 g
Agar	15 g
Water	500 ml

Sterilize by autoclaving for 20 minutes at 121°C. This medium has a water activity of 0.91, which is not always selective enough, the genus *Saccharomyces* in particular still being able to develop on this medium. A modification has thus been proposed which lowers the a_w to 0.86, making the medium more selective:

Leveau and Bouix modified Whalley medium (1979):

Yeast extract	5 g
Peptone	2 g
Soluble starch	2 g
Ammonium chloride	2 g
Glucose	20 g
Sucrose	20 g
Fructose	400 g
Glycerol	95 ml
Distilled water	405 ml

This medium allows the growth of xerotolerant yeasts only, particularly of *Zygosaccharomyces*.

15.7.2 Isolation and Detection

The isolation of total yeasts or of yeasts among a bacterial flora may be performed by using some of the previously described media, in particular Sabouraud, malt extract, or wort media.

In fermentation industries using pure cultures of a highly selected yeast, it is essential to be able to test the purity of the starter and therefore to be able to search for various types of contaminating yeasts within the cultured yeast population. Specific culture media for the detection of wild yeast in brewing have been proposed: actidione medium (Harris and Watson, 1968) and lysine medium (Morris and Eddy, 1957) inhibit the growth of yeasts of the genus *Saccharomyces* while brewery-contaminating yeasts such as *Brettanomyces*, *Candida*, *Kloeckera*, *Pichia*, and *Rhodotorula* develop normally. These media therefore detect contaminants other than *Saccharomyces*. Another medium developed by Taylor and Marsch (1984), is essentially an MYGP medium with 0.15 mg/ml of copper sulfate added allowing the detection of wild strains of *Saccharomyces* while the *S. cerevisiae* culture strain does not develop. Other media-containing antibiotic mixtures have been developed for the selective search for yeasts (Beech et al., 1980).

Isolation methods on selective media must be tested beforehand in each laboratory in order to produce a standard of responses as a function of the particular conditions of use. These methods have the disadvantage of requiring incubation time and therefore of yielding the response too late to be very useful.

In certain cases, immunofluorescence, which is a rapid technique, may be used for the ready detection of contaminants. A smear of the cultured yeast may be made fluorescent using a specific serum followed by a fluorescein isothiocyanate marked serum (the indirect immunofluorescence method) while coloring the contaminants nonspecifically using another fluorochrome, Evans blue. Contaminants thus appear red while the cultured yeast cells are fluorescent green when the preparation is examined by FITC fluorescent microscopy. The challenge of this technique is in the preparation of a serum specific to brewers' yeast. When a relatively specific serum is obtainable, a rapid technique for the evaluation of the purity of the starter culture is feasible (Bouix and Leveau, 1980).

Radiometric techniques may provide rapid detection (within 10 hours) of small numbers of yeasts in large volumes, for example, in the case of testing the effectiveness of filtration or pasteurization of beer (Bourgeois et al., 1973; Mafart et al., 1976).

More recently, flow cytometry has been applied to the analysis of microbial populations and particularly of contaminants in food products. The advantage of this technique is that it is able to analyze cells one by one using various marking systems, that is, nucleic acid stains, monoclonal antibodies, and, more recently, viability stains. The latter (ChemChromes Chemunex S.A.) provide an instant measurement of the metabolic activity of each cell by fluorescent intensity, the very active cells being very fluorescent and the dead cells appearing black (Fuhrmann et al., 1989). Variations in fluorescence reflect heterogeneity in the metabolic activity of a microbial population. This new system (ChemFlow) allows the simultaneous enumeration of microorganisms, the evaluation of their metabolic activity, and, in conjunction with monoclonal antibodies, the identification of those of interest. The results are displayed as a histogram. The speed of this analysis (15–90 minutes) and the simplification of the current devices makes this technique particularly well adapted for quality control applications in food industries, especially in the search for yeasts among other microorganisms (e.g., in yogurt) or for the evaluation of their physiological status during fermentation processes (Gatley, 1989).

In the case of yeasts selected for wine making, some have been genetically marked by imparting to them antibiotic resistance traits carried by mitochondrial DNA and acquired by mutation. It is therefore possible to detect these yeasts among populations of wild yeasts by isolation or enumeration on media such as those described earlier with the corresponding antibiotic added to them (Loiseau et al., 1987).

15.8 Identification Techniques

Identification must be done using a culture of a pure strain isolated on agar medium for yeasts. The identification criteria are of three types: (1) vegetative reproduction characteristics; (2) sexual characteristics; and (3) biochemical and physiological characteristics.

15.8.1 Vegetative Reproduction Characteristics

15.8.1.1 Mode of Vegetative Reproduction

Vegetative reproduction may be by budding or by simple fission or sometimes by a combination of these. Bud formation begins as an evagination of the cell, followed by swelling of the bud until its separation from the mother cell. Budding may be polar (at one end of the cell only), bipolar (at both ends), or multilateral when the buds are positioned all around the surface of the mother cell (Figure 15.1).

Reproduction by fission involves splitting the cell in two by forming a lateral division plane inside the cell, as is the case for the genus *Schizosaccharomyces.* Finally, the genus *Sterigmatomyces* forms a conidium on a stemlike tubular protuberance or sort of conidiophore (Figure 15.1).

15.8.1.2 Morphology of Vegetative Cells

The morphology of vegetative cells is examined microscopically between slide and cover-slip using cultures in liquid media or on solid media. First, the cultural characteristics are observed. This is a matter of examining the appearance of cultures in liquid media and on solid media after incubation at 28°C for at least 3 days.

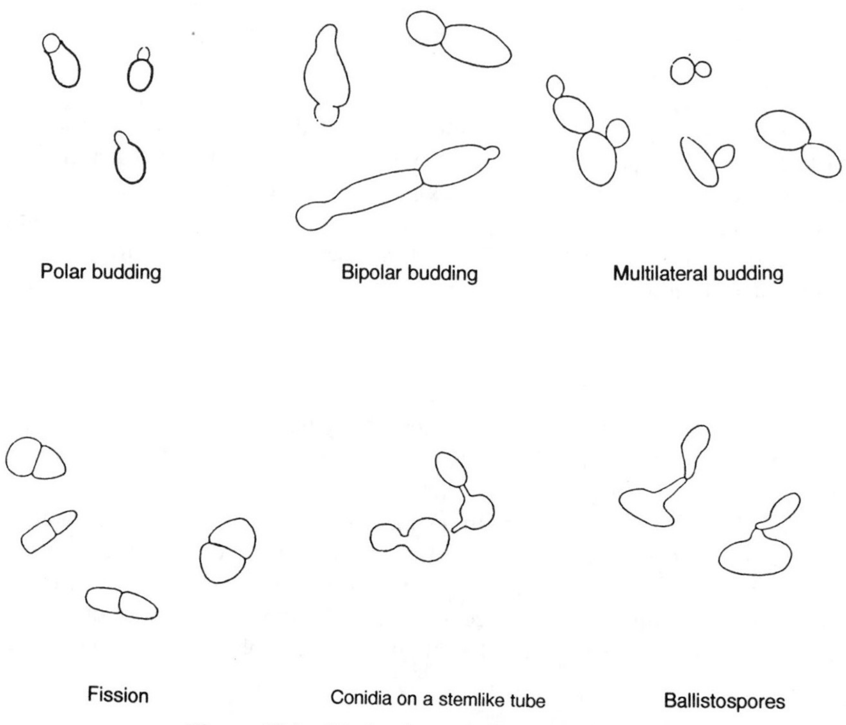

Figure 15.1 Mode of vegetative reproduction.

Growth in Liquid Media

Malt extract medium (2%) and glucose-yeast extract-peptone medium are suitable for culturing.

Glucose	20 g
Peptone	10 g
Yeast extract	5 g
Distilled water	1,000 ml

Sterilize by autoclaving for 15 minutes at 121°C.

The appearance of the cultures is noted:

- formation of sediment at the bottom of the tube and its texture (fine, coarse);
- skin at the liquid surface;
- gas formation.

Growth on Solid Media

Culturing is done using the same media as described earlier, with 2% agar and poured into Petri plates. The plates are inoculated by streaking. Visual observation is done to assess the following:

- size of the colonies;
- their shape (smooth or rough outline, convex or concave);
- their finish (dull or shiny);
- their pigmentation.

Cells may be spherical, ovoid, elongated, ellipsoidal, apiculate (lemon-shaped), bottle-shaped, triangular, and so on. Sometimes the shape is a characteristic of the genus (e.g., the bottle shape belonging to the genus *Malassezia* or the triangular shape of the genus *Trigonopsis*). More often, however, the size and shape of the cells of strains within a single species are subject to variation (e.g., *S. cerevisiae* (Figure 15.2).

15.8.1.3 Formation of Mycelium

The tendency of a culture to form filaments may be observed on a microscope slide. Melted potato dextrose agar is pipetted onto a sterilized slide placed in a Petri plate and left to solidify. The yeast to be examined is streaked onto the slide and the plate is incubated with a small amount of sterile water to prevent dessication of the medium. The growth is observed around the edges after 3–5 days at a magnification of 100–400X. Formation of filaments, the mycelial type (pseudomycelium or true mycelium), its abundance and degree of branching are all noted. A few examples of different morphologies are shown in Figure 15.2.

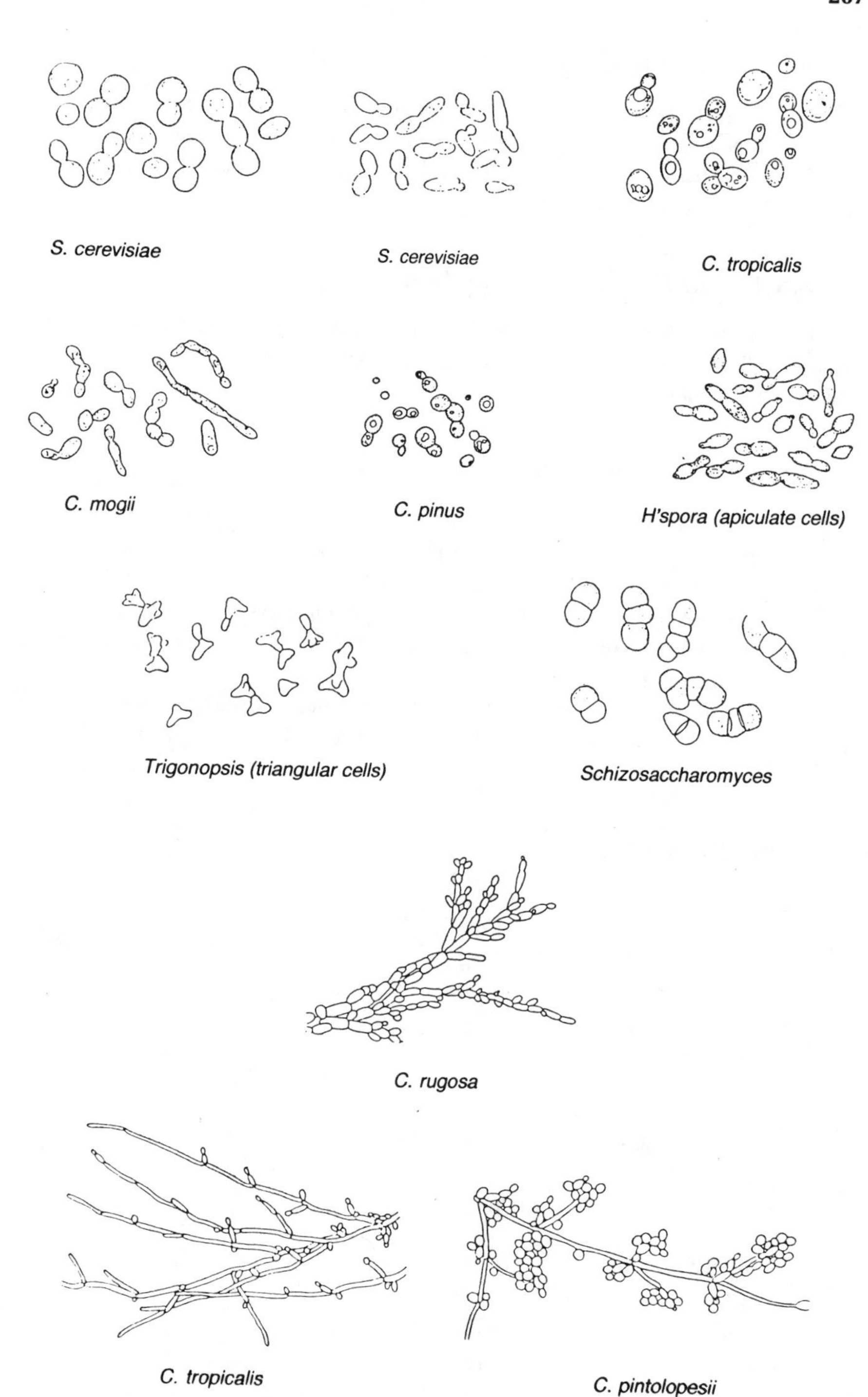

Figure 15.2 Morphology of yeast cells and mycelium.

15.8.1.4 Chlamydospore Formation

Chlamydospore formation is a characteristic of *Candida albicans* and *Metschnikowia* and may be observed in old cultures of *Trichosporon cutaneum* or *Cryptococcus*. These are thick-walled, lipid-rich asexual spores well adapated for periods of dormancy.

15.8.1.5 Ballistospore Formation

Ballistospores are spores produced at the end of protuberances from the mother cell and which are launched into the air at maturity. They characterize the genus *Sporobolomyces*.

15.8.2 Sexual Characteristics

The ascomycetous yeasts show sexual reproduction capabilities by forming an ascus containing ascospores resulting from meiosis. The yeasts being homothallic or heterothallic, stabilized in haplophase or/and diplophase, the asci may be formed by different processes. Regardless of the process, it is necessary to find asci and ascospores in order to identify a yeast, if it is capable of sexual reproduction.

Since not all yeasts have the same requirements, several sporulation media must be used simultaneously in order to make sure that sporulation will occur if the yeast is ascosporogenous. The most commonly used media are the following:

MacClary's medium

Glucose	1 g
KCl	1.8 g
Yeast extract	2.5 g
Sodium acetate	8.2 g
Agar	15 g
Distilled water	1,000 ml

Fowells' medium

Sodium acetate	4 g
Agar	15 g
Distilled water	1,000 ml

Gorodkowa's medium

Glucose	1 g
Peptone	10 g
NaCl	5 g
Agar	25 g
Distilled water	1,000 ml

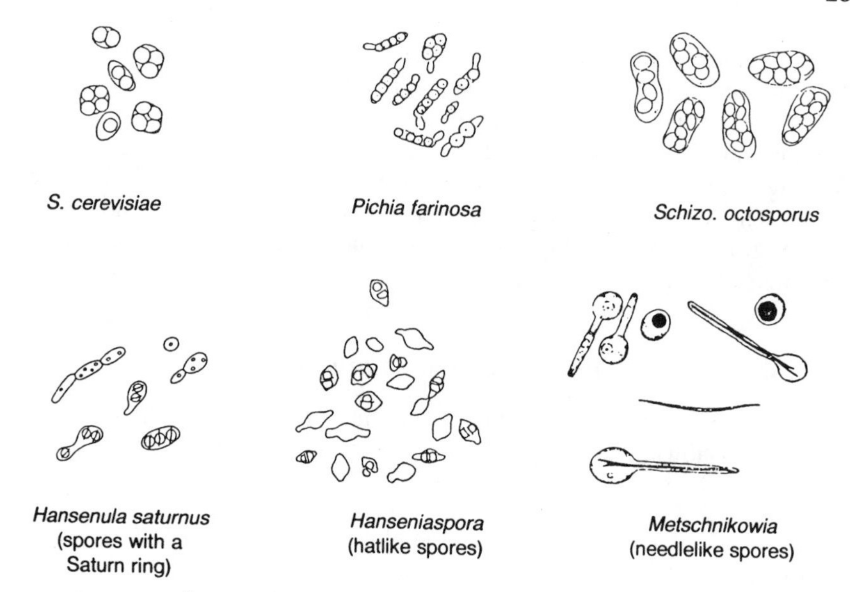

Figure 15.3 Asci and ascospores.

All three media are sterilized by autoclaving for 15 minutes at 115°C.

Other media may be used such as V8 or YM agar. All of these media are listed in the Kreger Van Rij (1984) manual.

Media poured into tubes as slants are surface inoculated using a 24–36 hour exponential phase liquid culture (e.g., malt extract) and incubated for 8 days to 1 month at 20°C. These cultures are monitored by regular microscopic observation between slide and cover-slip. In the case of ascosporogenous yeasts, the shape of the ascus and the ease with which it breaks are noted as well as the number, shape, and position of the ascospores inside the ascus (Figure 15.3).

The basidiomycetous yeasts show sexual reproduction capabilities by forming a basidium. These yeasts are rarely found in food products.

15.8.3 Biochemical and Physiological Characteristics

While morphological and sexual characteristics are generally sufficient for the identification of a genus, biochemical characteristics must be observed in order to define the species of a yeast. This refers to the use of carbon compounds, nitrogen compounds, growth at 37°C, growth in vitamin-free media, and cycloheximide resistance as principal traits.

15.8.3.1 Sugar Fermentation

The sugars usually tested in fermentation are glucose, galactose, sucrose, maltose, lactose, and raffinose. Others, such as trehalose, melibiose, and polysaccharides (i.e., inulin or starch) may also be tested.

The basal solution used for sugar fermentation is yeast water medium (0.5% yeast extract in water) in test tubes with inverted Durham tubes. Prior to inoculation, sterile solutions of the sugars to be tested are added to these tubes to a final sugar concentration of 2% except for raffinose which must be at 4%. The sugar solutions are sterilized by membrane filtration or flash pasteurization. The tubes are inoculated with a drop of test yeast suspension and the cultures are incubated at 25°–28°C for 48 hours to 3 weeks. A sugar has been fermented when gas appears in the inverted tube.

15.8.3.2 Assimilation of Carbon Compounds

For the assimilation tests, 18 compounds may be used in a first step from among the following compounds:

- Hexoses
  glucose, galactose, sorbose, methyl-D-glucoside, salicine, arbutine
- Pentoses
  D-xylose, arabinose, D and L ribose, L-rhamnose
- Disaccharides
  sucrose, lactose, maltose, melibiose, trehalose, cellobiose
- Trisaccharides
  raffinose, melezitose
- Polysaccharides
  starch, inulin
- Alcohols
  methanol, ethanol, glycerol, ribitol, galactitol, erythritol, mannitol, sorbitol, inositol
- Organic acids
  lactic, succinic, citric, pyruvic, malic, glucuronic, and 2-keto-gluconic acids
- Decane, hexadecane

The auxanographic method is used for the simple sugars, on Petri plates with yeast nitrogen base (YNB) (Difco) agar (2%). One ml of yeast suspension (10^6–10^7 cells/ml) from a 48-hour-old colony on agar medium is placed in a Petri plate to which melted YNB agar at 48°C is then added with mixing to ensure a uniform suspension. When the medium has solidified, the sugars to be tested are deposited on the agar surface either in the form of crystals or sugar-loaded paper disks such as are available from various suppliers including Difco. The plates are incubated at 25°C and may be checked after 24–48 hours. Assimilation is indicated by the development of colonies in the diffusion zone of the sugar.

The auxanographic method has the advantage of quick execution. For slowly assimilated substrates, however, Petri plates are not usable due to dessication. A liquid method is preferred in such cases, usable regardless of the carbon source being tested. Liquid YNB is distributed in 5 ml portions per tube and carbon source is added to a final concentration of 0.5%. The tubes are then inoculated with a drop of test yeast suspension and incubated at 25°–28°C on a shaker to facilitate oxygenation. After a few days to a few weeks, clouding may develop indicating assimilation of the compound.

The arbutine hydrolysis test is done using a 0.5% arbutine agar (2%) medium poured as slants. Hydrolysis of arbutine appears as a brown color after 2–7 days at 25°C.

15.8.3.3 Assimilation of Nitrogen Compounds

Generally, two nitrogen sources are tested: potassium nitrate and ethylamine chlorhydrate. This test is carried out according to the auxanographic method with yeast carbon base (Difco) or a nitrogen-free synthetic medium:

Glucose	20 g
KH_2PO_4	1 g
$MgSO_4$	0.5 g
Agar	20 g
Distilled water	1,000 ml

Once the cells are dispersed in the Petri plate, the nitrogen sources are deposited on the solidified agar surface in the form of a few crystals. A control using ammonium sulfate is also done. Other nitrogen sources (e.g., creatine, amino acids) may also be tested using this method.

15.8.3.4 Growth in Vitamin-free Medium

The determination of vitamin requirements is done in vitamin-free yeast extract base (Difco) which is obviously not supposed to contain vitamins. The medium is dispensed in 5 ml portions per tube, lightly inoculated, and incubated at 25°C. Since growth may be observed due to the transfer of small amounts of vitamin with the inoculum, it is necessary to transfer the culture at least once to an identical tube. If growth appears in the second tube or in a third tube, it may be concluded that the yeast does not require any vitamins for growth.

During routine identification, all tests are not carried out simultaneously but rather stepwise. In the first phase, the following are inoculated:

- six sugar fermentation tests: glucose, galactose, sucrose, maltose, lactose, raffinose;
- nine sugar assimilation tests: those for fermentation plus trehalose, melibiose, and cellobiose;
- one auxanogram plate for nitrogen sources.

15.8.3.5 Rapid Technique: Combined API Tests

The identification of yeasts thus requires a large number of tests and a considerable amount of work. API 20C galleries, originally designed for the identification of yeasts in hospitals, may be used to perform the biochemical tests for the identification of yeasts in food products. This gallery includes eight sugar fermentation tests, ten sugar assimilation tests, and a test for the detection of cycloheximide resistance (actidione).

Cuinier and Leveau (1979) have critically examined this technique applied to the identification of various genera and species of yeasts. They have shown that, compared with the conventional method, it is time saving, decreases the risk of contamination by molds, and often yields results more rapidly. In the case of slow or weak metabolism, however, the gallery may give a negative result. For delicate tests or ambiguous responses, it is therefore recommended that the conventional method be retained. The total time gained remains considerable.

15.8.4 Identification

Indentification is made possible using the following test results:

- cultural characteristics;
- morphological characteristics of the vegetative cells;
- sexual characteristics;
- the first results of biochemical tests done by conventional methods or with API galleries.

The Kreger Van Rij (1984) identification key may be used for the preliminary orientation of the diagnosis. The identification criteria do not all have the same importance and must be considered in a determined order. First, morphological and sexual traits aid in the definition of the genus (the first dichotomous key). Next, the biochemical characteristics determine the species within each genus (dichotomous key for each genus). The preliminary results thus lead generally to the genus and to a few species in this genus. Complementary tests indicated by the key are then done to narrow down the species diagnosis.

Another identification system proposed by Barnett and co-workers (1983) is based almost entirely on biochemical criteria. The latest edition lists 55–57 traits for species identification purposes.

15.9 Fine Differentiation Technique

It was mentioned in the first part of this chapter that the classification of yeasts was overhauled in 1984. This new classification has led to an important simplification for the genus *Saccharomyces*, since applying the cross-fertility criterion has resulted in 17 strains previously recognized as different species being grouped as members of a single species, *Saccharomyces cerevisiae*. Hence the names *S. cerevisiae*, *S. uvarum*, *S. bayanus*, and *S. diastaticus*, which designated strains having well-differentiated technological properties for brewers and enologists, have all become synonymous. It remains well known among the users of these strains that within a given species as defined by the preceding criteria are clones with different technological properties. These users therefore require fine techniques to differentiate strains of the same species but of different clones.

15.9.1 The Exocellular Fraction

The exocellular fraction is composed of all of the macromolecular compounds excreted by cells during their growth. Biochemically, this fraction is made up of several cell-wall related glycoproteins with a molecular weight of around 500,000. It has been shown by Bouix and Leveau (1983) that the electrophoretic patterns of these fractions differ from one clone to another within a given species. This technique has been applied to the determination of strain origin within one enological species (Cuinier et al., 1980; Bouix et al., 1984) and to testing of yeast efficiency (Bouix et al., 1980).

15.9.2 Mitochondrial DNA

It is possible to extract mitochondrial DNA from yeasts and split it up using restriction endonucleases. The fragments may then be analyzed by gel (agarose) electrophoresis. Using this technique, Aigle and colleagues (1984) have shown that different clones of yeast within the same species possess differences at the mitochondrial DNA level with respect to the specific restriction enzyme activity sites, resulting in differing electrophoretic profiles for different strains treated with the same enzyme. This technique has been applied to wine yeast strains (Hallet et al., 1988).

A simplified mitochondrial DNA mini-extraction method has been developed and applied on a routine basis to testing lot purity of active dried yeast produced for enological purposes (Bertin et al., 1989).

15.9.3 Pulsed Field Electrophoresis

Pulsed field electrophoretic techniques have recently been used to separate large DNA fragments such as whole chromosomes of yeasts (Schwartz and Cantor, 1984). Work by Blondin and Vezinhet (1988) on enological strains have established their karyotypes. Intravarietal and intervarietal chromosomal polymorphism is another important factor that has been well exploited by this technique for strain characterization. The purity of dried active yeast lots has been monitored in this manner (Poulard and Daniel, 1989).

References

Aigle M., Erbs D., Moll M., 1984. Some molecular structures in the genome of lager brewing yeast. *Am. Soc. Brew. Chem.*, **43**, 1–7.

Barnett J. A., Payne R. W., Yarrow D., 1983. *Yeasts: Characteristics and identification.* Cambridge University Press, Cambridge.

Beech F. W., Davenport R. R., Mossel D. A. A., Dijkmann K. E., Koopmans M., Jong J., Put H. M. C., Tilbury R. H., 1980. Media and methods for growing yeasts: proceedings of a discussion meeting. *In*: Skinner F. A., Passmore S. M., Davenport R. R. (eds). *Biology and activity of yeasts*, 259–293. Academic Press, London.

Bertin B., Debesque D., Francou F., Bouix M., Van Hoegaerden M., 1989. *Differenciation des souches de champignons de même espèce par analyse électrophorétique de leur génome.* Agoral 89, Paris.

Blondin B. and Vezinhet F., 1988. Identification des souches de levures œnologiques par leurs caryotypes obtenus en électrophorèse en champs pulsés. *Rev. Fr. Œnol.*, **115**, 7–11.

Bouix M. and Leveau J. Y., 1980. Mise en évidence des levures sauvages dans les levains de brasserie par une méthode de double fluorescence. *Bios*, **11**, 27–35.

Bouix M. and Leveau J. Y., 1983. Electrophoretic study of the macromolecular compounds excreted by yeasts: Application to differenciation between strains of the same species. *Biotech. Bioing.*, **25**, 133–142.

Bouix M., Leveau J. Y., Cuinier C., 1980. Application de l'électrophorèse des fractions exocellulaires de levures au contrôle de l'efficacité d'un levurage en vinification. Current developments in yeast research. Advances in biotechnology. Proceedings of the Fifth International Symposium, London, Canada, 87–92.

Bouix M., Leveau J. Y., Cuinier C., Vaterkowski A., 1984. Origine des levures de fermentation d'un vin de Montlouis. 5èmes Journées Scientiques de l'ENSIA, Lille.

Bourgeois C., Mafart P., Thouvenot D., 1973. Méthode rapide de détection des contaminants dans la bière par marquage radioactif. *Europ. Brew. Conv.*, **14**, 219–230.

Cuinier C. and Leveau J. Y., 1979. Méthode rapide d'identification des levures des vignobles et des vins à l'aide de la galerie API 20C. *Vignes et Vins*, **283**, 44–49.

Cuinier C., Bouix M., Leveau J. Y., 1980. Méthode d'étude de l'origine des levures en analogie: Identification des espèces et différenciation des clones. Current developments in yeast research. Adv. in biotechnology. Proc. of the Fifth International Symposium, London, Canada, 505–508.

Fuhrman B., Dumain P. P., Laplace-Builhe C., Van Hoegaerden M., 1989. Nouveau système de détection, d'identification et de numération des microorganismes viables. 2ème congrès SFM, Strasbourg.

Gatley S., 1989. Rapid process control with flow cytometry. *Food Technol. Int.* Sterling Publications Ltd., 323–329.

Geneix C., 1984. Recherche sur la stimulation et l'inhibition de la fermentation alcoolique du moût de raisin. Thèse Bordeaux II.

Gustafsson L. and Norkraus B., 1976. On the mechanism of salt tolerance. Production of glycerol and heat during growth of *Debaryomyces hansenii. Arch. Microbiol.*, **110**, 177–188.

Hallet J. N., Graneguy B., Zucca J., Poulard A., 1988. Caractérisation des différentes souches industrielles de levures œnologiques par les profils de restriction de leur ADN mitochondrial. *Prog. Agric. Vitic.*, **105**, 328–333.

Harris J. O. and Watson W., 1968. The use of controlled levels of actidione for brewing and non brewing yeast strain differentiation. *J. Inst. Brew.*, **74**, 286.

Kreger Van Rij N. J. W., 1984. *The yeasts, a taxonomic study.* 3d ed. Elsevier Science Publishers, Amsterdam, 1082 p.

Leveau J. Y. and Bouix M., 1979. Etude des conditions extrêmes de croissance des levures osmophiles. *Ind. Alim. Agric.*, **11** 1147–1151.

Leveau J. Y. and Bouix M., 1986. Les microorganismes producteurs d'éthanol. *In*: *La production industrielle d'éthanol de fermentation.* Colloque FCB. Ed. SEDA, 3–56.

Lodder J., 1971. *The yeasts, a taxonomic study*, 2d ed. North Holland, Amsterdam, London, 1385 p.

Loiseau G., Vezinhet F., Valade M., Vertes A., Cuinier C., Delteil D., 1987. Contrôle de l'efficacité du levurage par la mise en œuvre de levures œnologiques marquées. *Rev. Fr. Œnol.*, **27**, 29–36.

Mafart P., Bourgeois C., Duteurtre B., Moll M., 1976. Radiometric method for control of filtration and pasteurisation. *Tech. Quar. MBAA*, **13**, no. 157–160.

Morris E. O. and Eddy A. A., 1957. Method for the measurement of wild yeast infection in pitching yeast. *J. Inst. Brew.*, **63**, 34.

Parry J. M., Davies P. J., Evans W. E., 1976. The effect of "cell age" upon the lethal effects of physical and chemical mutagens in the yeast, *Saccharomyces cerevisiae. Mol. Gen. Genet.*, **146**, 27.

Poulard A. and Daniel P., 1989. Nouvelles techniques fines de caractérisation des souches en œnologie. *Vignes et Vins*, November, **4**, 71–73.

Rose A. H., 1987. Responses to the chemical environment. *In*: Rose A. H., Harrison J. S. (eds). *The yeasts, V2, yeasts and the environment*. Academic Press, London.

Russell I. and Stewart G. G., 1980. *Liquid nitrogen storage of yeast cultures compared to more traditional storage methods. Current developments in yeast research. Advances in biotechnology.* Pegamon Press, 123–128.

Schimz K. L. and Holzer H., 1979. Broad decrease of ATP content in intact cells of *Saccharomyces cerevisiae* of sulfite. *Arch. Microbiol.*, **121**, 225–229.

Schwartz D. C. and Cantor C. R., 1984. Separation of yeast cytochrome sized DNAs by pulsed-field gradient gel electrophoresis. *Cell*, **37**, 67.

Taylor G. T. and Marsh A. S., 1984. MYGP + copper, a new medium that detects both *Saccharomyces* and non-*Saccharomyces* wild yeast in the presence of culture yeast. *J. Inst. Brew.*, **90**, 127.

Walton E. F. and Pringle J. R., 1980. Effect of growth temperature upon heat sensitivity in *Saccharomyces cerevisiae. Arch. Microbiol.*, **124**, 285.

Watson K., 1987. Temperature relations. *In*: Rose A. H., Harrison J. S., (eds). *The yeasts, V2, yeasts and the environment.* Academic Press, London, 41–65.

Suggested Readings

Candish A. A. G., 1991. Immunological methods in food microbiology. *Food Microbiol.* London: Academic Press. March 1991. **8**, no. 1, 9–14.

Candish A. A. G., Smith J. F., Stimson W. H., 1990. Aflatoxin monoclonals: Academic development to commercial production. *Letters Appl. Microbiol.* **10**, no. 4, 167–169; 10 ref.

Deak T. 1991. Foodborne yeasts. *Adv. Appl. Microbiol.* San Diego: Academic Press. **36**, 179–278.

Fung D. Y. C. and Liang C., 1990. Critical review of isolation, detection, and identification of yeasts from meat products. *Cr. Rev. Food Sci. Nutr.* Boca Raton, FL: CRC Press. **29**, no. 5, 341–379.

Jewers K., 1988. Aflatoxin in food and feeds. *Food-Lab. Newsletter*, **11**, 30–34; 10 ref.

Niller R. and Galston G., 1989. Rapid methods for the detection of yeast and lactobacillus by ATP bioluminescence. *J. Inst. Brew.* London: The Institute. Sept./Oct. 1989. **95**, no. 5, 317–319.

Ozcelik S., Ozcelik N, Beuchat L. R., 1990. Toxin production by *Alternaria alternata* in tomatoes and apples stored under various conditions and quantitation of the toxins by high-performance liquid chromatography. *Int. J. Food Microbiol.* **11**, no. 3/4, 187–194; 33 ref.

Rohm H. and Lechner F., 1990. Evaluation and reliability of a simplified method for identification of food-borne yeasts. *Appl. Environ. Microbiol.* Washington, D.C.: American Society for Microbiology. May 1990. **56**, no. 5, 1290–1295.

Torok T. and King A. D. Jr., 1991. Comparative study on the identification of food-borne yeasts. *Appl. Environ. Microbiol.* Washington, D.C.: American Society for Microbiology. April 1991. **57**, no 4, 1207–1212.

CHAPTER

16

The Molds

C. Moreau

16.1 Molds in the Food Industry

The concept of "moldy food" is easy enough to grasp. Everybody has seen bread, cheese, or jams with molds on them. Molds are difficult to define, however, because they do not correspond to any homogeneous group but rather to a wide range of fungal families having few traits in common. The most frequently encountered are from the order Mucorales or are asexual forms belonging to the genera *Aspergillus*, *Penicillium*, *Fusarium*, and so on. Like the majority of yeasts, these microscopic fungi (the "Micromycetes") develop on dead substrates, that is, they are saprophytes. Whereas yeasts often occur as unicellular elements which give rise to very similar individuals, molds consist of a filamentous thallus or mycelium easily fragmented into autonomous shoots but which also differentiates into a variety of structures for multiplication (e.g., spores). Under favorable conditions, a spore germinates into mycelial hyphae which rapidly branch out to invade the substrate and extract from it the necessary nutrients for continued growth. Later on the species forms its spores for dissemination purposes, sometimes in very large numbers (a grain of wheat supports 25 million *Penicillium cyclopium* spores and an orange 15 billion *Penicillium digitatum* spores). Depending on the species, spores are easily dispersed into the air (xerospores) or water (myxospores). Molds may often form thick-walled resting spores (chlamydospores) which enable them to resist adverse conditions.

16.1.1 Useful Molds

Molds which are totally unwelcome in the food industry are actually very common in our environment and useful in agriculture. Their role in degrading organic wastes and in humification is predominant in soils. Furthermore, the cheese-making industry uses *Penicillium camemberti* for soft-curd cheese with powdery crusts and *Penicillium roqueforti* for speckled cheeses. Moreover, molds such as *Mucor miehei* provide rennin substitutes.

Enzymatic processing carried out by molds is the basis of culinary preparations which are greatly appreciated in the Far East. Rice and soya undergo long and complex fermentations in which *Aspergillus oryzae* and various Mucorales species are involved. Molds can also contribute to the upgrading of food industry wastes.

16.1.2 Undesirable Molds

More foodstuffs, during their production and especially during their storage, are vulnerable to deterioration brought about by molds, to which considerable losses are attributable. It is often a matter of a simple visual defect, harmless but sufficient to devalue or render a product unsalable. People simply refuse to buy cakes or loaves of bread with mold on them.

Spoilage of foodstuffs by molds may result in changes in the nutritional value of the product, the development of undesirable flavors, modification of organoleptic traits (cheese with an ammonia taste because of too much *Geotrichum candidum* activity, coffee turned bitter under the influence of *Aspergillus tamarii*, vegetable oil turned rancid by *Aspergillus glaucus*, meat with a "moldy" taste asociated with *Scopulariopsis brevicaulis*, wine spoiled by metabolites of some strains developing in the cork, etc.).

In other cases, the health of the consumer is at stake. Some molds produce toxic substances called *mycotoxins* which may diffuse into the food and if present in sufficient quantity may cause chronic or acute poisonings. These poisonings are rare among humans due to the variety of regularly consumed foods and due to the care with which food is usually stored. They are, however, more frequent among herds of farm animals which receive essentially the same rations every day (Table 16.1). For this reason, the detection of molds in grain, oilseeds and their derivatives, fodders, and silaged feeds is especially important.

A few molds may have allergenic effects on workers (handlers of hay, flour, cork, etc.) while others may sometimes be pathogenic to humans or animals (pulmonary aspergilloses).

Production defects due to molds represent the most widespread risk and make rigorous testing indispensable. The detection and identification techniques best suited to the mold or molds responsible, as well as the systematic tracing of sources of contamination, must be implemented. The most appropriate methods of prevention and maintenance must be adapted to each case.

This chapter will concisely review the various steps in the rational examination of production defects due to molds in the food industry and of preventive monitoring in cases where the presence of undesirable molds is presumed.

Table 16.1 Examples of mycotoxicoses

Predominant syndrome	Causative myotoxin	Mold responsible
Hepatotoxicoses	Aflatoxins	*Aspergillus flavus*
Nephrotoxicoses	Ochratoxins	*Aspergillus ochraceus* *Penicillium viridicatum*
Neurotoxicoses	Patulin	*Aspergillus clavatus* *Penicillium expansum*
Gastroenterotoxicoses Hemorrhages	Trichothecenes	various *Fusarium* ssp.
Œstrogen-like effects	Zearalenone	*Fusarium graminearum*

16.2 Methods of Analysis

16.2.1 Direct Microscopic Examination

Microscopic examination of material removed from a moldy product is of no interest unless a well-developed young thallus is present on the spoiled substrate (e.g., on the surface of a rotting fruit). In routine practice, molds are rarely at such a stage of development and precise identifications are illusory. By placing the food samples in a warm and moist enclosure, the evolution of mold development may sometimes be followed by normal microscopic examination procedures. More often it will be necessary to culture and purify the organism on suitable nutrient media at the most favorable incubation temperature. The microscopic techniques described later are equally applicable to molds removed directly from foodstuffs and to those developed under pure culturing conditions after isolation.

16.2.1.1 Unstained Wet Mount

A fragment of thallus is carefully placed with a drop of water in the well of a depression slide, covered with a cover-slip, and observed directly under the microscope. A nonstaining mounting solution which gives a sharper image is Amann's lactophenol solution:

Phenol, white crystalline	20 g
Lactic acid	20 g
Glycerin	40 g
Distilled water	20 ml

Stored in an amber bottle away from light.

16.2.1.2 Stained Mount

The best stain for fungi is methyl blue (Poirier C_4B cotton blue). This is an acidic aniline blue and a specific stain for callose. It is usually used mixed with other mounting liquids, for example, as a 0.5% solution in lactophenol. The microscopic slide preparation should be heated slightly, for instance, using a

Bunsen burner pilot flame. Other more specific stains may be used in order to reveal lipid bodies (Sudan III) or "amyloid" cell walls (iodine-iodide solution), and so on.

In order to improve the observation of fungal cells inside darkly pigmented organs (e.g., asci inside a perithecium), hot 10% potassium hydroxide solution may be used to decolorize followed by rinsing and staining with methyl blue.

16.2.1.3 *Storage of Preparations*

Lactophenol preparations may be kept for some time, especially if sealed with a cellulosic varnish. For longer lasting mounts, gelatin jellied glycerol is more suitable.

16.2.2 Culture Techniques

Laboratory techniques for fungal culture are similar to those used by bacteriologists with respect to the choice of culture vessels and sterilization procedures. The similarity ends there, however, and instant extrapolations to standard bacteriological techniques must be avoided since the two types of microorganisms have totally different morphology and physiology. Culturing is of great value to mycologists for the purposes of revealing species polymorphism and attempting to discern its complete developmental cycle as well as possibly separating several fungal species where the presence of only one was suspected.

16.2.2.1 *Isolation by Direct Inoculation*

Using a sterile inoculating loop or toothpick to remove portions of a thallus from a given substrate and inoculating onto a solid medium often gives good results. The portion of the thallus transferred should be as small as possible, such as may be taken with the aid of a stereoscopic-dissecting microscope or magnifying lens. In the case of particulate foods, it is sufficient to sprinkle the surface of the solid medium with a small quantity of the product under examination. When several molds, coexist on a substrate, this sort of direct inoculation, even if done in minute quantities, generally does not detect species other than the most competitive under the conditions employed.

16.2.2.2 *Isolations by Suspension-Dilutions*

When separating the various constituents of a mycoflora developing within a particulate substrate (flour, powdered milk, etc.) or on the surface of kernels of grain, the so-called suspension-dilution technique is used.

A known weight of sample (e.g., 10 g) is suspended in sterile water (190 ml). In some cases, the use of a special diluent such as the following is recommended in order to better dissociate the various elements:

tryptone: 1 g + NaCl: 8.5 g + Tween 80: 0.033 g + water: 1,000 ml

The suspension is vigorously mixed using stirring or homogenizing devices and a series of successive dilutions is then made aseptically (e.g., 1/100, 1/1,000, 1/10,000, etc.). Using a sterile pipette, 1 ml of each suspension is deposited per Petri plate into which 15 ml of agar-based medium kept liquified at 45°C is immediately poured. Each plate is carefully mixed (by sliding back and forth, etc.) before the agar solidifies and is then incubated. Several successive examinations at intervals of a few days are often necessary to follow the development of the thalli of the various species. They may be enumerated although a complete examination will require transfer to tubes and ultimately to the most well-adapted solid medium.

16.2.2.3 Culture Media

Since yeasts have varying nutritional requirements, they are not all apt to develop on one single culture medium. The so-called Sabouraud medium is an isolation medium for pathogenic dermatophytes of humans and animals and as such is normally unsuitable for most mold species.

The most usual isolation medium is malt extract agar (15–20 g of agar-agar per liter). However, some of the pioneering molds of prunes (*Aspergillus* of the *glaucus* group), jams (*A. restrictus*), dried sausages (*Wallemia sebi*), and many components of the mycoflora of grains, flours, powdered milks, and so on are incapable of development on common nutrient media because they tend towards xerophilia and require high osmotic pressures. In order to isolate these, sugar, salt, or glycerol-enriched media may be used. Malt extract: 50 g + NaCl: 50 g + water: 1,000 ml + agar-agar: 15 g is normally recommended.

Examination of the cultural traits of *Penicillium* (Pitt, 1979) is preferably done comparatively on three media:

MEA (malt extract agar)

Malt extract	20 g
Peptone	1 g
Glucose	20 g
Agar-agar	15 g
Water	1,000 ml

CYA (Czapek yeast autolysate agar)

K_2HPO_4	1.0 g
Czapek concentrate	10 ml
Yeast extract	5 g
Sucrose	30 g
Agar-agar	15 g
Water	1,000 ml

G 25 N (25% Glycerol nitrate agar)

K_2HPO_4	0.75 g
Czapek concentrate	7.5 ml
Yeast extract	3.7 g
Glycerol	250 g
Agar-agar	12 g
Water	1,000 ml

The "Czapek concentrate" for the latter two media is made up of the following:

$NaNO_3$	30 g
KCl	5 g
$MgSO_4 . 7H_2O$	5 g
$FeSO_4 . 7H_2O$	0.1 g
Water	100 ml

Fusarium develops and sporulates especially well on an oat flour medium:

Oat flour	50 g
NaCl	5 g
Agar-agar	15 g
Water	1,000 ml

Suspend the flour, add NaCl, cook for 20 minutes, filter, then dissolve the agar.

Among the conventional media worth mentioning is potato dextrose agar (PDA):

Potatoes	200 g
Glucose	20 g
Agar-agar	15 g
Water	1,000 ml

Wash, peel, and cut potatoes into cubes or thin slices. Cook for 15–20 minutes in 200 ml of water, filter through fine mesh cotton, and press. Add glucose to the filtrate and bring the volume up to 1,000 ml. Add and dissolve the agar-agar.

Numerous other vegetable-based media exist, made from corn, rice, various plant concoctions, and so on. All of these media must, in principle, be autoclaved for 20 minutes at 121°C, although slightly lower temperatures for longer times are usually preferred in order to minimize caramelization. For isolations from substrates on which bacteria are mixed in with molds, it is preferable to use very dry surface solid media. Incubation in a refrigerator may suffice to slow down the development of bacterial colonies to the benefit of mycelial filaments. It may be necessary to add a bactericidal agent to the medium, for example, chlortetracycline and rose bengal or, better yet, chloramphenicol (added to a concentration of 0.05 g/l just before sterilization). These should be used with discretion, however, and always in parallel with the equivalent antibiotic-free medium, since these agents may inhibit the development of some fungal species.

16.2.2.4 Incubation

In order to properly assess the full range of mold species which may be present in a foodstuff, it is advisable to incubate the cultures at the most suitable temperature for each species. It is well known, for example, that the extent of an infection by *Aspergillus flavus* or *Aspergillus fumigatus* (meaning a presumption of mycotoxosis) may be underestimated in cereal products, oilseed press cakes, silages, or powdered milk due to an error in the choice of the incubation temperature. At too low a temperature (22°C), these molds are uncompetitive and become dominated by common species of *Penicillium* and *Mucor*.

In practice, for foods which are likely to encounter higher temperatures, incubation at 25° or 35°C is used. For foods subjected to low temperatures, longer incubations at 11° or even 5°C are used in addition to 25°C. Pitt (1979) has thus recommended incubation at 5°C for the identification of *Penicillium*, a temperature which is selective for some species.

Since light has a stimulatory effect on sporulation for the majority of molds, enclosures which allow daylight to penetrate are preferred for incubation.

16.3 Types of Analyses Performed

16.3.1 Monitoring of Starting Material Contamination

Since many food products are the result of mixing or processing of some number of ingredients, the search for the source of contamination of a food requires mycological analysis of all of the starting materials used. In principle, each of these must be considered to be suspect since the possibility of harboring one of the elements of the final mycoflora exists. A single species, even though present in traces in one of the starting materials, may effectively find itself in an ideal environment for its development in the finished product, for example, in spices in fresh cheese or processed meats. Although it may be of some use to know that a given mold is very abundant, it is more important to enumerate species with precision.

For grains, in addition to the surface mycoflora which is revealed by means of suspension-dilutions, it is worthwhile to acquire some knowledge of the mycoflora lodged in the depths of the tissues. For this purpose, a surface disinfection may be carried out by immersing the grain samples in sodium hypochlorite (1%) for 10 minutes, rinsing three times in sterile water (first rinsing: 5 min.; second rinsing: 10 min.; third rinsing: 15 min.). Splitting in half as many kernels of grain as possible and inoculating solid medium by placing the freshly exposed face of each half-kernel down on the agar surface is also recommended.

Table 16.2 Critical growth temperatures (°C) for some molds in foodstuffs

	Minimal	Optimal	Maximal
Aspergillus candidus	10	28	55
Aspergillus clavatus	10	26	38
Aspergillus flavus	10	30	45
Aspergillus fumigatus	12	37	52
Aspergillus repens	5	30	38
Botrytis cinerea	−4	22	37
Cladosporium herbarum	−6	25	40
Helicostylum pulchrum	0	5	28
Humicola lanuginosa	30	48	60
Mucor pusillus	20	40	55
Neurospora sitophila	5	36	45
Penicillium aurantio-griseum	−2	24	29
Penicillium brevicompactum	−2	22	28
Penicillium citrinum	12	27	34
Penicillium expansum	−2	24	31
Penicillium glabrum	3	24	29
Penicillium roqueforti	2	22	35
Penicillium viridicatum	−2	24	35
Rhizopus stolonifer	1	28	34

16.3.2 Monitoring the Contamination of Work Areas

When faced with production anomalies occurring during the preparation or stocking of a food product, it is advisable to search for sources of contamination not only in the starting material entrants of the product but also in the areas where manufacturing and storage take place.

Taking into consideration the known spore dispersal mechanisms of molds (xerospores and myxospores), the contaminating source will be sought at the same time among airborne mycoflora and that which is present on surfaces (walls, floors, ceilings, various elements of the production line).

6.3.2.1 Airborne Mycoflora

In different areas of rooms and at various heights from the floor, Petri plates containing a solid culture medium (preferably two different plates, one with 2% malt extract and one with 5% malt extract + 5% NaCl) are opened side by side for a given time interval (10 minutes or less depending on how heavy an atmospheric contamination is expected). The plates are carefully closed and individually packaged for incubation in the laboratory. Each spore (in the case of xerospores) or clump of spores (in the case of myxospores, which are rarely airborne) will in theory give rise to a thallus. The two plates are to be considered complementary in the qualitative expression of the flora at a given location. This simple method of evaluation by sedimentation detects mold species which fall onto the surface of a food product in production or stockage during a time interval equivalent to that of the plate exposure. In smaller rooms or in those where rigorous hygiene is essential, a spore-detecting device may be preferred, although in the majority of industrial enclosures the faster, more flexible, and more economical sedimentation technique is an adequate reflection of reality when the plates are placed in locations chosen by a competent mycologist. Fungal contamination entering through openings into the room being monitored (doors, windows, filter vents, etc.) may also be tested in this manner.

16.3.2.2 Surface Mycoflora

The conventional swabbing technique practiced by bacteriologists is not without its critics. A technique is required which detects not only spore deposits on surfaces but especially of fungal thalli which may be deeply lodged in various substrata (paint, deglossed tiles, plaster, various porous materials). Besides, some of the more xerophilic of these species do not tolerate the wetting of their spores for germination.

Numerous scrapings at carefully selected sites are done using a scalpel regularly dipped in alcohol and flamed, each sampling used to inoculate tubes or Petri plates with a choice of selective media to be incubated in the laboratory in order to detect possible sources of contamination. All of the elements of a production line and all of the instruments used (especially those which are supposed to be used for cleaning and which are often themselves contaminated) will be examined in this manner. Roofs, eaves troughs, drain holes for stray water and so on will all be the subject of special scrutiny.

16.3.3 Evolution of Foodstuff Contamination

During the examination of a production line, sampling must be done at all levels. It may be discovered that production involving long periods in ovens or high temperature pelleting yield a product that is momentarily sterile even though the starting materials may have been heavily contaminated. This has been the case for bread and bakery products, cooked processed meats, pelleted filter press cakes, granulated animal feeds, and so on. In such cases, very rigorous hygienic standards must be enforced at all stages of production subsequent to this step in order to prevent recontamination of the product. In processes which do not involve such a sterilizing step (soft curd cheeses, dry sausages, etc.), the problem is obviously more complex. Moreover, examin-

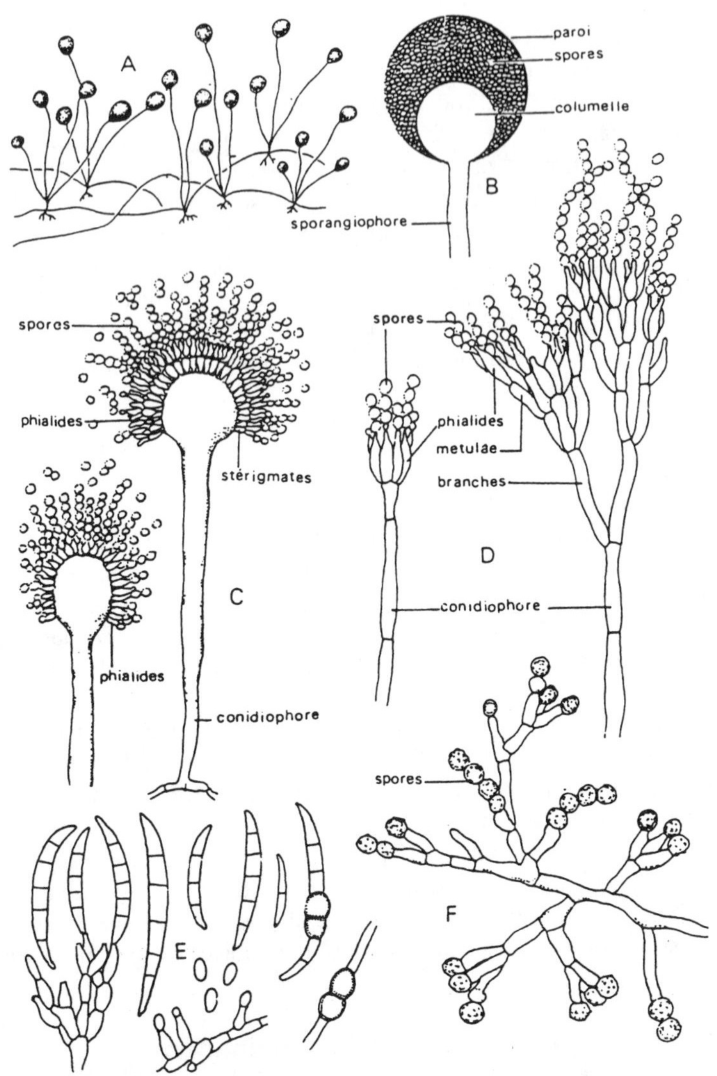

Figure 16.1 Some common genera of molds in the food industries. A, Mycelial stolons, sporangiophores, and sporangia of *Rhizopus* (Mucorales); B, Organization of the sporangium of a species of *Mucor*; C, Two types of spore heads of *Aspergillus* (*left*: uniseries; *right*: biseries); D, Two types of *Penicillium* spore brushes (*left*: monoverticillate; *right*: asymmetrical biverticillate); E, Macroconidia, microconidia, and chlamydospores of *Fusarium*; F, Sporiferous branch of *Scopulariopsis*.

ations at regular intervals throughout the period of stockage (as for the case of animal feeds) reveal floral successions which enable the subsequent evolution of fungal contamination to be predicted, thereby establishing storage life predictions taking into consideration the ambient conditions that are likely to prevail.

16.4 Identification of Molds

Due to the large number of different molds which exist (several tens of thousands of species), their precise identification requires the knowledge and skill of a specialist. Their classification is based essentially on the observation of morphological traits as revealed by careful microscopic examination at the various stages of development most often complemented by cultural traits which they display under specific conditions (e.g., growth rate, thallal texture, pigmentation, etc.). The following works dealing with the mycological systematics of the most common molds may then be consulted:

Booth C., 1971. *The genus* Fusarium. Kew, C.M.I.
Pitt J. I., 1979. *The genus* Penicillium *and its teleomorphic states* Eupenicillium *and* Talaromyces. Academic Press, London, New York.
Raper K. B. and Fennel D. I., 1965. *The genus* Aspergillus. Williams & Wilkins Co., Baltimore.
Raper K. B. and Thom C., 1968. *A manual of* Penicillia. Hafner Publishing Co., New York, London.
Zycha H. and Siepmann R., 1969. *Mucorales.* Lehre, Cramer.

It is also worthwhile to consult the "Mycological Papers" collection edited by the Commonwealth Mycological Institute of Kew as well as "Studies in Mycology" of the Centraal-bureau voor Schimmelcultures of Baarn.

Comparison with correctly identified strains deposited in the mycological libraries of various establishments may also be of use.

16.5 Expression of Results

The simple analysis of a moldy food product by the suspension-dilution method constitutes a conventional test of which the results may be contested. Although the enumeration of unicellular organisms such as bacteria and yeasts as colonies developing in Petri plates does provide a reasonable assessment of the quality of a contaminated product, in the case of molds, which are filamentous organisms, counting thalli only indicates the presence of strongly sporulating species or of a viable mycelium which has been broken up to some degree during grinding.

Species of *Penicillium* or *Aspergillus* may produce small but very prolific thalli easily detected by plating methods, while *Fusarium* or some strains of Mucorales may sporulate poorly or moderately and be underestimated even though they may be the predominant causal agent of the production anomaly under consideration.

A "total count" of molds without identification has no valid meaning. It is therefore recommended that the relative "degree of abundance" for each species identified be indicated in terms of numbers of germinating elements per gram of a given product, to distinguish them as follows:

+ + + +	very abundant species	($>$10,000 CFU/g)
+ + +	abundant species	(1,000–10,000 CFU/g)
+ +	moderately abundant species	(100–1,000 CFU/g)
+	nonabundant species	($<$100 CFU/g)

When sufficient numbers of analyses of the same product are performed, the "frequency" of the species may be noted, that is, the number of times the species is detected relative to the total number of samples examined:

+ + + +	characteristic contaminant	($>$80%)
+ + +	frequent contaminant	(50–80%)
+ +	occasional contaminant	(10–50%)
+	opportunistic contaminant	($<$10%)

16.6 Experimental Studies and Prospects for Improved Defense Against Mold Infections

All searches for the causes of production anomalies must include an experimental study once the mold or molds supposedly responsible have been detected and identified. It may be necessary for this study to focus on acquiring knowledge of the biological particularities of the undesirable strain, especially if these are not sufficiently documented in the literature (e.g., optimal and lethal temperatures, most favorable growth media, incubation time before onset of spore germination, growth rate, speed and intensity of sporulation, resistance to physical and chemical agents, etc.).

The study will also involve the experimental contamination of the food product with pure cultures of the isolated mold strain(s) and observation of its (their) evolution under varied ambient conditions.

Once equipped with reliable experimentally controlled examples of the foodstuff uniformly contaminated with the undesirable agents, appropriate measures may be sought to inhibit their normal course of development in the product. Treatments applied directly to the food may be used effectively to combat molds in the food industry as may indirect means designed to reduce the sources of contamination. The direct treatments consist essentially of sterilization by heat, use of low temperatures, dehydration-dessication-lyophilization, modification of the acidity, chemical sterilization, or use of chemical preservatives.

To prevent contamination, surface treatments (detergents, disinfectants, fungicidal paints, etc.) or environmental treatments (sanitizing or sterilizing mists, smoke, gases, etc.) may be used.

Testing of the effectiveness of specialty products marketed as industrial fungicidal or fungistatic agents is usually done by authorities (e.g., governmental ministries or departments of agriculture) within national jurisdictions who establish recommended methods for products intended for the sanitary treatment of work spaces, supplies, vehicles, buildings, and surrounding areas used for the harvesting, storage, industrial processing, and commercialization of products of animal or vegetable origin subject to regulatory approval. In France, only those products appearing in the October 27, 1975 (modified December 21, 1979) statement issued by the Comité d'homologation des Produits Phytosanitaires et Assimilés of the Ministère de l'Agriculture (175, rue du Chevaleret, 75646 Paris, Cedex 13) are authorized for cleaning and disinfecting in the food industries.

16.7 Conclusion

The difficulties in standardizing methods of testing for the presence of development of molds in the food industries should now be apparent. A specifically adapted method should be applied to each particular case.

Although the techniques must be performed carefully, they do not require very sophisticated devices or supplies. This does not mean, however, that simple and automated bacteriological methods can be applied to molds. The only valid basis for effectively combating undesirable molds is the correct identification of the suspects, a process whch can easily baffle beginners and which is therefore best left in the hands of a specialist with many years of experience.

At most, it may be hoped that techniques will become standardized for applications to a single product, a single type of processing, or the search for a single fungal species. Much remains to be learned, however, before this can become reality since very little work has been done on molds in the various food industries, especially when compared to the enormous volume of work dealing with bacteria.

PART III

EVALUATION METHODS FOR MICROFLORA OF SANITARY INCIDENCE

CHAPTER

17

Indicators of Fecal Contamination

M. V. Catsaras

Fecal contamination indicators are microorganisms which normally live in the large intestine of humans and animals. Their presence in foods may indicate that fecal contamination may have occurred, an event strongly correlated with the risk of pathogenic organisms being present. These indicators include the coliforms, *Escherichia coli* (or fecal coliforms) and the enterobacteria as a whole, but also fecal streptococci and sulfite-reducing species of *Clostridium*. The search for these indicators, especially of coliforms and of *E. coli*, is widely employed in the testing of food product quality.

17.1 Basis for the Use of Indicator Microorganisms and the Justification of their Choice

17.1.1 General Principle of the Detection of Contaminations

Every specimen is characterized not by a single microbial species but by a number of species which make up an association. As a result, if a food contamination from a given source is considered very serious because of the possible presence of dangerous species by association, the contamination may be detected by either searching directly for pathogenic species or searching for another species or group of species within the same association if advantageous in terms of convenience or sensitivity; such species in this case constitute a contamination index or barometer.

In some cases (e.g., for *Salmonella*), direct testing for pathogenic bacteria is carried out systematically, while for other pathogens this is done only rarely or not at all. Furthermore, since a negative finding is not always a guarantee of the total absence of pathogens, it follows that testing for indicators of fecal contamination is always necessary. This test is thus one of the most basic operations in food microbiology.

17.1.2 Characteristics of a Good Indicator of Contamination

The characteristics are as follows:

- The chosen microorganism should be specific to the contaminating source, and ideally an inherent characteristic of the source.
- It should provide high sensitivity (the greater its fraction of the microbial population in the source of contamination, the higher this sensitivity will be).
- It is obviously desirable that the resistance of the microbe—in terms of its survival time in the external environment and in the food, guaranteeing the reliability of the analysis—be at least equal to if not superior to that of the pathogens in the association.
- The techniques used for the detection and enumeration of the microbial barometer must be simple, rapid, and reliable.

17.1.3 Signs of Fecal Contamination

Examination for fecal contamination indicators is of generalized use in the testing of water and foods which may act as vehicles for the transmission of foodborne diseases and poisonings. The indicator organisms from among the intestinal microflora have been chosen as a function of the criteria cited earlier.

The most abundant microorganisms in the intestine are anaerobes, especially in the caecum (*Bifidobacterium*, *Ristella*), and microaerophiles (*Lactobacillus*), but these bacteria are not usable as indicators due to problems with their detection.

Next, in decreasing order, are the enterobacteria, of which *E. coli* is the most common, followed by the fecal streptococci and the sulfite-reducing species of *Clostridium*. These three groups have been studied as indicators.

This distribution is generally the case for humans but not necessarily for animals such as sheep and hogs, for which fecal streptococci are often much more numerous than enterobacteria.

17.2 Coliforms and *E. coli*

Among the gram-negative bacteria living especially in the large intestine of humans and animals are coliform bacteria characterized by the ability to ferment lactose at varied rates in the presence of bile salts. Fecal coliforms also possess a characteristic associated with their habitat, namely, the ability to grow at 44°C. In addition to these traits, *E. coli* has the property of producing indole from tryptophan at 44°C.

17.2.1 Value as Indicators of Fecal Contamination

The coliform group as a whole does not have very great specificity. They are intestinal bacteria but may also be found in other environments. Nevertheless, some representatives of the group distinguish themselves in this respect.

E. coli is probably the most specific of all the fecal contamination bacteria. Although its presence in water is clearly indicative of fecal contamination, the significance of its presence in food is less certain of recent fecal contamination since the possibility of its multiplication exists, for example, in certain poorly cleansed places in the processing plant which may thus become sources of contamination for products. The genus *Klebsiella* also has good specificity while *Citrobacter* and *Enterobacter* are not as specific. The sensitivity which *E. coli* brings to the detection of fecal contamination is also good since these bacteria are so numerous in the contents of intestines. Their resistance leaves something to be desired, however. *E. coli* is often less resistant than pathogens, particularly *Salmonella* both in external environments and in some raw foods (e.g., shellfish) as well as processed foods (e.g., dehydrated, frozen, or ionizing radiation treated). In any event, the conditions for the analyses of coliforms and *E. coli* (or fecal coliforms) are, on the whole, very satisfactory.

17.2.2 Conventional Techniques for Coliform, *E. coli*, and Fecal Coliform Analysis

Essentially their lactose utilizing ability is exploited, with or without gas production, detection being either in liquid media or on solid media.

17.2.2.1 Detection in Liquid Medium

Coliforms are distinguished from other enterobacteria by their ability to ferment lactose. Thus their detection consists of incubating the sample at 30°C for 48 hours (± 2 hours) in a lactose broth made selective by the addition of bile and Brilliant Green and containing an inverted Durham tube to detect fermentation (brilliant green lactose bile broth, BGLBB, see Appendix, medium 1). The technique has been standardized by the AFNOR (NF-V-08-016).

Three tubes each are inoculated with the original suspension and each dilution, each tube containing 10 ml of medium and the inoculum volume being 1 ml. The sensitivity can be increased by also testing 10 ml of inoculating suspension in 10 ml of double concentration medium. It is recommended for certain products (e.g., meats, eggs, dehydrated soups) that the tubes be shaken during incubation.

Reading of the results consists of observing whether or not sufficient gas evolution occurs. If at least 1/10 of the volume of the Durham tube is filled with gas, the result is positive. The raw data are processed by the most probable number (MPN) method for three tubes and expressed in terms of coliform/g or coliform/ml of product. Coliforms thus detected may be identified by isolating and applying the IMViC tests (indole, methyl red, Voges-Proskauer, citrate). Usually the detection of *E. coli* is considered sufficient. It is also possible to search directly for *E. coli* or fecal coliforms. For this, the same medium is inoculated in the same manner but the tubes are incubated at 37°C for 48 hours and then transferred as follows:

For *E. coli*

The tubes are transferred to BGLBB and to peptone water which are incubated for 48 hours at 44°C (±0.5); if gas evolution is observed in the BGLBB Durham tube and indole is detected (using the Kovacs reagent) in the peptone water, the presence of *E. coli* is confirmed. This is the Mackenzie test.

For Fecal Coliforms

The tubes are transferred to BGLBB only, incubated for 48 hours at 44°C (±0.5); gas evolution corresponds to a positive test.

17.2.2.2 Detection on Solid Media

Inoculation by the double agar layer is used, the most commonly utilized media being deoxycholate agar and crystal-violet neutral-red lactose bile agar (VRLB, see Appendix, media 2 and 3). The technique has also been standardized by the AFNOR (NF-V-08-015).

For the enumeration of coliforms, the plates are incubated at 30°C for 24 hours (±2 hours). Enumeration of fecal coliforms may be done directly after incubation at 44°C for 24 hours (±2 hours). Reading consists of counting red and violet colonies and so on, at least 0.5 mm in diameter for coliforms and 1 mm for *E. coli*.

The identification of various colonies may then be done by the IMViC tests or by detection of β-D-glucuronidase and tryptophanase (a recent technique from Fluorocult for solid media makes use of a fluorescent reaction).

17.2.2.3 False Reactions

Certain false reactions, such as those due to *Clostridium* and to *Bacillus polymyxa-macrans* are well known, while others are less known but nevertheless quite frequent. The genus *Aeromonas*, which is not a fecal bacterium, interferes with coliform detection in milk and drinking water in similar fashion. For a long time they were considered to be paracoli, since tests for oxidase were not done. Similarly, pigmented coliforms which are not of fecal origin also interfere with the enumeration of fecal coliforms whether on solid or liquid media.

17.2.3 Other Techniques

In fact, a great variety of both solid and liquid media are usable for the enumeration of coliforms and *E. coli* (or fecal coliforms). The main difference between them is the type of selective agent, coloring agent, and surface active agent used. Bile salts are obviously applicable to favoring bacteria which normally live in the intestine, although the variability of this particular physiological secretion has led some workers to propose the use of more specifically defined surface active agents such as lauryl sulfate salts.

Many efforts are currently focused on accelerating the detection of coliforms. The enumeration of fecal coliforms can now be done within 24 hours while the identification of *E. coli* requires an additional 48 hours in order to carry out the Mackenzie test. In any event, it is claimed by Mara (1973) that the time required for the confirmation test may be saved by passing directly to the use of a liquid medium to detect lactose fermentation with production of gas and indole. A membrane culture procedure allowing the characterization and enumeration of *E. coli* in food after as little as 24 hours was proposed by Anderson and Baird-Parker in 1975. Moreover, Francis and co-workers suggested in 1974 an accelerated version of fecal coliform enumeration on solid medium, giving an answer in 7 hours at 41.5°C.

Some of the more original techniques are able to give an even faster evaluation of these indicators. The detection of *E. coli* and fecal coliforms is possible by immunofluorescence after a few hours of enrichment (Abshire and Guthrie, 1973). Certain procedures based on the measurement of medium parameters may also be applied to the detection of coliforms or of *E. coli*. This is particularly true for impedance and radiometry. Bachrach and Bachrach (1974) applied a CO_2 radiometric technique for this purpose which enabled them to detect from one to ten coliforms in 6 hours. The more recent Petri Film gives results in 24–48 hours.

Today it may be said that the so-called rapid techniques are progressing very slowly. In fact, the nonprogressive conventional methods of enumeration continue to be references which although unsatisfactory are nevertheless inevitable.

17.3 The Enterobacteria

17.3.1 Justification for Their Use as Indicators

The principle of substituting the entire family of the Enterobacteriaceae for the coliform group as indicator organisms was proposed by Seeliger in 1952 for heat-treated milk; by Henriksen in 1955, Habs in 1958, and Kretzchmar in

1959 for chlorinated water; and especially by Mossel beginning in 1958 for the following reasons:

- Taxonomical definition of the coliform group is very imperfect.
- The false sense of security associated with detection of lactose-positive strains alone while pathogenic Enterobacteriaceae such as *Salmonella* are lactose-negative.
- The reduced sensitivity of the test when the number of lactose-positive coliforms is small compared to the number of total enterobacteria in a given product.

The enumeration of enterobacteria is recommendable when the equilibrium between the different flora has been disturbed due to any of the following:

- intrinsic antimicrobial properties of the food (low water activity or pH);
- storage under unfavorable conditions for nonsporulating (prolonged cold storage; freezing);
- heat treatment.

This analysis is currently recommended in France for pasteurized egg-based products.

17.3.2 Techniques of Analysis

The enumeration of enterobacteria may be done directly on a solid medium such as VRBG (crystal-violet, neutral-red, bile, glucose; Mossel et al., 1962) or in buffered glucose containing liquid broth medium with brilliant green and lauryl sulfate (Mossel et al., 1974).

17.4 The Fecal Streptococci

In the gram-positive, catalase-negative, facultative anaerobic Streptococcaceae family, the streptococci distinguish themselves by their coccoid form, their grouping into pairs or chains, and their homofermentative trait. In this group, the fecal streptococci are characterized by the Lancefield serological group D and by the fact that their normal habitat is the digestive tract of warm-blooded animals. They grow at 45°C and in media containing 40% bile. Among the fecal streptococci are certain enterococci (*Streptococcus faecalis* and its varieties as well as *S. faecium* and *S. durans*) which generally live in the human intestine, and two species (*S. bovis* and *equinus*) which live generally in the intestine of animals, although not as an absolute rule. These are generally innocuous in the intestines of animals, behaving as commensals, although their implication in certain food intoxications is not excluded.

17.4.1 Their Indicator Value

Fecal streptococci have long been used as indicators of fecal pollution of water, but this same use in the case of foods has met with much more debate.

Their specificity is satisfactory in spite of a few exceptions, certain enterococci multiplying on the surface of plants. In any event, it appears to be difficult to deduce from the number of fecal streptococci whether the contamination is of human or animal origin, although this distinction is of minor importance since some pathogens such as *Salmonella* are spread to humans by animals. Their widespread occurrence in the intestines of humans and animals does allow the detection of fecal contaminations with satisfactory sensitivity.

They have great resistance, which is both an advantage and a disadvantage. The vulnerability of *E. coli* to environmental conditions and to industrial treatments makes the hardier fecal streptococci a more certain indicator in the case of frozen, heated, or dehydrated foods. Conversely, however, their superior resistance compared to pathogens such as *Salmonella* makes their presence less indicative of real danger. Overall, there is agreement that the presence of excessive numbers is a sign of a manufacturing hygiene defect or of defective storage conditions. Their detection in the laboratory does not pose any special problems.

17.4.2 Enumeration Techniques

The techniques of analysis are comparable to those described for coliforms including the possibility of carrying out the analysis using either liquid or solid media.

17.4.2.1 Detection in Liquid Media

In the case of water, the official circular of January 21, 1960 (*Journal Officiel*, 15 March 1960) prescribes performing a presumptive test in Rothe medium followed by a confirmation test in Litsky medium (see Appendix, media 4 and 5). The same may be done for foods.

Each tube containing 10 ml of Rothe medium, of which the selective agent is sodium azothydrate, is inoculated with 1 ml of original suspension or decimal dilution. The original suspension is inoculated with 10 ml into 10 ml of double-strength medium. For water, inoculating five tubes per dilution is recommended. In the case of foods, usually three tubes per dilution are inoculated. The tubes are incubated at 37°C for 24–48 hours. Tubes showing microbial turbidity are considered positive and are subjected to tests of confirmation. The confirmation test is done in Litsky medium of which the selective agents are sodium azothydrate and ethyl-violet. The positive Rothe tubes are shaken and a loopful of culture is removed and transferred onto the Litsky medium which is then incubated at 37°C for 24–48 hours.

The presence of fecal streptococci appears as microbial turbidity throughout the medium and by the possible formation of a violet pellicle at the bottom of the tube (in the latter case, the turbidity of the medium may be very slight). In order to identify the streptococcal species thus detected, isolation on Barnes medium may be carried out by transferring a small quantity of positive Litsky medium with an inoculating loop.

17.4.2.2 Detection on Solid Medium

Numerous media may be used for this purpose. Slanetz medium (see Appendix, medium 6) is used for enumerating fecal streptococci in water by means of filtering membranes. It contains sodium azothydrate, sodium chloride, and 2,3,5-triphenyl tetrazolium chloride (TTC) as selective agents, the latter being added to the agar base just prior to use. After solidifying the medium, the membrane is placed on the surface for incubation at 37°C for 24–48 hours. Streptococcal colonies appear violet-red, with or without a white halo.

The medium Bea (Isenberg et al., 1970) (see Appendix, medium 7), which contains notably beef bile, esculine, sodium azide, and ferrous ammonium citrate, is surface inoculated and incubated for 24–48 hours at 37°C. Streptococcal colonies appear gray surrounded with a black halo (esculine positive).

Barnes medium with TTC (Barnes, 1956) (see Appendix, medium 8) allows identification. After incubation at 37°C for 24–48 hours, colonies of streptococci have the following appearances:

- *S. faecalis* and its varieties: center red surrounded by a white halo
- *S. faecium* and *S. durans*: white or pinkish
- *S. bovis*: very small, red or white

The media of Packer (Packer, 1943; Mossel, 1957) and Mead (1963) may also be used.

17.5 The Sulfite-Reducing *Clostridium* Species

Sulfite-reducing species of *Clostridium* are still sometimes used as fecal contamination indicators. Their sporulating ability obviously gives them great resistance. This will be discussed in further detail later in this book.

17.6 Conclusion

Relatively recent works (Miskimin et al., 1976; Solberg et al., 1976) show that the correlation between the presence and numbers of fecal contamination indicating bacteria and the actual risk which they represent is often weak and that, consequently, the aptness of the indicators to provide information about the actual health risk of a food to the consumer may be legitimately questioned. On the other hand, the monitoring of these indicators definitely allows testing of the value of the manufacturing practices with respect to hygiene and thereby represents an important factor in industrial progress.

In any event, it is now known that if indicators of fecal contamination are isolated, the measurement of the actual health risk of a given situation is revealed. In order to be really sure, all testing must be processed within an integrated evaluation system to reduce the risk to the consumer as much as possible. Fecal contamination indicators form an integral part of such systems.

17.7 Appendix: Culture Media

All formulas are expressed in grams per liter of distilled water.

1. BGLBB

Formula: brilliant green lactose bile broth

Peptone	10
Lactose	10
Beef bile	20
Brilliant green	0.0133
Final pH	7.2

Preparation

Completely dissolve 40 g of dehydrated medium in 1 l of distilled water by gentle stirring. Adjust the pH to 7.2, if necessary. Distribute 10 ml each into tubes with inverted Durham tubes. Sterilize by autoclaving at 115°C maximum for 20 minutes.

2. Deoxycholate agar

Formula

Peptone	10
Lactose	10
Sodium deoxycholate	1
Sodium chloride	5
Sodium citrate	2
Agar	12
Neutral red	0.03
Final pH	7.1

Preparation

Place 40 g of dehydrated medium in 1 l of distilled water previously heated to 100°C for 10 minutes and brought back to room temperature. Wait 5 minutes then mix until a homogeneous suspension is obtained. Heat slowly with frequent stirring and bring to boiling until totally dissolved. Do not autoclave. Allow to cool to about 50°C before use. Adjust the pH to 7.1 if necessary.

3. VRBL agar

Formula

Peptone	7
Yeast extract	5
Bile salts	1.5
Lactose	10
Sodium chloride	5
Agar	11
Neutral red	0.03
Crystal-violet	0.002
Final pH	7.4

Preparation

Place 38 g of dehydrated medium in 1 l of distilled water previously heated to 100°C for 10 minutes and brought back to room temperature. Wait 5 minutes then mix until a homogeneous suspension is obtained. Heat slowly with frequent stirring and bring to boiling until totally dissolved. Adjust the pH to 7.4 if necessary and distribute. This medium may be used as soon as it is prepared or sterilized by autoclaving at 110°C for 20 minutes.

4. Rothe medium

Formula

Peptone	10
Glucose	5
Sodium chloride	5
Dipotassium phosphate	2.7
Monopotassium phospate	2.7
Sodium azothydrate	0.2
Final pH	6.8–7

Preparation

To obtain single-strength medium, place 35.6 g of dehydrated medium in 1 l of distilled water. To obtain double-strength medium, place 71.2 g of dehydrated medium in 1 l of distilled water. Mix gently until totally dissolved. Adjust the pH to 6.8–7 if necessary. Distribute to tubes in 10 ml portions. Sterilize by autoclaving at 115°C for 20 minutes.

5. Litsky medium

Formula

Peptone	20
Glucose	5
Sodium chloride	5
Dipotassium phosphate	2.7
Monopotassium phospate	2.7
Sodium azothydrate	0.3
Ethyl-violet	0.0005
Final pH	6.8–7

Preparation

Place 35.7 g of dehydrated medium in 1 l of distilled water. Mix gently until totally dissolved. Adjust the pH to 6.8–7 if necessary. Distribute to tubes in 10 ml portions. Sterilize by autoclaving at 115°C for 20 minutes.

6. Slanetz medium

Formula

Peptone	20
Yeast extract	5
Glucose	2
Dipotassium phosphate	2
Sodium azothydrate	0.4
Agar	10
Final pH	7.2

Preparation

Place 41.5 g of dehydrated medium in 1 l of distilled water. To obtain double-strength medium, place 71.2 g of dehydrated medium in 1 l of distilled water. Wait 5 minutes then mix until a homogeneous suspension is obtained. Heat slowly with frequent stirring and bring to boiling until totally dissolved. Adjust the pH to 7.2 if necessary, distribute, and sterilize at 110°C for 20 minutes.

7. Bea medium

Formula

Peptone	20
Yeast extract	5
Beef bile	10
Esculine	1
Ferrous ammonium citrate	0.5
Sodium chloride	5
Sodium azide	0.15
Agar	15
Final pH	7.1

Preparation

Autoclave for 15 minutes at 121°C.

8. Barnes TTC medium

Formula

Peptone	10
Meat extract	10
Glucose	10
Agar	15
Final pH	6

Preparation

Distribute into 100 ml flasks. Autoclave for 20 minutes at 121°C. For use, melt in a boiling water bath and cool to about 50°C. Add to each 100 ml of basal medium 5 ml of 2% thallium acetate and 5 ml of 0.2% triphenyl tetrazolium chloride.

References

Abshire R. and Guthrie R., 1973. Fluorescent antibody as a method for the detection of fecal pollution: *Escherichia coli* as indicator organisms. *Can. J. Microbiol.*, **19**, 201–206.

Anderson J. and Baird-Parker A., 1975. A rapid and direct plate method for enumerating *Escherichia coli* biotype 1 in food. *J. Appl. Bacteriol.* **39**, no. 2, 111–117.

Bachrach V. and Bachrach Z., 1974. Radiometric method for the detection of coliform organisms in water. *Appl. Microbiol.*, **28**, 169–171.

Barnes M. J., 1956. *J. Appl. Bacteriol.*, **19**, 193.

Buttiaux R. and Mossel D. A. A., 1961. *The significance of various organisms of fecal origin in foods and drinking water*, **24**, 353.

Catsaras M., 1967. Les coliformes pigmentés des aliments déshydratés. *Ann. Inst. Pasteur de Lille*, **18**, 177–184.

Catsaras M. and Buttiaux R., 1965. Les *Aeromonas* dans les matières fécales humaines. *Ann. Inst. Pasteur de Lille*, **16**, 85–88.

Francis D., Peeler J., Twedt R., 1974. Rapid method for detection and enumeration of fecal coliforms in fresh chicken. *App. Microbiol.* **27**, no. 6, 1127–1130.

Isenberg et al., 1970. *Appl. Microbiol.* **20**, 433–436.

Mara D., 1973. A single medium for the detection of *Escherichia coli* at 44°C. *J. Hygiene*, **71**, 783–785.

Mead G. C., 1963. *Nature*, **197**, 1323.

Miskimin D., Berkowitz K., Solberg M., Riha W., Franke W., Buchanan R., 1976. Relationships between indicator organisms and specific pathogens in potentially hazardous foods. *J. Food Sci.*, **41**, no. 5, 1001–1006.

Mossel D. A. A. et al., 1957. *J. Appl. Bacteriol.*, **20**, 265.

Mossel D. A. A., Mengerink W. H. J., Scholts H. H., 1962. Use of a modified MacConkey agar medium for the selective growth and enumeration of *Enterobacteriaceae*. *J. Bacteriol.*, **84**, 381.

Mossel D. A. A., Harre Wijn G. A., Nisselrooy-Zadelhoff C. F. M. van, 1974. Standardization of the selective inhibitory effect of surface active compounds used in media for the detection of *Enterobacteriaceae* in foods and water. Processed for safety. *Health Lab. Sci.*, **11**, 260.

Solberg M., Miskimin D., Martin B., Page G., Goldner S., Libfeld M., 1976. What do microbiological indicator tests tell us about the safety of foods? *Food Prod. Dev.*, **10**, no. 9, 72–80.

Work of synthesis

Food-borne microorganisms of public health significance. AIFST-Food Microbiology group. CSIRO-Division of Food Research.

Thatcher F. S. and Clark D. S., 1968, 1973, 1975. *Micro-organisms in foods.* University of Toronto Press.

Suggested Readings

Barret T. J., Green J. H., Griffin P. M., Pavia A. T., Ostroff S. M., Wachsmuth I. K., 1991. Enzyme-linked immunosorbent assays for detecting antibodies to Shiga-like toxin I, Shiga-like toxin II, and *Escherichia coli* 0157:H7 lipopolysaccharide in human serum. *Curr. Microbiol.*, **23**, no. 4, 189–195; 26 ref.

Basem A., Gardini F., Paparella A., Guerzoni M. E., 1992. Suitability of a rapid gas chromatographic method for total mesophilic bacteria and coliform enumeration in hamburgers. *Letters Appl. Microbiol.*, **14**, no. 6, 255–259; 8 ref.

Bej A. K., DiCesare J. L., Haff L., Atlas R. M., 1991. Detection of *Escherichia coli* and *Shigella* spp. in water by using the polymerase chain reaction and gene probes for *uid*. *Appl. Environ. Microbiol.*, **57**, no. 4, 1013–1017; 14 ref.

Berger S. A., 1991. Ability of the Colibert method to recover oxidant-stressed *Escherichia coli*. *Letters Appl. Microbiol.*, **13**, no. 6, 247–250; 19 ref.

Brood M. G. and Bannister B. A., 1991. Verocytotoxin producing *Escherichia coli.*, *Br. Med. J.*, **303**, no. 6806, 800–801; 38 ref.

Bush C. E., Vanden-Brink K. M., Sherman D. G., Peterson W. R., Beninsig L. A., Godsey J. H., 1991. Detection of *Escherichia coli* rRNA using target amplification and time-resolved fluorescence detection. *Mol. Cell. Probes*, **5**, no. 6, 467–472; 22 ref.

Calabrese J. P. and Bissonnette G. K., 1990. Improved membrane filtration method incorporating catalase and sodium pyruvate for detection of chlorine-stressed coliform bacteria. *Appl. Environ. Microbiol.*, **56**, no. 11, 3558–3564; 45 ref.

Candrain U., Furrer B., Hoefelein C., Meyer R., Jermini M., Luethy J., 1991. Detection of *Escherichia coli* and identification of enterotoxigenic strains by primer-directed enzymatic amplification of specific DNA sequences. *Int. J. Food Microbiol.*, **12**, no. 339–352; 26 ref.

Clark D. L., Milner B. B., Stewart M. H., Wolfe R. L., Olxon B. H., 1991. Comparative study of commercial 4-Methylumbelliferyl-beta-D-glucuronide preparations with the Standard Methods membrane filtration fecal coliform test for the detection of *Escherichia coli* in water samples. *Appl. Environ. Microbiol.*, **57**, no. 5, 1528–1534; 22 ref.

Cook D. W. and Ruple A. D., 1989. Indicator bacteria and Vibrionaceae multiplication in post-harvest shellstock oysters. *J. Food Prot.* Ames, Iowa: International Association of Milk, Food, and Environmental Sanitarians. May 1989. **52**, no. 5, 343–349.

Covert T. C., Rice E. W., Johnson S. A., Berman D., Johnson C. H., Mason P. J., 1992. Comparing defined-substrate coliform tests for the detection of *Escherichia coli* on water. *J. Am. Water Works Assoc.*, **84**, no. 5, 98–104; 41 ref.

Curiale M. S., Fahey P., Fox T. L., McAllister J. S., 1989. Dry rehydrate films for enumeration of coliforms and aerobic bacteria in dairy products: Collaborative study. *J. Assoc. Off. Anal. Chem.* Arlington: The Association. March/April 1989. **72**, no. 2, 312–318.

Cryan B., 1990. Comparison of three assay systems for detection of enterotoxigenic *Escherichia coli* heat-stable enterotoxin. *J. Clin. Microbiol.*, **28**, no. 4, 792–794; 26 ref.

Entis, P., 1989. Hydrophobic grid membrane filter/MUG method for total coliform and *Escherichia coli* enumeration in foods: Collaborative study. *J. Assoc. Off. Anal. Chem.* Arlington: The Association. Nov./Dec. 1989. **72**, no. 6, 936–950.

Entis P. and Boleszczuk P., 1990. Direct enumeration of coliforms and *Escherichia coli* by hydrophobic grid membrane filter in 24 hours using MUG. *J. Food Prot.* Ames, Iowa: International Association of Milk, Food, and Environmental Sanitarians. Nov. 1990. **53**, no. 11, 948–952.

Franco B. D. G. M., Gomes T. A. T., Jakabi M., Marques L. R. M., 1991. Use of probes to detect virulence factor DNA sequences in *Escherichia coli* strains isolated from foods. *Int. J. Food Microbiol.*, **12**, no. 4, 333–338; 23 ref.

Frampton E. W. and Restaino L., 1993. Methods for *Escherichia coli* identification in food, water and clinical samples based on beta-glucuronidase detection. *J. Appl. Bacteriol.*, **74**, no. 3, 223–233.

Fratamico P. M., Schultz F. J., Buchanan R. L., 1992. Rapid isolation of *Escherichia coli* 0157:H7 from enrichment cultures of foods using an immunomagnetic separation method. *Food Microbiol.*, **9**, no. 2, 105–113; 26 ref.

Grabow W. O. K., de Villiers J. C., Schildhauer C. I., 1992. Comparison of selected methods for the enumeration of fecal coliforms and *Escherichia coli* in shellfish. *Appl Environ. Microbiol.*, **58**, no. 9, 3203–3204; 7 ref.

Green D. H., Lewis G. D., Rodtong S., Loutit M. W., 1991. Detection of faecal pollution of water by an *Escherichia coli uidA* gene probe. *J. Microbiol. Methods*, **13**, no. 3, 207–214; 14 ref.

Harris L. J. and Stiles M. E., 1992. Reliability of *Escherichia coli* counts for vacuum-packaged ground beef. *J. Food Prot.*, **55**, no. 4, 266–270; 28 ref.

Hornes E., Wasteson Y., Olsvik O., 1991. Detection of *Escherichia coli* heat-stable enterotoxin genes in pig stool specimens by an immobilized, colorimetric, nested polymerase chain reaction. *J. Clin. Microbiol.*, **29**, no. 11, 2375–2379; 26 ref.

Hsu H. Y., Chan S. W., Sobell D. I., Halbert D. N., Groody E. P., 1991. A colorimetric DNA hybridization method for the detection of *Escherichia coli* in foods. *J. Food Protect.*, **54**, no. 4, 249–255; 24 ref.

Jamiecon A. C. and Batt C. A., 1992. Fluorescent properties of the *Escherichia coli* D-xylose isomerase active site. *Protein Engineering*, **5**, no. 3, 235–240; 18 ref.

Kim M. S. and Doyle M. P., 1992. Dipstick immunoassay to detect enterohemorrhagic *Escherichia coli* 0157:H7 in retail ground beef. *Appl. Environ. Microbiol.*, **58**, no. 5, 1764–1767; 24 ref.

Kneifel W., Manafi M., Breit A., 1991. Adaption of two commercially available DNA probes for the detection of *Escherichia coli* and *Staphylococcus aureus* to selected fields of dairy hygiene—An exemplary study. *Zentralbl.-Hygiene Umweltmedizin*; **192**, no. 6, 544–553; 34 ref.

Law D., 1988. Virulence factors of enteropathogenic *Escherichia coli*. *J. Med. Microbiol.*, **26**, no. 1, 1–10; 82 ref.

Lee R. M. and Hartman P. A., 1989. Optimal pyruvate concentration for the recovery of coliforms from food and water. *J. Food Prot.* Ames, Iowa: International Association of Milk, Food, and Environmental Sanitarians. Feb. 1989. **52**, no. 2, 119–121. charts.

Loewenadler B., Lake M., Elmblad A., Holmgren E., Holmgren J., Karlstroem A., 1991. A recombinant *Escherichia coli* heat-stable enterotoxin (STa) fusion protein eliciting anti-STa neutralizing antibodies. *FEMS Microbiol. Lett.*, **82**, no. 3, 271–277; 20 ref.

Lortie L. A., Harel J., Fairbrother J. M., Dubreuil J. D., 1991. Evaluation of three new techniques for the detection of STb-positive *Escherichia coli* strains. *Mol. Cell. Probes*, **5**, no. 4, 271–275; 23 ref.

Matner R. R., Fox T. L., McIver D. E., Curiale M. S., 1990. Efficacy of Petrifilm *E. coli* count plates for *E. coli* and coliform enumeration. *J. Food Prot.* Ames, Iowa: International Association of Milk, Food, and Environmental Sanitarians. Feb. 1990. **53**, no. 2, 145–150.

Meyer R., Luethy J., Candrian U., 1991. Direct detection by polymerase chain reaction (PCR) of *Escherichia coli* in water and soft cheese and identification of enterotoxigenic strains. *Letters Appl. Microbiol.*, **13**, no. 6, 268–271; 7 ref.

Olsvik O., Wasteson Y., Lund A., Hornes E., 1991. Pathogenic *Escherichia coli* found in food. *Int. J. Food Microbiol.*, **12**, no. 1, 103–114; 29 ref.

Padhye N. V. and Doyle M. P., 1991. Rapid procedure for detecting enterohemorrhagic *Escherichia coli* 0157:H7 in food. *Appl. Environ. Microbiol.*, **57**, no. 9, 2693–2698; 20 ref.

Padhye N. V. and Doyle M. P., 1992. *Escherichia coli* 0157:H7: Epidemiology, pathogenesis, and methods for detection in food. *J. Food Prot.*, **55**, no. 7, 555–565; 126 ref.

Peterkin P. I., Conley D., Foster R., Lachapelle G., Milling M., Purvis U., Sharpe A. N., Malcolm S., 1989. A comparative study of total coliform recovery from foods by most probable number and hydrophobic grid membrane filter methods. *Food Microbiol.* London: Academic Press. June 1989. **6**, no. 2, 79–84.

Pollard D. R., Johnson W. M., Lior H., Tyler S. D., Rozee K. R., 1990. Rapid and specific detection of yerotoxin genes in *Escherichia coli* by the polymerase chain reaction. *J. Clin. Microbiol.*, **28**, no. 3, 540–545; 32 ref.

Preixens S. and Sancho J., 1987. [Evolution of coliform bacteria in natural yoghurt.] Evolution de los coliformes en el yogur natural. *Alimentaria.* Madrid: s.n., Nov. 1987. **24**, no. 187, 33–37.

Ray B., 1989. Enumeration of injured indicator bacteria from foods. Injured index and pathogenic bacteria: Occurrence and detection in foods, water, and feeds. Boca Raton, FL: CRC Press, 1989, 9–55.

Read S. C., Clarke R. C., de Grandis S. A., Hii J., McEwen S., Gyles C. L., 1992. Polymerase chain reaction for detection of verocytotoxigenic *Escherichia coli* isolated from animal and food sources. *Mol. Cell. Probes*, **6**, no. 2, 153–161; 44 ref.

Reid C. M., 1991. *Escherichia coli* 0157:H7—the "hamburger" bug: A literature review. Meat Industry Research Institute of New Zealand, no. 879, iv + 14 p.; 32 ref.

Roth J. N. and Bontrager G., 1989. Temperature-independent pectin gel method for coliform determination in dairy products: Collaborative study. *J. Assoc. Off. Anal. Chem.* Arlington: The Association. March/April 1989. **72**, no. 2, 298–302.

Schets F. M. and Havelaar A. H., 1991. Comparison of indole production and beta-glucuronidase activity for the detection of *Escherichia coli* in a membrane filtration method. *Letters Appl. Microbiol.*, **13**, no. 6, 272–274; 5 ref.

Sernowski L. P. and Ingham S. C., 1992. Low specificity of the HEC 0157 registered ELISA in screening ground beef for *Escherichia coli* 0157:H7. *J. Food Prot.*, **55**, no. 545–547; 11 ref.

Shrestha K. G. and Sinha R. N., 1987. Occurrence of coliform bacteria in dairy products. *Indian J. Dairy Sci.* New Delhi: Indian Dairy Association. March 1987. **40**, no. 1, 121–123.

Singleton E. R.; Hamdy M. K., McCay S. G., Zapatka F. A., 1991. Detection of fimbriae (CFA/I) on enterotoxigenic *Escherichia coli* using antibodies. *J. Food Safety*, **11**, no. 3, 215–225; 29 ref.

Smith H. R. and Scotland S. M., 1988. Vero cytotoxin-producing strains of *Escherichia coli*. *J. Med. Microbiol.*, **26**, no. 2, 77–85; 63 ref.

Smith H. R., Cheasty T., Roberts D., Thomas A., Rowe B., 1991. Examination of retail chickens and sausages in Britain for Vero cytotoxin-producing *Escherichia coli*. *Appl. Environ. Microbiol.*, **57**, no. 7, 2091–2093; 23 ref.

Smith J. L., 1988. Heat-stable enterotoxins: Properties, detection and assay. *Dev. Ind. Microbiol.*, **29**, 275–286; 89 ref.

Spangler B. D., 1992. Structure and function of cholera toxin and the related *Escherichia coli* heat-labile enterotoxin. *Microbiol. Rev.*, **56**, no. 4, 622–647; 297 ref.

Speirs J., Stavric S., Buchanan B., 1991. Assessment of two commercial agglutination kits for detecting *Escherichia coli* heat-labile enterotoxin. *Can. J. Microbiol.*, **37**, no. 11, 877–880; 13 ref.

Stavric S., Buchanan B., Speirs J., 1992. Comparison of a competitive enzyme immunoassay kit and the infant mouse assay for detection *Escherichia coli* heat-stable enterotoxin. *Letters Appl. Microbiol.*, **14**, no. 2, 47–50; 10 ref.

Sukupolvi S. and Vaara M., 1989. *Salmonella typhimurium* and *Escherichia coli* mutants with increased outer membrane permeability to hydrophobic compounds. *Biochimica Biophysica Acta*, **988**, no. 3, 377–387; 122 ref.

Tsuji T., Hjonda T., Miwatani T., Miyama A., 1991. Detection and purification of the free A subunit of heat-labile enterotoxin produced by enterotoxigenic *Escherichia coli*. *FEMS Microbiol. Lett.*, **77**, no. 2/3, 277–282; 24 ref.

Wasteson Y., 1991. Application of molecular biological methods in diagnosis of pathogenic *Escherichia coli*. Dissert.-Abs. Internat., -C; **52**, no. 4, 539.

Weeratna R. D. and Doyle M. P., 1991. Detection and production of verotoxin 1 of *Escherichia coli* 0157:H7 in food. *Appl. Environ. Microbiol.*, **57**, no. 10, 2951–2955; 23 ref.

Wernars K., Delfgou E., Soentorop P. S., Notermans S., 1991. Successful approach for detection of low numbers of enterotoxigenic *Escherichia coli* in minced meat by using the polymerase chain reaction. *Appl. Environ. Microbiol.*, **57**, no. 7, 1914–1919; 20 ref.

CHAPTER

18

The Genus *Salmonella*

J. Gledel, Béatrix Corbion

Bacteria of the genus *Salmonella* belong to the Enterobacteriaceae family, of which they share the principal traits.

18.1 Definition

According to Bergey's *Manual of Systematic Bacteriology* (9th edition, 1984), the genus *Salmonella* is defined "Gram-negative facultative anaerobic bacilli of a size in the range of 0.7–1.5 × 2.0–5.0 μm and usually motile by means of peritrichous cilia. Nonmotile mutants do occur and *S. gallinarum* is always nonmotile. *Salmonella* grows well on ordinary nutrient media producing colonies 2–3 mm in diameter within 18–20 hours except for certain serotypes which always give dwarf colonies (*abortusovis, abortusequi, typhisuis*). The genus *Salmonella* has a G + C% of 50–53."

18.2 Principal Biochemical Traits

Nitrate reduction	+	
Oxidase	–	
Catalase	+	
Carbohydrate fermentation	+	(with acid production)
Citrate utilization	+	(except *typhi* and *paratyphi* A)

Glucose fermentation with gas	+(*)	
Hydrogen sulfide production (H2S)	+(*)	
Lactose utilization	−(*)	
Sucrose utilization	−(*)	
Possession of various enzymes		
β galactosidase	−	(except subspecies III)
Urease	−	
Decarboxylases: lysine (LDC)	+	(*) (except *paratyphi* A)
ornithine (ODC)	d	(variable)
Dihydrolase: arginine (ADH)	−	(+ rare)
Deaminases: phenylalanine (APP)	−	
tryptophan	−	
Tetrathionate reductase	+	
Other traits:		
Mannitol	+	
Indole	−	
Acetylmethylcarbinol (VP)	−	

*The majority of strains are gas-producing and H_2S-producing, but some serotypes such as *typhi* and *Gallinarum* never produce gas. Nongas-producing variants of normally gas-producing serotypes may occur notably with *Dublin.* Most *Paratyphi* A strains and some *Cholerae-suis* strains do not produce H_2S. Some atypical strains may ferment lactose (e.g., *senftenberg*) or sucrose or may not decarboxylate lysine. *S. arizonae* may ferment lactose.

18.3 Antigenic Structure

The genus *Salmonella* is made up of a single species (see Section 18.4) which is divided into subspecies and serotypes based on the antigenic structure of the bacteria. These are distinguished as described below.

18.3.1 The Somatic Antigens ("O" Antigens)

Constituents of the bacterial cell wall and lipopolysaccharide (LPS) in nature make up the endotoxin of *Salmonella.* They are thermostable and alcohol-stable. They are made up of several elements, *lipid* A being responsible for the toxic effects while the "core" or basal portion and the *polysaccharide support* determine antigenic specificity, a trait associated with the nature of their constituent sugars and of their bonding (in the case of the S or smooth forms) (Le Minor and Veron, 1989).

The "O" antigenic factors are very numerous although only the major factors are retained for the characterization of the different antigenic groups (04 = group B, 09 = group D, etc.). Among the accessory "O" factors are those whose expression depends on the presence of a bacteriophage (lysogenic conversion). This is an example of interfering information having perhaps some epidemiological significance but which should not be taken into consideration for antigenic characterization of strains. This occurred particularly frequently in the former groups E_2 and E_3 of which the serotypes were carriers of phages, for example, the phage ε15 bringing about the transformation of O:3, 10 to O:3, 15 or the phase ε34 transforming O:3,*15* to O:3,*15*,*34*.

Agglutination obtained with anti "O" sera is slow, granular, and difficult to dissociate. It should be noted that some bacteria may lose their antigenic specificity and change over to the Rough (R) form as the result of a mutation. They thereby become self-agglutinatable in 2.0% salt water.

18.3.2 The Envelope Antigens (or Capsular K Antigens)

The only one of the envelope antigens recognized among *Salmonella* is antigen Vi (for *Vir*ulence) which may occur among *Typhi*, *Paratyphi C*, and *Dublin*. The presence of Vi antigen masks "O" agglutination. It is usually eliminated in order to detect somatic antigens (1 hour at 60°C). Two genes, Vi A and Vi B, must both be present in the bacteria in order for this antigen to be expressed.

In addition to envelope antigens, surface protein structures also exist, that is, pili which are differentiated into common pili (involved in mannose dependent hemagglutination) and sexual pili (involved in bacterial conjugation) and of which the presence is plasmid coded.

18.3.3 The Flagellar or "H" Antigens

Made up of a single protein, flagellin has a constant amino acid composition for a given antigenic type. It is dependent on two structural genes corresponding to phase 1 and phase 2. The majority of *Salmonella* are diphasic although a certain number appear to be monophasic, one of the phases being absent. Phase 1 is designated by a letter and phase 2 by a numeral. Dominance of one phase over the other may mask the expression of one phase. In such cases the masked phase may be detected by using the so-called phase inversion technique of Sven-Gard, that is, culture in medium with a low concentration of agar and containing antidominant phase serum. The "H" antigens are thermolabile and are destroyed by alcohol. Moreover, H agglutination is fast, flaky, and easily disrupted.

18.4 Taxonomy

New data have been acquired in this field which have led to the development of proposals of which the essential elements are summarized here.

18.4.1 Older Data

Until recently, the genus *Salmonella* was considered to be made up of a certain number of species which corresponded to serotypes. The genus itself was subdivided into four subgenera according to a scheme described by Kauffman

(1975) based on the following traits:

	SG I	SG II	SG III	SG IV
Dulcitol	+	+	−	−
Lactose	−	−	+/x	−
ONPG	−	−	+	−
Salicin	−	−	−	+
Gelatin	−	+	+	+
Malonate	−	+	+	−
d-Tartrate	+	−/x	−/x	−/x
KCN	−	−	−	−

The majority of the strains used to belong to subgenus I while subgenus III included the strains described under the name *S. arizonae*. The other two subgenera included only a few strains. + is positive, − is negative, x is variable.

18.4.2 Newer Data

Le Minor and co-workers (1982, 1986, 1987) proposed significant modifications to the classification based on studies focusing on phenotypic and genomic traits (by DNA/DNA hybridization) and which demonstrate that *the genus* Salmonella *is in fact made up of a single species*. This species, for which Le Minor et al. have suggested the name *Salmonella enterica*, is divided into seven subspecies according to the following taxonomic classification:

Subspecies I: *S. enterica* ssp. *enterica*
Subspecies II: *S. enterica* ssp. *salamae*
Subspecies IIIa: *S. enterica* ssp. *arizonae*
Subspecies IIIb *S. enterica* ssp. *diarizonae*
Subspecies IV: *S. enterica* ssp. *houtenae*
Subspecies V: *S. enterica* ssp. *bongori*
Subspecies VI: *S. enterica* ssp. *indica*

The resemblance to the Kauffmann scheme is quite obvious, the principal differences being the recognition of subspecies V and VI as well as the distinguishing of two subspecies within the former subgenus III, monophasic *arizonae* and biphasic *diarizonae*.

Since the use of this system turns out to be very cumbersome in practical terms for the expression of results and creates interpretation difficulties for bacteriologists, the authors recommend retaining the names of the serotypes for subspecies I, which represents 99.5% of the isolated strains, and to designate the others by their single antigenic formula. The names given to the serotypes no longer have any taxonomical status and may be written simply with a capital letter (Typhimurium, Dublin,...). It remains acceptable nevertheless to use the form *S. typhimurium*. The seven subspecies are differentiated by the traits listed in Table 18.1.

Table 18.1 Distinguishing traits of the seven subspecies of *Salmonella*

	Subspecies						
	I	II	IIIa	IIIb	IV	V	VI
Dulcitol	+	+	−	−	−	+	d
OPNG (2 hours)	−	−	+	+	−	+	d
Malonate	−	+	+	+	−	−	−
Gelatinase	−	+	+	+	+	−	+
Sorbitol	+	+	+	+	+	+	−
Growth on KCN	−	−	−	−	+	+	−
L(+)-tartrate (a)	+	−	−	−	−	−	−
Galacturonate	−	+	−	+	+	+	+
γ-glutamyltransferase	+(*)	+	−	+	+	+	+
β-glucuronidase	d	d	−	+	−	−	d
Mucate	+	+	+	−(70%)	−	+	+
Salicin	−	−	−	−	+	−	−
Lactose	−	−	−(75%)	+(75%)	−	−	d
Lysis by phage 01	+	+	−	+	−	d	+
Habitat of the majority of strains	Warm-blooded animals		Essentially cold-blooded animals and the environment				

(a) = *d*-tartrate; (*) *Typhimurium* d, *Dublin*-; + = at least 90% of results positive. − = at least 90% of results negative; d = results variable.

Source: Adapted from Le Minor and Popoff, 1987.

Remarks: Currently, more than 2,200 serotypes have been ratified, of which over 50% (1,299 of 2,213) belong to subspecies I.

Lysis by phage 01 (Felix and Callow), which may be used as the initial confirmation test for belonging to the species *Salmonella*, does not provide a positive result with all strains. Subspecies IIIa and IV are not lysed. Moreover, some strains of groups 3, 10 (E_1) and 1, 3, 19 (E_4) of subspecies I are not lysed by this phage (to be considered when dealing with atypical *Salmonella* such as *Senftenberg* lactose + found frequently among poultry).

18.5 The Kauffmann-White Scheme

The Kauffmann-White Scheme constitutes a nomenclature of the serotypes, classed according to their antigenic formulas. It currently includes only the factors necessary for practical identification and few accessory factors (Le Minor and Popoff, 1987). The classification is by antigenic group 0—group 02, 04, 07, etc.—and within each "O" group according to the phase 1 and phase 2 flagellar antigens. The antigenic formula characterizes a strain as belonging to a serotype, for example, *Typhimurium*:1, 4 [5], 12:i:1, 2.

The factors associated with lysogenization are underlined and the accessory factors without any relationship to the presence of a bacteriophage are given in brackets. This is to take into account the fact that the possession of factors associated with lysogeny were of little taxonomic interest, that some serotypes have been dropped and are now designated as variants of another serotype. This concerns the old groups C4 (6,7,*14*), E2 (3,*15*), and E3 (3,15,*34*). *Examples:*

- *Eimsbuettel* is included in *Livingstone* var 14+
- *Minneapolis* is included in *Anatum* var 15+34+

18.6 Complementary Distinctive Traits

These are especially of epidemiological interest.

18.6.1 Biotypes

Possession of certain biochemical traits allows a better characterization of some strains once it has been established that they belong to the genus *Salmonella* (e.g., biotypes of *Typhimurium*).

18.6.2 Lysotypes

The determination of lysotypes is done with the aid of specific bacteriophages. Phage systems exist for the principal serotypes (*Typhi*, *Paratyphi* A and B, *Typhimurium*, *Enteritidis*, *Dublin*, *Montevideo*, etc.). Some are recognized and used in all of the major centers of specialization while others are only complementary systems defined locally. In France, the lysotyping of *Salmonella* strains is ensured by the Centre National de Lysotypie Entérique at the Pasteur Institute in Paris.

18.6.3 Plasmid Profiles

In spite of the fact that the possession of plasmids is a transcient phenomenon (either loss or acquisition is possible), it is possible to turn to plasmid examination in order to establish a common origin for some strains (e.g., those isolated from a patient and found in a food). Threlfall and colleagues (1986) have made an excellent demonstration of this with the *Goldcoast* strains, as has Whiley and co-workers (1988) with *Typhimurium*. However, more importance must still be ascribed to chromosomal gene coded traits than to those coded by extrachromosomal factors.

18.7 Epidemiology-Pathogenic Potential

18.7.1 Epidemiology

Salmonella are associated naturally with the bodies of all domestic and wild animals, these along with the environment making up their actual reservoir. The role of healthy carriers in the dissemination of *Salmonella* is decisive. *Salmonella* are quite resistant in external media and the Netherlands school (Edel et al., 1975) has well demonstrated the existence of contaminating cycles. Food is the primary vector of *Salmonella* towards humans, especially food of animal origin (butchered meats, poultry, eggs, and derivatives thereof). Their contamination is most often very slight and superficial, which explains why *Salmonella* are easily destroyed during the normal preparation of foods. In the majority of cases, food intoxications or foodborne infections are the result of errors in handling which provide the bacteria with the opportunity to grow abundantly. The programs proposed thus far to combat the risks associated with the dissemination of these bacteria have not yet met with great success (Kampelmacher, 1983).

18.7.2 Pathogenic Potential

Aside from a few serotypes, each adapted to a particular species (*Typhi* in humans, *Abortusovis* in sheep), the majority of *Salmonella* are "ubiquitous" bacteria. Among these, only a limited number of serotypes show any pathogenic capability in humans or animals (an average of 200 serotypes are isolated each year), although some such as *Typhimurium* are found at a high and constant frequency. Others emerge more or less rapidly and establish themselves or decrease without any valid explanation. *Enteritidis* is currently showing up actively in human pathology, apparently in association with the consumption of eggs or egg-based products which are contaminated and insufficiently heated.

The symptoms associated with *Salmonella* food poisoning are those of gastroenteric fever, usually without serious consequences. Severe and possibly fatal forms may occur among weakened patients (the aged). The expression of pathogenic power appears to be associated with the possession of virulence factors carried by plasmids, among these a large 60 Mdal plasmid (Pardon et al., 1986; Helmuth et al., 1983; Popoff, 1984).

It is still important to consider, however, that the majority of serotypes isolated from foods (mainly subspecies I) may manifest a pathogenic nature as soon as the *Salmonella* are present in sufficient numbers (i.e., in the range of 10^6 cells). There are exceptions to this rule.

Remark: Salmonella frequently harbor plasmids which carry antibiotic resistance factors. This has been used to identify antibiotypes. It is considered wise to remain highly critical with respect to this sort of information especially for epidemiological purposes.

18.8 Detection of *Salmonella* in Foods

Principles: With the exception of analyses carried out for the investigation of food-poisonings, the purpose of *Salmonella* tests is to detect small numbers of this species usually dominated by associated flora which may be either co-contaminants or agents added for technological purposes and therefore much more abundant. For this reason, the methods implemented conventionally are intended to favor the proliferation of *Salmonella* and inhibit adventitious flora. Articles by Harvey and Price (1979), Fricker (1987), and Cox (1988) are worth consulting. The analysis consists of the following successive steps:

- preenrichment;
- enrichment in selective liquid media;
- isolation on selective solid media;
- identification.

The technical details appear in the AFNOR V-08-013, ISO/DIS 6579, and FIL 93A-1985 standards which codify the reference bacteriological methods of which the complete implementation may require 1 week.

Various new methods are currently being proposed in order to obtain the information more rapidly and to simplify the analytical procedures. These employ immunology and genetic engineering and permit screening of lots. In the event of positive reactions, they must be backed up by conventional tests allowing the final characterization the serotypes present.

18.8.1 Conventional Methods

18.8.1.1 Test Sample

The analysis of *Salmonella* is usually carried out using 25 g of food. There may be a variety of technical solutions depending on the purpose of the analysis, for example, analyzing for surface contamination of poultry by a rinsing technique or by skin analysis (neck).

When the food being analyzed is intended for consumption by a high-risk population, the volume or weight of the samples is increased. For some dairy products, 15, 30, or 60 samples of 25 g may be taken and pooled into composite samples (15×25 g) for analysis. The rate of recovery of *Salmonella* is affected by the mode of rehydration. On this subject it may be very useful to consult the *Bacteriological Analytical Manual* (BAM, 6th edition) of the U.S. Food and Drug Administration (FDA) or the FIL 93A-1985 standard.

18.8.1.2 Preenrichment

Preparation of the original suspension is usually done using peptone water (PW) in a 1:9 ratio (e.g., 25 g of product + 225 ml PW). Peptone-salt solution, Ringer's solution, or phosphate buffer may be used for dairy products

(FIL 122A, 1988). The preenrichment incubation phase at 37°C for 16–20 hours must allow the stressed bacteria to recover all of their growth potential. The length of this incubation may at times be reduced and the preenrichment step may even be eliminated (e.g., for unprocessed starting materials).

18.8.1.3 Enrichment

Enrichment is carried out by transferring a variable volume of preenrichment medium or original suspension to selective liquid media and incubating these either at 37° or 43°C for 24–48 hours. The following enrichment media are used:

- sodium tetrathionate and brilliant green broth (Muller-Kauffmann);
- sodium selenite broth (possibly with cystine and novobiocin);
- Rappaport-Vassiliadis (RV 10) with malachite green and magnesium chloride.

The first two media are inoculated either with 1 ml or with 10 ml (in a 1:10 ratio, 1 ml + 10 ml or 10 ml + 100 ml). Incubation may be done at 37° or 43°C for 24–48 hours (the ISO standard recommends 43°C for tetrathionate broth and 37°C for selenite broth).

Inoculation of RV 10 medium, on the other hand, should be done at a ratio of 1:100, that is, either 0.1 ml + 10 ml RV 10 or 1 ml + 100 ml RV 10. Incubation is at 43°C for 24 or 48 hours. This medium gives excellent results especially with meats and meat products. Its substitution for one of the other two media is being contemplated for international standards.

Note: Laboratories which analyze *Salmonella* only intermittently may also use a system of simpler presentation known as the "Salmosyst" (Merck). De Smedt and co-workers (1987) proposed using a modified semisolid Rappaport-Vassiliadis medium following a 6 hour preenrichment, from which a serological test may be carried out with detection of positive samples within 24 hours.

18.8.1.4 Isolation

Using the enrichment media incubated for 24–48 hours, the following two solid media are inoculated:

- an agar-based medium with phenol red and brilliant green (BGA, formula of Edel and Kampelmacher);
- a second agar-based medium from among the following: Hektoen agar, bismuth sulfite (BS) agar, XLD agar (xylose, lysine, deoxycholate), DCLS agar (deoxycholate, citrate, lactose, sucrose).

A new medium has been proposed, MLCB (mannitol, lysine, crystal-violet, brilliant green) by Van Schothorst and associates (1987). This medium is quite useful, particularly for the detection of atypical strains (it does not detect Typhi and Paratyphi A). Rambach medium (using propylene glycol) also appears to be of great value.

When there are grounds for presuming the possible presence of atypical strains (e.g., those fermenting lactose), attention should be given to colonial characteristics on media containing this sugar as well as to the simple presence of L+ colonies. For this, it may be useful to use a second medium detecting *Salmonella* on the basis of other biochemical capabilities.

Incubation of the selective media is done at 37°C for 24–48 hours.

Note: Entis (1985) recommended the use of hydrophobic membranes with the Isogrid system (HGMF: hydrophobic grid membrane filter, GA Laboratories Ltd.) onto which 1 ml of enrichment medium is filtered, the membranes then being placed on the selective solid media.

18.8.1.5 Identification

Several (at least five) characteristic and noncharacteristic colonies must be transferred. The greater the number of transferred colonies, the greater the chances of detecting the presence of several serotypes or atypical strains.

18.8.1.5.1 Presumptive Tests

Use of phage 01 with the restrictions just mentioned (response time 6–8 hours). MUCAP test (AES) using methylumbelliferone with observation of blue fluorescence in Wood light.

18.8.1.5.2 Confirmation of Biochemical Traits

The following media are to be inoculated using colonies recovered on the solid media:

- TSI agar or Kligler agar (glucose, lactose/sucrose, H_2S, gas);
- urea agar or urea-indole medium (urease −, indole −);
- Taylor lysine medium (decarboxylase +);
- mannitol-motility medium (also detecting nitrate reductase);
- Sodium malonate medium (used essentially to detect strains belonging to subspecies IIIa and IIIb);
- ordinary nutrient agar.

β-galactosidase testing (not to eliminate β-galactosidase + strains but those with the biochemical traits of *Salmonella*).

The production of H_2S by lactose or sucrose fermenting strains may not be observed on TSI or Kligler agars.

Note: Various miniaturized identification systems are available. The most widely used in France is the API 20E. Others include the Miniteck System (BEL), MicrolD (Organon-Teknica), Enterotube II (Roche). More recent is an identification micromethod developed by Diagnostic Pasteur under the name of MIS Enterobacteriaceae, which relies on reading absorbance on different substrates distributed to microplates and interpreting results using a bench-top computer (response in 16–24 hours).

18.8.1.5.3 Serological Confirmation

Strains isolated from selective media and having the essential biochemical traits of *Salmonella* are subjected to slide agglutination tests using agglutinating sera to narrow down their antigenic structure or simply to confirm the biochemical identification.

So-called mixed polyvalent anti "O" and anti "H" sera have been commercialized and confirm identification of *Salmonella*. It is therefore sufficient in most cases to transfer the strains and to send them to a specialized center in order to obtain a complete identification of the serotype (use of monovalent sera following the scheme of Kauffmann-White). Strains must always be accompanied by detailed records.

In France, strains isolated from foodstuffs may be sent to the Laboratoire Central d'Hygiène Alimentaire (LCHA), 43 rue de Dantzig, 75015 Paris. This epidemiological monitoring is carried out in association and in harmony with the Central National des *Salmonella* (Institut Pasteur in Paris). The LCHA publishes an inventory every 2 years of all *Salmonella* strains received during the period under consideration (last printing in 1988 for the period 1986–1987).

Note: Although priority must be given to the observation of biochemical traits, in the event of disagreement between biochemical and serological confirmations, a qualified bacteriologist must give special attention to the identification.

18.8.2 Nonconventional Methods

Articles in *Food Technology* (1987, 41, 7, 54–73) and by Ibrahim and coworkers (1985) may be consulted.

The development of new so-called rapid methods is currently very active. In fact, it is more appropriate to refer to these as "more rapid" methods with respect to conventional techniques.

Note: In the case of positive tests, these specific methods require that the analysis be pursued using the usual bacteriological techniques in order to be able to completely identify the serotypes.

This evolution in methods development is fueled primarily by immunology and genetic engineering.

18.8.2.1 Immunological Techniques (Ibrahim, 1986)

Immunofluorescence has not been effectively applied in this field. Sero-enrichment (Sperber and Deibel, 1969) relies on a pool of anti "H" sera and allows numerous tests to be carried out under satisfactory conditions.

Immunoenzymatic methods (EIA: Enzyme Immunoassay)

- The Salmonella Bio-Enzabead Test proposed by Organon-Teknica (Transia) consists of the use of two monoclonal antibodies specific for flagellar structures and which used to be attached to metallic beads but has been modified. This may yield results in 48 hours. Prior steps of preenrichment, enrichment, and postenrichment are required while the assay itself requires 1 hour and 30 minutes (Mattingly et al., 1985; Eckner et al., 1987; Flowers et al., 1986–1988a, b).
- The Tecra *Salmonella* visual immunoassay (Bioenterprises Pty Ltd.). The initial phases are similar to those of the previous test but the assay is carried out in microplates to which specific antibody is attached. Reading is done by direct visual comparison or with the aid of a reader (Flowers et al., 1988b).
- The Immunoband-*Salmonella* 1-2 Test (Biocontrol System Inc. AES). This system consists of a chamber containing a nonselective medium favoring motility and a second perpendicular chamber containing a selective enrichment medium. A polyvalent antiflagellar serum is added to the mobility chamber. If *Salmonella* are present, a white line of immunoprecipitation is observed. Only motile *Salmonella* may be detected.
- The Chemiluminescence Immunoassay: Lumi-Phage Testing System (MCLAS Technologies Inc.). Results may be obtained within 19 hours.

18.8.2.2 Hybridization Techniques (Tompkins, 1986)

Hybridization techniques are based on the use of DNA probes, made of single-stranded nucleic acid sequences labeled either with a radionuclide or an enzymatic system. These sequences are exactly complementary to the nucleic acids of the target organism. In the presence of the latter, the probe hybridizes, yielding a double-stranded system which is labeled and therefore detectable. Some hybridization techniques that are currently available:

The Gene-Trak *Salmonella* Assay (Gene Trak System-3M) which includes the use of a ^{32}P labeled DNA probe (maximum period usable = 14 days). A prior enrichment in target organisms is necessary (preenrichment, enrichment, postenrichment) and a nylon membrane is used for filtering. The bacterial DNA is treated to yield single strands and the probe is added. Radioactivity is measured with the aid of a liquid scintillation counter (Flowers et al., 1987, no. 2).

18.8.2.3 Other Techniques

Measurements of Bacterial Development in Selective Media (Impedance)

Measurements of conductance are often done, which is one of the factors which make up impedance (Ogden et al., 1987; Gibson, 1987). Selenite-cystine/trimethylamine oxide (TMAO)/dulcitol medium is used, which favors the multiplication of *Salmonella*. Mannitol has been proposed to replace dulcitol.

A prior preenrichment is done preferably. The measurement of impedance or conductance and the interpretation of their variations are performed using equipment which includes a desktop computer (Malthus) and which thus represent a significant investment.

Another system has been proposed by Bactomatic Inc. (Bactometer Microbiol Monitoring System).

Radiometry

Measurement of $^{14}CO_2$ formed from a labeled substrate (^{14}C dulcitol).

The list of "rapid" methods already given is not meant to be exhaustive and is certainly less so given that after a period of relative dormancy, specialized firms have once again begun to pursue the agrifood industrial testing and monitoring market with sustained interest.

The techniques and methods outlined in this chapter have, for the most part, been validated during collaborative studies carried out under the auspices of the U.S. Association of Official Analytical Chemists (AOAC) and are considered recognized as usable by the U.S. FDA. It remains nevertheless useful to verify whether or not techniques or methods have been used for the analysis of *Salmonella* in the type of food to be tested before implementing them and to not forget that even validated methods (AFNOR) include some element of uncertainty. The quality testing required in laboratories may, as far as the detection of *Salmonella* is concerned, benefit from the development of reference samples (Beckers et al., 1987).

References

Beckers H. J., Roberts D., Pietzsch O., Van Schothorst M., Vassiliadis P., Kampelmacher E. H., 1987. Reference samples for checking the performance *Salmonella* isolation methods. *Intern. J. Food Microbiol.*, **4**, 51–75.

Cox N. A., 1988. *Salmonella* methodology update. *Poult. Sci.*, **67**, 921–927.

De Smedt J. M., Bolderdijk R. F. 1987. Dynamics of *Salmonella* isolation with modified semi-solid Rappaport-Vassiliadis medium. *J. Food Protect.*, **50**, 379–385.

Eckner K. F., Flowers Russel S., Robison Barbara J., Mattingly J. A., Gabis D. A., Silliker J. H., 1987. Comparison of *Salmonella* bio-enzabead immunoassay method and conventional culture procedure for detection of *Salmonella* in foods. *J. Food Protect.*, **50**, 379–385.

Edel W., Von Schothorst M., Kampelmacher E. H., 1975. Epidemiological studies on *Salmonella* in a particular area (Walcheren Project). The presence of *Salmonella* in man, swine, insects and seagulls as well as in foods and effluents.

Entis P., 1985. Rapid hydrophobic grid membrane filter for *Salmonella* detection in selected foods. Collaborative study. *J. Assoc. of Anal. Chem.*, **68**, 555–564.

Flowers Russel S., Eckner K., Gabis D. A., Robison Barbara J., Mattingly J. A., Silliker J. H., 1986. Enzyme Immunoassay for Detection of *Salmonella* in Foods. Collaborative study.

Flowers Russel S., Klatt Mary J., Mozola M. A., Curiale M. S., Gabis D. A., Silliker J. H., 1987. DNA hybridization assay for detection of *Salmonella* in foods. Collaborative study.

Flowers Russel S., Klatt Mary J., Robison Barbara J., Mattingly J. A., 1988a. Evaluation of abbreviated enzyme immunoassay method for detection of *Salmonella* in low-moisture foods.

Flowers Russel S., Klatt Mary J., Keelan Susan L., 1988b. Visual immunoassay for detection of *Salmonella* in foods. Collaborative study. JAOAC., **71**, 973–980.

Fricker C. R., 1987. The isolation of *Salmonella* and *Campylobacter*. *J. Appl. Bacteriol.*, **63**, 99–116.

Gibson D. M., 1987. Some modification to the media for rapid automated detection of *Salmonellas* by conductance measurement.

Harvey R. W. S. and Price T. H., 1979. Principles of *Salmonella* isolation.

Helmuth R., Stephan R., Bunge C., Hoog B., Steinbeck A., Bulling E., 1985. Epidemiology of virulence-associated plasmids and outer membrane protein-patterns within seven common *Salmonella* serotypes.

Ibrahim G. F., 1986. A review of immunoassays and their application to *salmonellae* detection in foods.

Ibrahim G. F. and Fleet G. M., 1985. Detection of *salmonellae* using accelerated methods.

Kampelmacher E. H., 1983. 51e Session Générale de l'Off. Inter. Epizooties, Paris (May 22–28).

Le Minor L., Veron H., Popoff M., 1982. Taxonomie des *Salmonella*. *Ann. Microbiol.* (Institut Pasteur). **133B**, 223–243. Proposition pour une nomenclature des *Salmonella*. *Ann. Microbiol.* (Institut Pasteur). **133B**, 245–254.

Le Minor L. and Popoff M. Y., 1987. Designation of *Salmonella enterica* sp. nov. nom. rev. as the type and only species of the genus *Salmonella*.

Mattingly J. A., Robison Barbara J., Boehm Any, Ghele W. D., 1985. *Food Technol.*, **39**, no. 3, 90–94.

Ogden I. D. and Cann O. C., 1987. A modified conductance medium for the detection of *Salmonella* spp.

Pardon P., Popoff M. Y., Coynault C., Marly J., Miras I., 1986. Virulence-associated plasmids of *Salmonella* serotype typhimurium in experimental murine infection.

Popoff M. Y., Miras I., Coynault C., Lasseun C., Pardon P., 1984. Molecular relationships between virulence plasmids of *Salmonella* serotype typhimurium and Dublin and large plasmids of other *Salmonella* serotypes.

Rambach A., 1990. New plate medium for facilitated differentiation of *Salmonella* spp. from *Proteus* spp. and other enteric bacteria.

Sperber W. H. and Deibel A. H., 1969. Accelerated procedure for *Salmonella* detection in dried foods and feeds involving only broth cultures and serological reactions.

Threlfall E. J., Hall M. L. M., Rowe R., 1986. *Salmonella* Gold-coast from outbreaks of food-poisoning in the British Isles can be differentiated by plasmid profiles.

Tompkins Lucy S., Troup Nancy, Labigne Agnès, Mitchell L. Cochen, 1986. Cloned, random chromosomal sequences as probes to identify *Salmonella* species.

Van Schothorst M., Renaud A., Van Beck C., 1987. *Salmonella* isolation using RVS broth and MLCB agar.

Whiley Susan J., Lanser Janier A., Manning P. A., Murray C., Steele T. W., 1988. Plasmid profile analysis of a salmonellosis outbreak and identification of a restriction and modification system.

Works

FDA. *Bacteriological analytical manual.* 6th ed., 1984. Publ. and dist. by Association of Official Analytical Chemists. Arlington, VA, 22209, USA.

Kauffmann F., 1975. *Classification of bacteria.* Munksgaard, Copenhagen.

Le Minor L. and Veron H., 1989, 2d ed. *Bacteriologie médicale.* Flammarion, Medecine Sciences, Paris Ed.

Suggested Readings

Barker R. M. and Old D. C., 1989. The usefulness of biotyping in studying the epidemiology and phylogeny of salmonellae. *J. Med. Microbiol.*, **29**, no. 2, 81–88; 52 ref.

Beckers H. J., Tips P. D., Soentoro P. S. S., Delfgou van Asch E. H. M., Peters R., 1988. The efficacy of enzyme immunoassays for the detection of salmonellas. *Food Microbiol.*, **5**, no. 3, 147–156; 21 ref.

Blackburn C., 1991. Detection of *Salmonella* in foods using impedance. *Europ. Food Drink Rev.*, Winter, 35, 37, 39–40; 21 ref.

Cartwright K. A. V. and Evans B. G., 1988. Salmon as a food-poisoning vehicle—two successive salmonella outbreaks. *Epidemiology Infection*, **101**, no. 2, 249–257; 13 ref.

Comerchero V. 1988. Effect of chilling and/or freezing on the survival of salmonellae in poultry and on meat quality. *Food Sci. Technol. Today*, **2**, no. 4, 272–278; 48 ref.

Coombs P., Ligugnana R., Rovera E., 1991. [Rapid determination of salmonellae using the Malthus technique.], *Latte*, **16**, no. 1, 40–43; 11 ref.

D'Aoust J. Y., 1991. Pathogenicity of foodborne *Salmonella. Int. J. Food Microbiol.*, **12**, no. 1, 17–40.

D'Aoust J. Y., Sewell A. M., Greco P., 1991. Commercial latex agglutination kits for the detection of foodborne *Salmonella. J. Food Prot.*, **54**, no. 9, 725–730; 21 ref.

Eley A., 1990. New rapid methods for *Salmonella* and *Listeria. Br. Food J.*, **92**, no. 4, 28–31; 17 ref.

El-Gazzar F. E. and Marth E. H., 1992. Salmonellae, salmonellosis, and dairy foods: A review. *J. Dairy Sci.*, **75**, no. 9, 2327–2343; 140 ref.

Feng P., 1992. Commercial assay systems for detecting foodborne *Salmonella*: A review. *J. Food Prot.*, **55**, no. 11, 927–934; 45 ref.

Foster K., Garramine S., Ferraro K., Groody E. P., 1992. Modified colorimetric DNA hybridization method and conventional culture methods for detection of *Salmonella* in foods: Comparison of methods. *J. AOAC Int.*, **75**, no. 4, 685–692; 7 ref.

Jackson G. J., Langford C. F., Archer D. L., 1991. Control of salmonellosis and similar foodborne infections. *Food Control*, **2**, no. 1, 26–34; 72 ref.

Jetton J. P., Billgili S. F., Conner D. E., Kotrola J. S., Reiber M. A., 1992. Recovery of salmonellae from chilled broiler carcasses as affected by rinse media and enumeration method. *J. Food Prot.*, **55**, no. 5, 329–332; 20 ref.

Krueger G. and Fehlhaber K., 1992. [Immunological procedures for detection of Salmonella in foods.] Gegenwaertiger Stand der Entwicklung immunologischer Verfahren zum Salmonella-Nachweis aus Lebensmitteln. *Minats. fuer Veteri.*, **47**, no. 11, 578–586.

Manafi M. and Sommer R., 1992. Comparison of three rapid screening methods for *Salmonella* spp.: "MUCAP test, MicroScreen Registered latex and Rambach agar." *Letters Appl. Microbiol.*, **14**, no. 4, 163–166; 9 ref.

Marth E. H., 1993. Growth and survival of *Listeria monocytogenes*, *Salmonella* species, and *Staphylococcus aureus* in the presence of chloride: A review. *Dairy Food Environ. Sanitation*, **13**, no. 1, 14—18; 41 ref.

Miyamoto T., Yonemura K., Yoshimoto M., Morinaga Y., Hatano S., 1991. Rapid detection of *Salmonella* by fluorogenic assay. *Jpn. J. Food Microbiol.*, **8**, no. 3, 143–150; 14 ref.

Prusak-Sochaczewski E. and Luong J. H. T., 1989. Utilization of two improved enzyme immunoassays based on avidin-biotin interaction for the detection of *Salmonella. Int. J. Food Microbiol.*, **8**, no. 4, 321–333; 17 ref.

Roberts T., 1988. Salmonellosis control: Estimated economic costs. *Poult. Sci.*, **67**, no. 6, 936–943; 25 ref.

Scholl D. R., Kaufmann C., Jollick J. D., York C. K., Goodrum G. R., Charache P., 1990. Clinical application of novel sample processing technology for the identification of salmonellae by using DNA probes. *J. Clin. Microbiol.*, **28**, no. 2, 237–241; 8 ref.

Sharpe J. C. M., 1990. Salmonellosis. *Br. Food. J.*, **92**, no. 4, 6–12; 21 ref.

Silva-Nda, 1991. [Enzyme immunoassays applied to detection of Salmonella spp. in foods.], *Coletanea Instituto Tecnologia Alimentos*, **21**, no. 2, 173–186; 54 ref.

Slavik M. F. and Tsai H. C. S., 1992. A membrane blotting procedure for detection of *Salmonella* on chicken carcasses. *J. Food Prot.*, **55**, no. 7, 548–551; 23 ref.

Sukupolvi S. and Vaara M., 1989. *Salmonella typhimurium* and *Escherichia coli* mutants with increased outer membrane permeability to hydrophobic compounds. *Biochimica Biophysica Acta*, **988**, no. 3, 377–387; 122 ref.

Tsai H. C. and Slavik M. F., 1991. Rapid method for detection of salmonellae attached to chicken skin. *J. Food Safety*, **11**, no. 3, 205–214; 13 ref.

Tsang R. S. W. and Nielsen K. H., 1992. Immunoassays for *Salmonella. Genet. Engineer Biotechnol.*, **12**, no. 2, 14–18; 67 ref.

Wolber P. K. and Green R. L., 1990. New rapid method for the detection of *Salmonella* in foods. *Trends Food Sci. Technol.*, **1**, no. 4, 80–82; 16 ref.

CHAPTER

19

The Genus *Campylobacter*

M. Catteau

It was at the beginning of this century that the microorganism then known as *Vibrio fetus* was recognized as the agent responsible for the occurrence of spontaneous abortions in cattle and sheep. This microorganism of curved morphology is difficult to grow in the presence of oxygen and is classed in the group called *microaerophilic vibrions*. However, its particular traits (Table 19.1) distinguish it from *Vibrio* and in 1963, Sebald and Veron proposed the name *Campylobacter*. Since then, the genus *Campylobacter* has been recognized as the principal cause of infectious diarrhea in many countries.

19.1 Culture Conditions and Identification

Of curved or helicoidal morphology, possessing great motility provided by one or several polar flagella, gram-negative, microaerophilic, neither fermenting nor acidifying sugars, oxidase positive, the genus *Campylobacter* belongs to the Spirillaceae family. The principal species are listed in Table 19.2.

Culturing Atmospheres

Most strains require an atmosphere containing 3–6% oxygen for growth. The mixture usually used is composed of 5% oxygen, 10% CO_2, and 85% nitrogen.

Media

Culture media are generally made up of a Muller Hinton base, Columbia agar, Thioglycolate agar, Brucella agar, with 5–10% blood or 0.4% carbon.

Table 19.1 Vibrion–*Campylobacter* differentiating traits

	Vibrions	*Campylobacter*
Morphology	gram -ve, curved	gram -ve, curved or helicoidal
Oxidase	+	+
Respiratory type	facultative anaerobic	microaerophilic
Sugar fermentation	+	–
G: C%	40–52	30–38

Surface colonies after 48 hours of culture are small, rounded, grayish or yellowish, nonhemolytic, and tending to be motile along the inoculating streaks on very moist agar.

The media are made selective by the addition of antibiotics. Numerous mixtures have been proposed by various authors. Their formulas are given in Table 19.3.

Antigenic Structure

The analysis of antigens has been achieved for *C. jejuni/coli* and for *C. fetus*. Two schemes have been developed for *C. jejuni/coli*:

- the thermostable antigen based scheme of Penner and Lauwers which defines over 50 passive hemagglutination serotypes;
- the thermolabile antigen based scheme of Lior which defines 90 sero-agglutination serogroups.

These techniques have led to better understanding of the epidemiology of *Campylobacter* by revealing the significance of poultry as a source of contamination, for example.

19.2 Pathogenic Potential

Some species are nonpathogenic to slightly pathogenic (*C. sputorum*, *C. concisus*), while others are pathogenic to animals and opportunistically infect humans. *C. fetus fetus* may cause sporadic abortions in sheep and cattle while it may be the cause of septicemia in immuno-suppressed humans.

C. jejuni/coli are normal inhabitants of the small intestines of numerous animals (poultry, cats, dogs, pigs, cattle, etc.) They are also found in the intestine of healthy humans. Since 1975, however, *C. jejuni* is recognized as one of the principal agents responsible for acute diarrhea in humans. Its frequency today exceeds that of *Salmonella* and *Shigella* in many industrialized nations. In developing countries, enteritis caused by *Campylobacter* is extremely frequent. *C. jejuni* is now recognized as a cause of travel-related diarrhea.

Table 19.2 Principal traits of the genus *Campylobacter*

	Catalase	Culture 42°C	25°C	H_2S Kligler	Nitrate	Urea	Hippurate	Naladixic acid	Cefalotine
Campylobacter jejuni	+	+	–	–	+	–	+	S	R
Campylobacter coli	+	+	–	+W	+	–	–	S	R
Campylobacter laridis	+	+	–	–	+	–	–	R	R
Campylobacter fetus fetus	+	(+)	+	–	+	–	–	R	S
Campylobacter fetus venerealis	+	–	+	–	+	–	–	R	S
Campylobacter hypointestinalis	+	(+)	(+)	+	+	–	–	R	S
*Campylobacter pyloridis**	+	(+)	–	–	–	+	–	R	S
Campylobacter sputorum sputorum	–	+	–	+	+	–	–	S	S
Campylobacter sputorum bubulus	–	+	(+)	+	+	–	–	R	S
Campylobacter sputorum fecalis	+	+	–	+	+	–	–	R	S
Campylobacter concisus	–	+	–	+	+	–	–	R	R
Campylobacter mucosalis	–	+	+	+	+	–	–	S	S

**Helicobacter pylori*; (+) = variable depending on the strains; w = weak; R = resistant; S = sensitive.

Table 19.3 Some selective media for *Campylobacter*

	Skirrow		modified Butzler		Preston		Karmali CSM		Bolton CCDA	
Base	Proteose peptone	15 g	Peptone	23 g	Meat extract	10 g	Peptone	23 g	Peptone	10 g
	Liver extract	2.5 g	Starch	1 g	Peptone	10 g	Starch	1 g	Meat extract	10 g
	Yeast extract	5 g	Sodium chloride	5 g	Sodium chloride	5 g	Sodium chloride	5 g	Sodium chloride	5 g
	Sodium chloride	5 g	Agar-agar	12–18 g	Agar-agar	12–18 g	Agar-agar	12–18 g	Casein hydrolysate	3 g
	Agar-agar	12–18 g	Water	1,000 ml	Water	1,000 ml	Water	1,000 ml	Agar-agar	12–18 g
	Water	1,000 ml	Water	1,000 ml					Water	1,000 ml
Supplements	Lysed horse blood	50 ml	Sheep blood	50 ml	Lysed horse blood	50 ml	Carbon	4 g	Carbon	4 g
							Sodium pyruvate	0.1 g	Ferrous sulfate	0.25 g
							Hematin	0.032 g	Sodium pyruvate	0.25 g
Selective agents	Vancomycin	10 mg	Amphotericin	2 mg	Trimethoprime	10 mg	Vancomycin	20 mg	Cefoperazone	32 mg
	Trimethoprime	5 mg	Bacitracin	10,000 IU	Polymyxin	5,000 IU	Cycloheximide	100 mg	Sodium deoxychlorate	1 g
	Polymyxin	2,500 IU	Rifampicin	10 mg	Cycloheximide	100 mg	Cefoperazone	32 mg		
			Cefoperazone	15 mg	Rifampicin	10 mg				

Campylobacter enteritis

Campylobacter enteritis afflicts mostly children under the age of 5. The incubation period is usually 2–5 days on the average, the sickness beginning with a fever associated with general fatigue and headaches. Nausea then follows, along with abdominal cramps seeming like acute appendicitis. The latter are rapidly followed by diarrhea which is at first watery and copious and which may then become mucous, bloody, and purulent.

C. pyloridis is today suspected of being responsible for gastritis and for gastroduodenal ulcers. Its taxonomic position has recently been revised, belonging now to a new genus, *Helicobacter*.

19.3 Analysis of *Campylobacter* in Foods

Sample Transport

If the sample must be transported, this should be done at 4°C. Adding 0.01% sodium bisulfate and storing under nitrogen appear to provide favorable conditions for transportation. Freezing of the sample must be avoided since the genus *Campylobacter* is very sensitive to this. In all cases, inoculation must be done as quickly as possible after sample removal.

Direct Isolation

For certain foods rich in *Campylobacter*, direct isolation on a selective medium (Skirrow and/or Butzler) is sufficient. The plates are incubated at 42°C under a microaerophilic atmosphere (5% O_2) for 48 hours.

Enrichment

In the majority of cases, an enrichment is necessary. The most widely used medium is currently Preston broth (Table 19.4). The sample is used to inoculate a volume of enrichment medium ten times the inoculum size. This is then incubated at 42°C for 18 hours under a microaerophilic atmosphere and preferably with agitation. After incubation, isolation is done on two media: Skirrow agar and one other medium such as modified Butzler, Karmali, Bolton CCDA, or Preston agars. These agar-based media are incubated at 42°C for 24–48 hours under a 5% oxygen atmosphere. Characteristic colonies are examined between slide and cover-slip and are also Gram stained.

Observation of so-called insect flight motility and the characteristic comma-like or corkscrew morphology orients the diagnosis towards *Campylobacter*. After reisolation on Columbia blood agar, selected colonies will be subjected to biochemical confirmation.

One technique showing some originality consists of placing a filtering membrane on nutrient blood agar and depositing a drop of enriched suspension thereon. After 30 minutes of contact, the membrane is removed and the

Table 19.4 Some selective broths for *Campylobacter*

	Preston broth		Park and Sanders broth	
Base	Meat extract	10 g	Tryptone	10 g
	Peptone	10 g	Peptic meat peptone	10 g
	Sodium chloride	5 g	Glucose	1 g
	Agar-agar	1 g	Yeast extract	2 g
	Water	1,000 ml	Sodium citrate	1 g
			Sodium chloride	5 g
			Sodium bisulfate	0.1 g
			Sodium pyruvate	0.25 g
			Water	1,000 ml
Supplements	Lysed horse blood	50 ml	Lysed horse blood	50 ml
Selective agents	Polymyxin B	5,000 IU	Solution A	
	Rifampicin	0.01 g	Vancomycin	0.01 g
	Trimethoprine	0.01 g	Trimethoprine	0.01 g
	Cycloheximide	0.1 g	Solution B	
			Cefoperazone	0.032 g
			Cycloheximide	0.01 g

plate is incubated. *Campylobacter* are capable under these conditions of crossing the filter and usually giving rise to pure cultures.

Numerous other methods have been described. One worthy of mention is that of Park and Sanders intended for the best possible recovery of spoiled *Campylobacter* cells. This enrichment method begins by keeping the sample at 32°C in a nonselective medium (solution A) for 4 hours and then adding antibiotics (solution B) and continuing the incubation at 37°C for 2 hours and finally at 42°C for 40–42 hours under microaerophilic atmosphere (Figure 19.1).

19.4 Occurrence in the Environment and in Foods—Epidemiology

The genus *C. jejuni/coli* is extremely frequent in fecal matter from a great variety of animals (e.g., sheep, cattle, hogs, wild birds, rodents, dogs, cats, etc.). Around 70% of pigs, and according to studies, 30–100% of poultry harbor these microorganisms in their intestinal contents. Much the same can be said for numerous foods (that is, pork and poultry) as well as milk and water will be easily contaminated. The example of poultry is typical since *C. jejuni/coli* is constantly found in the scalding tanks of poultry kills and therefore very frequently on the carcasses.

Fortunately, these microorganisms are fragile. They require very special medium, temperature, and atmosphere in order to develop or even to survive. In many foods, at ambient temperatures they do not multiply but rather slowly

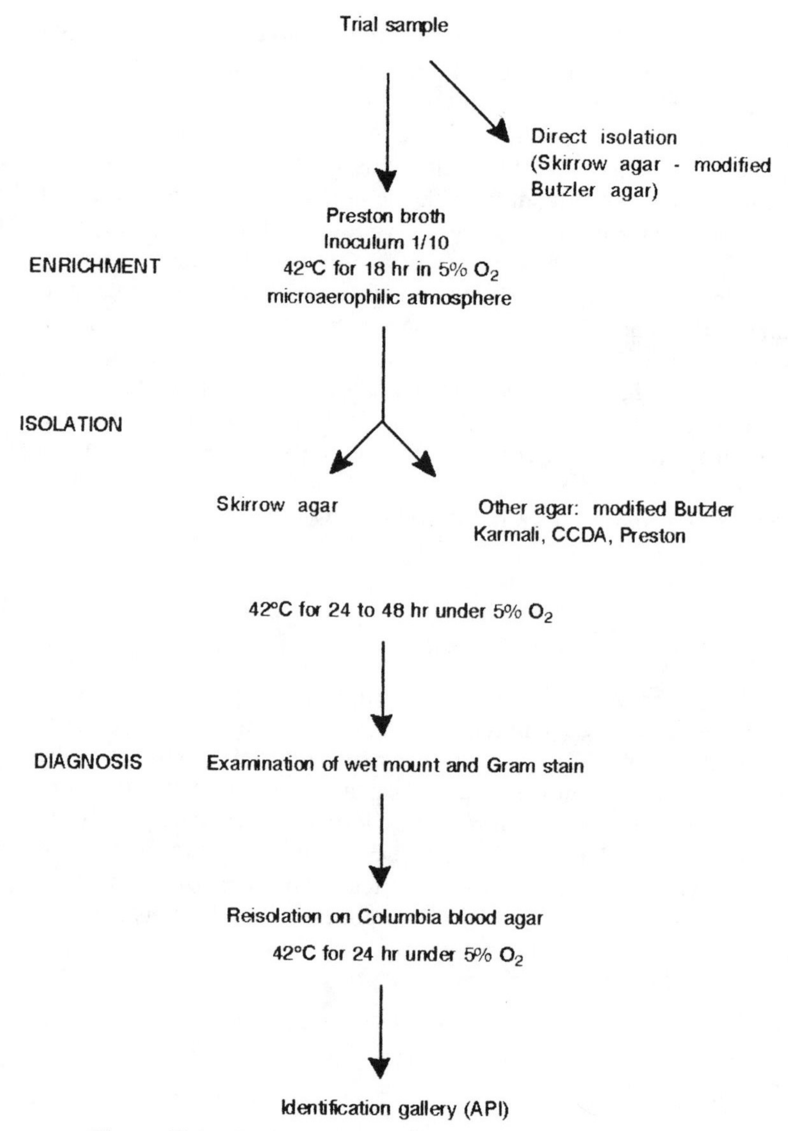

Figure 19.1 Testing for *Campylobacter jejuni/coli* in foods.

decrease in numbers. Their survival is likely under refrigeration, several weeks in water or milk having been observed although no multiplication occurs. Being very sensitive to the quantity of water available, their numbers decrease rapidly as soon as the slightest dessication occurs. This is frequently observed on pork carcasses (e.g., during holding in cold rooms).

In the freezer at −18°C, *C. jejuni/coli* disappears quite rapidly. They are very rarely found in frozen foods. This fragility makes the storage of strains difficult in laboratories. The characteristic morphology is lost and replaced by coccoid shapes which precede cell death. Generally, storage of strains at 4°C or at very low temperature (−80°C) if possible is preferable.

The mode of transmission of the sickness in humans therefore seems quite obvious. A few cases of direct transmission from livestock to humans have been described for persons in contact with the animals (animal breeders, slaughter-house workers). Cases of transmission from domestic animals to humans have also been reported as well as a few cases of human to human transmission at the cradle. Clearly, however, the principal mode of transmission may be summarized as animal → food → human.

The foods most often implicated are raw milk, red meat, and water. Curiously enough, poultry consumption is rarely involved. It is a fact that cooking easily destroys *Campylobacter* and that the outbreaks described resulted from inadequate cooking of poultry (barbecue) or recontamination after cooking. Water has been the source of the largest outbreaks: 2,000–3,000 sicknesses in Vermont in 1978 and 2,000 in Sweden in 1981.

19.5 Conclusion

A great number of studies have been carried out in recent years and have provided a better understanding of the conditions required for the isolation and identification of *Campylobacter*. Their frequency, particularly that of *C. jejuni/coli*, are better known today. These microorganisms are found routinely in our foods and the transmission to humans occurs frequently. In many countries (Canada, United Kingdom, USA, Japan, and others) *Campylobacter* is isolated two to three times more often than *Salmonella*. Today, they are considered to be the principal agents responsible for infectious diarrhea. There is certainly justification for analyzing for them in our foods on a more systematic basis.

References

Bolton F. J., Coates D., Hinchcliffe P. M., Robertson M., 1983. Comparison of selective media for isolation of *Campylobacter jejuni/coli*. *J. Clin. Pathol.*, **36**, 78–83.

Bolton F. J., Hutchinson D., Coates D., 1984. Blood-free selective medium for isolation of *Campylobacter jejuni* from feces. *J. Clin. Microbiol.*, **19**, 169–171.

Butzler J. P., De Boeck M., Goossens H., 1983. New selective medium for isolation of *Campylobacter jejuni* from faecal specimens. *Lancet*, **1**, 818.

Karmali M. A., Simor A. E., Roscoe M., Fleming P. C., Smith S. S., Lane J., 1986. Evaluation of a blood-free, charcoal based, selective medium for the isolation of *Campylobacter* organisms from feces. *J. Clin. Microbiol.*, **23**, 456–459.

Lauwers S. In: Newell D. G. (ed). 1982. Serotyping of *Campylobacter jejuni*: A useful tool in the epidemiology of *Campylobacter* diarrhea, 96–97. *Campylobacter: Epidemiology, pathogenesis and biochemistry*. MTP Press, Lancaster.

Lior H., Woodwiard D. L., Edgar J. A., Laroche L. J., Gill P., 1982. Serotyping of *Campylobacter jejuni* by slide agglutination based on heat-labile antigenic factors. *J. Clin. Microbiol.*, **15**, 761–768.

Penner J. L. and Hennessy J. N., 1980. Passive hemagglutination technique for serotyping *Campylobacter fetus* subsp. *jejuni* on the basis of soluble heat-stable antigens. *J. Clin. Microbiol.*, **12**, 732–737.

Skirrow M. B., 1977. *Campylobacter enteritis*: A "new" disease. *Br. Med. J.*, **2**, 9–11.

Steele T. W. and McDermott S. N., 1984. Technical note: The use of membrane filters applied directly to the surface of agar plates for the isolation of *Campylobacter jejuni* from feces. *Pathology*, **16**, 263–265.

Veron M. and Chatelain R., 1973. Taxonomic Study of the genus *Campylobacter* Selbald and Veron and designation of the neotype strain for the type species *Campylobacter fetus* (Smith and Taylor) Sebald and Veron. *Int. J. System Bact.*, **23**, 122–134.

Suggested Readings

Baggerman W. I. and Koster T., 1992. A comparison of enrichment and membrane filtration methods for the isolation of *Campylobacter* from fresh and frozen foods. *Food Microbiol.*, **9**, no. 2, 87–94; 6 ref.

Buck G. E., 1990. *Campylobacter pylori* and gastroduodenal disease. *Clin. Microbiol. Rev.*, **3**, no. 1, 1–12; 135 ref.

Cowden J., 1992. *Campylobacter*: Epidemiological paradoxes. *Br. Med. J.*, **305**, no. 6846, 132–133; 26 ref.

Dubreuil J. D., Kostrzynska M., Logan S. M., Harris L. A., Austin J. W., Trust T. J., 1990. Purification, characterization, and localization of a protein antigen shared by thermophilic campylobacters. *J. Clin. Microbiol.*, **28**, no. 1321–1328; 47 ref.

El-Shenawy M. A. and Marth E. H., 1989. *Campylobacter jejuni* and foodborne campylobacteriosis: A review. *Egypt. J. Dairy Sci.*, **17**, no. 2, 181–200; 89 ref.

Franco D. A., 1988. Campylobacter species: Considerations for controlling a foodborne pathogen. *J. Food Prot.*, **51**, no. 2, 145–153; 88 ref.

Griffiths P. L. and Park R. W. A., 1990. Campylobacters associated with human diarrhoeal disease. *J. Appl. Bacteriol.*, **69**, no. 3, 281–301; 249 ref.

Hoffman P. S. and Blankenship L. C., 1986. Significance of *Campylobacter* in foods. *Dev. Food Microbiol.*, **2**, 91–122; 147 ref.

Humphrey T. J., 1992. *Campylobacter jejuni*: Some aspects of epidemiology, detection and control. *Br. Food J.*, **94**, no. 1, 21–25; 27 ref.

Park C. E. and Sanders G. W., 1992. Occurrence of thermotolerant campylobacters in fresh vegetables sold at farmers' outdoor markets and supermarkets. *Can. J. Microbiol.*, **38**, no. 4, 313–316; 52 ref.

Park R. W. A., Griffiths P. L., Moreno G. S., 1991. Sources and survival of campylobacters: Relevance to enteritis and the food industry. *J. Appl. Bacteriol.* (symposium supplement), no. 20, 975–1065.

Peterz M., 1991. Comparison of Preston agar and a blood-free selective medium for detection of *Campylobacter jejuni* in food. *J. Assoc. Off. Anal. Chem.*, **74**, no. 4, 651–654; 13 ref.

Reid C. M., 1991. Evaluation of two selective media and a rapid latex test for the isolation/confirmation of *Campylobacter jejuni.* Meat Industry Research Institute of New Zealand, no. 870, iii + 10p.; 12 ref.

Rosef O., Kapperud G., Gondrosen B., Nesbakken T., 1986. (*Campylobacter jejuni* and *Campylobacter* coli—food hygiene and epidemiological aspects. II. Norwegian results.), *Norsk-Veterinaertidsskrift*l, **98**, no. 2, 111–119; 26 ref.

Skirrow M. B., 1989. Campylobacter perspectives. *PHLS Microbiol. Digest*, **6**, no. 4, 113–118; 27 ref.

Stelzer W., Jacob J., Schulze E., 1991. Environmental aspects of *Campylobacter* infections. *Zentralbl. Mikrobiologie*, **146**, no. 1, 3–15; 51 ref.

Sutcliffe E. M., Jones D. M., Pearson A. D., 1991. Latex agglutination for the detection of *Campylobacter* species in water. *Letters Appl. Microbiol.*, **12**, no. 3, 72–74; 5 ref.

Valentine J. L., Arthur R. R., Mobley H. L. T., Dick J. D., 1991. Detection of *Helicobacter pylori* by using the polymerase chain reaction. *J. Clin. Microbiol.*, **29**, no. 4, 689–695; 34 ref.

CHAPTER

20

The Genus *Yersinia*

M. Catteau

Although *Yersinia pestis* is of concern only in medical bacteriology and is being encountered less and less, *Y. pseudotuberculosis*, *Y. enterocolitica*, and its related species are of interest to the food bacteriologist. This is particularly the case for *Y. enterocolitica*, a psychrotrophic species more and more frequently isolated and of which some strains are responsible for human infections originating in foods.

20.1 Taxonomy

The name *Yersinia* comes from the French bacteriologist Yersin who was the first to isolate the agent which caused the bubonic plague in 1894. Formerly classified in the genus *Pasteurella*, the generic denomination *Yersinia* was proposed by Van Loghem in 1944 for the purpose of separating the species *Y. pestis* and *Y. pseudotuberculosis* from species of *Pasteurella*, strictly speaking.

20.1.1 The Genus *Yersinia*

Yersina is a member of the Enterobacteriaceae family. It is a gram-negative coccoid-shaped bacillus of a width between 0.5 and 0.8 μm and 1–3 μm in length. It forms neither capsules nor endospores, although an envelope may arise in the case of *Y. pestis* grown at 37°C or strains obtained from in vivo samples. The genus *Yersinia* is nonmotile at 37°C but motile at 30°C with peritrichous cilia, except for *Y. pestis*, which is always nonmotile.

The genus grows well on ordinary media, forming small colonies compared to other enterobacteria. Colonies are translucent to opaque, from 0.1–1 mm in diameter within 24 hours and 2 to 3 mm within 48 hours with elevated centers. Colonies are distinguishable as large (3 mm) or small (1 mm) on ordinary media within 48 hours. Culture is possible anywhere between 4° and 42°C, with a growth optimum of 25°–29°C, the biochemical traits being better expressed in this range than at 37°C. The optimal pH for all species is from 7.2 to 7.4, although growth is possible at any pH between 5 and 10.

Belonging to the Enterobacteriaceae family, *Yersinia* possesses all of its traits:

- gram-negative bacillus;
- facultative anaerobe;
- ferments glucose;
- reduces nitrates to nitrites (except certain biotypes of *Y. enterocolitica*);
- oxidase-negative;
- catalase-positive.

The % G + C in its DNA is in the range of 46–50.

As a facultative anaerobe, *Yersinia* has a respiratory type which is oxidative and fermentative at the same time. Glucose and other carbohydrates are fermented with production of acid but little or no gas.

Hydrogen sulfide is not produced, β-galactosidase is positive, urease is intensely positive (except for *Y. pestis*), phenylalanine deaminase (APP) and tryptophan deaminase (TDA) are negative. Most strains do not ferment lactose, although lactose positive strains are sometimes encountered, particularly in milk.

The genus *Yersinia* includes three principal species: *Y. pestis*, *Y. pseudotuberculosis*, and *Y. enterocolitica*, plus the related species *Y. intermedia*, *Y. frederiksenii*, *Y. kristensenii*, *Y. aldovae*, *Y. rohdei*, *Y. mollaretii*, and *Y. bercovieri*. The particular biochemical traits of *Y. ruckeri* should place it in a different genus. *Y. ruckeri* is responsible for diseases of certain fish and has been isolated only in North America.

20.1.2 *Yersinia pseudotuberculosis*

Y. pseudotuberculosis was discovered in 1883 by Malassez and Vignal in a guinea pig inoculated using a cutaneous nodule from a child who had died of tuberculous meningitis. Its biochemical traits are given in Table 20.1. Its antigenic traits allow the recognition of six principal serogroups: I–VI. Three new serogroups, IIC, VII, and VIII have been proposed by Tsubokora. Antigenic relationships exist with *Salmonella*, *E. coli*, and *Enterobacter cloacae*.

20.1.3 *Yersinia enterocolitica* and Related Species

Discovered in 1939 in the United States, this species was given the name *Y. enterocolitica* in 1964 by Frederiksen.

Table 20.1 Principal biochemical traits of *Yersinia pseudotuberculosis, enterocolitica*, and related species

Species designation	Tween 80	Esculine (24 hr.)	Pyrazin-amidase	Indole	Xylose	Sucrose	Rhamnose	Melibiose	V.P.	Citrate	Mucate
*Yersinia enterocolitica**	V	V	V	V	V	+/−	−	−	+	−	−
Yersinia intermedia	+	+	+	+	+	+	+	+	+	+	+
Yersinia frederiksenii	+	V	+	+	+	+	+	−	+/−	V	+
Yersinia kristensii	V	−	+/(+)	V	+	−	−	−	−	−	V
Yersinia mollaretii	−	−	+	−	+	+	−	−	−	−	+[†]
Yersinia bercovieri	−	−	+	−	+	+	−	−	−	−	+[††]
Yersinia aldovae	V	−	+	−	+	−	+	−	+	V	V
Yersinia rohdei	−	−	+	−	+	+	−	V	−	+	−

*See also Table 20.2.
V = variable; +/− = majority of strains +; (+) = reaction limited or weak; † = sorbose + fructose −; †† = sorbose − fructose +.

Biochemical traits are given in Table 20.1. Some of these traits appear only when the strain is incubated at a temperature below 30°C, confirming the psychrotrophic nature of the microorganism.

In 1980, Brenner showed by DNA/DNA hybridization that certain strains related to *Y. enterocolitica*—but having unusual traits such as fermentation of rhamnose or melibiose, the absence of sucrose fermentation, or citrate utilization—should belong to different species. The creation of the species *Y. intermedia*, *Y. frederiksenii*, and *Y. kristensenii* was therefore proposed. Bercovier in 1984 added the species *Y. aldovae* and more recently, Aleksic (1987) described *Y. rohdei*. Finally two new species were recognized, *Y. mollaretii* and *Y. bercovieri*.

Several biogroups have been described for *Y. enterocolitica*. Wauters proposed the scheme shown in Table 20.2.

The study of the antigens of *Y. enterocolitica* and related species has made possible the individualizing of over 70 O factors. Some correlations exist between biogroups, serogroups, and ecological origins of strains. These are shown in Table 20.3 (from Wauters).

In Europe, Japan, and South Africa, the principal type of pathogens belong to biogroup 2 serogroup O:9 and biogroup 4 serogroup O:3. However, factor O:3 is encountered frequently among strains apparently devoid of pathogenic capability such as *Y. frederiksenii*. Serotyping alone is therefore insufficient and must be preceded by a precise determination of species and biogroup. In North America, biogroup 1B serogroup O:8 is one of the most frequent pathogenic *Yersinia* species associated with humans.

20.2 Pathogenic Potential

20.2.1 *Y. pseudotuberculosis*

All animal species may be afflicted, in particular guinea pigs and hares. The sickness evolves over a period of a few weeks, causing diarrhea, weight loss, and terminal septicemia. Microabcesses occur principally in the liver and

Table 20.2 Characterization of different biotypes of *Yersinia enterocolitica*

Biotypes	1A	1B	2	3	4	5
Tween esterase	+	+	−	−	−	−
Esculine (24 hr.)	+/−	−	−	−	−	−
Indole	+	+	(+)	−	−	−
Pyrazinamidase	+*	−	−	−	−	−
Xylose	+	+	+	+	−	d
Trehalose/NO_3	+	+	+	+	+	−
Acetoin (VP)	+	+	+	+/−	+	(+)
Beta-D-galactosidase	+	−	−	−	−	−
Proline peptidase	d	−	−	−**	−	−

*Very rare negative strains exist.

**A few chinchilla strains are weakly positive.

Table 20.3 Correlation between biogroups, serogroups, and ecology of strains of *Yersinia enterocolitica*

Biogroups	Serogroups	Ecology
1A	Various	Ubiquitous: foods, water, animals, (humans)
1B	O:8, O:13, O:18	Humans (North America)
	O:20, O:21	
2	O:9, O:5, 27	Humans, swine
3	O:1, 2, 3	Rodents, chinchilla, (humans)
	O:5, 27, O:3	Humans, swine
4	O:3	Humans, swine
5	O:2, 3	Hares, goats
Y. intermedia		
Y. frederiksenii		
Y. kristensenii		
Y. aldovae		Ubiquitous, environmental
Y. rohdei		
Y. mollaretii		
Y. bercovieri		

spleen. Intestinal carrying is frequent among rodents, lagomorphs, and birds. Herbivores (except for horses and cattle) and carnivores (particularly the cat, which is often the source of human contamination) may be afflicted. The frequency of healthy carriers is high.

In humans, mesenteric adenitis simulating an acute appendicitis attack is more widespread and may evolve into knotty erythema. Generalized septicemic forms are, however, frequent and serious. Among the localized forms, ocular and pulmonary versions correspond to particular ports of infection. Contamination occurs almost always via the digestive tract, either by direct contact with infected animals (rodents, hamsters, guinea pigs, cats) or by foods soiled with animal excrement. The ground is a significant reservoir in the contamination of both animals and foods.

After having been localized for a long time in central and occidental Europe, which remains its predominant locale, *Y. pseudotuberculosis* infection has spread to the world, beginning slowly with the First World War and then more rapidly during the Second World War. Exporting of live animals plays a major role in the dissemination of the sickness.

In France, *Y. pseudotuberculosis* is the primary cause of mortality in hares and is widespread among small mammals and birds. In humans, *Y. pseudotuberculosis* is responsible for 5–10% of the cases of mesenteric adenitis diagnosed, not including the nonovert or undeclared forms.

20.2.2 *Y. enterocolitica*

In humans, enterocolitis remains the predominant form of infection, although infections due to *Y. enterocolitica* may assume a wide range of forms depending on the strain, the infectious dose, genetic factors, age, and the physical condition of the host. Gastroenteritis and mesenteric adenitis with right ilial syndrome simulating an attack of appendicitis are the predominating symptoms in children and adolescents. Abdominal troubles, diarrhea, and arthritis are the most common manifestations in adults from 20–60 years of age. Knotty erythema is observed principally in persons over 60 years old. In animals, *Y. enterocolitica* is spontaneously pathogenic only in chinchillas, hares, and monkeys, bringing about symptoms similar to those described for *Y. pseudotuberculosis*. Among other species, infection is generally limited to carrying for some variable period of time.

Contamination of humans and animals occurs essentially by the oral route and involves foods. Healthy human or animal carriers and soil play a major role in transmission.

20.2.3 Characterization of Potentially Pathogenic Strains

Several strains of *Y. enterocolitica* and related species produce thermostable enterotoxins (ST type). These enterotoxins, however, do not seem to have a significant role in vivo. The principal pathogenic power is of an invasive type. This invasive capability is coded by a so-called virulence plasmid of 40–45 megadaltons which also codes for a few properties, the laboratory detection of which is useful, such as the following:

- calcium dependence of strains at 37°C (the culture is totally or partially inhibited on oxalate medium at this temperature);
- self-agglutination of strains in broth at 37°C;
- cell wall hydrophobicity bringing about agglutination of latex beads at 37°C;
- binding of Congo red or crystal-violet by colonies at 37°C;
- synthesis at 37°C of certain so-called surface proteins easy to detect by electrophoresis.

Among these properties, we consider the calcium requirement at 37°C and the detection of the extracellular proteins to be the most reliable. The virulence plasmid, on the other hand, is easily lost during repeated transfers of the strains in the laboratory.

The “virulence” plasmid is observed in *Y. pestis*, *Y. pseudotuberculosis*, and strains of *Y. enterocolitica* belonging to biogroups 1B, 2, 3, 4, and 5. This corresponds well with the pathological manifestations observed.

Wauters demonstrated that the nonpathogenic strains of biogroup 1A possess a pyrazinamidase which the pathogenic strains of the other biogroups do not possess even when the plasmid has been lost. This test, associated with the hydrolysis of esculine, positive for some but not all nonpathogenic strains, is very useful as a method of orienting the diagnosis. It may be said in effect that the following are true:

- Pyrazinamide and esculine negative strains are potentially pathogenic.
- Pyrazinamide and/or esculine positive strains are nonpathogenic.

20.3 Analysis for *Yersinia enterocolitica* in Foods

20.3.1 Direct Isolation

If the food being analyzed is rich in *Yersinia*, direct isolation may be suitable. Schiemann (CIN) agar-containing Cefsulodine, Irgasan, and Novobiocine may be used, or that of Wauters (SSDC): *Salmonella-Shigella* (SS) medium with added deoxycholate and calcium chloride. Both media are incubated at a temperature below 30°C. The first, nonselective, allows the growth of all species and biogroups of *Yersinia* while the second, much more selective, is especially suitable for the strains O:3 of biogroup 4 (Table 20.4).

20.3.2 Enrichment

An enrichment step is usually necessary for most foods. Various techniques may be used:

- Enrichment at 4°C in phosphate buffer is often recommended. This requires incubation times of 2–3 weeks, however, and appears to be generally more effective for ubiquitous strains than for potentially pathogenic ones.
- Enrichment at 22–25°C for 48 hours on Wauters ITC medium (Irgasan, Ticarcilline, Chlorate). A small inoculum (1/100 by volume) seems to give better results (Table 20.5).
- Enrichment at 22–25°C for 3–5 days on PSB medium (Peptone, Sorbitol, Bile, Table 20.5).

Table 20.4 Formulas of isolation media

Schiemann CIN		Wauters SSDC	
Peptone	20 g	Peptone	5 g
Yeast extract	2 g	Yeast extract	5 g
Mannitol	20 g	Meat extract	5 g
Sodium pyruvate	2 g	Lactose	10 g
Sodium chloride	1 g	Bile salts	8.5 g
Magnesium sulfate $7H_2O$	0.01 g	Sodium deoxycholate	10 g
Sodium deoxycholate	0.5 g	Calcium chloride	1 g
Neutral red	0.03 g	Sodium citrate	10 g
Crystal-violet	0.001 g	Sodium thiosulfate	8.5 g
Agar-agar	12–18 g	Ferric citrate	1 g
Water	1,000 ml	Brilliant green	0.0003 g
pH 7.4. Autoclave at 121°C for 15 minutes.		Neutral red	0.025 g
Cefsulodine solution, 0.15 mg/ml	1 ml	Agar-agar	12–18 g
Irgasan solution (alcohol), 4 mg/ml	1 ml	Water	1,000 ml
Novobiocin solution, 2.5 mg/ml	1 ml	pH 7.4. Do not autoclave.	

Table 20.5 Formulas of enrichment media

ITC broth		PSB broth	
Tryptone	10 g	Peptone	5 g
Yeast extract	1 g	Sorbitol	10 g
Magnesium chloride	60 g	Sodium chloride	5 g
Sodium chloride	5 g	Na_2HPO_4	8.23 g
Malachite green, aqueous 0.2% solution	5 ml	$NaH_2PO_4H_2O$	1.2 g
Water	1,000 ml	Bile salts	1.5 g
pH 6.9. Autoclave at 121°C for 15 minutes.		Water	1,000 ml
Ticarcilline solution, 1 mg/ml	1 ml	pH 7.6.	
Irgasan solution (alcohol), 1 mg/ml	1 ml	Autoclave at 121°C for 15 minutes.	
Potassium chlorate solution, 100 mg/ml	10 ml		

The use of an alkaline shock according to the method recommended by Aulisio. This operation is carried out on the product itself before direct isolation or more often on the incubated enrichment broth. This consists of placing one volume of enriched broth in nine volumes of 0.25% KOH solution and after 10 seconds of contact, isolating on selective agar. The medium SSDC, itself very inhibitory, is not suited to this treatment, for which CIN medium is preferred.

20.4 Incidence of *Y. enterocolitica* in Food Products

Isolation of *Y. enterocolitica* and related strains from many foods is commonplace. We have frequently detected them, as have many other authors, in uncooked vegetables, and they are as frequently isolated from raw meats (pork, beef, poultry), raw milk, and water. In the vast majority of cases they are nonpathogenic strains most often belonging to biogroup 1A. These strains are naturally found episodically in the intestinal tract of many animals and humans.

The simple exercise of analyzing foods for *Y. enterocolitica* is therefore of limited use and their presence must not be construed as having a negative impact. Obviously, the same may not be said when potentially pathogenic *Y. enterocolitica* are detected. We have rarely isolated these from raw foods, although they are sometimes isolated from raw or pasteurized milk and some number of food intoxications have been cited following consumption of pasteurized plain or chocolate milk.

In animals, a few potentially pathogenic strains of serotype O:3 or O:9 have been isolated (cat, dog, monkey), but isolations from swine are the most frequent. They are present in 5–20% of fecal matter and much more often in the pharyngeal cavity. Wauters in Belgium has isolated them from practically

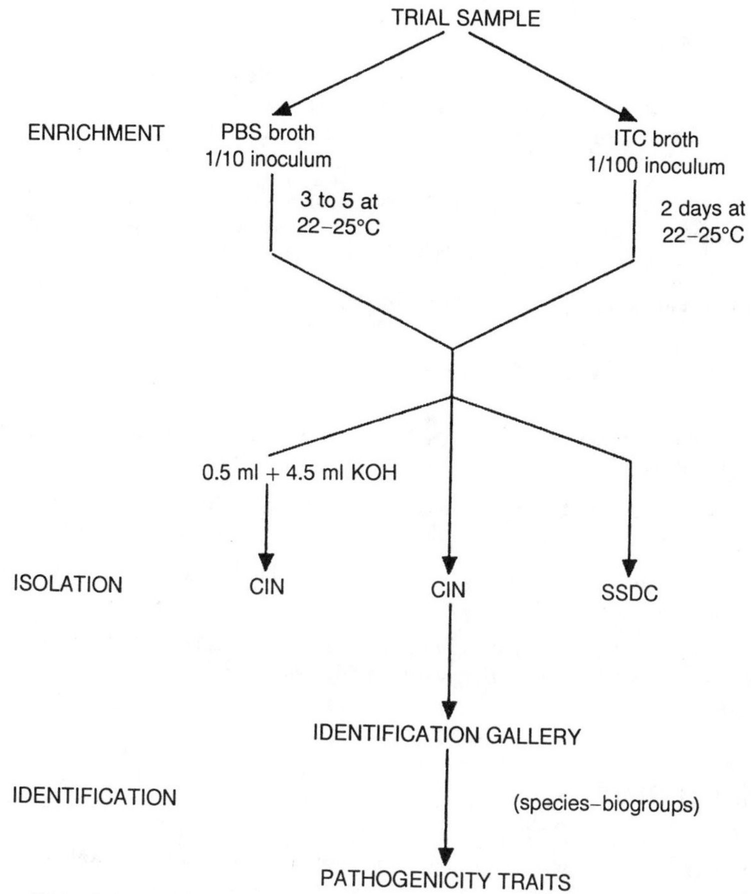

Figure 20.1 Scheme for the analysis of presumably pathogenic *Y. enterocolitica* in foods.

all pork tongues recovered from butchering, while their isolation appears to be slightly less frequent in France and other countries but remains frequent.

Ground pork often prepared from mixtures of meats including the masseter muscles and containing fragments of tonsils is frequently contaminated with *Y. enterocolitica* serotype O:3 (25% according to Wauters). Epidemiological studies carried out in Belgium have now clearly established that the consumption of these meats constituted the principal risk factor in *Y. enterocolitica* infections.

A more systematic search for these microorganisms in our foods thus appears worthwhile. Routine refrigeration likely selects psychrotrophic strains. It should be borne in mind that analyses must be brought to completion and must detect the potential pathogenic trait in the strains (Figure 20.1). Overall analysis for the simple presence of *Y. enterocolitica* is of little use.

References

Aleksic S., Steigerwalt A. G., Bockemuhl J., Huntley-Carter G. P., Brenner R. D. J., 1987. *Yersinia rohdei* sp. nov. Isolated from human and dog feces and surface water. *Int. J. System Bacteriol.*, **37**, 327–332.

Aulisio C. C. G., Mehlman I. J., Sanders A. C., 1980. Alkali method for rapid recovery of *Yersinia enterocolitica* and *Yersinia pseudotuberculosis* from food. *Appl. Environ. Microbiol.*, **39**, 135–140.

Bercovier H and Mollaret H. H., 1984. *Yersinia*. *In*: *Bergey's manual of systematic bacteriology*, vol 1, 498–505.

Catteau M., Krembel C., Wauters G., 1985. *Yersinia enterocolitica* dans les crudités. *Sci. Aliments*, **4**, 103–106.

Catteau M., Ringle P., Daunizeau A., Papierok G., 1987. Etude comparative de tests de pathogénicité de *Yersinia enterocolitica*. *Sci. Aliments*, 7 237–241.

Delmas C. L. and Vidon D. J. M., 1985. Isolement de *Yersinia enterocolitica* et de souches analogues à partir de produits alimentaires. *Appl. Environ. Microbiol.*, **50**, 767–770.

Wauters G., Kandolo K., Janssens M., 1987. Revised biogrouping scheme of *Yersinia enterocolitica*. *Contr. Microbiol. Immunol.*, **9**, 14–21.

Wauters G., Goossens V., Janssens M., Vandepitte J., 1988. New enrichment method for isolation of pathogenic *Yersinia enterocolitica* serogroup 0,3 from pork. *Appl. Environ. Microbiol.*, **54**, 851–854.

Suggested Readings

Anonymous 1991. USDA develops tests for quick detection of *Yersinia*. *Food Technol.*, **45**, no 5, 95–96.

de Boer E., 1992. Isolation of *Yersinia enterocolitica* from foods. *Int. J. Food Microbiol.*, **17**, no. 2, 75–84; 45 ref.

Fenwick S. G. and Murray A., 1991. Detection of pathogenic *Yersinia enterocolitica* by polymerase chain reaction. *Lancet*, **337**, no. 8739, 496–497; 1 ref.

Goverde R. L. J., Jansen W. H., Brunings H. A., Huis in't Veld J. H. J., Mooi F. R., 1993. Digoxigenin-labelled inv- and ail-probes for the detection and identification of pathogenic *Yersinia enterocolitica* in clinical specimens and naturally contaminated pig samples. *J. Appl. Bacteriol.*, **74**, no. 3, 301–313; 46 ref.

Halligan A. C., 1990. The emerging pathogens—*Yersinia*, *Aeromonas* and verotoxigenic *E. coli* (VTEC)—a literature survey. *Food Focus*, no. 11, 73 p.

Ibrahim A., Liesack W., Stackebrandt E., 1992. Polymerase chain reaction-gene probe detection system specific for pathogenic strains of *Yersinia enterocolitica*. *J. Clin. Microbiol.*, **30**, no. 8, 1942–1947; 29 ref.

Kapperud G., Vardund T., Skjerve E., Hornes E., Michaelsen T. E., 1993. Detection of pathogenic *Yersinia enterocolitica* in foods and water by immunomagnetic separation, nested polymerase chain reactions, and colorimetric detection of amplified DNA. *Appl. Environ. Microbiol.*, **59**, no. 9, 2938–2944; 40 ref.

Landgraf M., Iaria S. T., Falcao D. P., 1993. An improved enrichment procedure for the isolation of *Yersinia enterocolitica* and related species from milk. *J. Food Prot.*, **56**, no. 5, 447–450; 17 ref.

Nesbakken T., 1992. [Yersinia enterocolitica—epidemiological and food hygiene aspects. Are swine a source of infection for man?], *Norsk Veterinaertidsskrift*, **104**, no. 8/9, 667.

Nesbakken T., 1993. Epidemiological and food hygienic aspects of *Yersinia enterocolitica* with special reference to the pig as a source of infection. *Dissert. Absts. Internatl., -C*, **54**, no. 2, 537: ISBN 82 992573 0 1.

Nesbakken T., Kapperud G., Dommarsnes K., Skurnik M., Hornes E., 1991. Comparative study of a DNA hybridization and two isolation procedures for detection of *Yersinia enterocolitica* 0:3 in naturally contaminated pork products. *Appl. Environ. Microbiol.*, **57**, no. 2, 389–394; 30 ref.

Walker S. J. and Brooks J., 1993. Survey of the incidence of *Aeromonas* and *Yersinia* species in retail foods. *Food Control*, **4**, no. 1, 34–40; 42 ref.

CHAPTER

21

The Genus *Clostridium* and the Sulfite-Reducing Anaerobes

Martine Poumeyrol, J. Billon

These are gram-positive, endospore-forming anaerobic bacilli. Two species are responsible for food intoxications:

- *Clostridium perfringens:* nonmotile, encapsulated
- *Clostridium botulinum:* motile, ciliated

Generally, sulfite-reducing *Clostridium* species are considered to be indicators of contamination for purposes of assessing the hygienic quality of animal products and products of animal origin. The media used for enumeration and the incubation temperatures are currently more focused on *Clostridium perfringens* as indicator.

21.1 *Clostridium perfringens*

21.1.1 General

Clostridium perfringens belongs to group II of the genus *Clostridium*, bacilli which liquify gelatin, produce central or subterminal spores, are proteolytic, saccharolytic, and nonputrefying. This is a short, thick, nonmotile bacillus possessing a capsule which is polysaccharide in nature and easily seen in the fresh state. Spores are subterminal, appearing inconsistently, especially in the usual culture media. They appear more easily in VL media, glucose-free and enriched with meat extract.

C. perfringens is a strict anaerobe, strongly reducing, fermenting glucose, most hexoses and lactose very rapidly. It reduces sulfites in the presence of a hydrogen donor forming hydrogen sulfide. It is this biochemical property which was used first and foremost for the enumeration of spores and vegetative cells. It is due to an intracellular sulfite reductase which does not diffuse into the medium (Prevost, 1948). This activity is detected by including ferrous salts in the medium. The hydrogen sulfide produced reacts with the iron salt to form a black iron sulfide heavy precipitate which surrounds the colonies in deep agar.

One of the first media used was that of Wilson-Blair which is made from meat and yeast extracts, glucose, 1% agar, 0.25% sodium sulfate, and 0.05% ferrous aluminum sulfate. Other media less rich in sulfite (which may inhibit some strains of *C. perfringens*) were subsequently proposed and made selective by adding antimicrobial substances nontoxic for *C. perfringens* (i.e., polymyxin B, sulfadiazine, neomycin, and D-cycloserine).

Lactose fermentation in sulfite-containing medium ("L.S." medium) is currently proposed for the enumeration of spores and vegetative cells. Mossel and de Waart (1968) demonstrated that an incubation temperature of 46°C was specific for *C. perfringens* in selective media. Finally, for the enumeration of spores in sulfite-containing medium, the addition of lysozyme (2–20 units/ml of media) allows the recovery of those which are lysozyme dependent (Cassier and Sebald, 1969).

21.1.2 Indicator Value of *C. perfringens* and the Significance of Counts

C. perfringens is both a common microbe present in the intestine of healthy swine, cattle, and humans (although not preponderant) and a pathogenic microbe. In animals, it is responsible for enterotoxemias afflicting young sheep and cattle and in humans, it causes foodborne toxi-infections and myonecroses. Enterotoxemias due to type C have been described in certain primitive human populations.

Foodborne toxi-infections come from so-called atypical type A strains or FP strains (Hobbs, 1965) producing highly thermoresistant spores. Sporulation and enterotoxin production are apparently concomitant (Sebald and Cassier, 1970), although it now seems certain that the two phenomena are not associated (Goldner et al., 1986). The identification of the food responsible in the case of a foodborne toxi-infection must be based on the presence of a high number of spores in the food ($>10^4$ or 10^5/g).

21.1.3 Detection of *C. perfringens* in foods

In order to find *Clostridium perfringens*, which may be present in small number, PEM medium (perfringens enrichment medium) may be used.

Casein pancreatic digest	15 g
Yeast extract	5 g
Sodium chlorate	2.5 g
Sodium thioglycolate	0.5 g
L-cystine	0.5 g
Resazurine	0.001 g
Agar	0.75 g
Distilled water	1,000 ml

Adjust the pH to 7.1–7.2, distribute 10 ml per tube, and sterilize for 15 minutes at 121°C. Just prior to use, add D-cycloserine:

Crystalline D-cycloserine	4 g
Distilled water	100 ml

Dissolve and sterilize by filtration. Add 1 ml of solution to each 100 ml of medium, that is, 0.1 ml (or two drops) per tube containing 10 ml of liquid medium.

This medium is intended for the recovery of vegetative or sporulated forms which have suffered a loss of vitality due to a thermal treatment or due to the presence of inhibiting ions. It allows specification of the absence of *C. perfringens* in 1, 0.1, or 0.01 g of product.

Remark: Dextrose has been omitted from the conventional formula since the presence of lactobacilli in the analyzed product would result in lactic acid production from dextrose, which would lower the pH and inhibit *C. perfringens.*

21.1.4 Enumeration of Sulfite-Reducing Anaerobes and of *C. perfringens* in Foods

21.1.4.1 in Solid Media

The media used most often are agar based prepared from commercially available dehydrated media:

- SPS medium (Sulfite, Polymyxin, Sulfadiazine)
- TSN medium (Tryptone, Sulfite, Neomycin)
- SFP medium (DIFCO) used to make TSC medium (Tryptone, Sulfite, Cycloserine), adding cycloserine solution just prior to use as described earlier for the thioglycolate medium

Using a sterile pipette, transfer 1 ml of an original suspension or of the sample in the case of a liquid food into the medium liquefied in a water bath and cooled to 50°C (±2°C). Then carry out dilutions (decimal or 1/5°) in peptone water or tryptone salt and inoculate tubes of media as earlier in duplicata with 1 ml of suspension. In all cases, mix well to distribute the inoculum in the medium without introducing air. After solidification of the agar, incubate the inoculated tubes in a water bath at 37° or 46°C and observe after 24–48 hours.

The presence of colonies appearing blackened (which are in reality white and surrounded with a halo of precipitation of ferrous sulfate) in these selective media is presumptive of *C. perfringens* and the count is given by choosing a series of two tubes in which the number of colonies is between 10 and 30. The average is taken and multiplied by the dilution factor, thus yielding the number of sulfite reducers per gram of product.

This enumeration may also be carried out using Petri plates, incubation being done in anaerobic jars. Counting is made easier because the colonies are better isolated and more easily recoverable for subsequent identification tests; however, the use of 18 mm tubes containing 17–20 ml of agar is much more convenient for routine testing.

In order to obtain the number of *C. perfringens*, it is necessary to carry out identification of suspect colonies, enumerate the colonies corresponding in reality to *C. perfringens*, and apply a correction factor to the number of anaerobic sulfite-reducing colonies found.

An AFNOR standard (NF-V-08-019) appearing in December of 1985 and which completely revises ISO standard 7937-1985 describes "general directives for the enumeration of *Clostridium perfringens*; colony counting method." The medium used is TSC agar (Tryptose, Sulfite, Cycloserine) used in Petri plates and incubated in anaerobic jars. Incubation is at 35° or 37°C for 20 hours.

In France, as described in the AM of December 21, 1979 on microbiological criteria applicable to food products, incubation at 46°C has become more common since it has been documented that this temperature is more favorable to the growth of *C. perfringens* among the sulfite reducers. This practice has not yet been adopted in other countries (ISO standards) and consequently has not been incorporated into AFNOR standards.

For food gelatin, a specific AFNOR standard (NF-V-59-106) dated October 1982 describes the "enumeration of spores of anaerobic sulfite-reducing microorganisms, anaerobic 37°C colony counting method." The recommended medium is VL-sulfite ferric ammonium citrate agar.

Tryptic peptone	10 g
Bacteriological meat extract	3 g
Bacteriological yeast extract	6 g
Glucose	2 g
Sodium chloride	5 g
Cysteine chlorhydrate	0.30 g
Agar (by manufacturer's specifications)	9–18 g
Soluble starch certified free of bacterial growth inhibitors	5 g
Anhydrous sodium metabisulfite ($Na_2S_2O_5$)	1 g
Ferric ammonium citrate	1 g
Water	1,000 g

Dissolve the individual components of the dehydrated medium in boiling water. Adjust to have a final pH at 7.6 ± 0.1 at 25°C after sterilization.

Distribute 20 ml each into 20 × 200 mm test tubes. Sterilize by autoclaving at 115 ± 1°C for 30 minutes and allow the tubes to cool in an upright position.

This medium is used in the same manner as the other media but with heating of the original solution to 80 ± 1°C and holding for 10 minutes in a water bath in order to obtain the number of spores only. Five ml each of the 1/10 original solution is introduced into two tubes of medium. After incubation at 37°C for 72 hours, the black colonies are counted. The sum of the number of colonies in each of the two tubes represents the number of spores of sulfite-reducing anaerobic microorganisms present in a gram of gelatin.

21.1.4.2 In Liquid Medium

The count is done at 46°C in so-called L.S. lactose and sulfite-containing medium.

Composition

Casein tryptic peptone	5.0 g
Yeast extract	2.5 g
Sodium chloride	2.5 g
Pure lactose	10 g
Cysteine chlorhydrate	0.3 g
Distilled water	1,000 ml

Adjust the pH to 7.1–7.2, distribute 8 ml each to 16 mm tubes each containing a 5 × 35 mm Durham tube. Sterilize for 15 minutes at 121°C. When ready to use, add to each tube 0.5 ml of 1.2% anhydrous sodium bisulfite solution and 0.5 ml of 1% ferrous ammonium citrate solution prepared previously.

Inoculation

Prepare a 10% suspension or emulsion of the product being analyzed in the following solution, which is also used for the decimal dilutions:

Sodium chloride	2.25 g
Potassium chloride	0.105 g
Calcium chloride	0.12 g
Sodium carbonate	0.05 g
Cysteine chlorhydrate	0.3 g
Distilled water	1,000 ml

Sterilize for 20 minutes at 121°C and remelt prior to use. Inoculate three tubes of medium each with 1 ml of original solution or dilution. Incubate in a water bath at 46°C for 48 hours and observe after 18, 24, and 48 hours.

Positive tubes show a black iron sulfide precipitate which may be precocious (after 6 hours) and then progressively disappear and lactose fermentation which manifests itself as at least 1 cm of gas in the Durham tube. These reactions are specific for *C. perfringens.*

An AFNOR standard (NF-V-59-107) from March 1984 describes the use of L.S. medium for the detection of spores of *C. perfringens* in gelatins.

The number of vegetative cells or spores of *C. perfringens* is estimated by means of the most probable number (MPN) technique.

21.1.5 Identification of *C. perfringens*

The purpose of the biochemical confirmation is to specify the following traits: reduction of nitrates to nitrites, absence of motility, lactose fermentation (if liquid medium was not used for enumeration), and gelatin liquefaction. In order to do this, the following media are used:

Nitrate-Motility Medium

Peptone	5.0 g
Beef extract	3.0 g
Galactose	5.0 g
Glycerol	5.0 g
Potassium nitrate	1.0 g
Sodium dihydrogen phosphate	2.5 g
Agar	3.0 g
Distilled water	1,000 ml

Adjust the pH to 7.3–7.4, distribute 10 ml each to test tubes, sterilize for 20 minutes at 121°C, keep refrigerated at 4°–5°C, remelt and cool prior to use and do not use later than 4 weeks after preparation.

Reagent for the Detection of Nitrite

1. Dissolve 0.1 g of 5-2 ANSA (5-amino, 2-naphthylene sulfonic acid) in 100 ml of 15% (v/v) glacial acetic acid and filter on Whatman no. 41 paper.
2. Dissolve 0.4 g of sulfonic acid in 100 ml of the glacial acetic acid solution and filter on Whatman no. 41 paper.

The two solutions may be mixed and stored in amber flasks at 4°C, but the mixture does not keep for longer than 4 weeks.

After incubating the inoculated nitrate-motility medium at 37°C for 48 hours, 0.5 ml of the mixture or 0.2 ml of each separate reagent is added. The formation of a red coloring in less than 10 minutes signifies the reduction of nitrate to nitrite. In some cases, the reaction goes beyond the nitrite stage. If this has happened, adding zinc powder to reduce any nitrate which may be present does not produce a red color.

Lactose-Gelatin Medium

Tryptone	15 g
Yeast extract	10 g
Lactose	10 g
Sodium dihydrogen phosphate	15 g
Gelatin	120 g
Phenol red	0.05 g
Distilled water	1,000 ml

Dissolve all ingredients and distribute in 10 ml portions to 16 mm tubes and sterilize for 15 minutes at 121°C. Remelt and cool for use and do not use later than 3 weeks after preparation. After incubation at 46°C for 48 hours, a color change towards yellow is observed, and after cooling, liquefaction of the gelatin.

For expression of the results, a correction factor is produced based on the number of colonies (isolated on agar-based media and purified by transfer to a nonselective medium such as Columbia agar or TSC basal agar) of ten which meet the test criteria, for example, seven of ten means multiply the count by 70%.

21.1.6 Detection of Enterotoxin

Enterotoxin detection used to be done by means of the so-called ligated rabbit intestine loop technique (Sutton et al., 1971). Progress in the development of assays based on the immunological properties of enterotoxins of *C. perfringens* has resulted in detection kits which are currently available which make the examination of the enteropathogenic capability of strains much easier. This is the case for the latex test distributed by Oxoïd, for example (PET-RPLA, code DR 930).

21.2 *Clostridium Botulinum*

21.2.1 General

Like *C. perfringens*, *Clostridium botulinum* belongs to group II of the genus *Clostridium*. It is a bacillus which produces oval, subterminal spores, liquefies gelatin, possesses motility by means of peritrichous flagella, and does not reduce nitrates. In fact, the species *botulinum* regroups bacteria which are taxonomically different but which all produce the same types of neurotoxic toxin (Sebald, 1988). Seven types are distinguished, A, B, C, D, E, F, and G which correspond to seven specific toxins. All strains of group A are proteolytic while those of group E are not. Only strains of type G are not hemolytic and do not produce lipase.

The importance of these bacteria in food hygiene is due to their being the source of toxins capable of causing neurological lesions which may be very serious. The affliction in humans often arises from an intoxication (ingestion of preformed toxin in a food) but may also, on more rare occasions, result from an infection (as has been described in very young children).

It should be noted that a few cases of botulism have been described in which the toxin-producing bacteria were not taxonomically *C. botulinum* (*C. barati*, Hall et al., 1985 and *C. butyricum*, McCroskey et al., 1986).

21.2.2 Indicator Value of *C. Botulinum* and Significance of Counts

Clostridium botulinum, like *C. perfringens* is a soil bacterium very widespread throughout nature, even occurring at times in the intestines of healthy animals, particularly swine.

The counting of *C. botulinum* is generally not very useful. On the other hand, the detection of toxigenic spores in a food may sometimes provide very useful information. When toxigenesis occurs, the product may become toxic without the appearance of any organoleptic signs of spoilage.

C. botulinum is capable of synthesizing a toxin which will bring about a neurointoxication of a gravity proportional to the quantity present in the food.

The diagnosis of botulism, beginning with a suspect food, is always done by the seroneutralization test specific for the toxin.

When testing for the presence of spores or of vegetative forms of *C. botulinum* in a food suspected of being heavily contaminated (particularly by sporulating bacteria) is desired, the medium described by Dezfulian and co-authors in 1981 may be used, that is, CBI medium (*Clostridium botulinum* isolation).

Basal Medium

Trypticase peptone	40.0 g
Na_2HPO_4	5.0 g
NaCl	2.0 g
$MgSO_4$ (5% aqueous solution)	0.2 ml
D-glucose	2.0 g
Yeast extract	5.0 g
Agar	20.0 g
Distilled water	900 ml

Adjust the pH to 7.4. Sterilize by autoclaving at 121°C for 15 minutes. Just prior to use, add the following to the lukewarm medium:

Egg-yolk suspension	100 ml
Cycloserine (1% aqueous solution)	25 ml
Sulfamethoxazole (1.9% aqueous solution)	4 ml
Trimethoprime (0.1% aqueous solution)	4 ml

(These solutions are sterilized by filtration.) Distribute into Petri plates. After 24–48 hours of incubation at 37°C, colonies of *C. botulinum* are surrounded by an opaque, iridescent zone which is due to the action of a lipase. Type G *C. botulinum* is difficult to recognize on this medium because it does not possess a lipase. This is a very rare type, however.

21.2.3 Detection of Toxigenic Spores in Foods

The medium to be used depends on the nature of the product:

Meat, Dairy, and Egg-Based Products

CMM (Cooked Meat Medium)	
Cooked meat (DIFCO)	5.5 g
Glucose	5 g
Starch	2 g
Distilled water	1,000 ml

Autoclave for 15 minutes at 121°C.

Fish and Seafoods

TPGYC	
Trypticase	50 g
Peptone (bacto peptone)	5 g
Yeast extract	20 g
Glucose	4 g
Cysteine	1 g
Distilled water	1,000 ml

Autoclave for 12 minutes at 121°C, store refrigerated and use within 15 days after preparation.

For both media, just prior to use, prepare and filter sterilize a 1.5% aqueous solution of trypsin (DIFCO 1/250).

Remelt the medium and cool to ambient temperature, add sufficient trypsin solution to obtain a final concentration of 0.1%, and mix without incorporating air into the medium. Never shelve the medium after adding trypsin.

Mode of Operation

Introduce successively into three sterile plastic bags of 100–150 ml capacity, 10 g of food to be analyzed (ground or fractionated), add 50 ml of appropriate medium for the food type, carefully expel the air from the bag by squeezing, and seal with a heat press. Heat each bag to 80°C for 10 minutes in order to destroy vegetative cells and nonsporulating bacteria, and after cooling incubate for 5 days one bag at 12°C, one at 20°C, and one at 30°C. It has been demonstrated that depending on the serological type and sometimes for a single type, the optimal temperature for toxigenesis may vary between these limits.

At the end of the incubation, centrifuge 20–30 ml from each bag at 3,000 rpm for 5 minutes and then inoculate intraperitoneally two 20 g mice each with 0.5 ml of each supernatant. If two mice inoculated from the same bag or four mice or all six mice die, carry out the seroneutralization test for typing of the culture toxin. This technique detects *C. botulinum* spores but not their enumeration.

21.2.4 Enumeration of Spores and Vegetative Cells in Foods

VL agar with added lacquered (lysed) blood has given us the best results. The counting of *C. botulinum* in foods is done in the following manner:

Prepare a homogenate of the food at a 1/3 dilution in physiological saline or buffered peptone water and make decimal dilutions in buffered peptone water. Transfer successively 1 ml of each dilution to two sterile Petri plates, pour 15 ml of 15% remelted meat-yeast agar cooled to 50°C (±1°C) and then add 0.5 ml of lacquered (lysed) sheep blood. Mix the inoculum, agar and lacquered (lysed) blood by imparting a circular horizontal movement to the plates, leave them to solidify and incubate at 30°C in anaerobic culture jars. Lacquered (lysed) blood is prepared from sterile sheep blood by freezing and thawing it twice. After 5 days of incubation, select a pair of plates having a number of colonies between 20 and 120, take the average of the two counts and multiply by the dilution factor in order to obtain the number of spores and vegetative cells per gram of product. The colonies surrounded by a decolorized zone due to hemolysis are counted. The G strains are not hemolytic, but no case of intoxication due to this type has ever been reported, at least not in France.

CBI medium may be used to enumerate *Clostridium botulinum* by depositing no more than 0.1 ml of test food suspension or dilution thereof onto the surface of an agar plate and spreading with the aid of a bent glass rod. According to the authors who developed this medium, few lipase positive bacteria (of the human fecal microflora) grow on it (Dezfulian et al., 1981). Lipase positive colonies are counted after 48 hours of incubation at 35°C.

Remark: Clostridium botulinum is motile, and the surface of the agar may sometimes show spreading colonies.

21.2.5 Identification of *C. botulinum*

The enumeration methods described earlier are not absolutely specific. Counting must be complemented by the isolation of ten colonies and examination of their biochemical traits. The procedure is then done in the same manner as for *C. perfringens*.

Identification is based on the following characteristics: gram-positive bacilli, motile, fermenting glucose (except for type G), not fermenting lactose, liquefying gelatin, not reducing nitrates, producing lipase (except for type G).

21.2.6 Detection and Typing of Botulin Toxin

21.2.6.1 Detection and Typing of Toxin in Foods

Sampling Procedure

Remove 50 g quantities from various points in the suspect food (whenever possible). Particularly for dried hams, it is necessary to remove material from near the bone and especially from areas of muscle that appear grayish and split by gas swelling. Place fragments are thus removed in a sterile 250 ml flask.

Sample Treatment

For 50 g of material, add 100 ml of gelatin-phosphate buffer of the following formula:

Sodium dihydrogen phosphate	2 g
Gelatin (Bacto)	1 g
Distilled water	1,000 ml

Autoclave for 20 minutes at 115°C. Refrigerate and remove just prior to use in order to begin at a low temperature and avoid overheating during grinding. Add 0.1% pure trypsin (Difco 1/250), that is, 150 mg per 50 g of sample material and 100 ml of buffer. Homogenize for the minimum time necessary (30 seconds to 1 minute) and place the flask at 37°C for 30 minutes to allow diffusion of the toxin.

Centrifugation

Centrifuge all or part of the homogenate at 6,000–9,000 rpm for 10 minutes.

Distribution

Sterilize about 20 ml of the clear liquid phase using a 0.45 μ filter. Dispense 2.5 ml each of this filtrate into six flasks labeled A, B, C, D, E, and T.

Neutralization

Add to each of the first five flasks four drops of corresponding antiserum (A, B, C, D, E) and hold them at 37°C for 1 hour to allow the neutralization of any toxin which may be present. The last flask labeled T is a control and does not receive any serum.

Remark: Currently anti-C and anti-D sera are no longer commercially available in France.

Inoculations

Label successively five pairs of mice weighing 19–21 g. Inject 1 ml of neutralized suspension intraperitoneally into each mouse, that is, two mice per sample of suspension possibly neutralized by specific antiserum. The last two mice are injected with 1 ml each of suspension from unneutralized flask T and serve as controls.

Observation of Inoculated Mice

In the absence of botulin toxin, the twelve mice should behave normally. If a toxin is present, depending on the type, one pair of mice from among samples A to E should be normal while the others should all show symptoms of flaccid paralysis: dragging their bellies, stumbling, and difficulty breathing. Depending on the level of toxin in the food, the clinical signs will appear eventually as well as the death of the unprotected animal subjects.

Results

If one type of toxin is present (e.g., botulinum toxin A) after a maximum of 48 hours, eight mice (not protected by antibotulinum A serum) should be found dead and two normal (protected by botulinum A serum). The control mice (injected from sample T) should also be dead. Sometimes the result is obtainable within a few hours.

21.2.6.2 Typing of the Culture Toxin

Typing of the culture toxin is carried out under the same conditions after centrifugation of the medium, distribution into flasks, neutralization by the five specific sera, and injection into the mice.

21.2.6.3 Assay of Botulin Toxin

Assay of botulin toxin is done by establishing a minimum mortal dose (MMD) using lots of four mice each. Intraperitoneal injections of initial centrifuged suspension and successive dilutions are done and the result is expressed in units of mouse lots.

References

Cassier M. and Sebald M., 1969. *Ann. Inst. Pasteur*, **117**, 312.

Dezfulian M. et al., 1981. *J. Clin. Microbiol.*, **13**, 526.

Goldner S. B., Solberg M., Jones S., Post L. S., 1986. *Appl. Environ. Microbiol.*, **52**, 407.

Hall J. D. et al., 1985. *J. Clin. Microbiol.* **21**, 654.

Hobbs B. C., 1965. *J. Appl. Bacteriol.*, **28**, 74.

McCroskey L. M. et al., 1986. *J. Clin. Microbiol.*, **23**, 201.

Mossel D. A. A. and De Waart, 1968. *J. Ann. Inst. Pasteur de Lille*, **19**, 1–7, 13.

Prevost A. R., 1948. *Ann. Inst. Pasteur*, **75**, 571.

Sebald M., 1988. *In: Microbiologie alimentaire*, Vol. 1: *Aspect microbiologique de la sécurité et de la qualité alimentaires.* Lavoisier Editeur, 77–88.

Sebald M. and Cassier M., 1970. *Bull. Inst. Pasteur*, **68**, 7.

Sutton R. G. A., Chost A. C., Hobbs B. C., 1971. *Appl. Microbiol.*, **12**, 39.

Suggested Readings

Australia, Standards Association of Australia, 1991. Food microbiology. Method 2.7: Examination for specific organisms—*Clostridium botulinum* and *Clostridium botulinum* toxin. *Australian Standard*; AS 1766, Pt. 2.7–1991, 5p.

Censi A., 1990 (publ. 1991), (Clostridium botulinum type G in honey.), *Annali Facolta Medicina Veterinaria, Universita Parma*, **10**, 125–129; 16 ref.

Dezfulian M., 1993. A simple procedure for identification of *Clostridium botulinum* colonies. *World J. Microbiol. Biotechnol.*, **9**, no. 1, 125–127; 11 ref.

Ferreira J. L., Baumstark B. R., Hamdy M. K., McCay S. G., 1993. Polymerase chain reaction for detection of type A *Clostridium botulinum* in foods. *J. Food Prot.*, **56**, no. 1, 18–20; 12 ref.

Ferriera J. L., Hamdy M. K., McCay S. G., Zapatka F. A., 1990. Monoclonal antibody to type F *Clostridium botulinum* toxin. *Appl. Environ. Microbiol.*, **56**, no. 3, 808–811; 15 ref.

Gibson A. M., Modi N. K., Roberts T. A., Hambleton P., Melling J., 1988. Evaluation of a monoclonal antibody-based immunoassay for detecting type B *Clostridium botulinum* toxin produced in pure culture and an inoculated model cured meat system. *J. Appl. Bacteriol.*, **64**, no. 4, 285–291; 5 ref.

Goodnough M. C., Hammer B., Sugiyama H., Johnson E. A., 1993. Colony immunoblot assay of botulinal toxin. *Appl. Environ. Microbiol.*, **59**, no. 7, 2339–2342; 11 ref.

Gryko R., Goszczynski D., Lorkiewicz Z., 1990. A simple method of detection of toxigenic *Clostridium botulinum* type B strains. *J. Microbiol. Methods*, **11**, no. 3/4, 187–193; 14 ref.

Hauschild A. H. W., 1990. *Clostridium botulinum* toxins. *Int. J. Food Microbiol.*, **10**, no. 2, 113–124.

Jiangh, H. M. and Genigeorgis C. A., 1993. Modeling lag phase of nonproteolytic *Clostridium botulinum* toxigenesis in cooked turkey and chicken breast as affected by temperature, sodium lactate, sodium chloride and spore inoculum. *Int. J. Food Microbiol.*, **19**, no. 2, 109–122; 26 ref.

Katsaras K., 1980. (Clostridium perfringens toxins.) Clostridium perfringens-Toxine., *Archiv-f[r-Lebensmittelhygiene*, **31**, no. 4, 121–125.

Lambert A. D., Smith J. P., Dodds K. L., 1991. Effect of initial O_2 and CO_2 and low-dose irradiation on toxin production by *Clostridium botulinum* in MAP fresh pork. *J. Food Prot.*, **54**, no. 12, 939–944; 36 ref.

McClane B. A., 1992. *Clostridium perfringens* enterotoxin: Structure, action and detection. *J. Food Safety*, **12**, no. 3, 237–252; 35 ref.

Mehta R., Narayan K. G., Notermans S., 1989. DOT-enzyme linked immunosorbent assay for detection of *Clostridium perfringens* type A enterotoxin. *Int. J. Food Microbiol.*, **9**, no. 1, 45–50; 13 ref.

Miller A. J., Murphy C. A., Call J. E., Konieczny P., Uchman W., 1993. Enhanced botulinal toxin development in beef sausages containing decolourized red blood cell fractions. *Letters Appl. Microbiol.*, **16**, no. 2, 80–83; 12 ref.

Nakano H., Yoshikuni Y., Hashimoto H., Sakaguchi G., 1992. Detection of *Clostridium botulinum* in natural sweetening. *Int. J. Food Microbiol.*, **16**, no. 2, 117–121; 13 ref.

Ogert R. A., Brown J. E., Singh B. R., Shriver-Lake L. C., Ligler F. S., 1992. Detection of *Clostridium botulinum* toxin A using a fiber optic-based biosensor. *Anal. Biochem.*, **205**, no. 2, 306–312; 23 ref.

Park K. B. and Labbe R. G., 1991. Isolation and characterization of extracellular proteases of *Clostridium perfringens* type A. *Curr. Microbiol.*, **23**, no. 4, 215–219; 20 ref.

Sato H., Chiba J., Sato Y., 1989. Monoclonal antibodies against alpha toxin of *Clostridium perfringens*. *FEMS Microbiol. Lett.*, **59**, no. 1/2, 173–176; 13 ref.

Szabo E. A., Pemberton J. M., Desmarchelier P. M., 1992. Specific detection of *Clostridium botulinum* type B by using the polymerase chain reaction. *Appl. Environ. Microbiol.*, **58**, no. 1, 418–420; 16 ref.

Szabo E. A., Pemberton J. M., Desmarchelier P. M., 1993. Detection of the genes encoding botulinum neurotoxin types A to E by the polymerase chain reaction. *Appl. Environ. Microbiol.*, **59**, no. 9, 3011–3020; 37 ref.

Takahashi M., Kameyama S., Sakaguchi G., 1990. Assay in mice for low levels of *Clostridium botulinum* [type A, strain 97] toxin. *Int. J. Food Microbiol.*, **11**, no. 3/4, 271–277; 17 ref.

Thomas R. J., 1991. Detection of *Clostridium botulinum* type C and D toxin by ELISA. *Australian Vet. J.*, **68**, no. 3, 111–113; 22 ref.

CHAPTER

22

Coagulase-Positive Staphylococci

Marie-Laure De Buyser

The growth of staphylococci in foods constitutes a public health risk since some strains belonging mainly to the species *Staphylococcus aureus* produce enterotoxins, the ingestion of which provokes staphylococcal food poisoning. The symptoms include violent vomiting often accompanied by diarrhea, 2–4 hours after ingestion. It would appear to be preferable to detect the enterotoxins rather than the staphylococci in foods since it is the ingestion of the toxin, and not of the bacteria, which results in the disease. However, due to the technical difficulties imposed by the detection of enterotoxins in foods, *S. aureus* continues to be the object of analysis and enumeration as well as a criterion for the assessment of food safety.

22.1 Taxonomic Position

Staphylococci belong to the genus *Staphylococcus* of the Micrococcaceae family. They share some traits with the genus *Micrococcus*. These are gram-positive, catalase-positive, nonmotile and nonsporulating cocci. Staphylococci, however, are facultative anaerobes while micrococci are strict aerobes. Other tests which allow these two species to be easily distinguished are shown in Section 22.5.3. The classification of staphylococci has undergone significant modifications. Baird-Parker (1974) divided the genus *Staphylococcus* into three species: *S. aureus*, *S. epidermidis*, and *S. saprophyticus*. *S. aureus* was characterized by the production of coagulase, an enzyme whose association with "coagulase-reacting factor" catalyzes the conversion of plasma fibrinogen into fibrin. There are currently 26 individualized species (Kloos and Schleifer, 1986; Freney et al., 1988); 3 of these possess a coagulase: *S. aureus*, *S. entermedius*, and *S. hyicus*.

22.2 Significance of the Presence of Coagulase-Positive Staphylococci in Foods

The staphylococci are saprophytic parasites of humans and animals (skin and mucous membranes). Some species are opportunistic pathogens, responsible for local infections (abcesses) or general infections (septicemias). The most virulent are the three coagulase-positive species. Only certain strains belonging to the species *S. aureus* and *S. intermedius* are capable of producing enterotoxins (Fukuda et al., 1984).

S. aureus is isolated from man and animals. The sinuses and the skin constitute the principal reservoir of the microbe, which may be easily disseminated therefrom throughout the environment and contaminate foods. Strains of *S. aureus* of varied origin (animal, human, or environmental) may contaminate raw foods. Being thermolabile, they are generally destroyed in foods during pasteurization or cooking. The presence of *S. aureus* in warmed foods handled after cooking tends to be an indicator of human contamination. The level of contamination may be low, but if the food is kept in conditions which favor the multiplication of the organism ($>10^5$ *S. aureus*/g) and toxigenesis, enterotoxins may be produced in sufficient quantity to cause an enterointoxication (as little as 1 ng/g of food).

S. intermedius is isolated only from certain animal species (dog, cat, horse, pigeon, fox, mink). This staphylococcus is often confused with *S. aureus* and its actual incidence in foods is not known. Contamination of food by this species is considered unlikely, however.

S. hyicus is isolated from swine, cattle, poultry, and horses. It may therefore contaminate foods and pose identification problems for analysts because it shares some important traits with *S. aureus* (see Section 22.3).

22.3 Characterization of Enterotoxigenic Staphylococci

No test allows enterotoxin-producing strains to be distinguished from nonproducing strains other than detection of enterotoxigenesis capability itself (Bennett, 1986). Because this detection remains difficult, analysts usually settle for the detection of presumptive traits of enterotoxigenicity. The most important traits of enterotoxigenic staphylococci are coagulase production (free coagulase) associated with thermostable nuclease or thermonuclease (TNase) production although not all coagulase-positive and TNase-positive

Table 22.1 Major biochemical traits common to *S. aureus* and other staphylococcal species

	S. aureus	*S. intermedius*	*S. hyicus*	*S. schleiferi**	*S. lugdunensis**
Protein A	+	−	V	−	−
Bound coagulase	+	V	−	+	+
Free coagulase	+	+	V	−	−
Thermonuclease	+	+	+	+	−
Enterotoxin	V	V	−	−	−

**S. schleiferi* and *S. lugdunenis* are two recently identified species (Freney et al., 1988).
+ = over 90% of strains are positive; − = over 90% of strains are negative; V = variable.

staphylococci are enterotoxin producers. *S. aureus* and *S. intermedius* produce coagulase and TNase at the same time. *S. hyicus* produces a TNase but only a few strains of this species coagulate rabbit plasma and often only after 24 hours of incubation at 37°C (Table 22.1).

Bound coagulase or "clumping factor" is a bacterial cell wall constituent which reacts directly with fibrinogen. It is present in *S. aureus* but also in three other staphylococcal species (Table 22.1). For this reason the detection of bound coagulase cannot be substituted for the detection of free coagulase which remains an essential test.

The Baird-Parker classification (1974) until recently, has been retained in food hygiene by various authors cited in this chapter. According to this classification, all coagulase-positive staphylococci belong to the species *S. aureus*. This may cause confusion given that a newer and now solidly established classification distinguishes three coagulase-positive species (Kloos and Schleifer, 1986). The differentiation of these three species requires that extra tests be done (De Buyser, 1988) which are not necessary in routine practice since it is the possession of coagulase and TNase which matters. However, in the absence of a precise species identification, it is henceforth recommended that the expression *coagulase-positive staphylococcus* be used when reporting the results of the analysis, instead of specifying *S. aureus*.

22.4 Detection and Enumeration

In food bacteriology, selective media must be used to preferentially isolate *S. aureus* (according to the Baird-Parker classification) especially since this organism is often a minority within a spectrum of related flora. The selective agents are less toxic for *S. aureus* than for other bacterial species although the more selective the medium, the more inhibitory it is to *S. aureus*. There is therefore no ideal medium.

Furthermore, bacterial cells are often damaged by the treatments which the food undergoes (heating, freezing, acidifying, osmotic shock). For this reason, a "resuscitation" period of 45 minutes in the original suspension is essential before transferring to selective media. Liquid selective enrichment media as well as solid selective isolation media are used.

22.4.1 Enrichment Media

Enrichment media are used in the case of foods likely to contain only a small number of cells of *S. aureus* which have been "stressed" to some degree. Chopin and co-workers (1985) used Giolitti and Cantoni medium with Tween 80. Lancette (1986) recommended trypticase soy broth with added sodium chloride (10%) and sodium pyruvate (1%). The tubes of inoculated media are incubated at 37°C and isolations on selective solid media to detect colonies of *S. aureus* are done after 24 and 48 hours of incubation. The response obtained is of the "presence" or "absence" type for a given inoculum. Enumeration may be done by the most probable number method. Liquid media are rarely used, however. Analysts prefer direct isolation on solid medium which gives similar results, at least according to some authors (Chopin et al., 1985).

22.4.2 Selective Isolation Media

Selective isolation media provide enumeration when inoculated directly without an enrichment phase. It is most often sufficient to use 1/10 and 1/100 dilutions of the original suspension for purposes of verifying adherence to microbiological criteria.

Numerous selective isolation media are available. Baird-Parker medium (1962) is the most widely used in food bacteriology since it gives a better percentage of recovery of *S. aureus*, particularly when damaged cells are involved, as demonstrated by comparative studies of media (Chopin et al., 1985). It is recommended by AFNOR standard NF-V-08-014 of January 1984: Microbiologie alimentaire—Directives générales pour le dénombrement de *S. aureus*, which reproduces the international standard ISO 6888 of 1983. Baird-Parker medium is rich and complex. It contains nutritional elements, three selective agents (potassium tellurite, glycine, and lithium chloride), a growth activator (sodium pyruvate), and a protective agent against the toxic effect of tellurite (egg yolk). Egg yolk and tellurite are at the same time differential agents. The basal formula is commercialized in dehydrated form. Egg yolk-tellurite suspension is also commercialized in dehydrated form but results are better when a laboratory preparation with fresh egg yolk is used. Once poured into Petri plates, the medium does not keep beyond 24–48 hours refrigerated because sodium pyruvate degrades rapidly. One alternative is to leave this constituent out thus allowing the medium to be kept for 1 month at 4°C once poured into Petri plates. Just prior to use, 0.5 ml of 20% (w/v) sodium pyruvate solution is spread onto the surface of the medium and dried at 50°C (Bennett, 1986).

Inoculation is done by quickly spreading 0.1 ml of suspension over the agar surface using a bent glass rod. Plates are incubated at 37°C for 24–48 hours and examined twice, once after 24 hours to note colonies with the characteristic appearance and again after 48 hours. Colonies of *S. aureus* appear black (reduction of tellurite) shiny, convex, encircled with an opaque white trim, and surrounded by a halo of clearing (degradation of egg-yolk lipoprotein).

There are some inconveniences associated with Baird-Parker medium:

- It is troublesome.
- It is weakly selective. Numerous bacterial species (micrococci, enterococci, *Bacillus*, yeasts, etc.) grow on it, interfering considerably with the reading of the plates.
- Some organisms (*Proteus*, coagulase-negative staphylococci) produce colonies identical in appearance to those of *S. aureus*. The genus *Proteus* may be inhibited by the addition of sulfamethazine to the medium (to a final concentration of 50 μg/ml) before pouring the plates.
- Some strains of *S. aureus* do not produce clearing of egg yolk. Uncharacteristic colonies lacking the halo of clearing are frequent in milk products (Harvey and Gilmour, 1985) and rare in meat products (De Waart et al., 1968).

As a result, confirmation tests are indispensable when using Baird-Parker medium. Variations of this medium have been proposed for avoiding these laborious tests. Among the more recent ones worth citing are those of Lachica (1984) and Beckers and associates (1984).

Lachica replaces egg yolk with Tween 80 and magnesium chloride. Colonies of *S. aureus* are identified in situ by a simplified TNase test. They appear black and surrounded by a pink halo. It is nevertheless necessary to transfer a few colonies and subject them to the coagulase test because the TNase alone no longer constitutes a very specific test for *S. aureus*. TNase may be produced by other staphylococci (Table 22.1) and by *Bacillus* and enterococci.

Beckers (1984) replaces egg yolk with rabbit plasma and bovine fibrinogen. An opaque precipitate surrounds coagulase-producing colonies. This medium is useful because it provides complementary tests of identification. According to Sawney (1986), the tellurite concentration must be decreased with respect to the Baird-Parker formula in order to obtain the best yield. The plasma-fibrinogen-potassium tellurite mixture to be added to the basal formula is commercially available (RPF supplement, ref. SR122, Oxoïd).

Tatini and colleagues (1984) recommends a combination of the preceding formulas (plasma-containing Baird-Parker medium on which the simplified TNase test is done) which provides in situ detection of coagulase-positive and TNase-positive colonies.

22.5 Identification

22.5.1 Detection of Coagulase

Five characteristic and/or uncharacteristic colonies, as the case may be, are sampled from each plate and used to inoculate tubes containing brain-heart infusion (BHI) broth. This is a rich broth which may also be used for the detection of TNase. The tubes are incubated for 18 hours at 37°C. The BHI cultures thus obtained are then each mixed with rabbit plasma (or according to the manufacturer's directions) in sterile hemolysis tubes (0.5 ml of each liquid per tube) which are brought to 37°C and observed after 2, 6, and 24 hours of incubation. The reaction is considered positive when the plasma is coagulated enough for the tube to be inverted with no more than a slight flow occurring (strong reactions rated 3+ and 4+, Tatini et al., 1984). When citrate is added to the plasma addition of EDTA is also recommended (final concentration of 0.1%) in order to avoid pseudocoagulations. Each lot of plasma must be tested with coagulase-positive and coagulase-negative strains. When the observed response is dubious (weak reactions rated 1+ and 2+), complementary tests must be carried out.

22.5.2 Detection of TNase

TNase cuts DNA into mono- and oligonucleotides. Its presence is detected by means of a technique described by Lachica and co-workers (1971) using a medium containing DNA and toluidine blue (toluidine DNA agar or TDA) which is commercially available. Five milliliters of TDA medium is poured into 60 mm Petri plates, left to solidify and then six to eight wells 3 mm in diameter are cut in the agar. This medium may be stored cold for several weeks. The 18-hour cultures in BHI are placed in a boiling water bath for 15 minutes (to eliminate thermolabile nucleases) and left to cool. A one-drop sample (or 5 μl) of each broth is removed to be tested and deposited in a well. The plates are incubated for 4 hours at 37°C. A positive reaction appears as a pink halo with a sharp boundary around the well. The halo is developed to a greater or lesser degree depending on the quantity of enzyme produced which varies from one strain to the next. The change from blue to pink is associated with the metachromatic properties of the toluidine blue which is violet-blue in agar with DNA and turns pink in agar alone when the DNA is digested by TNase.

The antigenic properties of TNase of *S. aureus* are different from those of nucleases produced by other staphylococci or by other bacterial genera. It may be specifically distinguished from other nucleases by a seroinhibition test (Gudding, 1983). The nuclease activity of an 18-hour BHI culture is tested in parallel on a TDA medium with and without an anti-*S. aureus* nuclease antiserum. Alone, the *S. aureus* TNase is inhibited in the presence of immunoserum (Staphynuclease kit, bioMérieux). This test identifies the species *S. aureus* according to the classification of Kloos and Schleifer (1986).

22.5.3 Other Tests

When the results obtained from the coagulase and TNase tests are unclear, the suspect colonies selected on Baird-Parker medium should be isolated on a nonselective medium. Colonies thus produced may be examined by Gram stain and for catalase activity. Gram-positive and catalase-positive cocci are then subjected to the differentiation tests for staphylococci and micrococci (Table 22.2). The isolates belonging to the genus *Staphylococcus* are then once again subjected to the coagulase and TNase tests. Other traits such as anaerobic utilization of mannitol, acetoin, hemolysin, and pigment production should also be tested (Kloos and Schleifer, 1986; De Buyser, 1988).

22.6 Interpretation of Results

Microbiological criteria specify for the vast majority of products fewer than 100 *S. aureus*/g or less than 1,000 *S. aureus*/g (cf. interministerial decree* of 21 December 1979, printed in the *Journal Officiel*, 19 January 1980). In actual fact, however, the "less than 100 *S. aureus*/g" is more the result of analytical convenience than of any public health requirement. For solid products, selective media are surface inoculated with 0.1 ml of a 1/10 mother suspension which corresponds to 10 mg of product. Finding one characteristic colony is equivalent to an *S. aureus* content of 100/g in principle. Such use of bacteriological results assumes that they are 100% reliable, which, of course, they are not.

The AFNOR standard NF-V-08-014 specifies that for enumeration purposes only plates containing between 15 and 150 characteristics and/or noncharacteristic colonies should be considered and that five characteristic and/or noncharacteristic colonies should be chosen for confirmation by the coagulase test. If at least 80% of the selected colonies are coagulase-positive, the number of colonies obtained by plate counting is multiplied by the

---

* *Translator's note:* The English word *decree* is used in this chapter to translate the French words *arrêté* and *décret* which correspond to different administrative levels. An asterisk indicates that the French word used was *arrêté*, otherwise the word was *décret*.

Table 22.2 Routine tests for distinguishing staphylococci and micrococci

	Staphylococci	Micrococci	References
Anaerobic formation of acid from glycerol plus erythromycin*	+	–	Schleifer and Kloos (1975)
Furans (300 μg/disc)†	S	R	Hebert and Caillet (1982)
Bacitracin (0.02 IU/disc)†	R	S	Falk and Guering (1983)
Vibriostatic compound 0/129 (500 μg/disc)†	R	S	Bouvet et al (1992)

*Purple agar-based medium (Difco) + glycerol 1% (vol/vol) + erythromycin (final conc. 0.4 μg/ml).
†Pasteur Diagnostic Discs. Mueller-Hinton antibiogram floating on medium method.
S = sensitive; R = resistant.

reciprocal of the inoculum volume and by the reciprocal of the dilution factor in order to obtain the number of *S. aureus*/ml or /g of product. If there are fewer than 15 characteristic and/or noncharacteristic colonies on the plates inoculated with the lowest dilution, the number of confirmed *S. aureus* colonies may still be estimated. For example, the confidence interval for estimation at the 5% level is 1–2 when there is a single confirmed colony per plate and 2–9 when there are five confirmed colonies per plate (see the appendix to the AFNOR standard NF-V-08-014).

The analytical potential may be improved nevertheless by working with 1/3 or 1/5 original suspensions and increasing the inoculum volume (up to 0.3 ml). Furthermore, Petri plates 140 mm in diameter permit surface inoculation with 1 ml, although this is onerous (30 ml of medium per plate). Baird-Parker medium may also be inoculated by mixing in 1 ml of dilution while pouring the plate, but this method has met with criticism since the appearance of the colonies is altered. Media-containing plasma may be either surface inoculated or depth inoculated without any difference.

It is very important to bear in mind that the detection and enumeration of *S. aureus* in foods has its limitations. It should be remembered that only a major multiplication of enterotoxigenic *S. aureus* ($>10^5$/g) produces enterotoxin in quantities worth noting in a food and this only when conditions for toxigenesis are favorable. The presence of enterotoxin may thus be strongly suspected when a high number of *S. aureus* is detected in a food which has been implicated in an enterotoxicosis (De Buyser, 1988). However, finding a high number of staphylococci in a food not implicated in an enterotoxicosis does not constitute anything more than a presumption of danger to the consumer since not all strains of *S. aureus* are enterotoxin producers.

Conversely, it is possible to detect enterotoxins in heated foods containing few or no viable *S. aureus* because the organisms have been destroyed after having produced the enterotoxins. The latter are, in fact, much more thermostable than *S. aureus* itself. In such cases, only analysis for the toxins allows an assessment of the safety of the product (Lapeyre et al., 1988).

The detection of enterotoxins has been carried out up to the present in specialized laboratories using immunological methods. The challenge is to provide methods which are at the same time sensitive (i.e., capable of detecting low levels of enterotoxins in foods) and easy to interpret (not giving rise to nonspecific reactions) (De Buyser, 1988).

Immunoenzymatic methods are currently being developed (Lapeyre et al., 1988). Reagent kits are being marketed and should eventually enable enterotoxin detection in food bacteriology laboratories to become more widespread (Wienecke and Gilbert, 1987). This would provide a useful addition to the information obtained by bacteriological analysis. In spite of its limitations, the latter continues to be indispensable, as much in the context of routine testing of foods as in the analysis of foods implicated in staphylococcal enterotoxicoses.

References

Baird-Parker A. C., 1962. An improved diagnostic and selective medium for isolating coagulase-positive staphylococci. *J. Appl. Bacteriol.*, **25**, 12–19.

Baird-Parker A. C., 1974. *Micrococcaceae. In:* Buchanan R. E., Gibbons N. E. (eds). *Bergey's manual of determinative bacteriology*, 8th ed. 478–490. Williams & Wilkins, Baltimore.

Beckers H. J., Van Leusden F. M., Bindschedler O., Guerraz D., 1984. Evaluation of a pour-plate system with a rabbit plasma-fibrinogen agar for the enumeration of *Staphylococcus aureus* in food. *Can. J. Microbiol.*, **30**, 470–474.

Bennett R. W., 1986. Detection and quantitation of Gram-positive nonsporeforming pathogens and their toxins. *In:* Pierson M. D., Stern N. (eds). *Foodborne microorganisms and their toxins: Developing methodology*, 345–392. Marcel Dekker Inc., New York.

Bouvet P., Chatelain R., Riou J. Y., 1982. Intérêt de composé vibriostatique 0/129 pour différencier les genres *Staphylococcus* et *Micrococcus. Ann. Microbiol.* (Institut Pasteur), **133B**, 449–453.

Chopin A., Malcom S., Jarvis G., Asperger H., Beckers H. J., Bertona A. M., Cominazzini C., Carini S., Lodi R., Hahn G., Heeschen W., Jans J. A., Jervis D. I., Lanier J. M., O'Connor F., Rea M., Rossi J. Seligmann R., Tesone S., Waes G., Mocquot G., Pivnick H., 1985. ICMSF methods studies. 15: Comparison of four media and methods for enumerating *Staphylococcus aureus* in powdered milk. *J. Food Prot.*, **48**, 21–27.

De Buyser M. L., 1988. Les staphylocoques. *In:* Bourgeois C. M., Mescle J. F., Zucca J. (eds). *Microbiologie alimentaire*, vol. 1: *Les microorganismes des aliments*, 65–76. Technique et Documentation, Lavoisier, Paris.

De Waart J., Mossel D. A. A., Tenbroeke R., Van De Moosdijk A., 1968. Enumeration of *Staphylococcus aureus* in foods with special reference to egg-yolk reaction and mannitol-negative mutants. *J. Appl. Bacteriol.*, **31**, 276–285.

Falk D. and Guering S. J., 1983. Differentiation of *Staphylococcus* and *Micrococcus* spp. with the taxo A bacitracin disk. *J. Clin. Microbiol.*, **18**, 719–721.

Freney J., Brun Y., Bes M., Meugnier H., Grimont F., Grimont P. A. D., Nerui C., Fleurette J., 1988. *Staphylococcus lugdunesis* sp. nov. and *Staphylococcus schleiferi* sp. nov., two new species from human clinical specimens. *Int. J. System. Bacteriol.*, **38**, 168–172.

Fukuda S., Tokumo H., Ogawa H., Sasaki M., Kishimoto T., Kawano J., Shimizu A., Kimura S., 1984. Enterotoxigenicity of *Staphylococcus intermedius* strains isolated from dogs. *Zbl. Bakt. Hyg.*, **A258**, 360–367.

Gudding R., 1983. Differentiation of *Staphylococci* on the basis of nuclease properties. *J. Clin. Microbiol.*, **18**, 1098–1101.

Harvey J. and Gilmour A., 1985. Application of current methods for isolation and identification of *Staphylococci* in raw bovine milk. *J. Appl. Bacteriol.*, **59**, 207–221.

Hebert J. P. and Caillet R., 1982. Differenciation entre microcoques et staphylocoques par l'étude de la sensibilité à la nitrofurantoïne. *Path. Biol.*, **31**, 213–216.

Kloos W. E., Schleifer K. M., 1986. Genus IV *Staphylococcus* Rosenbach 1884. *In:* Sneath P. A., Mair N. S., Sharpe M. E., Holt J. G. (eds). *Bergey's manual of systematic bacteriology*, vol. 2, 1013–1035. William & Wilkins, Baltimore.

Lachica R. V., 1984. Egg yolk-free Baird-Parker medium for the accelerated enumeration of food-borne *Staphylococcus aureus. Appl. Environ. Microbiol.*, **48**, 870–871.

Lachica R. V. F., Genigeorgis C., Hoeprich P. D., 1971. Metachromatic agar-diffusion methods for detecting staphylococcal nuclease activity. *Appl. Microbiol.*, **21**, 585–587.

Lancette G. A., 1986. Current resuscitation methods for recovery of stressed *Staphylococcus aureus* cells from foods. *J. Food Prot.*, **49**, 477–481.

Lapeyre C., Janin F., Kaveri S. V., 1988. Indirect double sandwich ELISA using monoclonal antibodies for detection of staphylococcal enterotoxins A, B, C and D in food samples. *Food Microbiol.*, **5**, 25–31.

Sawney D., 1986. The toxicity of potassium tellurite to *Staphylococcus aureus* in rabbit plasma fibrinogen agar. *J. Appl. Bacteriol.*, **61**, 149–155.

Schleifer K. H. and Kloos W. E., 1975. A simple test system for the separation of *Staphylococci* from *Micrococci. J. Clin. Microbiol.*, **1**, 337–338.

Tatini S. R., Hoover D. G., Lachica R. V., 1984. Methods for the isolation and enumeration of *Staphylococcus aureus. In: Compendium of methods for the microbiological examination of foods*, 2d ed., 411–427. Speck M. L., Washington.

Wieneke A. A. and Gilbert R. J., 1987. Comparison of four methods for the detection of staphylococcal enterotoxin in foods from outbreaks of food poisoning. *Int. J. Food Microbiol.*, **4**, 135–143.

Suggested Readings

Avery S. M., 1991. A comparison of two cultural methods for isolating *Staphylococcus aureus*, for use in the New Zealand meat industry. *Meat Industry Research Institute of New Zealand*, no. 868, iii + 13p.; 25 ref.

Brehm R. D., Tranter H. S., Hambleton P., Melling J., 1990. Large-scale purification of staphylococcal enterotoxins A, B, and C2 by dye ligand affinity chromatography. *Appl. Environ. Microbiol.*, **56**, no. 4, 1067–1072; 38 ref.

Gaston M. A., Duff P. S., Naidoo J., Ellis K., Roberts J. I. S., Richardson J. F., Marples R. R., Cooke E. M., 1988. Evaluation of electrophoretic methods for typing methicillin-resistant *Staphylococcus aureus*. *J. Med. Microbiol.*, **26**, no. 3, 189–197; 15 ref.

Khalafalla G. M., Zahra M. K., El-Shenawy M., Mahmoud O. S., 1991. Enterotoxin producing staphylococci in some bakery products. *Ann. Agric. Sci.*, (Ain Shams University), **36**, no 2, 347–353; 18 ref.

Kneifel W., Manafi M., Breit A., 1991. Adaptation of two commercially available DNA probes for the detection of *Escherichia coli* and *Staphylococcus aureus* to selected fields of dairy hygiene—an exemplary study. *Zentralbl. Hygiene Unweltmedizin*, **192**, no. 6, 554–553; 34 ref.

Langlois B. E., Karim-Parlindungan A., Harmon R. J., Akers K., 1990. Biochemical characteristics of Staphylococcus species of human and bovine origin. *J. Food Prot.*, **53**, no. 2, 119–126; 130; 29 ref.

Ocasio W. and Martin S. E., 1990. New methods for staphylococcal enterotoxin detection. *Dev. Ind. Microbiol.*, **31**, (Suppl. 5), 175–184.

Park C. E., Aktar M., Rayman M. K. 1992. Nonspecific reactions of a commercial enzyme-linked immunosorbent assay kit (TECRA) for detection of staphylococcal enterotoxins in foods. *Appl. Environ. Microbiol.*, **58**, no. 8, 2509–2512; 18 ref.

Reynolds D., Tranter H. S., Sage R., Hambleton P., 1988. Novel method for purification of staphylococcal enterotoxin A. *Appl. Environ. Microbiol.*, **54**, no. 7, 1761–1765; 41 ref.

Shinagawa K., Omoe K., Matsusaka N., Sugii S., 1991. Immunological studies on staphylococcal enterotoxin D: Production of murine miniclonal antibodies and immunopurification. *Can. J. Microbiol.*, **37**, no. 8, 586–589; 24 ref.

Windemann H., Luethy J., Maurer M., 1989. ELISA with enzyme amplification for sensitive detection of staphylococcal enterotoxins in food. *Int. J. Food Microbiol.*, **8**, no. 1, 25–34; 9 ref.

CHAPTER

23

The Genus *Listeria*

M. Catteau

Listeria is a bacterium which has attracted a lot of attention in recent years. With several recent outbreaks in both Europe and North America, it has become the subject of a multitude of studies.

23.1 Taxonomy-Identification

Listeria had for a very long time been placed in the Corynebacteriaceae family. Various works in numerical taxonomy have shown that *Listeria* is in fact more related to lactobacilli or streptococci than to the coryneform bacteria. The G+C% is in the 36–42 range and 16S ribosomal RNA sequencing have confirmed the results: the genus most related to *Listeria* is the genus *Brochothrix*.

Listeria are small gram-positive bacilli, noncapsulating and nonsporulating. Motile at 20°C by means of a small number of peritrichous flagellae, their characteristic "pirouette" motility may be observed in a simple wet-mount. On motility agar, *Listeria* frequently show a peculiar "umbrellalike" appearance. They are catalase-positive and oxidase-negative. The genus *Listeria* ferments numerous carbohydrates without gas production (e.g., glucose, fructose, mannose, amygdalin, salicine, cellobiose, maltose, trehalose, gentobiose, D-arabitol).

Esculine is hydrolyzed, the Voges-Proskauer and methyl red reactions are positive, urease is negative, indole and H_2S are not produced (Table 23.1).

Table 23.1 General traits of the genus *Listeria*

Small bacillus, gram	+	Mannose	+	D-arabitol	+
Capsule	–	Amygdalin	+	Esculine	+
Spore	–	Calicine	+	V.P.	+
Motility at 22°C	+	Cellobiose	+	RM	+
Catalase	+	Maltose	+	Urease	–
Oxidase	–	Trehalose	+	Indole	–
Fructose	+				

Table 23.2 Differentiation of *Listeria* species

Species designation	Hemolysis (beta)	Nitrate reduction	Mannitol	RM/VP	Rhamnose	Xylose	Pathogenic capability in mice
L. monocytogenes	+	–	–	+/+	+	–	Yes
L. ivanovii	+	–	–	+/+	–	+	Yes
L. innocua	+	–	–	+/+	V	–	No
L. welshimeri	–	–	–	+/+	V	+	No
L. seeligeri	+	–	–	+/+	–	+	No
L. grayi	–	–	+	+/+	–	–	No
L. murrayi	–	+	+	+/+	V	–	No

The genus *Listeria* includes seven species organized into two distinct branches: (1) *L. monocytogenes, L. ivanovii, L. innocua, L. welshimeri, L. seeligeri* and (2) *L. grayi* and *L. murrayi.* The principal biochemical traits which differentiate these species are listed in Table 23.2. Rocourt (1984) (Figure 23.1) distinguishes seven species using five traits: hemolysis, the CAMP test with *Corynebacterium equi* (*Rhodococcus equi*), nitrate reduction, and production of acids from mannitol and D-xylose. *L. denitrificans* has recently been excluded from the genus *Listeria* and moved to a new genus called *Jonesia* (Rocourt et al., 1987).

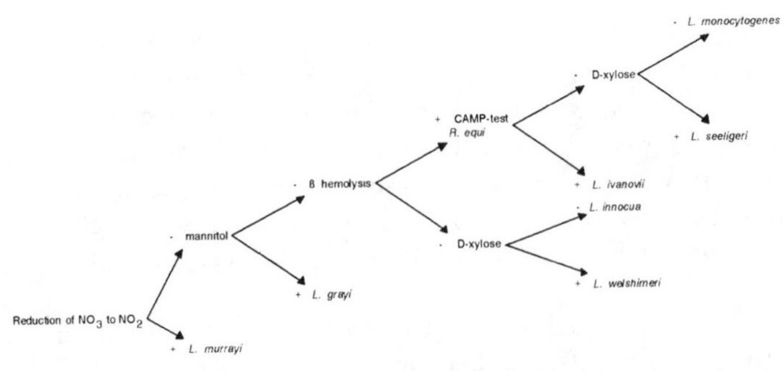

Figure 23.1 Biochemical differentiation of *Listeria* species.

23.2 Serotyping

Fifteen somatic and five flagellar antigens allow the definition of 16 serovarieties for *L. monocytogenes* and related species (Table 23.3). *L. grayi* and *L. murrayi* are antigenically identical but differ from other species.

23.3 Lysotyping

Nearly all of the strains isolated in human pathology belong to serogroups 1/2 and 4. The origin of strains solely on the basis of serotypes is therefore not easy to determine. It is hence useful to turn to lysotyping. Described by Audurier and his co-workers (1977) this system employs 28 bacteriophages and is recognized as an international standard. In conjunction with field enquiries, it enables the origin of the contamination to be traced.

23.4 Pathogenic Potential

Only *L. monocytogenes* appears to be pathogenic to humans. *L. ivanovii* has been isolated from rare cases of abortion in sheep, cattle, and goats. Human listeriosis preferentially afflicts pregnant women and infants as well as immunosuppressed adults. In pregnant women, the signs of infection remain discreet (e.g., pseudo-cold syndrome, lower back pain, urinary infection). Very often the evolution of the infection in the mother will result in interruption of pregnancy from the fifth or sixth month. In infants, listeriosis is a serious infection, fatal in one-third of all cases, or with aftereffects of varied seriousness. Septicemia and meningitis are the most common manifestations. In immunosuppressed adults, meningitis, encephalitis, and septicemia are the most frequent. The mortality rate remains high at 50%.

23.5 Epidemiology

Among the recent epidemics, the one in Canada in 1981 was caused by coleslaw kept at 4°C and sold in a supermarket. In 1983 in Boston, 49 people were stricken with listeriosis apparently caused by pasteurized milk. In 1985 in California, 142 cases of listeriosis were observed due to consumption of Mexican-type cheese and caused 48 fatalities. The strain responsible was a particular phage type of *L. monocytogenes* Type 4b. And in Switzerland from 1983 to 1987, 122 people were afflicted of which 33 died, the food responsible being a soft ripened cheese (Vacherin Mont-d'Or) contaminated by *L. monocytogenes* Serotype 4b and belonging to two particular lysotypes.

Table 23.3 Serovarieties of the genus *Listeria*

Species	Serovariety	O antigens															H antigens				
L. monocytogenes	1/2a	I	II	(III)*													A	B			
and related species	1/2b	I	II	(III)													A	B	C		
	1/2c	I	II	(III)														B		D	
	3a		II	(III)	IV												A	B			
	3b		II	(III)	IV								(XII	XIII)			A	B	C		
	3c		II	(III)	IV								(XII	XIII)				B		D	
	4a			(III)		(V)		VII		IX							A	B	C		
	4ab			(III)		V	VI	VII		IX	X						A	B	C		
	4b			(III)		V	VI										A	B	C		
	4c			(III)		V		VII									A	B	C		
	4d			(III)		(V)	VI		VIII								A	B	C		
	4e			(III)		V	VI		(VIII)	(IX)							A	B	C		
	5			(III)		(V)	VI		(VIII)		X						A	B	C		
	7			(III)									XII	XIII			A	B	C		
	6a			(III)		V	(VI)	(VII)		(IX)						XV	A	B	C		
	6b			(III)		(V)	(VI)	(VII)		IX	X	XI									
L. grayi				(III)									XII		XIV						E
L. murrayi				(III)									XII		XIV						E

*( ) = not always present.

Source: Adapted from Seeliger and Jones, 1986.

23.6 Contaminated Foods and Multiplication of *L. monocytogenes*

Although soft ripened cheeses have frequently been implicated in cases of transmission of listeriosis, it should be noted that *L. monocytogenes* may be isolated from numerous other foods:

- 85 out of 100 positive samples of poultry purchased commercially in Germany;
- 15 out of 50 positive samples of meats and meat products in Germany, 59 out of 306 products of the same type in France;
- 11 out of 68 positive samples of raw meats in Switzerland;
- 3 cases positive out of 91 samples of vegetables and salads in Switzerland, 11 cases positive out of 25 in the Netherlands;
- 10 cases positive out 190 samples of shrimp and crabmeat in the United States.

L. monocytogenes is also frequently encountered in the environment (soil, vegetables, human and animal fecal matter) and may therefore easily contaminate many raw foods. Proper pasteurization is sufficient to eliminate them, although cross-contaminations of *Listeria* may be found in treated foods.

L. monocytogenes is capable, when the other growth parameters are optimal, of developing at pH 5, in the presence of a water activity of 0.94, or at a temperature of 1°C. It may therefore multiply in numerous foods and numbers of *Listeria* as high as 10^6–10^7/g have been observed in some soft ripened cheeses.

23.7 Detection of *Listeria* in Foods

Various methods have been proposed to detect *Listeria*, particularly in foods. This is a relatively simple matter when *Listeria* are abundant in the sample being analyzed. Isolation on selective media is sufficient. In the majority of cases where foods are involved, the problem is of a different nature. *Listeria* generally occur in small numbers and are accompanied by an abundant associated microflora. A typical example would be the detection of from one to a few cells of *Listeria* in 25 g of cheese containing a very abundant lactic flora.

Selective enrichment methods are indispensable in such cases. A variety of recommendable enrichment methods exists. Some of these are based on the psychrotrophic nature of *Listeria* and consist of incubating inoculated enrichment broth at low temperature (4°C) for several days or several weeks. *Listeria* develops better than the accompanying flora, thus providing the basis for enrichment. The major inconvenience of these methods is obviously the long delay required, making them incompatible with the life spans of some products. These methods do, however, give satisfactory results. Other methods involve incubating the enrichment broth at 30°C. This is the FDA method or derivatives thereof currently recommended for the analysis of cheeses.

23.8 Method of Detection of *Listeria monocytogenes* in Cheeses

This is the method currently recommended by official testing agencies.

23.8.1 Sampling and Inoculation of the Enrichment Broth

Detection of *Listeria* is carried out on 25 g of cheese. The samples are removed from five cheeses of the same lot, that is, 5 g each using a cheese probe, throughout the entire diameter and including the rind. The 25 g are added aseptically to 225 ml of selective enrichment broth (see Appendix, Section 23.9) and homogenized for 2 minutes in a stomacher and then incubated at 30°C.

23.8.2 Isolation

Isolations are carried out on modified McBride agar (MMA, see Appendix, Section 23.9).

After 24 Hours of Enrichment at 30°C

Isolate on MMA. Remove 0.1 ml of incubated EB broth and inoculate 10 ml of fresh EB broth. Incubate for 24 hours at 30°C (first subculture). The MMA plates are incubated at 37°C, for 24 and 48 hours.

After 48 Hours of Enrichment at 30°C

Isolate on MMA. Isolate the first subculture on MMA. Carry out second subculture in EB broth. Examine the previous days' plates.

After 72 Hours

Isolate the second subculture on MMA. Examine the plates.

After 96 Hours

Examine the plates.

After 7 Days of Incubation of the EB Broth

Carry out an isolation on MMA. Carry out a third subculture on EB Broth.

After 8 Days

Examine the plates. Isolate the third EB Broth Culture.

After 9 Days

Examine the plates.

23.8.3 Reading of the Plates

The plates are read in oblique transillumination. *Listeria* appear as small, bluish-gray colonies. If suspect colonies appear, a minimum of five should be removed to a slide and tested for catalase followed by Gram staining. They should be catalase-positive, small gram-positive bacilli.

23.8.4 Identification

The following media are inoculated with suspect colonies, purified by isolation on trypticase soy agar:

- VL agar with glucose (0.6% agar) for respiratory type;
- blood agar, Oxoid blood no. 2 + sheep blood, inoculate by stab for hemolysis;
- the CAMP test on Oxoid blood no. 2 agar for β hemolytic staphylococci, *Corynebacterium equi*;
- nitrate nutrient broth for nitrate reduction;
- Sim agar (BBL) at 20° and 37°C for motility;
- Clark and Lubs medium for the methyl red and Voges-Proskauer tests;
- Christensen urea tests;
- glucose, xylose, rhamnose, mannitol, maltose, esculine (BCP broth) fermentations.

It is readily apparent that this method of *Listeria* detection in foods is very tedious, long (from 4 to 15 days), and costly. Simpler methods must therefore be developed.

Improvements to the isolation media may be made, making them more selective and facilitating the observation of colonies of interest. Numerous formulas are currently published, among the more interesting are the Oxford agar (Curtis et al., 1988), Palcam agar (Van Netten et al., 1988), and LPM agar (Lee et al., 1986). Newer methods using immunological methods or hybridization techniques are also being developed.

Immunoenzymatic methods are currently commercialized, allowing appreciable gains in time compared to conventional methods, although their specificity is not perfect.

The use of nucleic acid probes should develop considerably during the next few years. "Cold" probes (nonisotopic) and the recent development of amplification methods makes their use simpler and increases their sensitivity. The detection of a single DNA molecule and hence a single bacterial cell is now conceivable using probes. It should not be forgotten, however, that although the development of simple, sensitive, and rapid detection techniques is important, it is first and foremost the implementation of rigorous production practices which will reduce and ultimately eliminate the contamination of foods.

23.9 Appendix: Formulas of the Media Used

EB enrichment broth

Prepare trypticase soy broth according to the manufacturer's instructions. Add 0.6% yeast extract. Distribute into 225 ml flasks. Autoclave for 15 minutes at 121°C. The pH must be between 7.3 and 25°C after autoclaving. Just prior to use, add aseptically to each flask:

0.5% Acriflavine HCl solution	0.5 ml
2% Nalidixic acid solution	0.5 ml
2.5% Cycloheximide solution	0.5 ml

Dissolve separately the acriflavine HCl and nalidixic acid in distilled water and sterilize by filtering through 0.2 μm membrane. Dissolve the cycloheximide in a 40% ethanol in water solution and sterilize by filtration.

Modified McBride Selective Agar (MMA)

Phenyl ethanol agar (Difco)	35.5 g
Anhydrous glycine (Sigma)	10.0 g
Lithium chloride (Sigma)	0.5 g
Distilled water	1 l

Distribute into 100 ml flasks. Autoclave at 121°C for 15 minutes. Keep refrigerated. Just prior to use, remelt in a water bath, cool slightly and add 20 mg of cycloheximide per 100 ml (0.8 ml of a filter-sterilized 2.5% solution in 40% alcohol).

References

Audurier A., Rocourt J., Courtieu A., 1977. Isolement et caractérisation de bactériophages de *Listeria monocytogènes*. *Ann. Microbiol.* (Institut Pasteur), **128A**, 185–186.

Curtis G. D. W., Mitchell R. G., King A. F., Griffin E. J., 1988. A selective differential medium for the isolation of *Listeria monocytogenes*. *Letters Appl. Microbiol.*, **8**, 95–98.

Doyle M. P. and Schoeni J. L., 1987. Comparison of procedures for isolating *Listeria monocytogenes* of soft, surface-ripened cheese. *J. Food Prot.*, **50**, 4–6.

Lee W. H. and McClain D., 1986. Improved *Listeria monocytogenes* selective agar. *Appl. Environ. Microbiol.*, **52**, no. 5, 1215–1217.

Rocourt J., Venmeyer V., Stackebrandt E., 1987. Transfer of *Listeria denitrificans* to a new genus *Jonesia* gen. nov., as *Jonesia denitrificans*, comb. nov. *Int. J. System. Bacteriol.*, **37**, 226–270.

Seelinger H. P. R. and Jones D., 1986. *Listeria.* 94.

Sheath P. H. A., Mair N. S., Sharpe M. E. and Holt J. G., Eds. Bergey's manual of systematic bacteriology, vol. 2, Williams & Wilkins, Baltimore.

Van Netten P., Perales I., Van De Moosdijk A., Curtis G. D. W., Mossel D. A. A., 1988. Liquid and solid selective differential media for the detection and enumeration of *Listeria monocytogenes*. *Int. J. Food Microbiol.*, **8**, 299–316.

Suggested Readings

Beuchat L. R. and Brackett R. E., 1990. Survival and growth of *Listeria monocytogenes* on lettuce as influenced by shredding, chlorine treatment, modified atmosphere packaging and temperature. *J. Food Sci. Off. Publ. Inst. Food Technol.* Chicago, IL: The Institute. May/June 1990, **55**, no. 3, 755–758, 870.

Boerlin P. and Piffaretti J. C., 1991. Typing of human, animal, food, and environmental isolates of *Listeria monocytogenes* by multilocus enzyme electrophoresis. *Appl. Environ. Microbiol.*, **57**, no. 6, 1624–1629; 30 ref.

Cassiday P. K. and Brackett R. E., 1989. Methods and media to isolate and enumerate *Listeria monocytogenes*: A review. *J. Food Prot.*, **52**, no. 3, 207–214; 63 ref.

Chenevert J., Mengaud J., Gormley E., Cossart P., 1989. A DNA probe specific for L. monocytogenes in the genus *Listeria. Int. J. Food Microbiol.*, **8**, no. 4, 317–319; 5 ref.

Daneshvar M. I., Brooks J. B., Malcolm G. B., Pine L., 1989. Analysis of fermentation products of *Listeria* by frequency-pulsed electron-capture gas-liquid chromatography. *Can. J. Microbiol.*, **35**, no. 8, 786–793; 22 ref.

Doyle M. P., 1988. Effect of environmental and processing conditions on *Listeria monocytogenes. Food Technol.*, **42**, no. 4, 169–171; 29 ref.

Eley A., 1990. New rapid detection methods for *Salmonella* and *Listeria. Br. Food. J.*, **93**, no. 4, 28–31; 17 ref.

Farber J. M. and Peterkin P. I., 1991. *Listeria monocytogenes*, a food-borne pathogen. *Microbiol. Rev.*, **55**, no. 3, 476–511; 445 ref.

Fuchs R. S., 1991. *Listeria monocytogenes. ASEAN Food J.*, **6**, no. 1, 3–13.

Lacey R. W., 1992. *Listeria*: Implications for food safety. *Br. Food J.*, **94**, no. 1, 26–32; 31 ref.

McLauchlin J., 1989 (recd. 1990). Rapid non-cultural methods for the detection of *Listeria* in food. A review. *Microbiol. Alim. Nutr.*, **7**, no. 3, 279–284; 42 ref.

Mackey B. M. and Bratchell N., 1989. The heat resistance of *Listeria monocytogenes. Letters Appl. Microbiol.*, **9**, no. 3, 89–94; 31 ref.

Mafu A. A., Roy D., Goulet J., Savoie L., 1991. Characterization of physicochemical forces involved in adhesion of *Listeria monocytogenes* to surfaces. *Appl. Environ. Microbiol.*, **57**, no. 7, 1969–1973; 28 ref.

Mossel D. A. A., 1989. *Listeria monocytogenes* in foods. Isolation, characterization and control. *Int. J. Food Microbiol.*, **8**, no. 3, 183–195.

Ninet B., Traitler H., Aeschlimann J. M., Horman I., Hartmann D., Bille J., 1992. Quantitative analysis of cellular fatty acid (CFAs) composition of the seven species of *Listeria. System. Appl. Microbiol.*, **15**, no. 1, 76–81; 34 ref.

Palumbo S. A., 1991. A review of methods for detection of the psychrotrophic foodborne pathogens *Listeria monocytogenes* and *Aeromones hydrophila. J. Food Safety*, **11**, no. 2, 105–122; 61 ref.

Schuchat A., Swaminathan B., Broome C. V., 1991. Epidemiology of human listeriosis. *Clin. Microbiol. Rev.*, **4**, no. 2, 169–183; 179 ref.

Seeliger H. P. R. and Langer B., 1989. Serological analysis of the genus *Listeria*. Its values and limitations. *Int. J. Food Microbiol.*, **8**, no. 3, 245–248; 20 ref.

Slade P. J., 1992. Monitoring *Listeria* in the food production environment. I. Detection of *Listeria* in processing plants and isolation methodology. *Food Res. Int.*, **25**, no. 1, 45–56.

Wang R. F., Cao W. W., Johnson M. G., 1992. 16S rRNA-based probes and polymerase chain reaction method to detect *Listeria monocytogenes* cells added to food. *Appl. Environ. Microbiol.*, **58**, no. 9, 2827–2831; 14 ref.

Yu L. S. L. and Fung D. Y. C., 1991. OxyraseRegistered enzyme and motility enrichment Fung–Yu tube for rapid detection of *Listeria monocytogenes* and *Listeria* species. *J. Food Safety*, **11**, no. 3, 149–162; 24 ref.

Yu L. S. L. and Fung D. Y. C., 1991. Effect of OxyraseRegistered enzyme on *Listeria monocytogenes* and other facultative anaerobes. *J. Food Safety*, **11**, no. 3, 163–175; 22 ref.

PART IV

MICROBIOLOGICAL TESTING OF STARTING MATERIALS AND FINISHED PRODUCTS

CHAPTER

24

Water

A. Plusquellec

Water serves more than one purpose in our diets. First, as a basic food element, water for direct consumption must meet standards of potability which ensure the protection of the consumer. Water is also widely used in the preparation of foods (reconstitution, washing) and thus constitutes a possible vector for dangerous microbes. For these reasons, water has always received special attention from hygienists.

Contamination from water supplies is a major problem in developing countries where the struggle against diarrheic diseases responsible for mortality and morbidity (mainly of infants) is a priority (Leclerc, 1989). Of these diseases, cholera is one of the most serious. In countries where programs for treatment and testing of distributed water are routinely applied, the great waterborne epidemics have disappeared, although there is still the occasional bout with certain types of contamination (bacteria, viruses, protozoa).

24.1 Definition

Water analyzed in laboratories specializing in the area of food may come from a variety of sources.

24.1.1 Drinking Water

There are several categories of drinking water.

24.1.1.1 Publicly Distributed Water (Tap Water)

Tap water must conform to the definition of potable water. It is drawn from bodies of surface water (lakes, rivers) or from underground deposits or springs (which are generally better protected).

Before distribution, these waters may undergo several purification treatments such as filtration, decanting, or flocculation. The last operation is a disinfection step, capable of destroying pathogenic microorganisms, either by chlorination (with gaseous chlorine or a chemical chlorinated derivative) or sometimes ozonation.

24.1.1.2 Water Drawn by Independent Entities

Drawing of water from underground deposits or springs is generally for supplying a house, a hamlet, or a processing plant, mainly in rural areas not served by public water works. This type of water usually does not undergo any treatment prior to consumption.

24.2 Regulation of the Bacteriological Quality of Water

24.2.1 Waters Other than Natural Mineral Water

The quality of water for human consumption (excluding natural mineral waters) is regulated in France by decree no. 89, dated January 3, 1989. This includes the following:

- waters delivered for consumption, whether in containers or not;
- waters used in food companies for manufacturing, processing, or storage of foods;
- ice (water-derived) intended for food uses.

Verification of water quality must be ensured by the business owner at his or her expense and according to a defined program. The type of microbiological and physicochemical analyses depend on the point at which the water was sampled (resource, production, or distribution). The annual frequency of analysis depends on the daily flow rate for water sampled at the source or in the factory and on the size of the population served for samples taken at distribution points.

Supplementary analyses may be imposed by the local authority when quality requirements are not met, when quality criteria for crude water are not met, or when signs of deteriorating water quality or sanitary problems possibly associated with the water appear.

The quality criteria for water intended for human consumption are established in the January 3, 1989 decree. The microbiological parameters are as follows:

- The water must not contain any pathogens, especially not salmonella in any 5 l sample, no pathogenic staphylococci in a 100 ml sample, and no enterovirus in any 10 l sample.

- At least 95% of samples taken must not contain any coliforms per 100 ml.
- The water must not contain any thermotolerant coliforms or fecal streptococci in any 100 ml sample taken.
- There must not be more than one sulfite-reducing anaerobic bacterial spore per 20 ml of sampled water.
- When the water is delivered in closed containers, the enumeration of viable aerobic bacteria at 37°C after 24 hours must be less than or equal to 20 per ml of sample and less than or equal to 100 per ml of sample at 22°C after 72 hours. The analysis is to be done within 12 hours after bottling.

Aside from the microbiological requirements, the decree also establishes organoleptic and physicochemical parameters as well as parameters relative to undesirable substances, toxic substances, pesticides, and related products.

This decree also establishes the limits for the quality of crude water used for the production of water intended for human consumption (i.e., river or lake waters before purification). From a microbiological point of view, these waters must contain less than 20,000 thermotolerant coliforms and less than 10,000 fecal streptococci/100 ml.

24.2.2 Natural Mineral Waters

Testing of mineral water springs is regulated by the May 16, 1989 statement which deals mainly with the springs exploited in thermal establishments (hot spas). At all stages of use (spring, drinking fountains, pools, therapy), the water must meet the following criteria:

- absence of thermotolerant coliforms in 250 ml of water;
- absence of fecal streptococci (group D) in 250 ml of water;
- absence of spores of anaerobic sulfite-reducers in 50 ml of water.

24.3 Bacteriological Analysis

24.3.1 Types of Microbiological Analysis

The analysis of water intended for human consumption is done by certified laboratories, the certification being approved by ministerial decree.* Laboratories must use reference methods.

From a bacteriological point of view, three levels of analysis are defined: reduced analysis (B1), summary analysis (B2), and complete analysis (B3) (see Table 24.1).

For publicly distributed water, reduced analysis (B1) is done on water resources, summary analysis (B2) on water from the works, and complete analysis (B3) on treated water or water at the point of drawing in the absence

Translator's Note: The English word *decree* is used in this chapter to translate the French words *arrêté* and *décret* which correspond to different administrative levels. An asterisk indicates that the word used was *arrêté*, otherwise the word was *décret*.

Table 24.1 Bacteriological analyses

Reduced (B1)	Summary (B2)	Complete (B3)
Thermotolerant (44°C) coliforms Fecal streptococci	Thermotolerant (44°C) coliforms Fecal streptococci Enumeration of viable aerobic bacteria at 22° and 37°C	Thermotolerant (44°C) coliforms Fecal streptococci, coliforms Enumeration of viable aerobic bacteria at 22° and 37°C Spores of anaerobic sulfite-reducing bacteria

of treatment. For bottled water and ice for food use, complete analysis (B3) is done at the source before draining or freezing and after filling. The containers and filling equipment are also subjected to a summary analysis (B2). With respect to food industry, waters not taken from public utilities, complete analyses (B3) must be carried out at the source at a frequency which depends on the flow rate.

Monitoring analyses for natural mineral waters (May 16, 1989 decree*) include several physicochemical tests (conductivity, pH, temperature, alkalinity, assay of a characteristic ion) and several bacteriological tests which are as follows:

- enumeration of viable bacteria in 1 ml of water after 24 hours at 37°C and 72 hours at 22°C (inoculation within 8 hours following sampling);
- enumeration of sporulating anaerobic sulfite reducers in 50 ml of water;
- enumeration of coliforms at 37°C, thermotolerant (44°C) coliforms at 44°C, fecal streptococci and *Pseudomonas aeruginosa* in 250 ml of water;
- enumeration of *Legionella* in 1 l of water.

The techniques recommended for the microbiological analysis of water have been standardized by the AFNOR (standard NF-T-90-420). The techniques that will be described are based essentially on these recommendations.

24.3.2 Sample Removal

Sample removal must be done aseptically by qualified personnel. A sterile 500 ml glass flask is used. In the case of water delivered by public utilities works, it may sometimes be necessary to eliminate contamination due to pipes. The tap must be disinfected and flamed and the water must run for some time before sampling. For water treated with chlorine or derivatives thereof, flasks containing 5–10 mg of sodium thiosulfate are used. Samples must be brought quickly to the laboratory and refrigerated if the temperature exceeds 10°C. The analysis must be carried out within 12 hours of sampling.

Personnel responsible for sampling must gather as much information as possible relevant to the bacteriological quality of the water (e.g., the origin of the water, by what means it is drawn, the type of treatment to which it may have been subjected, probable causes of contamination, degree of rainfall prior to sampling, temperature during sampling).

24.3.3 Enumeration of the Viable Bacterial Flora

The January 3, 1989 decree recommends a double counting: first, colonies developing within 3 days at 20°C on ordinary counting medium (optimal conditions for most common saprophytic bacteria); second, colonies which develop within 24 hours at 37°C on the same medium (optimal conditions for bacteria of human or animal origin). This enumeration is of value insofar as it is repeated on a regular basis. Enumeration is done by the pour plate method using standard plate count agar (PCA).

For counting microorganisms viable at 37°C (AFNOR standard NF-T-90-401), 1 ml of water is used as inoculum or two times 2 ml for bottled water. If heavy contamination is expected, smaller inoculum volumes may be used. Plates are poured with ordinary PCA medium.

The same volumes of inoculum are used for enumeration at 20°C (AFNOR standard NF-T-90-402). A layer of sterile plain agar (4 ml) is overlaid in order to prevent surface spreading by the colonies.

24.3.4 Enumeration of Coliforms and of Thermotolerant (44°C) Coliforms

Regulatory criteria require the absence of thermotolerant (44°C) coliforms in 100 ml of water and the absence of coliforms in 100 ml for 95% of samples. Water samples of 100 ml are therefore treated in one of two ways: by filtration of 100 ml on a membrane or by inoculating 100 ml into liquid medium.

Note: The standardized technique for the analysis of water intended for human consumption is membrane filtration.

24.3.4.1 Membrane Filtration

Membrane filtration is the most widely used technique. A membrane of cellulose ester composition and 0.45 μ porosity is used to aseptically filter 100 ml of well-homogenized water. The membrane is then placed for incubation on a medium selective for coliforms. Two media are recommended by the AFNOR (standard NF-T-90-414): TTC and Tergitol 7 lactose agar and sodium lauryl sulfate agar. Incubation of these media at 37°C for 24 hours allows the enumeration of coliforms. Incubating at 44°C provides an enumeration of thermotolerant (44°C) coliforms.

On TTC and Tergitol 7 lactose agar, characteristic colonies are yellow or orange-stained (TTC-) with an underlying yellow halo (lactose fermented). On sodium lauryl sulfate agar, characteristic colonies are yellow with an underlying yellow halo.

After incubation at 37°C, those colonies which have the characteristic appearance and are oxidase negative are counted as coliforms. After incubation of 44°C, any colonies having the characteristic appearance are counted as thermotolerant (44°C) coliforms (the oxidase test is not necesary). When characteristic colonies appear at 37°C but not at 44°C, it is necessary to verify that the colonies are not thermotolerant (44°C) coliforms. For this purpose, the coliforms colonies are used to inoculate brilliant green lactose broth or Schubert's medium. After incubation at 44°C, gas-producing colonies are counted as thermotolerant (44°C) coliforms.

Precise identification as *Escherichia coli* may be done for colonies with the characteristic appearance, but this is optional, since such confirmations are not required by the regulatory criteria.

Membrane fecal coliform (MFC) medium also provides a selective enumeration of thermotolerant (44°C) coliforms (incubation at 44°C ± 0.5°C). The characteristic colonies have a bluish color.

Filtering of 100 ml of sample water constitutes a minimal protocol. In practice, it is wise (essentially in the case of nontreated water) to filter smaller volumes of sample (1–5 ml), principally for the enumeration of coliforms, where noncoliform common flora may otherwise render readings impossible.

24.3.4.2 Detection in Liquid Media

One alternative for processing 100 ml of sample consists of dividing it into ten 10 ml volumes and inoculating one tube each of coliform counting broth with these. Double-strength broth in 10 ml volumes is used since inoculation dilutes the broth by half.

The AFNOR recommended technique for public swimming water and for water to be made potable (standard NF-90-143) consists of first inoculating ten tubes of double-strength medium with 100 ml of sample water. The presumptive medium used at this stage may be BCP lactose broth or sodium lauryl sulfate broth. These media are incubated at 30°C for 24 or 48 hours. The tubes which indicate a presumption of coliforms are those with gas collecting in the inverted Durham tube.

The second step involves transferring the positive media onto selective confirmation medium in order to confirm coliforms or thermotolerant coliforms. Brilliant green lactose bile broth incubated at 37°C is used for coliforms and this medium or Schubert's medium incubated at 44°C ± 0.5°C allows the characterization of thermotolerant (44°C) coliforms.

Growth at 44°C associated with indole production at this temperature characterizes *Escherichia coli*.

This protocol does not constitute an enumeration but a simple detection in 100 ml of water. If a true count is desired, the multiple tube technique and a series of dilutions must be used and interpreted using most probable number tables.

24.3.5 Enumeration of Fecal Streptococci (Group D Streptococci)

Regulations require the absence of fecal streptococci in 100 ml of water, this being the volume which must be analyzed either by membrane filtration of by inoculating liquid media.

Membrane filtration is the most generally practiced (AFNOR standard NF-T-90-416). The medium most widely used for this purpose is Slanetz and Bartley agar (*M. enterococcus*) with sodium azide as selective agent and TTC reduction for the characterization of the colonies. Incubation is done at 37°C for 48 hours. After incubation, all red, pink, or brown colored colonies are counted.

A representative sampling (approximately the square root of the number of characteristic colonies counted) is taken for confirmation. The confirmation may be done by inoculating Litsky medium with sodium azide and ethyl violet. The appearance of growth with clouding confirms the presence of group D streptococci.

Confirmation may also be done by inoculating esculine-bile agar with characteristic colonies. Hydrolysis of esculine, producing a blackening of the agar, confirms that the colonies thus cultured are group D streptococci.

Detection in liquid media is done the same way as for coliforms, that is, a presumptive step which consists of distributing a 100 ml water sample into ten tubes of double-strength Rothe medium followed by a confirmation step involving the subculture of the positive tubes in Litsky medium.

24.3.6 Detection of Anaerobic Sporulating Sulfite-Reducing Bacteria (ASSR)

Regulations require the absence of ASSRs in 20 ml, which is the volume used to inoculate.

The AFNOR recommended technique (standard NF-T-90-415) consists of inoculating 20 ml of meat-liver agar de-aerated in a boiling water bath, cooled to 55°C, and supplemented with sodium sulfite and iron salt. After inoculating with 5 ml of water, the tubes are quickly cooled.

The selection of spores takes place before inoculation by heating the water sample to 80°C for 10 minutes.

After incubation at 37°C, colonies surrounded with a black halo are counted as having arisen from spores of anaerobic sporulating sulfite-reducers. Counting is done at 24 and 48 hours.

One alternative (NF-T-90-419) consists of filtering 20 ml of water (after heating to 80°C). The membrane is then placed onto a thin layer of meat-liver agar for sulfite reducers, the upper surface against the agar and overlaid immediately with enough of this same medium to completely fill the plate. It is also possible to inoculate the uninverted membrane containing agar plate and incubate it in an anaerobic jar.

In all cases, colonies with a black halo are counted after incubation at 37°C.

24.3.7 More Specific Detection

In some situations, such as the discovery of sanitary problems which may be water-related or in the case of defective analytical results, more specific analyses may be required (e.g., for pathogenic agents). Analysis for *Salmonella* then becomes a priority. This is a semiquantitative analysis intended to establish the presence of *Salmonella* in 5 l of water, a volume which is filtered on a cellulose ester membrane of 0.45 μ porosity which is then put through all of the conventional steps of *Salmonella* detection in food products: preenrichment, enrichment, isolation, characterization of suspect colonies, and serotyping.

Similarly, detection of *Staphylococcus aureus* may be done by filtering 100 ml of water followed by incubating the membrane in staphylococcal enrichment membrane followed by isolation on Baird-Parker medium and characterization. It is also possible to enumerate them directly by incubating the membrane on Chapman mannitol medium. The characteristic yellow colonies with a yellow halo are then further characterized by tests for catalase and coagulase.

Defection of enteroviruses generally employs the adsorption onto glass powder technique followed by elution at high pH. The eluate thus obtained is used to inoculate a culture of sensitive cells. This method allows a most probable number of cytopathogenic units to be obtained.

Detection of *Vibrio cholerae* in water is of paramount importance in countries where cholera outbreaks occur. This is also a semiquantitative analysis leading to isolation on TCBS medium. Pathogenicity is determined mainly by serum 01 agglutination.

Various other pathogens may be disseminated by water and will therefore be the object of occasional analyses. These include bacteria such as *Shigella*, *Yersinia enterocolitica*, *Campylobacter jejuni* and pathogenic leptospira, amoebae such as *Entamoeba histolytica*, *Naegleria* or various parasites (nematodes, etc.).

24.4 Interpretation of Results

Each of the bacteriological analyses already described has its own particular significance which must be taken into consideration when interpreting results.

Enumeration of total bacterial flora at 20° and 37°C provides a general indication of the level of contamination of the water by common bacteria. Variations from one sample to the next as well as variations in the equilibrium of these two flora must be taken into consideration. Large variations may reflect irregularities in the treatment process or poor protection of the water deposit in the case of untreated water drawn from wells.

Sanitary risks transmitted by the fecal-oral route are still the most important risks in waterborne pathology and analysis for fecal contamination indicators consequently forms the basis of the bacteriological testing of water. The thermotolerant coliform group in particular makes up the most specific indicator of fecal contamination. The survival of these bacteria in water networks is close to that of intestinal pathogens.

The fecal streptococcal group contains species which are specific to animals and which may thus indicate contaminations of animal origin. Species of human origin are persistent enough in water networks to make them useful for tracking down old or intermittent contaminations.

Anaerobic sporulating sulfite-reducing bacteria, and primarily sulfite-reducing *Clostridium* species have also been considered as indicators of fecal contamination. It is known today that their fecal specificity is poor, but the detection of this group in water remains useful for other reasons. It allows an assessment of the effectiveness of the treatments and of the state of cleanliness of the distribution network. Furthermore, the presence of these bacteria in water used in the food processing industry is undesirable due to the sanitary and organoleptic problems which may result from their introduction.

The conclusions which result from the microbiological analysis of water must be written in clear and precise terms on an analysis form which includes all details concerning the water analyzed and the analyses carried out.

CHAPTER

25

Milk and Dairy Products

A. Plusquellec

Since milk is such a favorable medium for the development of microorganisms, the dairy is an industrial sector in which microbiological testing plays a fundamental role. It may come into play at various stages anywhere from collection to the consumption of finished products. This testing may be implemented as part of efforts to meet various objectives. It may serve the purposes of quality control which may itself be implemented at various levels.

- Testing of freshly drawn raw milk is used for purposes of milk payment based on quality.
- Testing of raw milk during processing in dairies is intended to determine the suitability of the milk for one manufacturing process or another.
- Testing of finished products at distribution points or for exportation is hygienic testing intended to protect the health of the consumer.

Along with the quality testing which regroups all of the frequently performed analyses, the Veterinary Services ensure sanitary testing intended to protect livestock, and therefore indirectly the consumer.

25.1 Testing of Raw Milk at the Point of Collection

Testing of raw milk at the point of collection is carried out by laboratories certified within the milk payment infrastructure based on composition and hygienic and biological qualities, as legislated by the law of January 3, 1969 and the decree of November 16, 1970. The basis of quality-based payment varies depending on agreements between producers and professionals. Payment is based essentially on the following criteria:

- bacteriological quality;
- chemical quality, i.e., fat content and nitrogen content;
- presence of antibacterial agents.

Other parameters may influence payment depending on the intended use of the milk:

- the leukocytes concentration (which must meet European standards);
- the butyric spore concentration;
- lipolysis.

Each producer receives a monthly rating based on several analyses (usually two per month per producer), each analysis leading to the discerning of a grade:

Grade 3: good quality
Grade 2: fair quality
Grade 1: poor quality

The combination of the monthly grades allows the classification of the milk into three categories:

Class A: maximum price
Class B: intermediate price
Class C: minimum price

The determination of the bacteriological quality of milk consists of evaluating its bacterial concentration. This may be measured either by conventional enumeration methods of culturing aerobic mesophiles or by the epifluorescent microscopic counting technique. The thresholds used for grading are as follows:

Grade 3: $<10^5$ CFU/ml of milk
Grade 2: from 10^5 to 3×10^5 CFU/ml of milk
Grade 1: $>3 \times 10^5$ CFU/ml of milk

with the possibility of a bonus for milk of excellent quality (the sum of both analyses lower than 10^5 CFU/ml or both analyses below 5×10^4 CFU/ml).

As for leukocytes, the threshold of 3×10^5 units/ml of milk is the most often used.

When butyric spores are enumerated, the thresholds that must not be exceeded may differ from summer to winter (e.g., 370 butyric spores/l in the summer and 1,300 spores/l in the winter).

25.1.1 Enumeration of the Total Aerobic Mesophilic Flora

Counting total aerobic mesophilic flora is the method used for milk which has been refrigerated between milking and collection, that is, almost all milk. The enumeration is done by the pour plate method using plate count agar with incubation at 30°C for 72 hours. In practice, inoculating duplicate plates with a 1/1,000 dilution is sufficient.

Due to the large number of samples generally treated, the counting method has been simplified. Taking 1 ml of sample followed by three successive decimal dilutions has been replaced by taking 1 μl (0.001 μl) of milk directly. This sampling is done using a calibrated platinum loop (a Burri loop) which is fitted onto a syringe. Inoculation is done by rinsing the loop with 1 ml of sterile diluent, thereby inoculating directly with 1 ml of milk diluted by a factor of 1/1,000. This may be a manual operation, the loop being fitted onto an automatic syringe which takes up sterile diluent, or it may be automated when large numbers of samples are being tested. In the latter case, the loop is fitted onto a syringe which is in turn connected to the pump and valve network which usually also provides automation of pouring and mixing of the media. Counting of colonies after incubation is done by electronic reading.

Another possibility for the automated enumeration of viable microorganisms in milk is provided by the spiral inoculating technique (Chapter 1.3) followed by electronic counting with a specially adapted grid.

25.1.2 Reduction Methods

The reduction method applies only to milk which has not been subjected to refrigeration. It is an indirect measurement of bacterial density. As bacteria develop, they consume the dissolved oxygen and lower the redox potential of the medium. This may be detected by the use of colored redox indicators. This is an indirect, nonmicrobiological method which does not reveal the specific reducing activities of each bacterial group. Consequently, it is not always in agreement with the enumeration of colonies on agar but has the advantage of being rapid, simple, economical, and performed on large volumes.

Reducing activity may be measured by evaluating the time necessary to obtain a color change, as is the case for methylene blue. Another possibility consists of measuring the degree of coloring obtained after a predetermined time (the resazurin method).

25.1.3 Microscopic Counting

25.1.3.1 Microscopic Counting on a Slide (The Breed Method) (see Chapter 4, Section 4.1.2)

The Breed method, mentioned by the APHA (American Public Health Association), allows direct enumeration of bacteria or lymphocytes (detection of mastitis). The technique consists of spreading a small known volume of milk over a known surface. In practice, 0.01 ml is spread over 1 cm^2. Following staining, the average number of bacteria per field is determined. Using an objective micrometer, the diameter of the microscopic field is first measured, which provides the means of calculating the field surface area. Knowing the number of organisms per field and the surface area of the field, the number corresponding to 1 cm^2 of surface may be easily calculated, from which the number per milliliter may be deduced.

The mode of operation is to deposit and spread 0.01 ml of milk on a slide on which a 1 cm^2 surface is marked using the appropriate pipette tip. Dry the slide in a drying oven or on the edge of a hot plate. (Do not use a flame.) For raw milk, staining is done for 2 minutes using Newman's stain (see Chapter 33, Appendix, medium 1) in a Borrel tube. Dry the slide and rinse with water. Dry again and observe under magnification of 400X. Count twenty microscopic fields. If M is the average number of organisms enumerated per field and S (Πr^2) is the surface area of the field expressed in mm^2, the number of organisms per ml is equal to:

$$\frac{M \times 100 \times 100}{S}$$

The main advantage of this technique is that the microbial flora can be evaluated in a few minutes. Furthermore, its performance is simple and economical.

It gives different results from those of agar plate counting or reduction techniques, however, since dead microorganisms are also counted. Reproducibility is also not very good, due to the small volume of milk taken as sample and also due to variations attributable to the operator during counting. Moreover, the threshold of detection is high (in the 10^5 CFU/ml range) and incompatible with the thresholds used for milk payment based on quality.

25.1.3.2 Enumeration by the Direct Epifluorescence Filtration Technique (DEFT) (see Chapter 4)

The sensitivity and reproducibility limitations mentioned for the Breed method are raised by processing a larger sample volume. The DEFT technique described by Pettipher and co-workers (1980) consists of microscopic enumeration by epifluorescence on a membrane on which 2 ml of milk have been filtered. The filtering of this volume requires prior treatment of the sample due to the presence of large somatic cells and of fat globules. The two ml are incubated for 10 minutes with 0.5 ml of 20% trypsin and 2 ml of surfactant (i.e., 0.5% Triton × 100). The filtration is done on 25 mm diameter 0.6 μ porosity polycarbonate membranes. The membrane is then covered for 2 minutes with 2 ml of a fluorochrome solution (i.e., 0.025% acridine orange in citrate-NaOH buffer pH 6.6), which binds to nucleic acids.

Following rinsing with citrate-NaOH buffer pH 3 and 95% ethanol and drying, the membrane is mounted on a slide with immersion oil and observed under an epifluorescent microscope at a magnification of 400.

Actively growing bacteria fluoresce reddish orange (rich in RNA). Inactive bacteria containing mostly DNA appear green. Each single bacterial cell or clump is counted as one unit. The number of fields to be counted depends on the abundance of the microorganisms. Knowing the surface areas of the field of observation and of the membrane, the result may be expressed in terms of 1 ml of milk.

25.2 Microbiological Testing in Milk Processing

Most milk-processing plants have their own testing laboratory for monitoring the bacteriological status of their products at all stages of manufacturing.

Good quality obviously implies the absence of pathogenic microbes and very limited numbers of contamination indicating bacteria. All hygienic tests may therefore be done on this level.

Nevertheless, numerous technological problems associated with specific bacterial groups having no hygienic significance arise in manufacturing. It is consequently very important to evaluate these flora of technological significance in order to determine the suitability of the milk for undergoing one type of processing or another.

- Milk of which the storability is ensured by pasteurization must not have too high a thermoresistant microfloral content.
- The presence of a large number of butyric spores renders milk unsuitable for the production of some cheeses.
- Products involving a fermentation step require milk free of inhibitory substances.
- The quality of products with a high fat content (e.g., butter, cream, anhydrous milk fat) depends on having a low number of lipolytic organisms. Caseinolytic organisms are also undesirable in butter.
- In products inoculated with a starter culture, the lactic flora does not totally inhibit the contaminating flora, which makes it useful to know how to enumerate one versus the other by differential counting agars.

25.2.1 Enumeration of Thermoresistant Microorganisms

Several bacterial genera possess species capable of resisting relatively high temperatures and therefore of surviving pasteurization (*Micrococcus*, *Microbacterium*, *Streptococcus*, *Lactobacillus*, *Corynebacterium*). Sporulating bacteria are obviously capable of resisting even higher temperatures.

The same techniques as for enumerating aerobic, mesophilic flora are used here, but with the sample having undergone a prior heating treatment: either at 63°C for 30 minutes for enumerating organisms resistant to low temperature pasteurization or at 80°C for 10 minutes (or 85°C for 5 minutes) for enumerating sporulating microbes.

25.2.2 Enumeration of Butyric Microbes

Clostridia, particularly those of the butyric group such as *Clostridium tyrobutyricum, Clostridium butyricum* (but also *Clostridium sporogenes*), are responsible for late swelling of certain cheeses (especially cooked curd cheeses). This is the result of fermenting lactate to butyric and acetic acids with gas production (carbon dioxide and hydrogen) and it is accompanied by the development of an undesirable taste.

Spores of butyric bacteria are abundant in soil, which results in the frequent contamination of silages which, in turn, contaminate animals either by contact or following ingestion. Milk may be contaminated, mainly by contact with fecal matter from silage-fed animals. A small number of spores in milk (anywhere from a few hundred spores per liter) is sufficient to cause manufacturing processes to fail. The enumeration of butyric spores in milk intended for cheese manufacturing is therefore an essential precaution. This enumeration may also be done on the product during production or on the finished product, as well as on silages in the context of early prevention of process failures.

The count is done using the multiple tube method with a liquid medium, usually that of Bryant and Burkey modified by Bergere (see Chapter 33, Appendix, medium 2). Prior to inoculation, the medium must be de-aerated by boiling in a water bath for 20 minutes. The medium is inoculated with 1 ml of milk or of dilution. For milk, inoculating five tubes each with dilutions of 0, 10^{-1}, and 10^{-2} is recommended, the detection of small concentrations with sufficient precision being a requirement. For cheese, dilutions from 10^{-2} to 10^{-6} are inoculated in triplicate. All tubes are covered with sterile paraffin. The inoculated tubes are brought to 75°C for 15 minutes in a water bath which provides the thermal treatment intended to select for spores. Incubation at 37°C is extended to 7 days, with readings after 48 hours and then each day thereafter. The positive tubes are those in which the paraffin plug has risen due to gas formation. The number of butyric spores is determined by MPN tables. This enumeration may also be done in skim milk acidified to pH 5.8 by adding lactic acid (the Weinzirl technique). This MPN technique is far from satisfactory due to the difficulties in implementation, the imprecise results, and the long delay in obtaining them. Newer techniques have been proposed to replace it.

A technique based on culturing on Petri plates after filtering the milk has been proposed by Abgrall and Bourgeois (1985). The milk undergoes a thermal treatment and is then mixed with an equal volume of 1% Triton × 100 surfactant and a quarter volume of 1% trypsin in 0.1M Tris HCI buffer at pH 7.8. This treatment consists of filtrating the milk after homogenization and incubation at 55°C for 15 minutes. The filtration is done under pressure (3 bars) on membranes of 0.8 μ porosity. From 100 to 400 ml of milk may be filtered this way. The membrane is incubated on RCM medium (reinforced clostridium medium) with 0.1% acid fuschin and 0.02% cycloserine. The medium is incubated anaerobically at 37°C for 48 hours. A count of all colonies present corresponds to the enumeration of *Clostridium. Clostridium tyro-*

butyricum in particular gives convex or dome-shaped, smooth, shiny, yellowish pink or pink colonies.

An immunoenzymatic technique uses the same preliminary steps, that is, filtering 10 ml of milk after treatment and culturing of the membrane of RCM medium at 37°C for 24 hours under anaerobic conditions. The membrane is then immersed in a solution of anticlostridial polyclonal antibody and then in a second solution of peroxidase conjugated antibody. The presence of a *Clostridium* colony appears as a colored spot. A response may be obtained after about 24 hours, but the preparation process of the procedure requires a multitude of steps.

25.2.3 Analyses of Dairy Products for Antibacterial Substances

Legislation requires that all milk delivered by producers be free of inhibitory substances. This condition is verified at pickup by dairy testing services. Milk containing inhibitors results in the imposition of penalties. This requirement for the absence of antibacterial agents is justified by the following:

- The presence of antibacterial agents biases microbiological testing, making it possible to misrepresent the quality of the milk.
- Some antibodies such as penicillins cause allergic reactions or intolerance in some people.
- Antibiotic use promotes the selection of antibiotic resistant strains with the possibility of pathogenic strains acquiring resistance traits through genetic exchange in foods.
- The presence of inhibitory substances in raw milk makes it unsuitable for fermentation. Dairies which produce fermented milks, yogurt, butter, or cheese must systematically carry out analyses for inhibitors on all shipments of milk received.

Inhibitors that may be present include antibodies, sulfanilamides, and cleaning antiseptics. Their presence in milk may be the result of a deliberate attempt to increase storability or may be of accidental origin, for example, improper rinsing of equipment or accidental mixing of antibiotic or sulfanilamide-treated cow milks.

The principle of the techniques used to detect inhibitors is to test the ability of a sensitive organism to develop in the milk being analyzed.

Easy-to-use field tests are available in which the lyophilized test strain is mixed with a detector of culture activity.

The official methods for the detection of antibiotics or sulfanilamides in milk intended for consumption by humans or by livestock are regulated by the decree* of September 2, 1983 (*Journal Officiel*, 6 October 1983). These apply to

---

* *Translator's note:* The English word *decree* is used in this chapter to translate the French words *arrêté* and *décret* which correspond to different administrative levels. An asterisk indicates that the French word used was *arrêté*, otherwise the word was *décret*.

the detection of residues of antibiotics such as penicillins, tetracyclines, aminosides, macrolides, and chloramphenicol and of sulfamide residues in raw milk, heat-treated milk, and dehydrated milk. The analysis consists of two phases: nonspecific detection of milk acidification tests and confirmation tests on positive or dubious samples thus found.

25.2.3.1 Acidification Method

The test organism used is *Streptococcus thermophilus* strain TJ (Laboratoire de Technologie Laitière, CNRZ, 78350 Jouy-en-Josas). This strain is cultured in sterile skim milk. These cultures may be frozen (e.g., a reserve in the form of 2% culture in sterile skim milk). For the analysis, a reserve culture is thawed at 40°C and incubated at 37°C for 16–18 hours. The test culture is made with the following mixture:

- 10 ml of *Streptococcus thermophilus* culture (16–18 hours at 37°C)
- 5 ml of sterile 10% yeast extract
- 10 ml of sterile 0.25% bromocresol purple solution

For detection by the acidification test, the following three tubes are prepared:

- the first contains 2 ml of homogenized test milk
- the second contains 2 ml of control milk without antibacterial agents
- the third contains 2 ml of milk containing 0.01 IU/ml of penicillin

These three tubes are heated to 100°C for 5 minutes and cooled after which 0.1 ml of test culture and 0.02 ml of 0.01% trimethoprime are successively added. The trimethoprime is added for its synergistic action with sulfamides which are thereby more easily detected. The tubes are incubated at 45°C in a temperature-controlled bath for $2\frac{1}{2}$ hours. After this incubation, the control tube without antibiotic must show acidification (coagulation with a color change to yellow) while the control tube with penicillin must remain blue. If the tube with the test milk remains blue or an intermediate color, the test is considered to be positive or dubious and the milk must be subjected to confirmation tests.

25.2.3.2 Confirmation Methods Using Agar-Based Media

These confirmation methods employ agar diffusion techniques. Agar-based medium in Petri plates is inoculated with the test organism and paper disks impregnated with the milk being tested are placed on the surface of the inoculated medium. The presence of antibiotics or of sulfamide inhibits the growth of the test organism giving a transparent zone of inhibition. The complete method uses the following sensitive strains:

- *Bacillus stearothermophilus* var. *calidolactis* C953
- *Bacillus subtilis* ATCC 6633
- *Bacillus megaterium* ATCC 9885

Galesloot and Hassing agar (see Chapter 33, Appendix, medium 3) is inoculated with *Bacillus stearothermophilus*, Grove and Randal agar no. 11 with *Bacillus subtilis*, and Mueller-Hinton agar containing trimethoprime with *Bacillus megaterium.* The sterile 9 mm diameter disks impregnated with the milk being analyzed and the control milks are placed on the media thus inoculated. The test milk is previously heated to 80°C for 10 minutes. Incubation is at 55°C for 2 hours and 30 minutes for *B. stearothermophilus*, 30°C for 16–18 hours for *B. subtilis*, and 37.5°C for 5 hours for *B. megaterium.* The disks, which are surrounded by a zone of inhibition of at least 10 mm in diameter, are considered positive.

B. stearothermophilus is especially sensitive to penicillins and tetracyclines. *B. subtilis* is particularly useful for the detection of aminosides (streptomycin, neomycin, kanamycin, etc.) and macrolides (erythromycin, spiramycin, etc.). *B. megaterium* has particular sensitivity to sulfamides and to certain antibiotics (chloramphenicol).

25.2.3.3 Methods of Rapid Detection

Several manufacturers have commercialized very easy-to-use detection processes. These are preparations of sensitive strains with an indicator of culture activity either in ready-to-use ampules or absorbed into a filter paper support. The procedure consists of simply inoculating these with a sample of milk being tested (previously pasteurized) and noting the presence or absence of color change after incubation.

These kits have limited storage life and it is wise to do controls using milk without inhibitors. They are often used for routine testing, the positive results being confirmed by a more precise method.

The "Penzym" test is used for rapid and easy detection (20 minutes) of β-lactams (penicillin, oxacillin, cephalosporin). It is based on the inactivation of a D.D. carboxypeptidase by β-lactams.

25.2.4 Enumeration of Lipolytic Microbes

Lipolytic flora is often responsible for the rancidification of fats. Development of rancidity is associated with the appearance of malodorous compounds (acids, aldehydes, ketones) resulting from the hydrolysis of lipidic substances.

Lipases may also be of mammary origin (in milk near the end of lactation). The consequences of lipolytic activity may be evaluated by chemical methods (measurement of the lipolytic index).

Microorganisms that produce lipases are numerous and varied. They include bacteria, including many psychrotrophic genera (*Pseudomonas*, *Achromobacter*, *Micrococcus*, *Bacillus*); yeasts (*Candida*, *Torula*, some *Saccharomyces*); and molds (*Aspergillus*, *Cladosporium*, *Geotrichum*, *Mucor*, *Penicillium*).

The storage capacity of animal or vegetable fats (cream, butter, margarine, oils, lard, shortening) depends therefore directly on their lipolytic microbe content. Enumeration is thus very important.

Testing may be carried out on finished products or on starting materials. Several techniques are available which are able to enumerate lipolytic flora.

25.2.4.1 Tributyrin Agar Count

The medium is composed of a basal agar (see Chapter 33, Appendix, medium 4) melted and held at 80°C to which glycerol tributyrate (neutral tributyrine) is added to a concentration of 1%. The medium is cooled to 45°C and poured with 1 ml of product suspension/dilute. Plates are incubated at 30°C for at least 72 hours.

Lipolytic organisms produce colonies surrounded by a clear halo.

25.2.4.2 Tween 80 Medium (De Sierra)

Sterile Tween 80 is added to the basal medium (see Chapter 33, Appendix, medium 5) to a concentration of 1%. After inoculation by the pour plate method and incubating at 30°C for 3 days, colonies surrounded by an opaque halo are counted (due to the formation of insoluble calcium salts of free fatty acids liberated).

25.2.4.3 Victoria Blue Agar

An ordinary basal medium such as nutrient agar or PCA agar is melted and combined with a Victoria blue-fat mixture. The stain may also be incorporated into the basal agar medium. The fat used corresponds to the product being analyzed (e.g., butter oil for testing butter). The stain must be obtained as a stabilized powder (see Chapter 33, Appendix, medium 6). The fat-stain mixture is obtained by adding 75 mg of stain powder to 100 g of fat and sterilizing the mixture by autoclaving. The final medium is the result of adding this mixture to the melted basal agar held at 45°C to a final concentration of 5%. The pour plate method is used. The medium must be vigorously homogenized immediately before pouring. This is best done by mechanical means (e.g., mixer or homogenizer). The media are incubated for 72 hours at 30°C.

Lipolytic microorganisms produce free fatty acids which combine with the stain giving colonies impregnated with deep blue colored salts.

25.2.4.4 Nile Blue Agar

Plate count agar may be used as a basal medium. Added to 100 ml of base are 0.5 ml of 1% Nile blue sulfate and 1.5 ml of sterile fat.

The medium must be homogenized (see Section 25.2.4.3). Lipolytic organisms produce greenish-blue to blue colonies after 3 days of incubation at 30°C.

25.2.4.5 Copper Sulfate Technique

To a basal medium such as PCA, sterilized lipid phase of the product being analyzed is added to a concentration of 5%. This is homogenized and poured into Petri plates which are inoculated by spreading and incubated for 5 days at 30°C. The plates are then read by flooding the surface with a saturated copper sulfate solution for 15 minutes and then carefully rinsing. Lipolysis appears as colonies surrounded by bluish-green zones due to the formation of insoluble copper salts of free fatty acids liberated.

25.2.4.6 Comparison of the Methods

Techniques using fat from the analyzed product as substrate have the inconvenience of requiring a difficult homogenization step, except in the case of testing cream. The time which elapses between homogenizing and solidifying of the agar must be as short as possible. The quantity of medium per plate is limited by faster solidification.

These methods, however, have the advantage of detecting microbes which are capable of lipolytic activity in the product itself. Microorganisms able to hydrolyze artificial substrates such as Tween or Tributyrin are not necessarily capable of lipolysis in the food being tested.

25.2.5 Enumeration of Caseinolytic Flora

Proteolytic microflora is undesirable in most food products. Particularly in butter, caseinolytic microbes may be responsible for an aging of residual casein present, producing a butter that tastes like rotten cheese. Combinations of products of caseinolysis and lipolysis may yield compounds with highly objectionable odors.

Counting of caseinolytic microbes is a relatively simple matter. Calcium caseinate agar is used, obtained by adding one volume of calcium caseinate (this is commercially available in a ready-to-use form) warmed to 45°C to an equal volume of double-strength melted agar (24–30 g of agar per liter, depending on its gelifying power) cooled to 45°C. This mixture is poured into Petri plates with 1 ml of product suspension or dilution. Plates are incubated for 3 days at 30°C. Hydrolysis of casein appears as a total and sharp clearing around the colonies while the rest of the medium remains opaque. These colonies are counted. This count may also be done on skim milk agar (15 ml of 2.5% agar with 5 ml of sterile hot skim milk). The appearance of the plates after incubation is essentially the same as for the caseinate medium.

25.2.6 Enumeration of Contaminating Flora of Fermented Products

Counting contaminating flora (common bacteria) in products inoculated with lactic ferments presents practical problems. On media generally used for this purpose (plate count agar), lactic acid bacteria develop slowly, producing small colonies and hindering the culture of the contaminating microbes. Counts of contaminating flora therefore require culture media which impede development of lactic bacteria. Carbohydrate-free media, such as Sugar Free Agar (see Chapter 33, Appendix, medium 7) meet this requirement. The medium is inoculated by the pour plate method and plates are incubated at 30°C for 3–5 days. Gelysate agar (tryptic gelatin peptone) (see Chapter 33, Appendix, medium 8) may be used under the same conditions. Colonies counted on these media correspond to the nonlactic contaminating flora.

25.3 Microbiological Testing of Distributed Dairy Products

25.3.1 Raw Milk Sold as Such

The quality of raw milk must fulfill provisions established by the decree of May 21, 1955 (and its June 18, 1955 rectification). For centers of more than 20,000 inhabitants, the milk must come from registered producers protected from tuberculosis and brucellosis. In all cases, raw milk sold must be

- cooled and kept below 15°C;
- recognized as clean by the filtration test;
- kept in closed containers until sold;
- unable to decolorize methylene blue (reductase test) in less than 3 hours at the time of sale.

Sale must take place within 24 hours of milking.

25.3.2 Pasteurized Milk

Pasteurized milk may be sold either in bulk or in containers. In the latter case, two standards are acceptable: packaged pasteurized milk or high quality pasteurized milk. Hygienic quality criteria have been defined for these three categories.

25.3.2.1 Pasteurized Unpackaged Milk

Pasteurized unpackaged milk is usually intended for collectivities or industries (for subsequent processing). On leaving the processing plant, it must not contain more than 100,000 CFU/ml. At the time of sale, it must contain less than 200,000 CFU/ml. In addition, it must meet the following criteria:

- phosphatase negative;
- clean according to the filtration test;
- free of pathogens destroyed by pasteurization;
- kept at a temperature below 10°C.

If the milk is intended for export, it must also contain less than 100 coliforms/ml.

25.3.2.2 Packaged Pasteurized Milk

The quality of packaged pasteurized milk is regulated by the decree of January 4, 1974. This milk has been low-temperature pasteurized at 63°C for 30 minutes or flash pasteurized at 95°C.

Packaged pasteurized milk must fulfill the same provisions as milk sold in bulk with respect to cleanliness, pasteurization-sensitive pathogens, storage temperature, and phosphatase. But, at the time of sale, it must contain less than 30,000 aerobic mesophilic bacteria/ml and less than 10 coliforms/ml.

25.3.2.3 High Quality Pasteurized Milk

High quality pasteurized milk must meet certain conditions defined by the decree of October 9, 1965.

25.3.2.3.1 Criteria for Raw Milk Intended for Preparation of Pasteurized High Quality Milk

Only raw milk of good quality should be used to produce this type of milk. It must in particular

- contain less than 50,000 aerobic mesophilic bacteria/ml;
- not reduce methylene blue in less than 5 hours;
- not contain any antibiotic or antiseptic.

25.3.2.3.2 Pasteurization Conditions

High quality pasteurized milk must have undergone a single pasteurization carried out at 72°–75°C for 15 seconds.

25.3.2.3.3 Qualities Required at Time of Sale

High quality pasteurized milk must have a negative phosphatase test and a positive peroxidase test. In addition, it must also

- not contain any antibiotic or antiseptic;
- not contain any pathogens destroyed by pasteurization;
- not contain over 30,000 bacteria/ml;
- contain less than 1 coliform/ml after packaging and less than 10 coliforms/ml at the time of sale;
- be packaged in single-use containers;
- be kept at a temperature below 8°C;
- be sold no later than 2 days after packaging;
- satisfy the same chemical requirements as other pasteurized milks (fat content >36.05 g/l, protein content >28 g/l).

25.3.2.4 *Analysis of Pasteurized Milk*

25.3.2.4.1 Bacteriological Testing

The required bacteriological tests are:

- aerobic mesophilic bacterial count on PCA (inoculating with dilutions of 1/100 and 1/1,000 is sufficient);
- a coliform count on lactose and deoxycholate containing agar (inoculating directly with 1 ml of milk);
- detection of bacterial inhibitors (for high quality pasteurized milk).

Due to the requirement for the absence of pathogens, the detection of salmonella and of staphylococci must be carried out by complete examination.

25.3.2.4.2 Enzymatic Testing of Heating

In order to verify that the heating step was done as required, detection procedures are performed for enzymes present in raw milk which are supposed to be destroyed by pasteurization. Enzymes for which the thermal destruction curve is known are used.

Alkaline Phosphatase Test

Alkaline phosphatase hydrolyzes phosphoric esters. It is destroyed by all types of pasteurization. It is detected by the Aschaffenburg-Muellen method which measures the decomposition of para-nitrophenyl phosphate (PNPP) liberating yellow colored para-nitrophenol.

The mode of operation is to add 1 ml of test milk and 1 ml of boiled test milk as a control to tubes containing 5 ml of 0.15% PNPP in pH 9.6 buffer (see Chapter 33, Appendix, medium 9). After 30 minutes and 2 hours, yellowing due to phosphatase is evaluated using the Lovidond comparator fitted with a special filter (APTW). The result is given in micrograms of nitrophenol:

	Reading at 30 *minutes*	*Reading at* 2 *hours*
Negative	0	<10
Dubious reaction	6	10–18
Positive reaction	$\geqslant 10$	$\geqslant 18$

Peroxidase Test

Peroxidase is less sensitive. It is not destroyed by pasteurization at 72°–75°C for 15 seconds, allowing testing of high quality pasteurized milk. Peroxidase decomposes hydrogen peroxide in the presence of a reduced organic substrate yielding water plus the oxidized substrate.

The Dupouy reaction uses gaïacol as substrate, which oxidases to a salmon pink color. The reaction is simple. The following are added to a tube:

- 2 ml of test milk
- 2 ml of 2% gaïacol
- one drop of 3% hydrogen peroxide solution

The tube is warmed to 37°C and the coloring appears in less than 1 minute.

25.3.3 Sterilized Milk—UHT Milk—Concentrated Milk—Flavored Sterilized Milk

25.3.3.1 Definitions and Regulation

The August 10, 1967 decree* regulates the quality of sterilized milk intended for export. Sterilized milk is to have undergone a heat such that it contains no pathogenic or taxinogenic germs or any other microorganisms able to proliferate. Packaging must be in containers which are watertight, gastight, and impervious to microorganisms. The sterilized milk must remain stable after incubation at 31°C for 21 days and 55°C for 10 days.

According to the September 7, 1977 decree, UHT milk is to have been treated by a process of heating, direct or indirect, with continuous throughput, applied a single time for an uninterrupted and very short time at a temperature at least equal to 140°C. This treatment must destroy or totally inhibit enzymes, microorganisms and their toxins of which the presence could otherwise spoil the milk or render it unfit for human consumption. A small number of nonspoiling spore-formers may persist. Furthermore, UHT milk must be packaged aseptically in hermetic containers impenetrable by liquids, microorganisms, and light. Unsweetened concentrated milk must not contain more than a small number of viable microorganisms (at most 5/ml) and must remain stable after incubation. Sweetened concentrated milks must be free of microbes capable of spoilage and must contain fewer than 30,000 bacteria/ml and less than 1 coliform/ml.

25.3.3.2 Techniques of Analysis

The preceding definitions are similar to those for canned goods. As a general rule, analytical protocols may be transported from tests for the stability of canned goods (Chapter 32).

After incubating the samples, the following comparison is made with an unincubated control:

- Organoleptic examination (odor, appearance). This may be complemented by the alcohol test. An equal volume of alcohol added to the milk at 68°C must not bring about coagulation.
- Miscroscopic examination. Counting (by the Breed method [Section 25.1.3.1]) of microbes present before and after incubation.
- pH measurement. Incubation must not bring about a variation greater than 0.5 pH unit.

- Bacteriological examination. Nonselective media for aerobes and anaerobes (VL starch agar, Rosenow's medium) are inoculated and incubated at optimal temperatures for mesophiles and thermophiles in order to test for bacterial proliferation induced by the curing oven.

25.3.4 Dry Milk

Dry milk is defined by the decree of August 24, 1961 and the decree* of August 13, 1963. Regulations stipulate that in addition to being pathogen-free, dry milk for human consumption (these criteria also apply to category A whey) must contain

- less than 50,000 aerobic mesophilic bacteria/g;
- less than 5 coliforms/g (absence in 0.2 g).

Milk intended for food industry use is subject to less strict constraints (these criteria also apply to category B whey):

- less than 200,000 aerobic mesophilic bacteria/g;
- less than 25 coliforms/g.

Routine analyses must therefore include the enumeration of aerobic mesophilic flora and coliforms. For the latter, the liquid medium enumeration technique (MPN) is necessary due partly to the small number to be detected and partly to the physiological state of the bacteria which have been stressed by thermal treatments undergone. Group D streptococci are more resistant to these treatments and may persist within the installation (sprayers, piping). They may be considered as indicators of the effectiveness of the cleaning procedure within the installation. A complete analysis will also include detection of pathogenic germs (staphylococci, salmonella, *Clostridium perfringens*) and potential spoilage agents (sulfite-reducing anaerobes, molds). Analysis begins with the preparation of a 1/10 original solution (10 g brought to 100 ml with sterile diluent after dissolving). Stirring with glass beads or mechanical agitation may be necessary in order to facilitate dissolution.

25.3.5 Fermented Milks—Yogurts

Yogurt, which makes up most of the production of fermented milks, is defined as "fermented milk obtained by the exclusive development of the lactic acid bacteria *Lactobacillus bulgaricus* and *Streptococcus thermophilus*, with which the milk must be inoculated simultaneously and which must be alive in the sold product" (decree of July 10, 1963).

The product must contain a concentration of free lactic acid of at least 0.8 g/100 g. The December 21, 1979 decree* stipulates that it must contain fewer than 10 coliforms/g and less than 1 fecal coliform. *Salmonella* must be absent in 25 g.

Yogurt must be free of contaminating microbes. The fungal flora (yeasts and molds) must be very sparse. The lactic flora must be tested. Microscopic observation (e.g., after staining with methylene blue) permits estimation of the proportion of each of the ferments. In order to obtain the number of live organisms, a method of dilution followed by inoculation of Petri plates must be used.

MRS medium acidified to pH 5.4 with acetic acid allows the enumeration of *Lactobacillus*, after 3 days of incubation at 37°C in a jar with a CO_2 enriched atmosphere. The enumeration of *Streptococcus thermophilus* is done on M17 medium (see Chapter 33, Appendix, medium 10) which is incubated at 37°C for 48 hours. For these counts, it is wise to inoculate plates with dilutions up to 10^{-7}.

25.3.6 Cheeses

The microbiological testing of cheeses has two purposes: first, to verify that pathogens are absent and that hygienic indicators are present in limited numbers and then, to test for the absence of microbes having unfavorable technological implications.

The enumeration of coliforms is useful from this point of view since aside from their significance as indicators, coliforms with their gas-producing tendencies are also responsible for the spoilage of certain cheeses which acquire what is called an *open texture* (thousands of holes).

Other flora which may be of interest include fungal flora, common aerobic mesophilic flora, lactic flora, sulfite-reducing anaerobes, caseinolytic and butyric microbes.

Soft-curd cheeses must meet certain criteria for export (March 5, 1973 decree*):

- Coliforms: absence in 1 cg (< 100/g)
- Presumed pathogenic staphylococci: absence in 1 cg
- *Salmonella*: absence in 24 g

Fresh cheeses have greater limitations on storage and the December 21, 1979 decree* establishes the following criteria:

- Coliforms: 10/g
- Fecal coliforms: 1/g
- *Staphylococcus aureus*: 10/g
- *Salmonella*: absence in 25 g

Recent epidemics of listeriosis have uncovered the significance which cheese may have in the spreading of this disease. Semiquantitative analyses for *Listeria monocytogenes* must be performed on cheeses that may be at risk.

25.3.7 Caseins—Caseinates

Caseins and caseinates are widely used in the food industry and frequently in products which are very sensitive to contamination. They must therefore be of excellent bacteriological quality.

The EEC has established the bacteriological criteria for casein of quality A (regulation of January 31, 1973):

- Aerobic mesophilic bacteria: 30,000/g
- Thermophilic bacteria: 5,000/g
- Coliforms: absence in 0.1 g

25.3.8 Butter

25.3.8.1 *Definition and Regulations*

Butter is defined as a dairy product of the water-in-oil emulsion type, obtained by physical processes, and of which the constituents are of dairy origin. It must contain a maximum of 2 g of nonfat matter and 16 g of water per 100 g of product (decree* of March 25, 1924). The microbiological quality of butter is regulated by the April 15, 1986 decree* which establishes the following standards:

	Aerobic microorganisms at 30°C (per g)	Coliforms at 30°C (per g)	*Staphylococcus aureus* (per g)	Phosphatase	*Salmonella* in 25 g
Raw butter*			10^2	Positive	Absence
Butter†	10^2‡§	10	10‖	Negative	Absence‖
Butterfat based fat or oil†		10	10‖	Negative	Absence‖
Butter concentrates	5×10^2	Absence	Absence		

*Butter obtained from cream which has not undergone any heat treatment.
†Products obtained from components which have undergone a heat treatment at least equivalent to pasteurization.
‡Other than lactic species.
§Provisional standard.
‖In the event that results are to be considered unsatisfactory for any of the following criteria: phosphatase, coliforms, or aerobic microorganisms at 30°C other than lactics, these criteria should be tested on additional samples.

25.3.8.1.1 Basic Microbiological Analysis

Microbiological analysis includes verification of the preceding microbiological criteria, in accordance with the stipulations of the December 21, 1979 decree.* The enumeration of aerobic microorganisms at 30°C is done using media for nonlactic flora (see Section 25.2.6).

25.3.8.1.2 Enumeration of the Lipolytic Flora (see Section 25.2.4)

Lipolytic flora may be responsible for the developement of rancidity of the butter.

25.3.8.1.3 Enumeration of the Caseinolytic Flora (see Section 25.2.5)

Caseinolytic flora may produce objectionable tastes and odors in butter (cheesy tastes).

25.3.8.1.4 Enumeration of Fungal Flora

Among the yeasts and molds, those which may be responsible for organoleptic spoilages are numerous.

25.3.8.2 Peculiarities of the Analysis

Microbiological testing requires a sample preparation that is somewhat out of the ordinary. The examination is performed on the aqueous phase obtained after melting and centrifuging the sample. The microorganisms are supposedly partial to the aqueous phase which represents about 1/6 of the weight of the butter (the March 25, 1924 decree setting at 18 g the maximum amount of nonfat substances allowable in 100 g of butter). The result found for 1 ml of aqueous phase must therefore be divided by six in order to obtain the value corresponding to 1 g of butter. The sample is taken with the aid of a sterile instrument (knife, spatula, etc.) after removing a superficial layer. A probe may also be used, for example, a sterile glass tube which removes a long cylinder of product. The butter thus sampled is introduced into a sterile centrifugation tube with a screw cap and placed in a water bath at 45°C until completely melted. A low-speed centrifugation (a few minutes at 2,000–3,000 rpm) brings about phase separation. As much of the oil phase as possible is removed by siphoning and the aqueous phase is sampled and placed in a sterile container. The oil phase may be saved for incorporation (after sterilization) into the medium used for the enumeration of lipolytic bacteria (see Section 25.2.4).

25.3.9 Creams

Pasteurized cream must meet the following criteria (decree* of December 21, 1979):

- phosphatase test negative;
- activity less than or equal to 2.5 g lactic acid per 1,000 g;
- *Salmonella*: absence in 25 g.

	Criteria for 1 g
Aerobic mesophilic microorganisms	3×10^4
Coliforms	10 for packaged cream 100 for cream in bulk
Fecal coliforms	1
Staphylococcus aureus	10

The spoilage agents in cream are essentially the same as those in butter. The same tests may be carried out but no special preparation of the sample is necessary.

25.3.10 Ice Milk and Ice Cream

Ice milk and ice cream are media favorable to the growth or survival of microorganisms. Their bacteriological quality must be closely monitored. The components from which they are manufactured must be of good hygienic quality.

- Water used must meet potable water criteria.
- Milk used must be of the same quality as packaged pasteurized milk.
- Cream used must be of good bacteriological quality.

The rooms, the equipment, and the handling conditions are subject to strict hygienic requirements (September 13, 1967 decree*).

The quality of the finished product must correspond to the following criteria for 1 g:

- Aerobic mesophilic flora: 3×10^5
- Coliforms: 100
- Fecal coliforms: 1
- *Staphylococcus aureus*: 10
- *Salmonella*: absence in 25 g

The results are expressed in terms of the thawed product (placed in a water bath at 30°C for 1 hour).

25.3.11 Anhydrous Dairy Fat (Butter Oil)

Anhydrous dairy fat presents certain difficulties for bacteriological analyses, being rather difficult to incorporate into an aqueous medium. A protocol quite similar to that for butter is used (10 g of fat are added to 10 ml of sterile diluent and brought to 45°C). After vigorous shaking, the mixture is centrifuged. The bacteria are considered to be drawn to the aqueous phase on which the analysis is performed. One milliliter of aqueous phase therefore corresponds to 1 g of fat.

The best results are obtained using melted agar-based media cooled to 45°C. The inoculum is added and the mixture is homogenized using a Waring blender and quickly poured into Petri plates placed on a cold surface. The use of a reduced quantity of medium (10 ml) speeds up its solidification.

The most commonly practiced tests are as follows:

- enumeration of aerobic mesophilic flora
- enumeration of coliforms and fecal coliforms
- enumeration of presumed pathogenic staphylococci
- enumeration of yeasts and molds.

CHAPTER

26

Beer and Soft Beverages

A. Plusquellec

26.1 Introduction

Hygienic testing is of no use in brewing since pathogenic microbes do not develop in beer but die rapidly in it. Microbiological testing is of technological interest. The finished product must meet a certain number of physical, chemical, and organoleptic criteria. Contaminations that may spoil these characteristics must be detected early and rapidly.

26.2 Microbial Contaminations of Beer

Beer is a rich medium which during the various stages of its manufacture may be used as a growth medium by a variety of organisms.

Several selective factors intervene throughout the manufacturing steps, such as temperature, pH, a greater or lesser degree of anaerobiosis, deficiency of assimilable nutrients, and appearance of inhibitory compounds (alcohol, carbon dioxide). Nevertheless, the selection of the cultured yeast is not always absolute, and several contaminating microbes may develop and impart undesirable characteristics to the beer.

26.2.1 Wild Yeasts

Several genera of yeasts found in the environment are capable of multiplication in beer and overtaking that of the cultured yeast. This occurs generally after decanting and shows up as cloudiness, clotting, and taste defects.

26.2.2 Wort Bacteria

Wort bacteria is capable of contaminating wort during production and giving it abnormal odors. It is made up of gram-negative bacilli that brewers call *thermobacteria.* The genera most frequently encountered belong to the enterobacteria (75%) and especially to the coliforms. Forty percent of the strains isolated belong to the genus *Enterobacter*. The other genera belong to the Pseudomonaceae (21%): *Pseudomonas*, *Xanthomonas*, and to the Achromobacteraceae: *Achromobacter*, *Flavobacterium.*

26.2.3 Lactic Acid Bacteria

Lactic acid bacteria which are able to contaminate beer come from two genera: the lactobacilli and the pediococci.

26.2.3.1 The Genus Lactobacillus

Some of these bacteria develop in wort while others do so in beer. They bring about clouding and spoiling of the taste (which may be associated with acidification). Their development requires the presence of numerous growth factors.

26.2.3.2 The Genus Pediococcus

Bacteria from *Pediococcus* are characterized by homolactic fermentation. They produce turbidity and a sediment accompanied by acidification and a honey-like odor.

26.2.4 Acetic Acid Bacteria

Two genera are encountered: *Acetobacter*, which oxidizes ethanol all the way to carbon dioxide and water, and *Acetomonas* (or *Gluconobacter*), which oxidizes it only to acetic acid. These bacteria have limited effects due to their inability to develop anaerobically.

26.3 Microbiological Testing

Microbiological testing may be carried out at several stages of the manufacturing process:

- on the wort, before and during fermentation;
- on the beer before and after filtration and after bottling;
- on the yeast harvested after the fermentation.

Testing of starter culture purity is discussed in detail in Chapter 34.

26.3.1 Direct Microscopic Examination

Direct microscopic examination is applicable only when spoilage has been ascertained. It may be used, for example, as a rapid test for characterizing the cause of turbidity arising in the beer during storage. The examination may be done by wet-mount between slide and cover-slip or by either simple staining or Gram stain. The test may be given a quantitative aspect by spreading a known volume of sample over a known surface area.

26.3.2 Microscopic Examination After Filtration

The process of microscopic examination after filtration consists of concentrating the microorganisms in beer on a membrane which is subsequently stained (with Loeffler blue or fuchsin-methylene blue). After rinsing and drying, the membrane is made transparent by dipping in immersion oil, and observed using an immersion objective. By filtering 100 ml of beer, contaminations greater than 10^3 cells/ml can be detected (Bourgeois-Malcoste, 1980).

The use of fluorochromes instead of the stains mentioned earlier makes for better detection of the microorganisms with a lower detection threshold. In this case, filtering is done on polycarbonate membranes which are well suited to microscopic examination.

Microscopic examination may also be done after incubating the beer to allow the development of possible contaminants. The incubation may be done simply at a temperature conducive to such development or it may follow degassing and addition of a nutrient supplement such as sterilized yeast extract solution.

26.3.3 Microscopic Examination After Culturing

Microbiological testing of pasteurized beer requires concentration of the microorganisms. This is brought about generally by filtering on a membrane which serves two purposes: separating the cells from the liquid and providing a culture support (Bourgeois, 1983). The membrane is contacted with a culture medium by placing it on an agar surface or on a medium-soaked absorbant pad. The number of colonies resulting after incubation is counted. Culturing may allow characterization of the contaminants. This general technique may be used to enumerate total or specific contaminants.

26.3.3.1 Detection of Total Contaminants

The media used are not selective. The most commonly used are the wort agar, a complex medium, and the WL medium of Green and Gray (see Chapter 33, Appendix, medium 11), a semisynthetic medium.

26.3.3.2 Selective Culture of Lactics

Selective culture of lactics is done on media with greater or lesser degrees of selectivity and which are presented in Chapter 34, Section 34.2.3.1.

26.3.3.3 Detection of "Wild" Yeasts

Chapter 34, Section 34.2.2 describes detection of "wild" yeasts.

26.3.3.4 Other Analyses

Enterobacteria are detectable using the usual media for coliforms (i.e., Endo agar, Chapman's TTC-lactose agar, MacConkey agar).

Acetic acid bacteria require special media such as Williamson's low pH medium (see Chapter 33, Appendix, medium 12) with 2–4% ethanol.

The size of the inoculum for all of these media varies (between 1 and 500 ml) depending on the production stage from which the sample was taken.

The main disadvantage of membrane culture methods is the requirement for long incubation periods. Bourgeois (1983) envisioned a few possible improvements to the basic technique:

- The early enumeration of microcolonies after a brief incubation (4–6 hours) of the membrane on culture medium. The membrane is treated with a stain or a fluorochrome prior to microscopic counting at low magnification.
- The use of HGMF (hydrophobic grid membrane filter) membranes allowing a more rapid and reliable enumeration.

Aside from the conventional methods of enumeration after culturing, a wide variety of modern and rapid techniques has been proposed for testing beer and starter cultures for contamination (Bourgeois, 1983) including turbidometry, enzymatic methods, ATP assay, radiometric methods, pH, and impedance measurements.

26.4 Analysis of Soft Beverages

26.4.1 Characteristics

This designation regroups the following:

- fruit or vegetable juices:
- fruit juice concentrates;
- fruit juice beverages;
- natural extract based beverages, i.e., lemonade (acidified water + sugar + lemon flavor), sodas (carbonated beverages with fruit or plant extract including "bitters"), colas (cola extract, caramel as coloring, phosphoric acid, caffeine), tonics (bitter extract + quinine).

In general, these beverages constitute a medium which is hostile to microorganisms.

Sodas and lemonades in particular are characterized by the following conditions:

- high acidity (pH 3–4);
- lack of oxygen;
- high carbon dioxide content;
- B vitamin and nitrogenous substance content very low;
- high sugar concentration.

Fruit juices are richer in nitrogenous substances and vitamins. Moreover, most fruit juices are noncarbonated. These beverages are much more sensitive to contaminations than sodas and lemonades are. In syrups, the main selective factor in additon to the pH is the osmotic pressure.

26.4.2 Spoilage Agents

Among the microorganisms which spoil soft beverages, the most significant are yeasts (90% of cases) for the following reasons:

- They tolerate low pH.
- They utilize inorganic nitrogen.
- They do not require B vitamins.
- They tolerate a high carbon dioxide content.
- They are resistant to high sugar concentrations.

The most commonly encountered yeasts belong to the genera *Candida* (especially in clear carbonated beverages) and *Saccharomyces* (especially in fruit beverages). Bacteria are not as well adapted to life in this type of beverage. Only lactic acid bacteria have the necessary properties allowing their development. These organisms require amino acids and vitamins, however, and they are therefore encountered specifically in fruit juices. The spoiling lactic acid bacteria belong to the genera *Lactobacillus* and *Leuconostoc*, the latter being capable of creating viscous formations in the beverage.

Among the other bacteria which may be encountered are acetic acid bacteria (in noncarbonated beverages), butyric species of *Clostridium* (in tomato juice) and coliforms.

The majority of molds are aerobic organisms and are therefore rarely encountered in carbonated drinks. They may, however, appear in fruit juices. This occurs when the beverage has lost its selective properties, particularly when packaging leaks are involved. In this case, molds which are normally inhibited by the anaerobic conditions may also be able to develop (*Penicillium*, *Aspergillus*, etc.).

26.4.3 Testing

Testing is a matter of searching for contaminants which may be capable of spoiling the quality of the finished product. Several levels of microbiological testing may be envisioned:

- minimal testing, i.e., testing the stability of the product after storage under conditions which favor development (warm incubation);

- systematic microbiological testing of finished products (quantitative testing);
- testing as above for preventive purposes at the potential sources of spoilage, i.e., starting material, equipment, atmosphere, personnel, packaging (testing of bottle-cleaning processes is discussed in Chapter 35, Section 35.3.3.1).

Testing by microscopic observation may be done in the case of visible spoilage (clouding, clotting, gelifying) which may appear either spontaneously or after incubation (clouding corresponds to a microorganism concentration of 10^7–10^8/ml). This provides some characterization of the contaminant and may also suggest a culturing technique for it. When the microbial concentration is too low for direct microscopic examination, the observation may be done after filtering some volume of the beverage, staining the filtration membrane, drying, and rendering it transparent.

Nevertheless, the most widely used technique for testing beverages is counting contaminants after filtering on a membrane and culturing. The volume filtered depends on the expected contamination. A volume of 50 ml is often used. This technique allows the enumeration of total flora on rich media approximating as closely as possible the composition of the beverage or the enumeration of specific flora using adapted selective media. The two methods are to be considered complementary.

Testing for microorganisms of sanitary incidence is performed mainly at the starting material level (water). Coliforms and thermotolerant (44°C) coliforms are enumerated by means of the techniques described in Chapter 17. The media are incubated at 25°–30°C for total contaminants and lactic acid bacteria and at 20°–22°C for fungal contaminants. Contaminants thus cultured must be characterized.

There are no regulatory criteria governing the microbiological quality of beverages. It is generally accepted that production entrants are not supposed to contain more than one fungal contaminant per milliliter. For fruit juices, the following criteria may be retained:

osmophilic yeasts < 20/l
molds < 10/100 ml
coliforms < 1/10 ml
butyric Clostridia < 1/100 ml

Total contaminants may be enumerated on Mossel's medium or Orange Serum Broth (see Chapter 33, Appendix, media 13 and 14) depending on the type of beverage. Dextran-producing *leuconostocs* are detected on medium with a high sucrose content (see Chapter 33, Appendix, medium 15). Fungal contaminants may be enumerated on yeast malt medium or OGA medium.

CHAPTER

27

Meat and Meat Products

A. Plusquellec

27.1 Introduction

Animal flesh has always been one of the most fundamental components of the human diet. The meat industry represents one of the principal sectors of the food industry. Meat is being consumed more and more and in an increasingly diverse range of forms. It is considered to be the food of choice due largely to its nutritional value. Its richness in proteins and the quality of these has made it a difficult food to replace by any single substitute. However, due precisely to its nutritional qualities, meat constitutes a very favorable substrate for the majority of microbial contaminations (principally proteolytic organisms). It is therefore a difficult food to preserve. Yet, even meats consumed "fresh" require a certain storage period in order to acquire certain desirable organoleptic properties (tenderness, color). This is the maturation phenomenon. As soon as meat is butchered (ground or cut into units for sale), it becomes even more vulnerable. It is generally consumed rapidly or stabilized by freezing or refrigeration and often packaged under vacuum. The majority of meat-based preparations produced by industries which cure meats or produce heat-and-serve prepared dishes provide a medium which is well suited to bacterial proliferation. Moreover, these products during preparation undergo numerous handlings all likely to introduce at some point or other the majority of the undesirable microorganism in the food microbiological repertoire.

27.2 Characteristics of Animal Muscle

Animal muscle is made up of a bundle of contractile fibers surrounded by an envelope of conjunctive tissues. These tissues are made of fibrous proteins (collagen, reticulin, elastin) with little or no food value. They constitute effective protection against external contamination. On the average, the flesh of mammals contains 18% proteins and 75.5% water. The carbohydrate content is low or zero. The fat content is quite variable.

After slaughtering, muscle assumes a rigid state or *rigor mortis*. This transformation is accompanied by a progressive drop in the redox potential (the rH of normal muscle is approximately +250 mV) down to about −50 mV, and a drop in the pH from 7–7.4 to 5.5–5.7 as glycogen is converted into lactic acid. This occurrence plays an important role in preservation.

27.3 Microbiological Characteristics of Meat

At the moment of slaughter, muscle is practically free of microorganisms, at least in the case of healthy animals. One organism per 10 or 100 g of flesh may be counted. Lymphatic ganglions, on the other hand, are often contaminated and may be the source of infections of muscle tissue. The microbes likely to be found in meat originate in the intestine where they cross the intestinal barrier and are carried by the blood. They may cause deep contaminations.

As for the surface of the carcass, a large number of microbes of various types may always be found. This number varies from one location to another on the animal and with the hygienic conditions in the slaughtering plant.

The genera or bacterial families that predominate on fresh meats are *Pseudomonas*, *Acinetobacter*, Micrococcaceae, enterobacteria, *Flavobacterium*, *Microbacterium*, *Lactobacillus*. Most of these groups are responsible for spoilage, that is, putrefaction, greening, souring, lipolysis.

Other bacteria may be present on a more sporadic basis, for example, *Bacillus*, *Alcaligenes*, *Streptococcus*, *Aeromonas*, *Corynebacterium*, *Arthrobacter*, *Clostridium*.

27.4 Endogenous Infections

Meat may be a vector of infectious agents. Some of these are not foodborne infections although contaminated carcasses may be at the origin of the spreading of the disease. These include *Bacillus anthracis* (anthrax), *Mycobacterium tuberculosis*, and *Brucella*.

Other diseases may be contracted by consuming contaminated meats. In this category, the *salmonella* are the most prevalent, poultry being the product in which these bacteria are most frequently encountered.

Meat is often also responsible for the transmission of parasitic worms. *Trichinella spiralis* (agent causing trichinosis) is a small nematode disseminated by pork. *Tænia saginata* may be harbored by beef.

27.5 Spoilage of Meat

27.5.1 Origin of Contaminations

Between slaughtering of the animal and consumption of the meat product, the points where the introduction of contaminating microorganisms is likely are numerous. During slaughtering itself, microbes may cross the intestinal barrier and reach muscle via the circulatory system, particularly in the case of stressed animals. The device used for slaughtering may bring microbes from the skin surface into the depths of the meat. Evisceration is a critical operation which may be the cause of various contaminations. Butchering of the carcass may spread microorganisms from the environment (organisms from the ground and airborne organisms) or from personnel (staphylococci, fecal microorganisms). Contamination also occurs during transport or stocking under insufficiently hygienic conditions.

27.5.2 Factors Related to Spoilage

Due to its chemical composition, meat will always present an inviting medium for microbial contamination. Whether or not proliferation becomes possible will depend on intrinsic and extrinsic factors.

Intrinsic factors:

- pH: animals which have been fatigued immediately prior to slaughter have insufficient glycogen reserves for a normal lowering of the pH;
- Eh: this increases when the meat is butchered or fabricated.

Extrinsic factors:

- ambient humidity: and atmosphere which is too humid favors more intense development of the surface microflora;
- temperature: this is the predominating factor. Upon slaughter, the carcass must be refrigerated. Butchering must take place in a refrigerated room. This cold chain must not be interrupted.

27.5.3 Principal Spoilages of Carcasses

Spoilages observed during the storage of carcasses are of five major types.

27.5.3.1 Surface Spoilage

Surface spoilage appears as a slimy layer (fishy meat) accompanied by a nauseating odor. The odor becomes apparent when the number of bacteria surpasses 10^7/p cm^2. The slimy layer becomes visible when the concentration reaches 10^8/p cm^2. The agents which cause this putrefaction generally belong to the genera *Pseudomonas* and *Achromobacter*. These microorganisms are hydrophilic and ambient humidity plays a significant role in this type of spoilage. Since they are also psychrotrophic, the development of these contaminations is not stopped by cool temperatures.

This type of bacteria develops poorly at acidic pH values, and poor slaughtering conditions may partly explain these developments.

These are aerobic microorganisms which develop better on smaller cuts of meat or on ground meat than on carcasses.

Surface spoilage caused by other bacteria may also be found, for example, *Micrococcus*, *Lactobacillus* (particularly in the case of vacuum packaging), yeasts, and molds.

27.5.3.2 Deep Putrefaction

Meats with deep putrefaction are swollen by gas, their color is abnormal (gray or greenish), and they give off a highly objectionable odor. The odor and gas are the result of breakdown of the meat by proteolytic organisms. The microorganisms responsible are putrefying *Clostridium* species, particularly *Clostridium perfringens*, present in large numbers (10^9/g).

This type of spoilage occurs on unrefrigerated or poorly refrigerated carcasses. The bacteria responsible are inhibited at temperatures below 20°C. Meat in which the pH has not decreased to a low enough value (above 6.2) are most vulnerable to this type of putrefaction.

27.5.3.3 "Bone Taint"

"Bone taint" is the development of putrid odors from the depths of the meat generally around the bone. The bacteria responsible are from lymphatic ganglions.

Spoilage arises in cases of poor refrigeration of meat which has a high pH.

27.5.3.4 Spoilage of Fungal Origin

Meat from freshly slaughtered animals is a favorable substrate for hygroscopic molds, especially the Mucorales (*Mucor racemosus*). During cold storage, the growth of most molds is inhibited although a few cryophilic species may be encountered such as *Cladosporium herbarum* which causes black-spot on meats.

Mycotoxins are detected only rarely in meat products even though a significant proportion of the strains they harbor are toxigenic species.

27.5.3.5 Spoilage by Brochothrix Thermosphacta

Brochothrix thermosphacta (*Microbacterium thermosphactum*) represents a major portion of the bacterial flora of refrigerated meat products. Packaging under film decreases the gram-negative flora to the benefit of the genera *Lactobacillus* and *Brochothrix*. The latter bacterium, having psychrotrophic tendencies, is cultured between 0° and 25°C at pH values between 5 and 9. It is a facultative anaerobe capable of developing in vacuum packaged meats and is sensitive to heat and to nitrite.

Brochothrix is responsible for meat spoilage which appears as souring due to lactic acid production and as odors associated with volatile acids (acetic, butyric).

Brochothrix thermosphacta may be enumerated on a selective medium containing streptomycin, thallium acetate, and actidione (STAA) (Gardner, 1966).

27.5.4 Spoilage of Various Meat Products

Products obtained by processing of meats are exposed to a wide range of contaminations.

27.5.4.1 Brined Meats, Hams

Brined meats and hams may develop surface spoilages attributable to *Micrococcus*, *Microbacterium*, or yeasts. Spoilages of the muscle tissue, with gas formation and greenish discoloring, are due to clostridia or to streptococci. Other genera, such as *Pseudomonas*, *Achromobacter*, *Bacillus*, and *Lactobacillus* may cause souring in the meat or around the bone.

27.5.4.2 Brined Sausages

Brined sausages may develop a surface discoloration caused by *Micrococcus* or *Leuconostoc*. In the case of vacuum-packaged sausages, greenish discoloring and gas formation may be produced by lactobacilli. Surface spoilage of the mucilagenous type may be caused by *Micrococcus* or yeasts.

27.5.4.3 Dry Sausages, Fermented Products

The contaminations observed are mainly superficial, such as formation of slimy layers by yeasts, development of molds producing abnormal coloring. Fresh or dry sausages of the bratwurst, salami, pepperoni, or wiener type are all frequently invaded by a variety of molds, such as Mucorales, *Penicillium*, *Aspergillus*, *Scopuliarosis*, *Paecilomyces*, and so on which often come from manufacturing ingredients (spices, powdered milk, lactose, etc.) or are introduced by atmospheric pollution during drying.

27.5.4.4 Uncooked Ham

When packaged under vacuum, this product may develop a souring due to lactobacilli, micrococci, or streptococci. Superficial spoilages due to streptococci or molds may also be observed, mainly *Aspergillus* or *Penicillium*-type molds which form spots.

27.5.4.5 Smoked Meats

Smoked meats are vulnerable to contamination by molds. The most frequently encountered agents are the genera *Aspergillus* and *Penicillium.*

27.6 Microbiological Testing

Given the great sensitivity of the product to microbial contamination, it must be subjected to very close monitoring with tests at all levels. During slaughtering, carcasses are subjected to a veterinary examination. In the event of the discovery of lesions, this examination may be completed by microbiological analyses.

The processing work areas (quartering, butchering, etc.) must fulfill very strict hygienic conditions. It must be continually verified by means of microbiological testing that these conditions are satisfied at the level of the equipment, the rooms, the personnel, and the meat throughout the processing operation Monitoring must be especially rigorous in poultry kills which present some very particular hygienic problems.

Products requiring a large amount of handling such as ground meats, pâtés, sausages, and so on as well as meat-based prepared dishes are liable to contain large numbers and a large variety of microorganisms, including organisms responsible for food intoxications. These products must consequently be the object of absolutely complete and thorough microbiological testing.

27.6.1 Sampling for Analysis

The method of sample removal differs depending on the nature of the product being analyzed. For testing carcasses, the sample is removed from the depths of the tissue by using a sterile scalpel after treating the surface by burning with a bunsen flame. In the case of cuts of meat, sample removal must address the surface and the depths at the same time. Sample removal from ground meats, wieners, sausages, pâtés, and so on does not present any problem. For whole poultry, detection of salmonella is done on 25 g of pectoral muscle taken after removing the skin.

27.6.2 Analysis

The detections and enumerations done for routine testing purposes are entirely conventional and address the following:

- aerobic mesophilic flora;
- fecal indicators (coliforms, fecal coliforms);
- *Staphylococcus aureus*;

- sulfite-reducing anaerobes;
- *Salmonella*;
- yeasts and molds.

Chemical analyses are often associated with these tests. The ABVT assay in particular is a spoilage index.

More specialized microbiological techniques must be implemented in some cases, for example, identification of microorganisms responsible for a given manufacturing defect in salted products.

Among the more specific tests done on meats, antibiotic detection must be mentioned, which is done by the technique of diffusion in agar from disks. Two bacterial test strains are used: a strain of *Bacillus subtilis* (strain BGA) and a strain of *Sarcina lutea* (*Micrococcus luteus*), ref. ATCC 9341. Ordinary nutrient agar is surface inoculated. One plate of nutrient agar at pH 7 is inoculated with *Sarcina lutea.* For *Bacillus subtilis*, two plates are used: one of nutrient agar at pH 6 and one of nutrient agar at pH 8. The disks are impregnated with an exudate of the meat obtained by freezing a strip of it and thawing quickly at 37°C. Alternatively, the disks may be placed between two cubes of meat which are frozen followed by thawing. A disk thus prepared is placed at the center of each of three Petri plates. The media are incubated at 30°C for *Bacillus subtilis* and at 37°C for *Sarcina.* The diameter of the zones of inhibition is measured after incubation.

27.7 Regulations, Microbiological Standards

French and European regulations govern the conditions under which meat must be handled, transported, and stocked. The European regulations impose early refrigeration below 7°C, which must be maintained during butchering. French regulations establish the conditions for killing and sanitary inspection (law of July 8, 1965, decree of July 21, 1971) as well as for transportation and stocking (February 1, 1974 decree*).

With respect to microbiological criteria, the December 21, 1979 decree* attaches great importance to all meat products and derivatives.

27.7.1 Fresh or Frozen Meat, With or Without Fabrication

The criteria[1] are as follows:

**Translator's note*: The English word *decree* is used in this chapter to translate the French words *arrêté* and *décret* which correspond to different administrative levels. An asterisk indicates that the French word used was *arrêté*, otherwise the word was *décret*.

[1]These criteria indicate the average value, m, which should normally be encountered. The value which must not be exceeded is equal to:

- 10 m if working with agar-based media,
- 30 m if working with liquid media.

Carcasses or sides, refrigerated or frozen

aerobic mesophilic flora	5×10^2/g
anaerobic sulfite-reducers	2/g
Salmonella	absence in 25 g

Packaged parts, vacuum or otherwise, refrigerated or frozen

aerobic mesophilic flora	5×10^2/g
fecal coliforms	10^2/g
anaerobic sulfite-reducers	2/g
Salmonella	absence in 25 g

Packaged single portions, refrigerated or frozen, and single portions for retail distribution, refrigerated or frozen

fecal coliforms	3×10^2/g
Staphylococcus aureus	10^2/g
anaerobic sulfite-reducers	10/g
Salmonella	absence in 25 g

27.7.2 Meats Having Undergone Various Types of Processing

Meats ground in advance or by customer request

aerobic mesophilic flora	5×10^5/g
fecal coliforms	10^2/g
Staphylococcus aureus	10^2/g
anaerobic sulfite-reducers	2/g
Salmonella	absence in 25 g

Uncooked delicatessen products, ground

	Dessicated for consumption as such	To be consumed after cooking
fecal coliforms	10^2/g	10^3/g
Staphylococcus aureus	10^2/g	10^3/g
anaerobic sulfite-reducers	2/g	10^2/g
Salmonella	absence in 25 g	absence in 25 g

Uncooked salted products, salted and/or dried, sliced or not

fecal coliforms	10^3/g
Staphylococcus aureus	5×10^2/g
anaerobic sulfite-reducers	50/g
Salmonella	absence in 25 g

Cooked delicatessen products, sliced or not, meat loaves

aerobic mesophilic flora	3×10^5/g
coliforms	10^3/g
fecal coliforms	10/g
Staphylococcus aureus	10^2/g
anaerobic sulfite-reducers	30/g
Salmonella	absence in 25 g

Cooked whole ham

aerobic mesophilic flora	10^4/g
coliforms	10/g
fecal coliforms	1/g
Staphylococcus aureus	1/g
anaerobic sulfite-reducers	1/g
Salmonella	absence in 25 g

27.7.3 Poultry

Whole birds, refrigerated, frozen, or deep-frozen

Salmonella	absence in 25 g of pectoral muscle

Roasts, scallopini, meat rolls, uncooked, breaded or not

aerobic mesophilic flora	5×10^5/g
fecal coliforms	10^3/g
Staphylococcus aureus	10^2/g
anaerobic sulfite-reducers	10/g
Salmonella	absence in 1 g

Cooked roasts, whole or sliced, cooked or precooked meat rolls

aerobic mesophilic flora	3×10^5/g
fecal coliforms	10/g
Staphylococcus aureus	10^2/g
anaerobic sulfite-reducers	10/g
Salmonella	absence in 25 g

Mechanically de-boned uncooked meat

aerobic mesophilic flora	10^6/g
fecal coliforms	5×10^3/g
Staphylococcus aureus	10^3/g
anaerobic sulfite-reducers	10^2/g
Salmonella	absence in 1 g

Mechanically de-boned cooked meat

aerobic mesophilic flora	3×10^5/g
fecal coliforms	50/g
Staphylococus aureus	10^2/g
anaerobic sulfite-reducers	30/g
Salmonella	absence in 25 g

In addition to these regulatory tests for flora of sanitary significance, it is very useful to evaluate flora responsible for the main types of spoilage. Selective enumeration of the *Pseudomonas* group in particular yields an index of the risk of spoilage by surface putrefaction. This enumeration is done by pour plate inoculation of a selective solid medium, such as CFC medium (Mead and Adams medium). Three inhibitors are added to the basal medium, that is, cetrimide, fucidine, and cephalosporin. After incubation at 22°C for 48 hours, the cultured colonies are confirmed (oxidase, Kligler medium) according to AFNOR standard NF-V-04-504.

CHAPTER

28

Eggs and Egg Products

A. Plusquellec

28.1 Contamination of Eggs

Eggs are well protected from contaminations by natural physical and chemical defenses although this protection has its limitations. The successful storage of eggs depends on the following:

- initial contamination;
- stocking conditions;
- factors intrinsic to the egg.

28.1.1 Origin of Contaminations

The egg before laying is protected against invasion by microbes in the cloaca. Contamination of the shell occurs mainly during laying. The flora found at this stage is essentially gram-positive (Micrococcaceae).

Congenital contamination of eggs may be of some significance (spreading of *salmonella*), but this is quite exceptional (highly contaminated foods or drinking water).

28.1.2 Characteristics of Spoilage

Penetration may occur at the level of the shell. Certain factors such as thinness of the shell and high moisture levels favor this event. Colonization of the membranes and of the albumin is marked by a selection of microorganisms.

While the bacteria present on the shell are mainly gram-positive, spoiled eggs have a flora composed mainly of gram-negative organism (sometimes a few gram-positive organisms).

The contaminating flora is quite constant, its main representatives belonging to the genera *Alcaligenes*, *Achromobacter*, *Pseudomonas*, *Serratia*, *Hafnia*, *Citrobacter*, *Proteus*, and *Aeromonas*. The presence of numerous psychrotrophs among these genera should be noted.

Contamination depends largely on stocking conditions. Under normal conditions, a considerable delay (at least 20 days) is necessary in order for a large number of bacteria to be found in the albumin. As soon as the eggs are broken, however, conditions for microbial development become very favorable. They must be used very rapidly or stabilized (pasteurization, freezing, concentration, dehydration).

28.2 Bacteriological Analysis

Bacteriological analysis is performed mainly on frozen or dehydrated eggs or egg products and is generally accompanied by an organoleptic examination based essentially on olfactive tests. The odor may be normal, sour or sharp, moldy or putrid.

28.2.1 Sample Removal

Sampling is done according to the following plan:

Number of samples:

- equal to 10% if the lot consists of less than 100 packages;
- equal to the square root of the number of packages, if this number is between 100 and 1,000 (above 1,000 packages one sample is added per each additional 1,000).

If the lot is made up of a single package, the amount taken is

- two samples if the lot is smaller than 100 kg;
- three samples if the lot is bigger than 100 kg.

Samples of fresh products must be analyzed very soon after sampling. If transportation requires more than 20 minutes, freezing the sample is recommended.

For powdered eggs, sample preparation is very straightforward. An original solution in tryptone salt is made.

Sample removal from eggs or frozen egg products is a more delicate matter. It may be done using a sterile gouge or with the aid of a 3/32″ sterile drill bit.

28.2.2 Bacterial Groups Analyzed

28.2.2.1 Enumeration of Aerobic Mesophilic Flora

A representative sample (25 g) is placed into a stomacher bag and then 22 ml sterile phosphate buffer are added into the stomacher bag. The contents are placed into a stomacher instrument and "massaged" for 2 minutes (see Chapter 1.2). The sample (1:10 dilution) is further diluted to 10^{-2}, 10^{-3} and 10^{-4}, or higher if necessary. Duplicate samples (1 ml each) from each dilution (usually 3 dilutions are used) are then placed into sterile Petri dishes. Tempered standard plate agar (48°C) is then poured into the Petri dishes. After solidification the plates are inverted and are incubated for 48 h at 35°C. The plates are then removed and counted. This is the standard plate count method or the aerobic mesophilic count method widely used internationally (FDA, 1992).

28.2.2.2 Enumeration of Enterobacteria

Counting enterobacteria is done on VRBG agar (crystal-violet, neutral red, bile, glucose) (see Chapter 33, Appendix, medium 16) poured with an overlay creating a double layer and incubated at 30°C for 24 hours.

28.2.2.3 Enumeration of Staphylococcus aureus

After food is homogenized as described in Section 28.2.2.1 for aerobic mesophilic count the sample can be plated on Baird-Parker medium. For each dilution 1 ml sample suspension is distributed to 3 plates (e.g., 0.4 ml, 0.3 ml, and 0.3 ml). The liquid is then spread by a sterile bent glass rod. Plates are then inverted and incubated at 35°C for 45–48 h. "Typical *S. aureus* colonies are circular, smooth, convex, moist, 2–3 mm in diameter on uncrowded plates, gray to jet-black, frequently with light-colored (off-white) margin, surrounded by an opaque zone and frequently with an outer clear zone" (FDA, 1992). Further confirmation tests include coagulase, catalase, anaerobic utilization of glucose and mannitol, sensitivity to lysostaphin, and thermostable nuclease production.

28.2.2.4 Detection of Salmonella *in 25 g*

There are many methods for the detection of *Salmonella*. A detailed review on rapid methods for the detection of *Salmonella* and other food pathogens can be found in Chapter 1 and Feng (1992). The one described herein is the most commonly used for poultry products (FDA, 1992) and is considered as the "Conventional Method."

Preenrichment Step. First, 25 g of the sample is added to a stomacher bag containing 225 ml of sterile lactose broth. The contents are "stomached" for 2 min. The "stomached" solution is then incubated for 24 h at 35°C.

Enrichment Step. Transfer 1 ml mixture of the preenriched sample to 10 ml of selenite cystine (SC) broth and another 1 ml to tetrathionate (TT) broth. Incubate these tubes at 35°C for 24 h.

Selective Isolation Step. To isolate *Salmonella* streak for isolation from SC and TT broth tubes onto bismuth sulfite agar, xylose lysine desoxycholate agar and Hektoen enteric agar. Incubate plates for 24 h at 35°C.

Biochemical Reaction Step. After growth on these plates typical colonies are inoculated (streak/stab) into triple sugar iron (TSI) agar slant and lysine iron agar (LIA) slant. Incubate TSI slant for 24 h and LIA slant for 48 h at 35°C.

Typical reactions of *Salmonella* in TSI tube are acid butt, alkaline slant, and blackening of the tube. All cultures that give an alkaline butt in LIA should be considered suspect *Salmonella.* Suspect *Salmonella* isolates can be further tested by one of many diagnostic kits (such as API, MicroID, Minitek, Enterotube, etc.) available in the market (see Chapter 1.5).

Serological Testing Step. All suspect *Salmonella* must be confirmed by serological tests.

28.2.2.5 Testing of the Effectiveness of Pasteurization

As is the case for milk, testing for pasteurization effectiveness is based on the detection of enzymes which are naturally present in eggs and normally destroyed by heating. Phosphatase is not usable since it is not destroyed by the pasteurizations normally applied to egg products.

Alpha-amylase is destroyed in $2\frac{1}{2}$ minutes by heating to 64°C. Its detection is applicable to products processed at this temperature. For products heated to lower temperatures, catalase is detected. It is destroyed by processing at 54°C.

28.3 Regulation of Bacteriological Quality

The decree of June 15, 1939 regulates the grounds for the seizure of eggs. The decree* of December 21, 1979 establishes the following criteria and sets the mode of analysis.

**Translator's note:* The English word *decree* is used in this chapter to translate the French words *arrêté* and *décret* which correspond to different administrative levels. An asterisk indicates that the French word used was *arrêté*, otherwise the word was *décret*.

Pasteurized egg products	
aerobic mesophilic bacteria	10^5/g
enterobacteria	10/g
Staphylococcus aureus	100/g
Salmonella	absence in 25 g
Unpasteurized egg white	
Salmonella	absence in 25 g

CHAPTER

29

Fisheries Products: Fish, Crustaceans, and Shellfish

A. Plusquellec

29.1 Bacteriological Quality of Fish and Fisheries Products

29.1.1 Types of Contamination

The following two important aspects are to be considered with respect to the microbiological quality of fish and fisheries products:

- The sanitary aspect. The consumer must be protected from the presence of pathogenic microorganisms. In this respect, sea products are associated with a great variety of risks depending on their type, their original habitat, and the processing they undergo.
- The economic aspect. Sea products are fragile. They must respond to a variety of organoleptic criteria (appearance, taste, odor) at the market level.

The spoilage of fisheries products also varies widely depending on the type and origin of product considered. This spoilage is associated to a large extent with the presence of a common bacterial flora which does not represent any direct risks from a sanitary point of view.

29.1.2 Contamination by Pathogens

The composition of the microbial flora of fisheries products is generally quite similar to that of their natural environment. The species found in the intestines of fish are essentially the same as those isolated from the water in which they were caught.

Contaminating microorganisms found in fisheries products may be of two main origins.

29.1.2.1 Contamination of the Waters Being Fished

Littoral zones are especially subject to pollutions which may be quite heavy. Pathogens introduced by this route are generally fecally disseminated (*Salmonella*, viruses, parasites). The detection of fecal indicators is thus very useful in this particular case.

Aside from these transient pathogens in the fisheries environment, the existence of a halophilic bacterium adapted to the fisheries environment should be mentioned, that is, *Vibrio parahaemolyticus* which is responsible for a type of gastroenteritis.

The wide range of products of the fisheries have different sensitivities to environmental pollution, which depend on their physiology. In the fish themselves, contaminating microbes may be found in the gills, in the intestine, and on the skin. The flesh which is consumed is free of microorganisms in the live state. In crustaceans, the bacterial flora is found in the head (gills, digestive system) and below the abdomen. Shellfish, essentially filtering bivalves (mussels, oysters), are particularly at risk. In order to feed themselves, these animals filter large volumes of water and as a result accumulate both chemical and bacterial pollutants in their digestive tracts (Prieur et al., 1990).

It must be emphasized that conchological (shellfish-growing) zones are often situated in geographical sites which, being sedimentation basins, are especially subject to pollution (bays, estuaries).

These shellfish are often consumed raw or after rapid cooking, which increases sanitary risks considerably.

All of these factors make shellfish a product representing potential sanitary risks and which must be subjected to close and continuous monitoring.

It must be emphasized, however, that the majority of sanitary problems attributable to shellfish which have been observed were due to toxic dinoflagellates (*Dinophysis*, *Gonyaulax*, *Alexandrium*) and viruses (hepatitis, Norwalk virus).

29.1.2.2 Contamination Subsequent to Fishing

A product uncontaminated at the origin may become soiled during any of the various stages leading up to its distribution in the marketplace. The contamination may have occurred on the boat by contact with soiled equipment or material (containers, ice of poor bacteriological quality). Washing with contaminated water may at times explain the introduction of dangerous microbes. Processed products such as fish fillets are subject to even higher contamination risks. Tools, cutting tables, and personnel may all act as vectors for the introduction of microbes carrying hygienic risks (fecal bacteria, staphylococci, *Clostridium*). Furthermore, the product is stripped of its natural barriers (skin, scales), making it easier for contaminants to penetrate and attach. Finely ground products heighten the risk of proliferation even more and must be very closely monitored. Fish meals and concentrates intended for animal feeds may also represent risks. The presence of *Clostridium perfringens* and *Salmonella* in these products is frequent. Among the products of fish processing, some may undergo treatments intended to inhibit microorganisms, such as salting, drying, smoking, pickling. These processes are not equivalent to sterilization and microbiological testing remains necessary.

29.2 Spoilage Processes of Fisheries Products

29.2.1 Microflora of Fish and Crustaceans

The microflora of live fish depends largely on the environment. Microorganisms isolated from the gills, intestines, and the skin belong mainly to the genera *Pseudomonas* and *Acinetobacter* (for 60% of isolates) and *Corynebacterium*, *Flavobacterium*, and *Micrococcus* (for 20% of isolates). Alcaligenous bacteria such as *Bacillus*, *Proteus*, and *Serratia* are also encountered. Crustaceans have essentially the same flora but with a higher proportion of corynebacteria. These bacteria are also those which are the most frequently isolated from marine environments.

29.2.2 Spoilage of Fresh Fish

As soon as a fish dies, a spoilage process begins which involves mainly autolysis. The digestive enzymes of the fish destroy the intestinal barrier allowing the dissemination of the microbes normally found there. The microorganisms present are mostly psychrotrophs, among which many are producers of proteolytic and lipolytic enzymes. Thus, even at low temperatures, preservation is of very limited duration.

Decomposition compounds such as free amino acids (which themselves break down into amines and ammonia) very rapidly appear and impart objectionable tastes and odors to the flesh.

Bacterial spoilage of a fish, particularly of certain types (tuna, mackerel), may lead to the production of a toxic breakdown product (i.e., histamine, responsible for histamine intoxication). The bacterial responsible are those which are capable of decarboxylating histamine (*Proteus*, *Klebsiella*, *Photobacterium*, etc.).

The anatomy and chemical composition of the animal determine its susceptibility to spoilage to a large extent. Thus, selachian fish (sharks) rapidly develop a higher total basic volatile nitrogen (TBVN) content than do teleosts (bony fish).

29.3 Testing

29.3.1 Fresh Fish

Bacteriological analysis is rarely justifiable in the case of fresh fish for two essential reasons:

- The delays involved in market distribution are very short and almost always shorter than the delay in response of bacteriological analyses.
- Fish from the high seas, sold fresh, are at minimal risk of contamination by microorganisms of hygienic significance.

Spoilage may be due to bacteria which are difficult to culture on the usual media and the bacteriological results are not always indicative of the observed spoilage. Conventional food microbiological criteria are therefore quite poorly adapted to this type of product testing.

Evaluation of the freshness of fish and crustaceans is done mainly by inspection of organoleptic traits such as odor, general appearance, firmness, observation of the skin, scales, eyes, gills, anus, viscera, and flesh. It may be important, nevertheless, to have some knowledge of the bacteriological quality of fresh fish intended for freezing and transportation.

Testing may be done by swabbing of the skin or on a freshly prepared homogenate. The principal criterion studied is the aerobic mesophilic flora.

Enumeration of microbes capable of breaking down histidine to histamine may be performed on fresh or frozen fish. A differential solid medium has been proposed by Niven and co-workers (1981) (see Chapter 33, Appendix, medium 17) containing L-histidine and bromocresol purple (pH 5.3) and inoculated by the pour plate technique with a top agar overlay. After incubation at 25–30°C for 36–72 hours, purple colonies are counted and sampled to verify the histidine + trait.

29.3.2 Processed Fish

Slicing or filleting of fish generally introduces a contaminating microflora. In addition, fragmenting of the product renders it more accessible to microbial proliferation. These products must be monitored very attentively and must meet the sanitary criteria defined by the decree* of December 21, 1979.[1]

**Translator's note:* The English word *decree* is used in this chapter to translate the French words *arrête* and *décret* which correspond to different administrative levels. An asterisk indicates that the French word used was *arrête*, otherwise the word was *dêcret*.

[1]Amending decree* of March 13, 1989.

Fish steaks, coated or uncoated, fish fillets, fresh or refrigerated

aerobic mesophilic flora — 10^5/g
fecal coliforms — 10/g
Staphylococcus aureus — 10^2/g
anaerobic sulfite-reducers — 10/g
Salmonella — absence in 25 g

Fish steaks, coated or uncoated, fish fillets, frozen, or deep-frozen

aerobic mesophilic flora — 5×10^4/g*
fecal coliforms — 10/g[2]
Staphylococcus aureus — 10^2/g
anaerobic sulfite-reducers — 2/g
Salmonella — absence in 25 g

Ground, raw fish-based preparations

aerobic mesophilic flora — 5×10^5/g
fecal coliforms — 10^2/g
Staphylococcus aureus — 10^2/g
anaerobic sulfite-reducers — 2/g
Salmonella — absence in 25 g

Fresh, frozen, or deep-frozen frogs legs[2]

aerobic mesophilic flora — 10^5/g
fecal coliforms — 10^2/g
Staphylococcus aureus — 10^2/g
Salmonella — absence in 25 g

29.3.3 Crustaceans

Crustaceans may be fished near the coasts and thus exposed to sedimentation basin contaminants. The detection of fecal organisms is therefore carried out.

For products requiring human handling, mainly deshelled crustaceans, testing is completely by analysis for staphylococci which may be introduced during handling.

The following microbiological criteria have been retained (decree* of December 21, 1979). Whole crustaceans, cooked, refrigerated, other than shrimp

aerobic mesophilic flora — 10^5/g
fecal coliforms — 1/g
anaerobic sulfite-reducers — 2/g
Salmonella — absence in 25 g

[2]See footnote 1, p. 440.

All crustaceans, including shrimp, cooked or raw, frozen or deep-frozen

aerobic mesophilic flora	10^3/g
fecal coliforms	1/g
anaerobic sulfite-reducers	2/g
Salmonella	absence in 25 g

Cooked shrimp, deshelled, refrigerated and deshelled, frozen or deep-frozen

aerobic mesophilic flora	10^5/g
fecal coliforms	10/g
Staphylococcus aureus	10^2/g
anaerobic sulfite-reducers	10/g
Salmonella	absence in 25 g

29.3.4 Shellfish

Bacteriological testing of shellfish may come into play within the framework of the sanitary inspection of distribution outlets carried out by the Veterinary Services. This testing is intended to protect consumers against sanitary risks posed by contaminated shellfish and is therefore oriented towards the detection of fecally spread germs. The criteria retained by the decree* of December 21, 1979 are as follows[3]:

fecal coliforms	3×10^2/100 ml
fecal streptococci[4]	2.5×10^3/100 ml
Salmonella	absence in 25 g

Bacteriological analysis of shellfish is also used to indirectly evaluate the hygienic safety of conchological zones.[5]

The monitoring of targeted zones requires frequent sampling (26 per 12 months) and a single criterion is retained for testing, that is, the most probable number of fecal coliforms in 100 ml of flesh. A zone is classed as safe when the MPN of fecal coliforms is below 300 in 100 ml of flesh. Counts exceeding this amount are tolerated if there are no more than five of these in any period of 12 consecutive months and if there have not been more than two counts over 3,000 in 100 ml and the other two are below 1,000 in 100 ml of flesh.

Bacteriological testing of shellfish requires a special sample preparation. The examination requires a number of live individuals not smaller than 4 for large shellfish and not smaller than 12 for all others (a flesh + intrashell liquid volume $>$25 ml must be obtained). The shellfish are freed of all foreign bodies and then carefully cleaned by brushing under running water. After briefly

---

[3]Expressed per volume of flesh + intrashell water.

[4]The usefulness of analysis for fecal streptococci is associated with their high resistance. They may still be present when other fecal indicators have disappeared.

[5]Decree* of October 12, 1976 (*Journal Officiel*, 23 November 1976).

flaming the hinge, the shells are opened with a sterile knife. The body of flesh is then detached from the shell using a scalpel and the mixture of flesh + intrashell liquid is collected in a sterile graduated cylinder. Grinding with added sterile diluent yields an original suspension, generally a 1/3 dilution.

The AFNOR technique for the enumeration of fecal coliforms in shellfish is the MPN method in liquid medium (standard NF-V-45-110). The 1/3 suspension and its dilutions are used to inoculate brilliant green lactose bile broth (three tubes per dilution) for incubation at 37°C. The positive tubes are subcultured and incubated at 44°C in the same medium for the detection of indole production among others.

Production of gas simultaneously with the presence of indole is supposed to indicate the presence of coliforms, although this does not correspond to the conventionally accepted definition of fecal or thermotolerant (44°C) coliforms.

The detection of *Vibrio parahaemolyticus* in live marine shellfish is the object of a standardized technique (standard NF-V-45-111). This is a semiquantitative technique which consists of three successive phases:

- An enrichment step in selective medium which may be salted peptone water (30 g/l of NaCl) or salted sodium lauryl sulfate glucose broth. The pH of these media is adjusted to 8.6. This enrichment is done using a large volume of shellfish flesh (25 or 50 ml).
- An isolation step which is done using the enrichment incubated for 7–8 hours at 37°C. Conventionally used thiosulfate citrate bile sucrose (TCBS) medium for the detection of *Vibrio* is most often chosen. After incubation at 37°C for 18–24 hours, colonies characteristic of *Vibrio parahaemolyticus* appear smooth, green, and 2–3 mm in diameter.
- A confirmation step in which suspect colonies must be confirmed either on conventional media such as TSI, nitrate medium, indole medium, or LDC medium, or using API 20E or API 20 NE strips. In all cases, the media or suspension diluent must have salt (NaCl) added to a concentration of 30 g/l.

CHAPTER

30

Vegetable Products

A. Plusquellec

30.1 Introduction

Fruits and vegetables represent a rather particular case in terms of microbial contamination. In general, it may be said that because of their chemical and biological characteristics, these foodstuffs are susceptible only to certain types of contamination. The types of spoilage observed may be serious from an organoleptic point of view and are therefore of considerable economic interest. The contaminations usually encountered are not, however, of great significance in terms of health and sanitation. Microorganisms that represent a risk to the health of the consumer are quite exceptional. Nevertheless, the development of refrigerated storage of vegetables has allowed new risks to arise, particularly those associated with the contamination of vegetables by *Yersinia enterocolitica*. Furthermore, the processing of vegetables and fruits often brings about a loss of natural protective properties and makes these products subject to the sort of contamination normally associated with handling. New packaging techniques for vegetables (packaged minimally processed vegetables) have brought about new possibilities for the storage of vegetables but have also given rise to new microbiological factors.

30.2 Characteristics of Fruits and Vegetables

Spoilage microflora will be entirely dependent on the intrinsic parameters of the product, which in the case of fruits and vegetables, present certain peculiarities.

30.2.1 Vegetables

30.2.1.1 Biological Peculiarities

Vegetables make up a heterogeneous group defined mainly in terms of consumption habits (whether the product is consumed warm or in salads). They include, in fact, a wide variety of plant organs, that is, fruiting bodies (tomatoes, peppers), seeds (peas, beans), flowers (artichoke, cauliflower), tubers (potato), leaves (lettuce, cabbage, spinach), roots (carrots, beets), and stems (asparagus).

Some vegetables are therefore buried and subject to contamination by soil which may contain contaminating microorganisms, possibly even pathogens, in addition to the usual microflora, particularly where spreading of animal manure is involved. Other vegetables are above ground and the contamination of these may occur by various routes (air, insects, birds, sprinkler water).

With respect to protection against external contamination, vegetables as a group present a wide range of factors. For some vegetables, the edible part is very fragile (tomatoes, lettuce), while in other cases good protection is provided by a nonedible portion (peas, beans).

30.2.1.2 Natural Bacterial Flora

The flora normally found obviously depends on the peculiarities of each vegetable. The bacterial groups encountered are therefore highly varied.

Gram-negative bacilli are the most frequently found.

- *Pseudomonas*
- *Alcaligenes*
- *Chromobacterium*
- *Flavobacterium*
- *Enterobacter*

Lactic germs are equally abundant

- *Lactobacillus*
- streptococci
- *Leuconostoc*

as well as other Gram-positive genera.

- *Bacillus*
- *Micrococcus*
- *Corynebacterium*

This natural bacterial flora is accompanied by a wide variety of yeasts and molds.

30.2.1.3 Chemical Characteristics

The heterogeneity from this point of view is also considerable. On the average, vegetables contain 88% water and 9% carbohydrates. This is entirely compatible with the development of bacteria, yeasts, and molds. The pH is quite variable, with the average value being around 6. The redox potential is also quite high and favors aerobes.

30.2.2 Fruits

Inherent protection of fruits against external contamination is very unequally distributed. Some are protected by a thick skin (citrus fruits, bananas), others have a thin skin (apples, pears, peaches, etc.), while still others have only very feeble protection (strawberries).

The chemical composition of fruits includes a slightly lower water content (85%) than for vegetables and a higher carbohydrate content (13%). This composition is compatible with the growth of all microorganisms although the pH intervenes as a selective factor. It is generally below the values suited to bacterial life, that is, as low as pH 3 for citrus fruits and up to pH 5 for fruits such as bananas and melons.

30.3 Spoilage of Fruits and Vegetables

30.3.1 Spoilage of Fresh Products

Products sold fresh have already undergone some handling. During harvesting, certain numbers of fruits or vegetables fall victim to shocks or injuries. Transportation may also bring about numerous spoilages. The product is subjected to shocks, piling effects, and warming as well as an increase in surface moisture. Furthermore, spreading of agents from already contaminated units to healthy units is likely.

30.3.1.1 Spoilage of Fresh Vegetables

The storage of fresh vegetables is quite limited. Between harvest and sale, significant losses may occur as spoilages visible to the naked eye arise, generally referred to as *rotting*, which render the product unsalable.

One of the most widespread spoilages, although not the major cause of vegetable spoilage, is bacterial soft rot. This is facilitated by the presence of lesions on the vegetable and is associated with small numbers of bacteria. The most often cited is *Erwinia carotovora*. This bacterium produces a pectinase which breaks down pectins giving a soft mash having the appearance of wet soap, often accompanied by a foul odor. *Erwinia* is capable of developing by fermenting sugars or alcohols present in the vegetables. Furthermore, most strains of *Erwinia* are able to develop at low temperature. Almost all vegetables are vulnerable to this type of rotting.

Various species of the genus *Pseudomonas* are also responsible for soft rot, mainly in leafy vegetables (*Pseudomonas marginalis, P. viridiflava, P. cicorii, P. cepacia*).

Other bacteria cause "necrotic" spoilages, but the majority of spoilages of vegetables are of the fungal type, that is, caused by molds. Some of these lead to post-harvest rotting while others are responsible for plant diseases. *Botrytis cinerea* is responsible for gray rot. The development of this begins on the surface, especially if there are lesions, and produces a prominent gray mold. It is favored by high moisture levels. "Sour rot" is caused mainly by *Geotrichum candidum.* Insects play a major role in its transmission. The genus *Rhizopus* produces a rot having a cottony appearance with black spores. Anthracnosis is a plant disease caused by *Colletotrichum* which manifests itself as black spots on the stem, the fruiting bodies, and the seed envelopes. The genera *Phytophthora* and *Bremia* are agents of mildew-type diseases on living plants. Among the other genera responsible for spoilage, worth mentioning are *Penicillium* (blue rot), *Sclerotinia* (brown rot), *Trichothecium roseum* (pink rot), and *Alternaria* (black rot).

30.3.1.2 Spoilage of Fresh Fruits

Apart from pears, which may develop a rot caused by *Erwinia,* fruits are not sensitive to bacteria. Molds, however, may cause very economically significant damage. Citrus fruits are contaminated mainly by the genera *Penicillium* (*P. Italicum, P. digitatum*), *Alternaria,* and *Colletotrichum.* Fleshy fruits are infected by *Botrytis cinerea, Gloeosporium, Monilia fructigena,* and *Penicillium expansum.* Yeasts are found in abundance on fruits. They may precede molds in some spoilage processes.

30.3.2 Processing Related Contaminations

Fruits undergo rather limited processing, that is, production of juices, compotes (fruits preserved in syrup), jams, and so on. These products are characterized by acidic pH and a high sugar content. Problems encountered will therefore be related primarily to yeasts, molds, and lactic acid bacteria.

Vegetables which are not sold fresh are marketed as canned or deep-frozen goods. A variety of steps, depending on the type of product, come into play prior to processing for storage. Initial washings remove a portion of the contamination, but also stir up microbes which may be present. Blanching is the brief passage of the product through boiling water, the purpose of which is to destroy enzymes that contribute to spoilage. This also destroys a large portion of the microflora. It does break down the plant infrastructure as well, however, making the product much more sensitive to subsequent contamination. The processing steps occurring between blanching and the preserving process (cooling, packaging) will therefore be critical in determining the contamination of the final product.

The vegetables which turn up the most contaminated are products which are handled to a greater degree, such as chopped spinach or products favorable to bacterial development (peas).

Deep-freezing may bring about substantial reductions of the number of microbes present, but this should not be considered a process intended to destroy the microflora.

30.4 Microbiological Testing

30.4.1 Tests on Fresh Products

Since microorganisms of hygienic importance are very rarely encountered on fresh fruits and vegetables, microbiological testing is highly exceptional. The identification of agents responsible for rotting may be done by microscopic observation, either directly from the product or after culture.

If the rot is of fungal origin, it is wise to attempt culture in several conditions (i.e., on ordinary medium and on medium for osmophilic molds) and at several temperatures.

Counts of pectinolytic microorganisms are possible on polypectate gel medium (see Chapter 33, Appendix, medium 18) as recommended by the American Public Health Association. The medium may be made selective for gram-negative microorganisms by adding crystal-violet. The medium is inoculated by spreading. After inoculation at 25°–30°C, colonies forming depressions on the gel surface are counted.

30.4.2 Tests on Processed Products

Processing of vegetables is likely to introduce various contaminations at several levels. Microbiological testing should therefore be complete, that is,

- enumeration of aerobic mesophilic flora;
- enumeration of fecal indicators (coliforms, fecal coliforms, fecal streptococci);
- enumeration of staphylococci (particularly for handled products);
- enumeration of sulfite-reducing *Clostridium* species (especially for products soiled by dirt);
- enumeration of fungal flora;
- detection of pathogens if relevant (*salmonella*, *Yersinia enterocolitica*);
- enumeration of flat-sour spores for products to be heat-sterilized. This is done by inoculating dextrose tryptose agar containing bromocresol purple after heating the sample or previously inoculated medium to 100°C. Flat-sour spores produce acidifying colonies (yellow).

The preservation of deep-frozen vegetables depends on the effectiveness of the blanching step. This may be tested by detecting residual peroxidase activity.

This enzyme, which is present in all vegetables, must be destroyed during blanching.

In the case of mashed or cut vegetables, solutions of gaïacol (0.5%) and hydrogen peroxide (1.5%) are added in equal quantities. After 1 minute of contact, no change in color should be observed.

30.5 Regulation

Fresh products are not subjected to any particular specification. The only established criteria concern deep-frozen vegetables. These are as follows:

	Deep-frozen vegetables	Unbalanced deep-frozen vegetables or those handled between blanching and deep-freezing
aerobic flora	5×10^5/g	1.5×10^5/g
coliforms	5×10^5/g	5×10^5/g
fecal coliforms	15/g	15/g
anaerobic sulfite-reducers	10/g	10/g
Staphylococcus aureus	10/g	10/g
Salmonella	absence in 25 g	absence in 25 g
yeasts	1,000/g	2,000/g
molds	500/g	1,500/g

It must be emphasized that the enumeration of molds is meaningless unless followed by identification. It merely demonstrates the aptness of certain species to sporulate on the substrate considered and under the conditions used.

30.6 Dehydrated Plant Products

The preservation of plant products may be obtained by dehydration. This is the case for dried fruits and spices which are consumed as such. Vegetables sold dehydrated, however, are intended for use in the rehydrated form. Preparations sold in this form are mainly the following:

- soups, in which the vegetables are associated with other foodstuffs;
- vegetable purées for infants;
- instant mashed potatoes.

Preservation of the product requires storage away from moisture.

Microbiological testing requires the preparation of an original suspension. This is done using peptone water, since the microbes present must be resuscitated before proceeding with the analysis.

The microbiological criteria applicable to dehydrated soups are as follows:

aerobic mesophilic flora	5×10^5/g
coliforms	10^3/g
fecal coliforms	10/g
Staphylococcus aureus	10^2/g
anaerobic sulfite-reducers	30/g
Salmonella	absence in 25 g

In addition, this type of product is rich in starch, which makes it a medium favorable to the proliferation of *Bacillus cereus*. Cases of intoxications associated with this bacterium in vegetable soups and purées have been reported. It is therefore recommended that the enumeration of this bacterial group be carried out for dehydrated vegetables. This is done on the solid medium of Mossel for *Bacillus cereus* (see Chapter 33, Appendix, medium 19). The medium is inoculated by spreading 0.1 ml of a dilution of the product. After incubation for 18–40 hours at 32°C, colonies of *Bacillus cereus* are enumerated. They appear as dry and rough pinkish growths (mannitol is not decomposed) surrounded by a precipitation halo associated with the presence of lecithinase.

30.7 Minimally Processed Vegetables

30.7.1 General

Minimally processed vegetables are ready-to-use fresh products, that is, peeled, cut, washed, and packaged. Their popularity in the market in the form of chopped leafy vegetables, carrots, and celery in various mixtures is rapidly increasing. Vegetables thus treated are deprived of their natural protections and are thus much more vulnerable than fresh vegetables to physicochemical spoilage (drying up, discoloration) and microbial spoilage. With respect to the latter, minimally processed vegetables are more vulnerable than fresh vegetables both to contaminations having public health implications and to flora responsible for organoleptic spoilage. The formulation of this type of product therefore requires strict hygienic measures at all levels of processing (cutting, washing, packaging) as well as a starting material of good microbiological quality.

The level of contamination of starting materials is frequently high and disinfecting treatments have therefore been envisioned for the purpose of reducing the initial flora. Mazollier (1988) thus considers treating vegetables with a chlorinated solution (50 ppm) effective in providing a significant albeit limited improvement in the microbiological quality of minimally processed vegetables (the authorized concentration is from 4–10 ppm). Similarly, Strugnell (1988) proposed a process consisting of several steps, that is, chlorination followed by neutralizing the chlorine and immersing in a sugar/potassium sorbate solution, which leads to prolonged preservation (up to 20 days at 3°–5°C).

How well minimally processed vegetables fare during distribution is directly related to the packaging of the product. Picoche and Denis (1988) concluded that packaging under air appears to be preferable both for grated carrots and chopped leafy vegetables, but that the type of film to use depends on the vegetable packaged (semipermeable for carrots, low permeability for leafy vegetables).

Maintenance of the cold chain at all distribution stages is absolutely essential for minimally processed vegetables. Significant differences are observed between salads stored at 4° and 12°C both organoleptically and microbiologically (Scandella, 1988).

30.7.2 Microbiological Characteristics

The popularity of minimally processed vegetables is still relatively recent and the microbiology of this type of product is only partially understood. A study dealing with prechopped salad and grated carrots (Denis and Picoche, 1986) reveals that the aerobic mesophilic flora is quite abundant in these products from the first day of storage. This aerobic psychrotrophic flora is quite similar to the mesophilic flora, suggesting that there may be some problem of differentiation between the two groups. This flora is made up essentially of gram-negative bacteria which is in turn made up of two groups: enterobacteria and pseudomonads (in proportions which vary depending on the vegetable under consideration). The gram-positive flora consists essentially of lactic acid bacteria (*Lactobacillus* and *Leuconostoc*) with smaller numbers of corynebacteria and gram-positive aerobic and anaerobic spore-formers. Yeasts are also present (especially on carrots) while molds are encountered in very low proportions and do not pose problems during storage.

The same authors have analyzed for bacteria of sanitary importance on vegetables. Fecal contamination indicators (fecal coliforms and streptococci) are present. Among the potential pathogens, only *Yersinia enterocolitica* is encountered (on 75% of vegetables, serotypes not specified).

This study is corroborated by Nguyen The and Prunier (1989) who isolated from minimally processed salads mainly enterobacteria (30–40%) and fluorescent *Pseudomonas* species (40–50%). The latter group is widely responsible for soft rot-type spoilages observed on salad, particularly the species *Pseudomonas marginalis* which possesses pectinolytic activity. A clear relationship exists between the frequency of this species and the spoilage of the salads.

After *Pseudomonas*, *Erwinia herbicola* is the most commonly isolated species.

These findings for the microbiology of minimally processed vegetables serve as the basis for the following microbiological enumerations which are of use in evaluating the microbiological quality of these products (according to Denis and Picoche, 1986).

Spoilage flora:

mesophilic aerobic flora	PCA 30°C 3 days
psychrotrophic flora	PCA 4°C 10 days
gram-negative flora	PCA + crystal-violet (2 mg/l) 30°C 3 days
pectinolytic flora[1]	PT medium 30°C 5 days
Erwinia[2]	D_3 medium 30°C 2 days
Pseudomonas	Centrimide agar 30°C 2 days
fluorescent pseudomonads	King's B medium 30°C 2–3 days
lactic flora	MRS 30°C 2 days
lactobacilli	Rogosa agar, pH 5.5 30°C 5 days
Leuconostoc	Mayeux agar 22°C 4 days
lactic streptococci	Elliker medium 20°C 1 day
aerobic sporulating forms	BPC tryptose dextrose agar 30°C 3 days after treatment for 10 min. at 80°C

Flora of importance to public health:

Salmonella
Yersinia enterocolitica
fecal coliforms
fecal streptococci

There are currently no criteria governing the microbiological quality of minimally processed vegetables.

[1]The pectinolytic flora may be enumerated on polypectate gel medium or selectively on PT medium (see Chapter 33, Appendix, medium 20) in which polygalacturonic acid is the sole carbon source.

[2]*Erwinia* are selectively enumerated on D_3 medium (see Chapter 33, Appendix, medium 21) (Kado and Heskett, 1970). They appear as colonies surrounded by a red halo.

CHAPTER

31

Prepared Meals

A. Plusquellec

Prepared meals are often microbiologically complex. In order to obtain a product of good bacteriological quality, manufacturing under very strict conditions organized according to a very rational work plan are required. This must be carried out by competent and informed personnel in specially designed work areas. Testing must be of the same rigor as manufacturing (i.e., control at all levels and rationally planned) in order to ensure adequate safety in terms of public health.

31.1 Definition

Prepared meals are very diverse in nature. This group includes meals to be consumed cold or after warming. They may be sold in the form of fresh refrigerated products, deep-frozen products, semipreserved or canned. The latter two categories are the subject of Chapter 32.

Prepared meals may be defined as cooked or prepared meat, poultry, giblet, game, fish, crustacean, molluscan, or egg recipes accompanied by sauce, dressing, mash, and vegetables. Delicatessen and pickled or brined products are excluded from this category. The range of products under consideration here is therefore very broad and continues to expand.

31.2 Microbiological Problems Posed by Prepared Meals

Much more so than for other food products, two important aspects must be considered.

31.2.1 Sanitary Aspect

Prepared meals have the potential to represent risks to the consumer. In fact, almost all of the known microbial agents of food intoxication may be encountered in these products. Prepared meals represent one of the categories most frequently incriminated in documented cases of food poisonings (second after meat products). Cases of intoxication involving *Salmonella*, *Staphylococcus*, and *Clostridium perfringens* have all been reported. This problem of intoxications associated with prepared meals must be taken very seriously since the manufacturing of these products has passed directly from the kitchen level to the industrial stage. They are being consumed on an increasingly larger scale and the factors representing the potential for resulting in large-scale food poisonings with very serious consequences have all converged.

31.2.2 Organoleptic Aspects

The consumer demands an attractive product presentation which must be preserved during the delay allowed by regulations governing sale. Thus, aside from public health protection, the product must be kept in an organoleptically satisfying state throughout the entire duration of its shelving. This is according to regulations for a maximum of 6 days at a temperature which must be below 3°C. Certain groups of microorganisms are nevertheless capable of multiplying during this sale period and bringing about changes in the appearance, taste, or odor of the product. These are mainly psychrotrophic bacteria (*Pseudomonas* and related bacteria), yeasts, and molds.

31.3 Conditions Required for Good Bacteriological Quality

Good microbiological quality of a finished product requires uncontaminated starting materials. During processing, microbial entrants must be blocked and the proliferation of any microbes already present must be prevented. The finished product must be stored under proper conditions.

31.3.1 Starting Materials

A wide variety of starting materials may enter into the production of a given prepared meal. Each of these is capable of inoculating the meal with undesirable microbes. Surveillance with respect to all components of the product is therefore essential. Even secondary components such as spices may sometimes be vectors of contamination. It is also important to not mix starting materials of differing quality.

31.3.2 Proliferation of Microbes Present

The processing of the product must be carried out rationally, taking special precautions to avoid all causes of microbial proliferation.

The speed of the processing is important. The product must be brought to its formulated stage as quickly as possible and without any needless lags.

Since temperature has a very key role in the multiplication of microorganisms, holding at temperatures favorable to their development will have disastrous consequences for preservation. Cooling after cooking in particular must be completed as quickly as possible. Regulations require that after cooking, a temperature of 10°C at the center of the mass must be reached within no more than 2 hours. Beginning at the end of the cooling phase, products must be stored either in a refrigerated room ensuring a temperature below 3°C, or in a freezer at a temperature no higher than −18°C.

31.3.3 Microbial Entrants

In addition to avoiding the multiplication of microorganisms already present, contamination with external microbes must be minimized during the processing steps. Contaminations at this level may have various origins.

31.3.3.1 Equipment and Work Areas

The equipment and work areas must be of a special design in order to not permit harboring and multiplication of microorganisms (smooth surfaces free of crevices). In addition, tools, machines, floors, and walls must readily lend themselves to effective cleaning and disinfection (no porous or oxidizable materials).

31.3.3.2 Personnel

Workers may be direct sources of foodstuff contamination, introducing microbes which may pose dangers (fecal organisms, staphylococci) or may be indirectly responsible by failing to observe the principles stated earlier.

31.3.3.3 The Atmosphere

The atmosphere of an active processing facility may spread large numbers of microorganisms and in particular so-called dry spores (xerospores) of molds. It is therefore necessary to:

- not let the atmosphere become polluted;
- limit air circulation as much as possible;
- protect the food from atmospheric contact as much as possible. Cooling in particular must take place in an enclosure isolated from the rest of the processing plant, in accordance with hygienic standards.

31.3.4 Storage of the Finished Product

The formulated product is generally fragile, and its proper storage requires the maintenance of a constant low temperature and packaging protecting it from external contamination.

31.4 Microbiological Testing

In order to obtain prepared meals of suitable bacteriological quality, the conditions described earlier must be satisfied, but it is also necessary to verify that the precautions taken have been effective. This requires close and constant microbiological testing to be carried out at several stages.

31.4.1 Testing of Finished Products

The regulatory criteria are applicable at the selling stage. In order to guarantee an adequate safety margin, the product at the end of the manufacturing process must show contamination levels vastly lower than those imposed by regulations.

31.4.2 At All Stages of Manufacturing

In order to acquire true control of manufacturing hygiene, all elements which enter as contaminating factors must be tested. This work requires implementing a microbiological monitoring plan aimed at starting materials, equipment, work areas, personnel, and the atmosphere. These testing methods are presented in Chapter 35.

31.4.3 Microbiological Criteria Evaluated

Due to the diversity of the possible sources of contamination, the microbiological examinations employed must be among the most complete. These generally include:

- Evaluation of the common contaminating flora by counts of aerobic bacteria grown at 30°C. This criterion gives an indication of the general level of hygiene and also allows an evaluation of the capacity for preservation of the product without risk of organoleptic breakdown.
- Detection and enumeration of yeasts and molds on the same hygienic level. The effects of these microorganisms is felt particularly in certain types of products (sauces with low pH, products containing cheese).
- Evaluation of fecal contamination indicating flora (coliforms, fecal coliforms, fecal streptococci). Fecal microorganisms may come from starting materials or may be associated with manufacturing hygiene.

- Microbes responsible for intoxications (*Salmonella*, staphylococci, *Clostridium perfringens*). Staphylococci are of particular importance in human-handled products. They are frequently encountered and often attributable to the human factor associated with food processing.

The usual analytical mode is used. Nevertheless, it is important that the sample analyzed be representative of the product as a whole, that is, that it include all of the different components in proportions close to that which they each represent in the prepared meal.

31.5 Regulatory Criteria

The microbiological standards which must be satisfied by prepared meals, whether refrigerated or deep-frozen, have been established by the decree* of June 26, 1974 and updated by the decree* of December 21, 1979.

aerobic mesophilic flora[1]	3×10^5/g
coliforms	10^3/g
fecal coliforms	10/g
Staphylococcus aureus	10^3/g
anaerobic sulfite-reducers	30/g
Salmonella	absence in 25 g

**Translator's note:* The English word *decree* is used in this chapter to translate the French words *arrêté* and *décret* which corresponds to different administrative levels. An asterisk indicates that the French word used was *arrêté*, otherwise the word was *décret*.

[1]This criterion must sometimes be interpreted, particularly in the case of products containing cheese.

CHAPTER

32

Preserves and Semipreserves

A. Plusquellec

The bacteriological testing of preserves or canned goods represents a very particular case requiring an examination procedure which differs from that of conventional analyses carried out in food microbiology. The role of the laboratory is to verify that the canned products adhere well to their definition.

32.1 Definition

The February 10, 1955 decree defines a *preserve* as "a perishable foodstuff of animal or vegetable origin of which the preservation is ensured by the combined use of the following techniques:

- packaging in a water-tight and air-tight container which also acts as a physical barrier against microorganisms at any temperature below 55°C;
- processing by heating or by any other authorized treatment, the purpose of which is to destroy or totally inhibit enzymes and microorganisms as well as their toxins whose presence or proliferation would otherwise spoil the foodstuff or render it unfit for human consumption."

This definition is quite general and not all preserves are to be considered on the same level. First, from a technological and testing point of view, the following are differentiated:

- nonacidic preserves, with a pH higher than 4.5;
- acidic preserves, with a pH below 4.5. These do not allow the development of sporulating microorganisms. A limited thermal treatment is usually applied (up to about 100°C), sufficient to destroy the acidophilic flora (yeasts, molds, acidophilic bacteria).

Furthermore, the same guarantee of sterility is not required for all preserves. Although some must be sterile (meat, fish), so-called commercial sterility, often called *stability*, is sufficient for others, that is, they need only be free of microbes capable of proliferating in the product.

32.2 Role of the Testing Laboratory

This is applicable at two levels:

- Routine control intended to verify that the pressure does not show any modification after having undergone an incubation. This is stability testing. It is quite simple and may be done in the processing plant.
- Complete testing, which becomes relevant when a problem has been encountered or for the purpose of developing a reference standard. This is sterility testing, a sensitive examination which must be performed in a specialized laboratory.

32.3 Stability Testing

Stability testing is a simplified test to analyze the largest number of samples. It has been the object of standardization by the AFNOR:

- Standard NF-V-08-401 deals with the stability of preserves having a pH higher than 4.5.
- Standard NF-V-08-402 applies to the stability of preserves having a pH lower than 4.5.

The decree* of December 21, 1979, addressing the microbiological criteria applicable to foodstuffs of animal origin, describes in Article 8 the essential elements of these AFNOR protocols.

The test consists of subjecting a sample of the preserve to an incubation and verifying whether or not this has brought about any notable transformations compared to an unincubated control sample. This involves detecting

- variations in the appearance of the packaging;
- variations in pH;
- modifications of the microbial flora (as revealed by microscopic examination);
- possible variations in internal pressure.

**Translator's note:* The English work *decree* is used in this chapter to translate the French words *arrêté* and *décret* which correspond to different administrative levels. An asterisk indicates that the French word used was *arrêté*, otherwise the word was *décret*.

32.3.1 Incubation

The samples are chosen for their normal appearance, that is, showing no swelling, leakage, nor dents. The containers are carefully cleansed, labeled, and placed on Kraft paper in order to detect leaks. For acidic preserves, a single incubation at 30° or 32°C for 21 days is sufficient. Microorganisms which may develop under these conditions will not be thermophilic.

Nonacidic preserves are subjected to a double incubation: at 30°–32°C for 21 days[1] or at 55°C for 7 days. A control package is kept at ambient temperature.

32.3.2 Examinations After Incubation

When the incubation period has elapsed, the cans are stabilized at ambient temperature. The following are then examined:

- The appearance of the package, that is, the possible presence of swelling, leaks, or deformation is noted.
- The pH. The determination of the pH of preserves has been standardized (AFNOR standard NF-V-08-406). The measurement of pH is intended for liquid or homogenous products and for heterogenous products. The pH of the following is measured:
  - the liquid or the aqueous phase for liquid products or products considered homogenous (beans, peas, etc.);
  - grindings in the case of mushy or heterogenous products (such as corned beef);
  - each component or the mixture thereof for products with coarse components (sauerkraut, stew);
  - the aqueous phase or solid product/aqueous phase mixture for products in an oil-in-water emulsion.
- Microscopic appearance. A slide is made from grindings.

After Gram staining, the preparation is examined for the average number "n" of microbial elements per field. The nature of the elements encountered is also noted. The same assay is done for the unincubated control sample, the average number "n_0" of microbial elements per field being noted in this case.

32.3.3 Interpretation of the Stability Test

The analyzed pressure is considered stable if it satisfies all of the following criteria:

- absence of modification of the appearance of the package and of the product after incubation;

[1]The December 21, 1979 decree mentions incubation at 37°C for 7 days or at 35°C for 10 days, but it applies only to preserves of animal origin.

- variation in pH with respect to the unincubated control less than or equal to 0.5 pH unit;
- absence of variation of the microbial flora, either qualitatively or quantitatively. The n/n_0 ratio must remain below 100.[2]

32.4 Sterility Testing

32.4.1 Objectives of Testing

This is a complete examination which is delicate and must be done in a properly equipped and specialized laboratory. It is done mainly when the stability test gives positive or dubious results or in the case of a preserve manifesting spontaneous fluctations in storage characteristics.

The objectives of this analysis are first to establish that the product defect observed is in fact of microbiological origin, as is most frequently the case, and then to determine the origin of the stability defect. The defect may be due to either of the following:

- insufficient sterilization. The contaminants present will be high temperature resistant microbes in this case (sporulating);
- or a poststerilization contamination through leaks in the package. In this case, a wide range of microorganisms may be involved, coming from the environment or from cooling water.

The examination consists of aseptically removing a portion of the product and inoculating media to be cultured under a variety of conditions. These cultures are then examined microscopically. The following may thus be determined:

- the microbiology of the microbes present (shape, Gram staining trait, pure or mixed culture);
- thermoresistance, half of the media are subjected to heating to destroy the vegetative forms, i.e., for 10 minutes at 100°C for preserves having a pH > 4.5 and for 10 minutes at 80°C for preserves having a pH < 4.5;
- possible production of gas;
- respiratory type, a series of media is incubated anaerobically, another aerobically;
- their optimal growth temperature, a double incubation being done, i.e., at 30°C (culturing of mesophiles) and at 55°C (culturing of thermophiles).

A complete microbiological examination will employ two media for culturing aerobes and two for anaerobes. Each medium is inoculated in quadruplicate. Two from among these are heated in order to select for spores, one of these to be incubated at 30°C and the other at 55°C. The two unheated media are incubated similarly.

---

[2]This value is not meant to imply that small variations are tolerated, but that variations giving a ratio below 100 are normal due to the lack of precision of the technique.

32.4.2 Sampling

Sampling must be carried out under very securely aseptic conditions. The introduction of microorganisms during sampling may lead to conclusions based on false-positive results and have serious consequences. Correct sampling is therefore of paramount importance and for this reason has been the object of standardization. AFNOR standard NF-V-08-403 gives the directives to be followed for aseptic removal of samples (Figure 32.1).

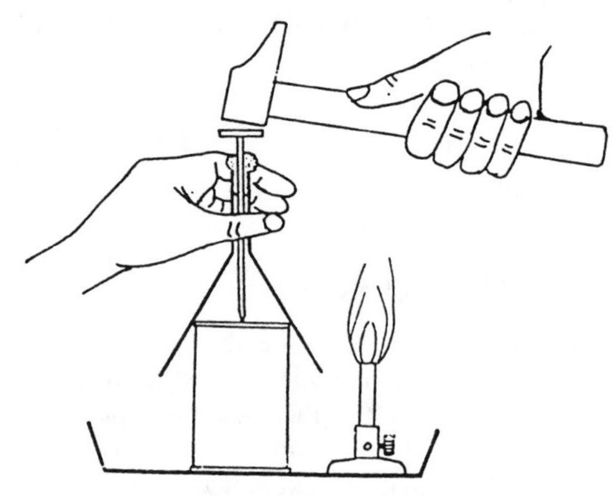

Figure 32.1.

All labels must be removed from the package which must then be carefully cleaned and disinfected with alcohols. Opening must be performed in a sterile enclosure such as a laminar flow hood. The package (e.g., can) is pierced using a sterile punch introduced through the stem of a sterile funnel covering the top of the package. This device protects the worker against possible spurting of the contents and helps maintain asepsis.

The product is then sampled with the aid of sterile instruments (tweezers, pipette, scalpel) and ground, mashed, or blended if necessary, in order to yield the original solution required for inoculations.

32.4.3 Inoculation

The culture media chosen must be nutritionally very rich in order to allow optimal growth of the largest number of microorganisms possible. The choice of media must be adapted to the product. Media may also be prepared from the product being analyzed (fruits, fish, etc.). The pH of the media must be adjusted to the normal pH of the product. The volume of the inoculum must be about one-tenth of that of the medium. Among the media used for the culture of aerobes are nutrient broth and agar, tryptose dextrose broth, trypticase soy agar, eugonobroth, and so on. The conventionally used media for the culture of anaerobes are:

- VF agar and broth;
- VL agar and broth;
- Rosenow cysteine medium;
- Brewer broth and agar.

All media for anaerobes must be de-aerated by prolonged passage in a boiling water bath. The solid media are inoculated by the pour plate method. Liquid media must be overlaid with a paraffin plug.

For acidic preserves ($pH < 4.5$), the media best adapted to the flora which may be present are used:

- MRS medium;
- yeast malt medium;
- tomato juice of tomato juice agar;
- Tryptose dextrose broth or agar.

32.4.4 Interpretation of Results

The first point which must be noted is the absence or presence of culture. When a positive stability test or a spontaneous storage fluctuation is not accompanied by positive culture results, other possible explanations must be sought (medium poorly adapted, pH poorly adjusted, self-destruction of the microbes in the product before inoculation). Nonmicrobiological causes must also be investigated (e.g., chemical swelling).

Once microbial growth is established, examination of the cultural and morphological characteristics of the microbes which have developed must provide an explanation for the stability defect.

The presence of thermoresistant species appearing in the form of gram-positive bacilli, sometimes sporulating, strictly anaerobic (*Clostridium*), or aerobic, sometimes facultative (*Bacillus*), is a sign of insufficient sterilization.

The presence of thermosensitive flora is associated with poststerilization contamination likely due to a container sealing defect. This flora is generally mesophilic, aerobic, or facultative anaerobic and composed most often of gram-positive cocci or gram-negative bacilli.

Acidic preserves undergo much more limited heating which may not destroy nonsporulating germs (lactobacilli, yeasts).

It is possible to search more specifically for thermophilic species of *Bacillus* (spores and vegetative forms) according to the AFNOR NF-V-08-404 protocol and to specifically detect thermophilic *Clostridium* species (*Clostridium thermosaccharolyticum, Clostridium nigrificans*) according to AFNOR standard NF-V-08-405.

32.4.5 Conclusion

The detection of contaminating microorganisms rules out other causes of spoilage (physical or chemical). Analysis of the cultures obtained determines the nature of the defect (e.g., poor sterilization or recontamination). Complementary examinations must be undertaken in order to root out the origin of the defect:

- in the case of insufficient sterilization, testing of tabled reference values, testing the proper functioning of the autoclaves, testing of the initial contamination of the product;
- in the case of recontamination, testing of cans and seams.

Careful attention must also be given to the possibility of false-positive results due to accidental contamination during the analysis. The incidence of this type of false-positive must be known and laboratories which perform the microbiological examination of preserves must test the asepsis of the operations by means of control preserves. These consist of a control nutrient product (meat peptone), packed into a container identical to the type used for the preserves usually analyzed, and sterilized at 121°C for 30 minutes. These controls are slipped in among the usual analyses. These tests must result in the establishment of a known characteristic proportion of false-positives to be associated with laboratory procedures and which will influence the decision to reject or accept the preserves subjected to the sterility test.

32.5 Semipreserves

With reference to preserves, semipreserves are defined as "foodstuffs of animal or vegetable origin, packaged in water-tight containers, having undergone an authorized treatment in order to ensure a more limited preservation" (decree of February 10, 1955).

	Aerobic microorganisms 30°C (per g)	Coliforms (per g)	*Staphylococcus aureus* (per g)	Anaerobic sulfite-reducers 46°C (per g)	*Salmonella* in 25 g
Pasteurized semipreserves*	10^4	Absence	Absence	Absence	Absence
Unpasteurized semipreserves*					
rollmop or smoked herring, anchovies in salt or oil	10^5	Absence	Absence	Absence †	Absence
smoked salmon, haddock, and other lightly salted and smoked fish	10^6‡	Absence	Absence	Absence	Absence

*Resuscitation of the original suspension for 2 hours at laboratory temperature for semipreserves and for 30–45 minutes for unpasteurized semipreserves.

†Special case for anchovies in brine: anaerobic sulfite-reducers 46°C: < 10/g.

‡Enumeration in seawater medium or else 3.5% saline and an incubation temperature of 20°C for 5 days.

For practical purposes, semipreserves must be stocked under refrigeration and consumed before an expiry or "best before" date.

Testing consists of verifying the absence of undesirable microorganisms in the product.

Sample removal must be carried out under aseptic conditions comparable to those implemented for preserves.

The analysis requires the preparation of an original suspension which must undergo a resuscitation step before proceeding with incubation and detection.

These products must be free of pathogens. The decree* of December 21, 1979 establishes more precise microbiological criteria applicable to semi-preserves of animal origin.

CHAPTER

33

Testing of Intermediate Moisture Foods

Laurence Lesage, D. Richard-Molard

33.1 Intermediate Moisture Foods

33.1.1 The Intermediate Moisture Concept

Given the evolution in consumption habits in the food business, the food industries have developed during the past several years a new class of products based on an American model, ready-to-eat and preservable for long periods of time without the aid of low temperature as for frozen products or for sterilization by thermal processing as for canned goods. These so-called intermediate moisture products have been defined from a technological point of view by the Centre National de Coordination des Etudes de Recherches sur la Nutrition et l'Alimentation (CNERNA) as follows (Multon, 1981):

> An intermediate moisture food is a soft-textured food product having undergone one or more processing technologies, usually consumable without culinary preparation and of which the preservation is ensured for several months without heat-sterilization, freezing or deep-freezing, but by suitable adjustment of its formulation, i.e., composition, pH, additives and especially water activity[1] which must be between 0.6 and 0.84.

The seemingly contradictory association of the soft texture and low a_w properties is the basis of the originality of this class of products, even though industrially it is also a source of limitations, according to Torres (1987). This

[1]The water activity of a product may be expressed in terms of the relative humidity in equilibrium with the product, according to the relation $a_w = {}^{\infty}R.H._{(eq)}/100$.

Table 33.1 A few examples of humectants and their roles in Intermediate Moisture Foods

	Decrease in a_w	Softening of texture	Antimicrobial action
Hexoses: glucose, fructose	++	++	−
Disaccharides: sucrose, lactose, maltose	++	++	−
Dextrins: pectin, cellulose	++	++	−
Polyols:			
ethanol	++		++
sorbitol	++	++	
glycerol	++	++	+
Mineral salts: NaCl, KCl, $CaCl_2$	++	+	−
Proteins and derivatives:			
amino acids and salts	++	++	−
protein hydrolysates	++	++	−

++ Principal role; + secondary role.

Source: Adapted from Taoukis et al., 1988.

problem is partially resolved by adding humectants (Taoukis et al., 1988) which both lower the a_w and impart a softer texture to products. A few examples are given in Table 33.1.

Numerous traditional food products, known since antiquity, are already classed as intermediate moisture foods; these include dried fruits, jams, candies, salted, dried and/or smoked fish and meats. In all foods, a decrease in a_w is obtained by adding traditional and ancestral (and therefore well accepted) additives such as salts, sugars, and the like.

The current trend resides in more elaborate formulations which include preservatives, a_w suppressors of additives which impart new nutritional qualities (amino acids, vitamins, etc.) and yield products that are better adapted to present needs (Guilbert, 1984). The development of these new technologies has been mainly in the sectors of industrial pastries, prepared meals, and pet foods. This has resulted in the design of foods adapted to difficult environmental conditions (developing countries, exploration expeditions, camping, boating) characterized by the absence of means for heating and the requirement for lightweight and low-cost packaging. It should be mentioned that at the regulatory level, these intermediate moisture products still have no legal status and no regulations deal specifically with them at the present time. The U.S. Food and Drug Administration has, however, defined in 1986 conditions indicating a "safe moisture level" for intermediate moisture foods or IMFs. The data must be sufficient to demonstrate that the a_w of an IMF can impede the development of undesirable microorganisms during its manufacture, storage, and distribution (Taoukis et al., 1988).

33.1.2 The Microbiological Status of Intermediate Moisture Foods

The IMF concept is therefore of interest to many sectors of the food industries. Nevertheless, in spite of this diversity which applies as much to starting materials used as to manufacturing processes, there is a definite microbiological common denominator due to the water activity values between 0.60 and 0.84 selecting the microorganisms which are the sources of spoilage. The vast majority of microorganisms require water activities higher than 0.90 for their development. This is especially the case for the majority of pathogenic or toxigenic bacteria, which will therefore, in principle, not represent any problem for the preservation of IMFs, except for *Staphylococcus aureus* (Scott, 1957; Corry, 1973; Leitsner and Rodel, 1978; Beuchat, 1981).

Selected by high osmotic pressure, the filamentous fungi (molds), yeasts with xerophilic tendencies, and in some particular cases halophilic bacteria will all find enough water in IMFs to develop and produce spoilage (Pitt and Hocking, 1985). Microbiological testing of IMFs must consequently be aimed at these groups of microorganisms. These tests, the only ones of interest in this context, will be performed especially during the development stages of an IMF, by the manufacturer. When this development phase is completed, the intermediate moisture food will be by definition microbiologically stable.

Storage mishaps such as re-wetting cannot be excluded, however. Temperature fluctuations to which products are exposed during manufacturing, storage, distribution, and use may bring about migration of water and hence surface condensation. In such cases, the IMF is destabilized and conventional food microbiological methods will apply, as will the standards and testing procedures for the most applicable category of food product based either on product composition or on type of packaging.

33.2 Microbiological Testing of IMFs

33.2.1 Microbiological Quality Criteria

In the field of food products, the microbiologist is preoccupied above all with the inocuity of foodstuffs and is interested mainly in pathogenic or toxigenic microorganisms. With respect to IMFs, the quality criteria may be considered to be more strict (i.e., consumer demands) since in addition to the required guarantees of product inocuity, no perceptible manifestation of any microbial entity will be tolerated no matter how "insignificant" it may be. Although today's customer may still accept that a few days after purchase a fruit may have a bit of mold, a cheese may not age well, or that milk may start to turn, the appearance of the slightest bit of mold on a product which is *supposed* to keep for a long time will turn the customer against the manufacturer. It is therefore unrealistic to refer to "insignificant microorganisms" in the field of IMFs. All molds, whether or not they are capable of producing mycotoxins, must be considered to be of decisive importance, at least on the commercial level.

33.2.2 Starting Materials, Formulations

At the starting material level, the microbiological quality criteria do not differ in any way from those generally applied in the food industry. Existing standards and their corresponding enumeration techniques will be applied to the evaluation of total aerobic mesophilic flora and to the detection of indicator organisms, pathogens, yeasts, and molds as a function of the components involved.

At the IMF formulation level, it should be emphasized that in most cases, and especially when the upper limit of the water activity range is approached ($a_w = 0.85$), the use of an antifungal additive will almost always be necessary. This will be chosen from among the products authorized by the legislation in effect (Dehove, 1984) and the effectiveness of incorporating different doses of the chosen agent, notably the homogeneity of distribution, will be determined.

33.2.3 Tests of the Stability of an IMF

33.2.3.1 Principle

Testing stability of an IMF is a matter of assessing as rapidly as possible the aptness of a food for prolonged preservation. The product will therefore be placed in conditions which accelerate microbiological processes and therefore show, within a reasonable delay, spoilage which would require much more time to appear under normal conditions.

Considering the tests as they are currently practiced, the experimenter may choose between two possibilities, either storage at an experimentally elevated temperature such as 37°–38°C (Johnson et al., 1972), in which case acceleration is due to the thermal effect, or subjecting the product to temperature shifts such as 12 hours at 12°C followed by 12 hours at 24°C, which are likely to produce water migrations and hence microbiological destabilization of the product. The latter procedure seems to approach actual conditions to which IMFs are likely to be subjected. Trials at elevated temperature also have certain inconveniences which will be examined later (Section 33.2.3.4).

In order to increase the significance of the test and to free it from the uncertainty due to the initial microflora (which may be very low or may not consist of characteristic microbes), the tested product may be artificially contaminated by inoculating it with various test microorganisms during manufacturing, whose trends will be monitored throughout the test (Plitman et al., 1973; Lee and Kraft, 1977).

Strains isolated from media with high osmotic pressure will be chosen, for example, microorganisms which are known to have caused fluctuations in

Table 33.2 Microorganisms likely to develop in intermediate moisture foods

a_w	Microorganisms generally inhibited at a_w below the indicated range	Examples of traditional foods in the indicated range
0.87–0.91	Most yeasts (*Candida, Torulopsis, Hansenula*)	Salami, dry cheeses, margarine, foods at 65% (w/w) sucrose or 15% (w/v) NaCl
0.80–0.87	Most xerotolerant molds (*Penicillium, Aspergillus*)	Fruit juice concentrates, chocolate syrup, flour
	Bacteria: *Staphylococcus aureus* Most osmotolerant yeasts (*Saccharomyces rouxii, Debaryomyces*)	Rice, fruit syrups, fruitcakes
0.75–0.80	Halophilic bacteria: a few toxigenic xerotolerant molds (*Aspergillus flavus*)	Jams, marshmallows, salted fish
0.65–0.75	Xerophilic molds (*Eurotium chevalieri, Wallemia sebi*)	Molasses, cane sugar, dried fruits, nuts
0.60–0.65	Osmophilic yeasts (*Saccharomyces rouxii*) and xerophilic molds (*Monascus bisporus*)	Dried fruits, honey, caramels

preservability in the IMF field. All precautions will be taken to avoid exposing these microorganisms to osmotic shocks during handling, transfers, culturing, suspension, and so on by using adapted culture media and diluents supplemented with sugar or salt. The enumeration media have recently been redefined by Hocking and Pitt (1987). In general, an antibiotic must be added for the selective detection of yeasts and molds while an antifungal agent is used when bacteria are to be detected. The flora likely to be encountered in IMFs is given in Table 33.2 from Beuchat (1981).

33.2.3.2 Bacteriological Analysis

33.2.3.2.1 Detection of Halophilic Bacteria

Halophilic bacteria constitute one source of spoilage of IMFs stabilized by high salt concentrations such as salted meats and fish. The genera *Halobacterium* and *Halococcus* (so-called red halophiles due to the presence of bacteriorhodopsin giving them a red color) develop exclusively in media containing 20–30% NaCl having a_w values vetween 0.87 and 0.75 (Hocking, 1988).

Several media have been recommended by Hocking and Pitt (1987), their common characteristic being the presence of salt in the culture and the diluent. The example of Harrigan and McCance (1976), which recommends adding 15% (w/v) NaCl, is typically cited.

33.2.3.2.2 Detection of *Staphylococcus aureus*

Among the toxigenic bacterial genera conventionally detected in food microbiology (AFNOR standard NF-V-08-014), *Staphylococcus aureus* is the most tolerant of low a_w given that it may still be detected in salty media in which the a_w is close to 0.85 at 30°C, aerobically with $pH > 5$. Consequently, in the field of IMFs, *S. aureus* is a particularly useful test microorganism and its nondevelopment in a product constitutes a reasonable guarantee of good bacteriological quality (Pawsey and Davies, 1976). Specific salt-based media (potassium tellurite) are currently widely used for the isolation and enumeration of *S. aureus* (media of Baird-Parker, 1962 and Tatini et al., 1984).

33.2.3.2.3 Detection of Lactic Acid Bacteria

Although in most cases, the limiting a_w thresholds for the growth of lactic acid bacteria is situated around 0.93, some strains (particularly *Lactobacillus*) appear capable of tolerating water activities as low as 0.90 (Leitsner and Rodel, 1976) or even lower and of approaching the upper limit corresponding to IMFs (Richard-Molard, 1990). In some cases (fermented delicatessen products), analyses for these bacteria using Rogosa's selective medium (with actidione as antifungal agent, using the double layer technique and incubating at 30°C for 2–5 days) may therefore be justified.

33.2.3.3 Mycological Analysis

33.2.3.3.1 Isolation Methods—Enumeration

In general, the "suspension-dilution" method applied to the enumeration of molds and described in AFNOR standard NF-V-18-301 is used. The assay sample must be adapted to the type of product, especially if the product is of a heterogenous nature (10–50 g weighed to the nearest 0.1 g). The diluent used will have to contain a small amount of Tween 80 (0.03 g/l) due to the hydrophobicity of many mold spores. Depending on the type of product analyzed, it may be necessary to either settle for homogenization by vigorous mixing of the original suspension or to use a Waring blender or similar device to make a purée of the sample with the diluent. Mixing times of 1–2 minutes at 25,000 rpm are used. After homogenization, the original suspension is left to "rest" for 20–25 minutes to allow adequate resuscitation of the microorganisms.

The choice of culture media is obviously guided by the variable degree of the xerophilic trait characterizing the molds and yeasts likely to develop in IMFs. A medium having a high osmotic pressure will be employed in parallel with a conventional malt extract solid medium of which the water activity is close to 1. For example, if glycerol is used to lower the water activity of the culture medium, an a_w of 0.95 will be obtained by adding 250 g of glycerol to 1 l of conventional medium containing 20 g/l of malt extract.

For the enumeration of osmophilic yeasts, known to be contaminants especially of highly sweetened food products (syrups, fruit pastes, etc.) (Tilbury, 1976), a medium with a high osmotic pressure obtained using sugar is preferred (malt extract 20 g, NaCl 70 g, sucrose 50 g, agar 14 g, water 1 l, pH 7, sterilized by autoclaving for 30 minutes at 104°C). The use of an antibiotic is almost always necessary when analyzing for yeasts, of which the slower growth is hindered by the development of other microorganisms such as sporulating bacteria.

Numerous other media allow slightly more specific isolation of IMF mycoflora. Yeasts such as *Saccharomyces bailii*, able to develop in products containing preservatives, are isolated on malt extract medium with 0.5% acetic acid (Pitt and Hocking, 1985).

33.2.3.3.2 Reading of Results—Identification

Two Petri plates should be used for each medium and for each dilution of the original suspension. After incubation (5–7 days at 25°C, 15 days for slow-growing osmophiles) the pair of plates containing no more than 50 colonies are used for the count. After calculating the average of the appropriate pair of plates, the results are expressed as the number of microbes per gram of product for each species thus found.

Species identification may represent a considerable obstacle for a non-specialized laboratory. Given the relatively low number of species likely to be found in IMFs, however, turning to the services of a mycological specialist should not represent an excessive expense.

To the extent that the question essentially concerns product stability over a period of time, identification may be temporarily put aside in favor of carefully noting the different genera or, in some cases, species counted (cultural and morphological traits in the case of yeasts). It should be noted that the identification of yeasts may require more time than for molds (biochemical tests are necessary) and is more delicate, and that the specialized literature is more difficult to grasp for nonspecialists (Barnett et al., 1983; Kreger-Van Rij, 1984). The identification of species present is necessary, however, since following the trend of each population is the only way to distinguish a mold which is developing, perhaps slowly, from others which are present but not developing.

33.2.3.3.3 Interpretation of Results

The suspension-dilution method applied to solid or semisolid substrates provides results of which the repeatability suffers very much from the heterogeneity of the distribution of the microorganisms. For this reason, the interpretation of the figures obtained must be advanced cautiously. In practice, only numbers of microbes per gram differing by tenfold will be considered significantly different. For example, if two consecutive tests give results of 100 and 500 microbes/g for a given species, it is wise to continue the analysis in order to confirm the observed trend in a third test. In some cases, if judged necessary, better precision may be obtained by increasing the number of analyses (several assays carried out separately for a given sample) which reduces, in exchange for reduced testing lattitude, the uncertainty due to sampling and also of the errors inherent in the method itself.

By definition, the results for a sterility test will be easily interpreted. If microbial growth is detected, regardless of the type, either the formulation of the product is to be revised or the conditions under which the test was performed were too severe.

Given current knowledge it is practically impossible to correctly predict the actual product life span of an IMF under normal conditions based on its apparent stability during an accelerated test. The most that can be said, which is often sufficient, is that under real conditions, its life span will be somewhat longer, at least from a microbiological point of view. Only long-term experimentation under actual storage conditions may provide unrefutable values.

33.2.3.4 Remarks concerning hydrothermal conditions of tests

Given the great diversity of situations which may be encountered—types of foods, initial contamination of starting materials, mode of packaging, and so on—it is rather difficult to establish rules governing the conditions under which stability tests should be carried out. In most cases, as a simple matter of convenience, storage at elevated temperature will be used as the spoilage process accelerating factor. Two comments, which may aid in the choice of a temperature, are relevant.

- Within the a_w range considered, for many products, notably cereals, an increased temperature brings about an increase in water activity which may be estimated at 0.002 per degree. This implies that an IMF prepared at a water activity of 0.85 and tested at 38°C, for example, will have its a_w rise to nearly 0.88 and may therefore become unstable with respect to various microorganisms, particularly *S. aureus*. This may lead to the appearance of spoilages which would not show up after any length of time in the same IMF under actual storage conditions.
- Although some molds are very tolerant of high temperatures (*Aspergillus candidus*, *A. fumigatus*), the majority have a growth optimum around 25–30°C and are somewhat hindered, indeed inhibited, as the temperature reaches 40°C, especially the penicilliums. Consequently, from a microbiological point of view, the accelerating effect of a higher storage temperature may end up being strongly counterbalanced due to the temperature requirements of molds.

For these reasons, it does not appear to be desirable to exceed a temperature of 30°C, even if the duration of the test must be lengthened as a result.

With respect to the frequency of the analyses to be carried out, the hydration factor serves as a guide. Products with higher water activity require more frequent testing. As a general indication, one test per week maximum seems suitable for products with a water activity of 0.85. One test per month is probably a reasonable minimum. Comparing the microflora only at the end of the storage trial to the initial microflora has certain risks, insofar as some species may undergo only a momentary development and then disappear, leaving a product spoiled perhaps organoleptically, for example.

33.2.4 Limitations of the Fungal Enumeration Method

Although the estimation of fungal populations by the suspension-dilution method is by far the most widely used, it is not without its critics, and situations are encountered, especially in the IMF field, where it does not give satisfactory results. Two examples demonstrate these inadequacies and the reasons for which alternative techniques of analysis are being sought.

33.2.4.1 Assays of Antifungal Additives

As mentioned earlier, IMFs by definition are not stabilized with respect to xerotolerant molds and yeasts with osmophilic tendencies. At best, it may be hoped that for low a_w values ($a_w < 0.75$), microbiological spoilage processes will be so slow that they need not be taken into consideration. More frequently, however, it will be necessary to resort to the use of antifungal agents, such as sorbic acid or sorbate salts (K^+, Na^+, Ca^{2+}). If, during a stability test, mycological analysis reveals a progressive increase in the number of fungal cultures (spores or mycelial fragments), it may be safely concluded that insufficient antifungal agent has been added.

It should be noted that very often an insufficient dosage will mean a protective effect of insufficient length of time, the increase in the fungal population appearing after a few weeks whereas stability lasting several months is expected. The reason for this is that the authorized substances are often fungistatic and not fungicidal. On the other hand, if for some reason (poor distribution of the antifungal agent, cooking losses, water migrations, condensations, etc.), the fungistic effect is locally decreased or eliminated, mold colonies will develop, and if the assay sample removed does not include this area, this localized development will, most likely, not be taken into consideration. For this reason, whenever possible, the size of the assayed portion should be maximized (e.g., up to 100 g) as is done for grains in which the distribution of microbes is most often very heterogenous. In many cases (i.e., for IMFs which are often packaged in small volumes), the analysis can be performed on the entire portion of the food being examined by adjusting the volumes of the diluent and of the glassware used to make the original suspension.

33.2.4.2 The Role of Packaging—Availability of Oxygen

Most IMFs are marketed in very light and low-cost packaging whose purpose is to protect the product from the surrounding air (water vapor, oxygen) and to display the manufacturer's label. This product protection may result in oxygen limitation in the interstitial atmosphere of the IMF, which will be of some consequence to microbial behavior. In anaerobiosis, the multiplication of yeasts is considerably slowed while their fermentative activity becomes preponderant, which may lead to spoilage (acidification, gas production, etc.), even though microbial enumeration do not indicate that this happens. For molds, known to be strict aerobes, some works have shown that anoxia limits, indeed stops sporulation, but that under some conditions mycelial growth and certain metabolic processes may occur (Tabak and Cooke, 1968).

In the IMF field, experimental data are still too rare and fragmentary to be able to make a definitive judgment of the relative importance of these spoilage processes which, moreover, although of microbial origin, are easily dismissable as being due to purely biochemical processes.

Recent studies carried out in the field of storage of moist grains have, however, established clearly that the majority of frequent fungal species in low a_w media are remarkably tolerant of rarified oxygen in the surrounding atmosphere. Packaging permitting maintenance of an oxygen concentration somewhat below 1% would be required in order to significantly inhibit most yeasts, *Penicillium* and *Aspergillus* species (Diawara et al., 1987; Richard-Molard et al., 1987). This is technically possible today thanks to progress which has been made in composite film technology. The cost of these materials, however, remains a major obstacle.

33.3 Biochemical Tests

Food product degradation processes which, in spite of being of a microbiological nature, are not traceable by the usual microbiological techniques for reasons just mentioned, have been studied using various biochemical methods. At least a few of these deserve mentioning here because of the applications which they may soon have in IMF field, as much for stability tests as for routine quality testing.

33.3.1 Analysis of Specific Volatile Compounds by Gas Chromatography

All metabolic activity of microbial origin results in the formation of volatile compounds foreign to the composition of the product being examined and likely to modify its flavor. These compounds may be detected with good sensitivity by high resolution gas chromatography using glass or silica capillary columns coupled to a flame ionization detector.

It has been shown that molds in particular (*Aspergillus* spp., *Penicillium* spp.) develop on grain-based media, kernels, and flours, producing very characteristic volatile compounds responsible for "moldy" or "musty" odors.

Several of these compounds have been identified, notably, 1-octene-3-ol, cis-2-octene-1-ol, 1-octanol, and 3-octanone (Stawicki et al., 1973; Richard-Molard et al., 1976) or 2-methylfuran, 2-methyl-1-propanol, and 3-methyl-1-butanol (Borjesson et al., 1989). Their detection or their evolution in an analyzed product therefore enables the characterization of a fungal spoilage in progress. Similarly, the evolution of higher alcohols, short-chain organic acids, and their corresponding esters may be monitored in order to detect possible fermentations indicating yeast activity.

Until recently, the extraction and concentration of these specific molecules were conceivable only by laborious methods in laboratories specializing in flavor chemistry. Today, systems automated to varying degrees are offered by various manufacturers for the rapid analysis of the vapors surrounding the product after a brief concentration step on adsorbants (Chromosorb, Tenax, Carbopack, etc.).

In spite of the relatively high cost of the equipment used, the development of these detection techniques may accelerate in the years to come, because of their speed (a detailed chromatographic profile may be obtained in less than 1 hour) and also because they are able, at the same time, to analyze a large number of volatile substances and therefore can test a product not only in microbiological terms but also in terms of the entire range of spoilage processes likely to arise.

33.3.2 Analysis for Fungal Ergosterol

As in the previous example, this method for fungal ergosterol analysis was developed and tested for use primarily in the field of cereal products, where molds and yeasts are particularly serious. Ergosterol is considered to be the major component of the sterol fraction of most fungi (Weete, 1974). Its detection in a food product normally free of this substance may therefore be exploited to provide evidence of fungal attack. The method described by Cahagnier and co-workers (1983)—which includes methanol extraction, saponification, and purification steps followed by an assay by high performance liquid chromatography with UV detection at 282 nm—is likely to be applied unmodified to a large number of IMFs. This method is obviously nonspecific but obtains precise information in less than 2 hours about the general level of fungal contamination and its possible trend compared to a control sample. The sensitivity and the repeatability of the HPLC assay appear to be good. The values measured may extend from 0.1–0.2 μg/g for a rich medium to values in the range of 10–20 μg/g for the onset of visibility of molds to the unaided eye and to over 200 μg/g for a massive invasion (Seitz et al., 1977; Cahagnier, 1988). This method has recently been the subject of standardization.

33.4 Conclusion

This quick inventory of the principal problems of microbiological origin which may be encountered in the very vast field of intermediate moisture foods, and of the analytical methods and techniques which provide the means of confronting them, does not call for a conclusion in the usual sense. Many questions about the physiology of microorganisms which develop in media of low water activity remain without answers. Although the hydration factor is certainly one of the most important elements to consider, other parameters such as the pH, temperature, oxygen tension, and the product composition certainly play a very important role (Richard-Molard et al., 1985). These various parameters all interact and it is obviously of the utmost importance to learn the effect of these interactions; for example, on the a_w thresholds for the development of microorganisms, or the minimal water activities required for the synthesis of mycotoxins or of staphylococcal enterotoxins. Leitsner and Rodel (1976) introduced the concept of barriers, according to which all of the previously cited parameters are additive obstacles to microbial development. Growth may thus be inhibited by an appropriate combination of barriers, any of which would have no effect alone.

The true microbiological methods and the new methods based on the detection of specific metabolites indicating spoilage in progress have been presented in opposition within this chapter to underscore the notion that if there is a field in which these two technological approaches are complementary, it is certainly that of intermediate moisture foods. Rapid examination of stable samples by a biochemical method lending itself to automation, with the fine techniques of conventional analyses reserved for samples with a suspect nature, may be easily conceived. The latter procedures have the disadvantages of taking a long time and of requiring experienced operators who should really be spending their time doing more research and development. At the present time, however, the conventional methods remain the only means of determining the actual causes of a spoilage of microbial origin and of orienting the technologist towards the remedy to the situation.

33.5 Appendix: Culture Media

1. Newman's stain

Methylene blue	1 g
95% Ethanol	54 ml
Tetrachlorethane	40 ml
Glacial acetic acid	6 ml

2. Bergere modified Bryant-Burkey medium

Casein peptone	15 g	
Yeast extract	5 g	
Meat extract	10 g	
Cysteine hydrochloride	0.5 g	
Sodium acetate	5 g	
Sodium lactate	5 g	(10 ml of 50% solution)
Water	1 l	

Adjust to pH 5.8–6 with HCl.

3. Galesloot agar

Tryptone	5 g
Yeast extract	2.5 g
Glucose	1 g
Agar	15 g
Distilled water	1 l

pH 7.0

4. Tributyrin agar

Special peptone	5 g
Yeast extract	3 g
Agar	12 g

pH 7.5

5. Tween 80 medium

Peptone	10 g
Sodium chloride	5 g
Calcium chloride	0.1 g
Agar	15 g
Water	1 l

pH 7.4

6. Stabilized powdered Victoria blue

Prepare a 1% solution of Victoria blue (boil to dissolve), add 10% sodium carbonate solution until decolorized (the base precipitates). Filter. Rinse with water slightly alkalinized with ammonia. Dry in an oven at 30°C.

7. Sugar-free agar

Gelatin peptone	7.5 g
Casein peptone	7.5 g
Sodium chloride	5 g
Agar	15 g
Distilled water	1 l

pH 7.5

8. Gelysate agar

Tryptic gelatin peptone	5 g
Carbohydrate-free agar	15 g
Distilled water	1 l

pH 7.3–7.6

9. Buffer for alkaline phosphatase

Anhydrous sodium carbonate	3.5 g
Sodium bicarbonate	1.5 g
Distilled water	1 l

10. M 17 medium

Tryptic casein peptone	2.5 g	
Peptic meat peptone	2.5 g	
Papainic soy peptone	5 g	
Dehydrated yeast extract	5 g	
Sodium glycerophosphate	19 g	
Magnesium sulfate · 7 H_2O	0.25 g	
Ascorbic acid	0.5 g	
Agar	9–18 g	(depending on the gelling power)
Water	950 ml	

pH 7.1–7.2

11. WL medium

Dextrose	50 g
Monopotassium phosphate	0.55 g
Potassium chloride	0.425 g
Calcium chloride	0.125 g
Magnesium sulfate	0.125 g
Manganese sulfate	2.5 mg
Ferric chloride	2.5 mg
Casein hydrolysate	5 g
Water	1 l

12. Williamson medium (for acetic spoilers)

This is obtained by combining the following three factors:

- Basal solid medium: 1 l of hopped beer is boiled down to 0.5 l then adjusted to pH 7 and brought to 800 ml. After adding 5 g of yeast extract and 20 g of agar, the solution is sterilized by autoclaving.
- Citrate-phosphate buffer: 47 g of citric acid and 43.8 g of disodium phosphate is dissolved in 1 l of distilled water. Sterilization is by autoclaving.
- Actidione solution: 0.1 g of actidione is dissolved in 100 ml of sterile water. The final medium is obtained by adding 16 ml of buffer, 1 ml of actidione solution, and 4 ml of alcohol to 80 ml of melted basal medium cooled to 60°C.

13. Mossel's medium for enumeration of total contaminants in soft drinks

Glucose	1 g
Meat extract	3 g
Yeast extract	5 g
Tryptone	15 g
Peptonized milk	15 g
Agar	15 g
Water	1 l

14. Orange serum broth

Casein peptone	10 g
Yeast extract	3 g
Orange extract	5 g
Glucose	4 g
Dipotassium phosphate	3 g
Water	1 l

15. Hypersucrose medium

Tryptone	10 g
Yeast extract	5 g
Dipotassium phosphate	5 g
Ammonium citrate	5 g
Sucrose	50 g
Water	1 l

16. VRGB agar

Peptone	7 g
Yeast extract	5 g
Bile salts	1.5 g
Glucose	10 g
Sodium chloride	5 g
Agar	11 g
Neutral red	0.03 g
Crystal violet	0.002 g

pH 7.4

17. Differential medium for the detection of histamine producing germs (cf. Niven et al.)

Tryptone	5 g
Yeast extract	5 g
L-histidine·2HCl	27 g
NaCl	5 g
$CaCO_3$	1 g
Agar	20 g
Bromocresol purple	0.06 g
Water	1 l

pH 5.3

18. Polypectate gel medium

Peptone	5 g
Dipotassium phosphate	5 g
Monopotassium phosphate	1 g
Calcium chloride	0.6 g
Sodium polypectate (salt of polygalacturonic acid)	70 g
Distilled water	1 l

Add the polypectate gradually after the other components. Sterilize by autoclaving.

19. Mossel medium for *Bacillus cereus*

Meat extract	1 g
Special peptone	10 g
D-mannitol	10 g
Sodium chloride	10 g
Phenol red	0.025 g
Agar	12 g
Water	1 l

pH 7.1

Just prior to use, add 2 ml of 50% egg yolk emulsion and 0.2 ml of 1 mg/ml polymyxin sulfate to 20 ml of melted basal medium held at 50°C.

20. PT medium for selective enumeration of pectinolytic flora

Polygalacturonic acid	5 g
$NaNO_3$	1 g
K_2HPO_4	4 g
$MgSO_4 \cdot 7H_2O$	0.2 g
Tergitol	0.1 ml
Agar	11 g

21. D_3 medium for selective enumeration of *Erwinia*

Sucrose	10 g
Arabinose	10 g
Casein hydrolysate	5 g
Lithium chloride	7 g
Glycine	3 g
NaCl	5 g
$MgSO_4 \cdot 7H_2O$	0.3 g
Sodium dodecyl sulfate (SDS)	50 mg
bromothymol blue	60 mg
Acid fuchsin	100 mg
Agar	15 g

pH 6.9–7.1

References, Part IV

Abgrall B. and Bourgeois C., 1985. Dénombrement des spores de *Clostridium tyrobutyricum* par filtration sur membrane et culture sur milieu gélosé. *Le Lait*, **65**, no. 1, 45–53.

AFNOR (Association Française de Normalisation), Tour Europe, cedex 7, Paris la Défense, France.

AFNOR, CNERNA, 1982. Défauts et altérations des conserves. Réf. ISBN 212 355 812 5.

Alais C., 1975. *Science du lait. Principes des techniques laitières*. 3d ed. Société Edition Publicité.

Baird-Parker A. C., 1962. An improved diagnostic and selective medium for isolating coagulase positive *Staphylococci*. *J. Appl. Bacteriol.*, **25**, no. 1, 12–19.

Barnett J. A., Payne R. W., Yarrow D., 1983. *Yeasts. Characteristics and identification*. Cambridge University Press.

Beerens H. and Luquet F. M., 1987. *Guide pratique d'analyse microbiologique des laits et produits laitiers*. Technique et Documentation, APRIA, Paris.

Beuchat L. R., 1981. Microbial stability as affected by water activity. *Cereal Food World*, **26**, no. 7, 345–349.

Borgstrom G. *Fish as food*. Academic Press, New York, 1961.

Borjesson T., Stollman U., Adamek P., Kaspersson A., 1989. Analysis of volatile compounds for detection of molds in stored cereals. *Cereal Chemistry*, **66**, no. 4, 300–304.

Bourgeois C., 1983a. Méthodes rapides de contrôle microbiologique en brasserie, vol. 1. *Bios*, **14**, nos. 1–2, 30–38.

Bourgeois C., 1983b. Méthodes rapides de contrôle microbiologique en brasserie, vol. 2. *Bios*, **14**, no. 11, 23–29.

Bourgeois C. and Malcoste R., 1980. Techniques classiques de détection et de dénombrement; leurs variantes modernes. *In: Techniques d'analyses et de contrôle dans les IAA*. Technique et Documentation, APRIA Eds, 3–11.

Cahagnier B., 1988. Qualité microbiologique des grains et teneurs en ergostérol. *Industries Alimentaires et Agricoles*, **1**, 5–16.

Cahagnier B., Richard-Molard D., Poisson J., Desserme C., 1983. Evolution de la teneur en ergostérol des grains au cours de la conservation. Une possibilité d'évaluation quantitative et rapide de leur mycoflore. *Sci. Aliments*, **3**, 219–244.

Cheftel H. Aspects techniques de l'expertise des conserves. Bull. 13. Etablissements J. J. Carnaud et Forges de Basse-Indre, 71 avenue Edouard Vaillant, Billancourt.

Cogitore A. *Traité pratique de réglementation laitère*. 2d ed. Edition du Sapin d'Or, Epinal.

Corry J., 1973. The water relations and heat resistance of microorganisms. *Prog. Industr. Microbiol.*, **12**, 73–108.

Dehove R. A., 1974. *La réglementation des produits alimentaires et non alimentaires*. 8th ed. Commerce Editions, Paris.

Dehove R. A., 1984. *La réglementation des produits et services: Qualité, consommation et répression des fraudes*. Commerce Editions, Paris.

Denis C. and Picoche B., 1986. Microbiologie des légumes frais prédécoupés. *Industr. Agric. Alim.*, June, 547–553.

Diawara B., Cahagnier B., Richard-Molard D., 1987. Oxygen consumption by wet grain ecosystem in hermetic silos at various water activities. *In:* Donahaye E., Navarro S. (eds.), Proceeding of the 4th International Working Conference on Stored-Product Protection, Tel Aviv, 77–84.

Food and Drug Administration, 1992. Bacteriological Analytical Manual, 7th ed. AOAC International, Arlington, VA.

Feng P. 1992. *Rapid methods for detecting foodborne pathogens.* In FDA. Bacteriological Analytical Manual, 7th ed. AOAC International, Arlington, VA. Appendix 1.

Gardner G. A., 1966. A selective medium for the enumeration of microbacterium thermosphactum in meat and meat products. *J. Appl. Bacteriol,* **29**, 455–460.

Guilbert S., 1984. Additifs et agents dépresseurs de l'a_w. *In:* Multon J. L., *Additifs et auxiliaires de fabrication dans les industries agro-alimentaires.* Technique et Documentation, APRIA, Paris, 199–227.

Guiraud J. and Galzy P., 1980. *L'analyse microbiologique dans les industries alimentaires.* Les Editions de l'Usine.

Harrigan W. F. and McCance M. E., 1966. *Laboratory methods in microbiology.* Academic Press, New York.

Harrigan W. F. and McCance M. E., 1976. Laboratory methods in food and dairy microbiology. Academic Press, New York.

Herschdoerfer, 1967. *Quality control in the food industry.* Academic Press, New York.

Hobbs E. C. and Christian J. H. B., 1973. *The microbial safety of foods.* Academic Press, New York.

Hocking A. D. and Pitt J. I., 1987. Media for detection and enumeration of microorganisms with consideration of water activity requirements. *In:* Rockland L. B., Beuchat L. R. (eds.). *Water activity. Theory and applications to food.* Marcel Dekker Inc., New York and Basel, 153–169.

Jay J. M., 1970. *Modern food microbiology.* D. Van Nostrand Company, New York.

Kado L. J. and Heskett M. G., 1970. Selective media for isolation of *Agrobacterium, Corynebacterium, Erwinia, Pseudomonas and Xanthomonas. Phytopathology,* **60**, 969–976.

Kreger-Van Rij N. J. W., 1984. *The yeasts: A taxonomic study,* 3d ed. Elsevier, Amsterdam.

Johnson R., Sullivan D., Socrist J., Brockmann M., 1972. The effect of high temperature storage on the acceptability and stability of intermediate moisture food. Natick, U.S. Army Natick Laboratories, 19 p. (Technical Report 72–76-FL).

Lawrie R. A., 1974. *Meat science,* 2d ed. Pergamon Press.

LeClerc, 1988. Eaux de consommation. *In: Microbiologie alimentaire,* vol. 1. Technique et Documentation, 189–200.

Lee B. B. and Kraft A. A., 1977. Microbiology of intermediate moisture pork. *J. Food Sci.,* **42**, no. 3, 735–737.

Leitsner L. and Rodel W., 1976. The stability of intermediate moisture foods with respect to microorganisms. *In:* Davies R., Birch C., Parker K., *Intermediate moisture foods,* 120–137. Applied Sciences Publishers Ltd., London.

Martin G. (coord). *Point sur l'épuration et le traitement des effluents. Bacteriologie des millieux aquatiques.* Technique et Documentation.

Mazollier J., 1988. IVe gamme: Lavage et désinfection des salades. *Infos CTIFL,* 41, 20–23.

Multon J. L., 1981. L'état actuel des travaux de la Commission "aliments à humidité intermédiaire" du CNERNA, *Industries Alimentaires et Agricoles,* **98**, 291–302.

Nguyen The C. and Prunier J. P., 1989. Involvement of *Pseudomonas* in deterioration of "ready to use" salad. *Int. J. Food Sci. Technol.*, **24**, 47–58.

Niven C. F., Jeffrey M. B., Corlett D. A., 1981. Differential plating medium for quantitative detection of histamine producing bacteria. *Appl. Environ. Microbiol.*, **41**, no. 1, 321–322.

Pawsey R. and Davies R., 1976. The safety of intermediate moisture foods with respect to *Staphylococus aureus*. *In:* Davies R., Birch G., Parker K. (eds.). *Intermediate moisture foods*, 182–202. Applied Sciences Publishers Ltd., London.

Petitpher G. L., Mansell R., McKinnon C. H., Cousins C. M.., 1980. Rapid Membrane Filtration-Epifluorescent Microscopy Technique for direct enumeration of bacteria in raw milk. *Appl. Environ. Microbiol.*, **39-2**, 423–429.

Pitt J. I. and Hocking A. D., 1985. *Fungi and food spoilage*. Academic Press, Sydney.

Plitman M., Park Y., Gomez R., Sinsey A., 1973. Viability of *Staphylococcus aureus* in intermediate moisture foods. *J. Food Sci.*, **6**, no. 38, 1004–1008.

Price J. F. and Schweigert B. S. *The science of meat and meat products,* 2d ed. W. H. Freeman and Company.

Prieur D., Mevel G., Nicolas J. L., Plusquellec A., Vigneulle M., 1990. Interactions between bivalve molluscs and bacteria in the marine environment. *Oceanogr. Mar. Biol. Ann. Rev.*, **28**, 277–352.

Richard-Molard D., 1990. Conservation of humid grains in controlled atmosphere storage. *In:* Calderon M., Barkai-Golan R., (eds.). *Food preservation by controlled atmosphere,* 57–82. CRC Press, Boca Raton, FL.

Richard-Molard D., Cahagnier B., Poisson J., Drapron R.., 1976. Evolutions comparées des constituants volatils et de la microflore de maïs stockés. *Annales de Technologie Agricole*, **25**, no. 1, 29–44.

Richard-Molard D., Lesage L., Cahagnier B., 1985. Effect of water activity on mold growth and mycotoxin formation. *In:* Simatos D., Multon J. L., (eds.). *Properties of water in foods.* Martinus Nijhoff Publishers, The Hague, 273–292.

Richard-Molard D., Diawara B., Cahagnier B., 1987. Susceptibility of cereal microflora to oxygen deficiency and carbon dioxide concentration. *In:* Dona-Haye E., Navarro S., (eds.). Proceeding of the 4th International Working Conference on Stored-Product Protection, Tel Aviv, 85–92.

Rodier J., 1975. *L'analyse de l'eau.* Dunod Technique.

Scandella D., 1988. IVe gamma: La qualité à l'étalage. *Infos CTIFL*, **39**, 24–26.

Scott W., 1957. Water relations of food spoilage microorganisms. *Adv. Food Res.*, **7**, 83–127.

Seitz L. M., Mohr H. E., Burroughs R., Sauer D. B., 1977. Ergosterol as an indicator of fungi invasion in grains. *Cereal Chemistry*, **54**, 1207–1217.

Serres L., Amariglio S., Petransxienne D. Contrôle de la qualité des produits laitiers, vol. 2. Direction des Services Sanitaires.

Stadelman and Cotterill, 1977. *Egg science and technology*, 2d ed. AVI Publishing Company.

Stawicki S., Kaminski E., Niewiarowicz A., Trojan M., Wasowicz E., 1973. The effect of microflora on the formation of odors in grains. *In:* Multon J. L., Guilbot A., Comptes-rendus du Symposium International sur la Conservation des Grains Récoltés Humides, Paris. *Annales de Technologie Agricole*, n.s., 309–336.

Strugnell C., 1988. Increasing the shelf-life of prepacked vegetables. *Irish J. Food Sci. Technol.*, **12**, nos. 1–2, 81–84.

Tabak H. H. and Cooke W. B., 1968. Growth and metabolism of fungi in atmosphere of nitrogen. *Mycologia*, **60**, 115–140.

Tatini S. R., Hoover D. G., Lachica R. V. F., 1984. Methods for the isolation and enumeration of *Staphylococcus aureus. In:* Speck M. L. (ed.). *Compendium of methods for the microbiological examination of foods*, 2d ed. American Public Health Association, Washington, D.C.

Taoukis P. S., Breene W. M., Labuza T. P., 1988. Intermediate moisture foods. *In:* Pomeranz Y. (ed.). *Advances in cereal science and technology*, vol. 9, 91–128. American Association of Cereal Chemists, St. Paul, MN.

Tilbury R., 1976. The microbial stability of intermediate moisture foods with respect to yeast. *In:* Davies R., Birch G., Parker K. (eds.). *Intermediate moisture foods*, 138–165. Applied Science Publishers Ltd., London.

Tirilly Y. and Thouvenot D., 1988. *Fruits et légumes dans microbiologie alimentaire.* Technique et Documentation, 265–278.

Torres J. A., 1987. Microbial stabilization of intermediate moisture food surfaces. *In* Rockland L. B., Beuchat L. R. (eds.). *Water activity: Theory and applications to food*, 329–368. Marcel Dekker, Inc., New York and Basel.

Veisseyre R., 1975. *Technologie du lait.* La Maison Rustique, Paris.

Weete J. D., 1974. *Fungal lipid biochemistry; distribution and metabolism.* Plenum Press, New York.

Weiser E. H., Mountney G. J., Gould W. A., 1976. *Practical food microbiology and technology*, 2d ed. AVI Publishing, Westport, Conn.

Code des usages en charcuterie et conserves de viandes. Méthodes de contrôle. Centre Techniques de la Salaison, Charcuterie et Conserves de Viandes, 1973.

Journal Officiel. Hygiène alimentaire. Textes généraux no. 1488-1.

Suggested Readings, Chapters 24–26

Abbaszadegan M., Gerba C. P., Rose J. B., 1991. Detection of Giardia cysts with a cDNA probe and applications to water samples. *Appl. Environ. Microbiol.*, **57**, no. 4, 927–931; 19 ref.

Anderson K., Walker R. L., Wesen D. P., 1990. Microbiological analysis of bulk-tank milks on blood agar: Comparison with regulatory methods and influence on sample collection and handling factors. *Dairy, Food Environ. Sanitation*, **10**, no. 4, 213–217; 25 ref.

Barer M. R. and Wright A. E., 1990. Cryptosporidium and water. *Letters Appl. Microbiol.*, **11**, no. 6, 271–277; 17 ref.

Bej A. K., DiCesare J. L., Haff L., Atlas R. M., 1991. Detection of *Escherichia coli* and *Shigella* spp. in water by using the polymerase chain reaction and gene probes for *uid. Appl. Environ. Microbiol.*, **57**, no. 4, 1013–1017; 14 ref.

Byrne R. D. Jr. and Bishop J. R., 1990. The Limulus amoebocyte lysate assay and the direct Epifluorescent filter technique as estimators of potential shelf-life of pasteurized fluid milk. *J. Food Prot.* Ames, Iowa: International Association of Milk, Food, and Environmental Sanitarians. Feb. 1990. **53**, no. 2, 152–153.

Byrne R. D., Bishop J. R., McGilliard M. L., 1989. Selective preliminary incubation for psychotrophic bacteria. *J. Food Prot.* Ames, Iowa: International Association of Milk, Food, and Environmental Sanitarians. June 1989. **52**, no. 6, 396–398.

Campbell I., 1987. *Microbiology of brewing: Beer and lager.* Prev. Emmaus, PA: Rodale Press, 261–290. ill., charts.

Cinvert T. C., Rice E. W., Johnson S. A., Berman D., Johnson C. H., Mason P. J., 1992. Comparing defined-substrate coliform tests for the detection of *Escherichia coli* in water. *J. Am. Water Works Assoc.*, **84**, no. 5, 98–104; 41 ref.

Clark D. L., Milner B. B., Stewart M. H., Wolfe R. L., Olson B. H., 1991. Comparative study of commercial 4-methylumbelliferyl-beta-D-glucuronide preparations with the Standard Methods membrane filtration fecal coliform test for the detection of *Escherichia coli* in water samples. *Appl. Environ. Microbiol.*, **57**, no. 5, 1528–1534; 22 ref.

Curiale M. S., Fahey P., Fox T. L., McAllister J. S., 1989. Dry rehydrate films for enumeration of coliforms and aerobic bacteria in dairy products: Collaborative study. *J. Assoc. Off. Anal. Chem.* Arlington: The Association. March/April 1989. **72**, no. 2, 312–318.

Deak T. and Beuchant L. R., 1987. Identification of foodborne yeasts. *J. Food Prot.* Ames, Iowa: International Association of Milk, Food, and Environmental Sanitarians. March 1987. **50**, no. 3, 243–264.

Dicsmore R. P., English P. B., Matthews J. C., Sears P. M., 1990. Effect of milk sample delivery methods and arrival conditions on bacterial contamination rates. *Cornell Vet.*, **80**, no. 3, 243–250; 7 ref.

Driessen F. M. and van-den Berg M. G., 1992. Microbiological aspects of pasteurized cream. *Bull. Int. Dairy Fed. no.* 271, 4–10; 33 ref.

Geldreich E. E., 1989. Drinking water microbiology—new directions toward water quality enhancement. *Int. J. Food Microbiol.*, **9**, no. 4, 295–312.

Guerzoni M. E., Gardini F., Cavazza A., Piva M., 1987. Gas-liquid chromatographic method for the estimation of coliforms in milk. *Int. J. Food Microbiol.*, **4**, no. 1, 73–78; 15 ref.

Jermini M. F. G., Geiges C., Schmidt-Lorenz W., 1987. Detection, isolation and identification of osmotolerant yeasts from high-sugar products. *J. Food Prot.* Ames, Iowa: International Association of Milk, Food and Environmental Sanitarians. June 1987. **50**, no. 6, 468–472.

Johnson E. A., Nelson J. H., Johnson M., 1990. Microbiological safety of cheese made from heat-treated milk. II. Microbiology. *J. Food Prot.*, **53**, no. 6, 519–540.

Johnson E. A., Nelson J. H., Johnson M., 1990. Microbiological safety of cheese made from heat-treated milk. III. Technology, discussion, recommendations, bibliography. *J. Food Prot.*, **53**, no. 7, 610–623; 378 ref.

Jordano R., Medina L. M., Salmeron J., 1991. Contaminating mycoflora in fermented milk. *J. Food Prot.* Ames, Iowa: International Association of Milk, Food, and Environmental Sanitarians. Feb. 1991. **54**, no. 2, 131–132.

Lawrence D. R., 1988. Spoilage organisms in beer. *In:* Robinson, R. K., Barking, U. K. (eds.). *Developments in food microbiology*, vol. 3. Elsevier. ISBN 1 85166 131 X [see FSTA (1988) 20 12B3].), *Dev. Food Microbiol.*, **3**, 1–48; 182 ref.

Macedo A. C., Xavier-Malcata F., Oliveira J. C., 1993. The technology, chemistry, and microbiology of Serra cheese: A review. *J. Dairy Sci.*, **76**, no. 6, 1725–1739; 42 ref.

Macler B. A. and Regli S., 1993. Use of microbial risk assessment in setting US drinking water standards. *Int. J. Food Microbiol.*, **18**, no. 4, 245–256; 26 ref.

Meyer R., Luethy J., Candrian U., 1991. Direct detection by polymerase chain reaction (PCR) of *Escherichia coli* in water and soft cheese and identification of enterotoxigenic strains. *Letters Appl. Microbiol.*, **13**, no. 6, 268–271; 7 ref.

Parish M. E. and Higgins D. P., 1990. Investigation of the microbial ecology of commercial grapefruit sections. *J. Food Prot.* Ames, Iowa: International Association of Milk, Food, and Environmental Sanitarians. Aug. 1990. **53**, no. 8, 685–688. charts.

Peladan F. Erbs D., Moll M., 1986. Practical aspects of the detection of lactic bacteria in beer. *Food Microbiol.* London: Academic Press. Oct. 1986. **3**, no. 4, 281–288.

Preixens S. and Sancho J., 1987. [Evolution of coliform bacteria in natural yoghurt.] Evolucion de los coliformes en el yogur natural. *Alimentaria.* Madrid: s.n. Nov. 1987. **24**, no. 187, 33–37.

Roberts D., 1992. Growth and survival of Vibrio cholera in foods. *PHLS Microbiol. Digest*, **9**, no. 1, 24–31; 35 ref.

Roy D. and Ward P., 1990. Evaluation of rapid methods for differentiation of *Bifidobacterium* species. *J. Appl. Bacteriol.*, **69**, no. 5, 739–749; 10 ref.

Schooner F., Simard R. E., Pandian S., 1991. Colorimetric assay for free fatty acids in butter using flow-injection and immobilized enzymes. *J. Food Sci.*, **56**, no. 5, 1229–1232; 33 ref.

Shrestha K. G. and Sinha R. N., 1987. Occurrence of coliform bacteria in dairy products. *Indian J. Dairy Sci.* New Delhi: Indian Dairy Association. March 1987. **40**, no. 1, 121–123.

Taguchi H., Ohkochi M., Uehara H., Kojima K., Mawatari M., 1990. KOT medium, a new for the detection of beer spoilage lactic acid bacteria. *J. Am. Soc. Brew. Chem.* St. Paul: The Society. **48**, no. 2, 72–75.

Toranzos G. A. and Alvarez A. J., 1992. Solid-phase polymerase chain reaction: Applications for direct detection of enteric pathogens in waters. *Can. J. Microbiol.*, **38**, no. 5, 365–369; 23 ref.

Tzeng C. H., 1985. Applications of starter cultures in the dairy industry. *Dev. Ind. Microbiol.*, **26**, 336–338; 39 ref.

Vasavada P. C., 1988. Pathogenic bacteria in milk—a review. *J. Dairy Sci.*, **71**, no. 10, 2809–2814; 83 ref.

Wartburton D. W., 1993. A review of the microbiological quality of bottled water sold in Canada. II. The need for more stringent standards and regulations. *Can. J. Microbiol.*, **39**, no. 2, 158–168.

Zottola E. A. and Smith L. B., 1991. Pathogens in cheese. *Food Microbiol.*, **8**, no. 3, 171–182; 45 ref.

Suggested Readings, Chapters 27–33

Baranowski J. D., Frank H. A., Brust P. A., Chongsiriwatana M., Premaratne R. J., 1990. Decomposition and histamine content in Mahimahi (*Coryphaena hippurus*). *J. Food Prot.* Ames, Iowa: International Association of Milk, Food, and Environmental Sanitarians. March 1990. **53**, no. 3, 217–222. charts.

Bell R. G., 1993. Development of the principles and practices of meat hygiene: A microbiologist's perspective. *Food Control*, **4**, no. 3, 134–140; 25 ref.

Beuchat L. R. and Brackett R. E., 1990. Survival and growth of *Listeria monocytogenes* on lettuce as influenced by shredding, chlorine treatment, modified atmosphere packaging and temperature. *J. Food Sci. Off. Publ. Inst. Food Technol.* Chicago, IL: The Institute. May/June 1990. **55**, no. 3, 755–758, 870.

Cartwright K. A. V. and Evans B. G., 1988. Salmon as a food-poisoning vehicle—two successive salmonella outbreaks. *Epidemiology Infection*, **101**, no. 2, 249–257; 13 ref.

Cole M. T., Kilgen M. B., Reily L. A., Hackney C. R., 1986 Detection of enteroviruses and bacterial indicators and pathogens in Louisiana oysters and their overlying waters. *J. Food Prot.* Ames, Iowa: International Association of Milk, Food, and Environmental Sanitarians. Aug. 1986. **49**, no. 8, 596–601.

Cook D. W., Ruple A. D., 1989. Indicator bacteria and Vibrionaceae multiplication in post-harvest shellstock oysters. *J. Food Prot.* Ames, Iowa: International Association of Milk, Food, and Environmental Sanitarians. May 1989. **52**, no. 5, 343–349. charts.

Dantas R. A. and Silva M. C. C., 1988. Microbiological analysis of meals prepared in canteens. *Catering & Health,* **1**, no. 1, 23–49; 4 ref.

de Mesquita M. M. F., Evison L. M., West P. A., 1991. Removal of faecal indicator bacteria and bacteriophages from the common mussel (*Mytilus edulis*) under artificial depuration conditions. *J. Appl. Bacteriol.*, **70**, no. 6, 495–501.

Dodd C. E. R. and Waites W. M., 1991. The use of toluidine blue for in situ detection of micro-organisms in foods. *Letters Appl. Microbiol.*, **13**, no. 5, 220–223; 7 ref.

Fliss I., Simard R. E., Ettriki A., 1991. Comparison of three sampling techniques for microbiological analysis of meat surfaces. *J. Food Sci. Off. Publ. Inst. Food Technol.* Chicago, IL: The Institute. Jan./Feb. 1991. **56**, no. 1, 249–250, 252.

Golden D. A. and Beuchat L. R., 1992. Effects of potassium on growth patterns, morphology, and heat resistance of *Zygosaccharomyces rouxii* at reduced water activity. *Can. J. Microbiol.*, **38**, no. 12, 1252–1259; 29 ref.

Grabow W. O. K., de Villiers J. C., Schildhauer C. I., 1992. Comparison of selected methods for the enumeration of fecal coliforms and *Escherichia coli* in shellfish. *Appl. Environ. Microbiol.*, **58**, no. 9, 3203–3204; 7 ref.

Gram L., 1992. Evaluation of the bacteriological quality of seafood. *Int. J. Food Microbiol.*, **16**, no. 1, 25–39.

Grant I. R. and Patterson M. F., 1991. Effect of irradiation and modified atmosphere packaging on the microbiological safety of minced pork stored under temperature abuse conditions. *Int. J. Food Sci. Technol.*, **26**, no. 5, 521–533, 28 ref.

Harris L. J. and Stiles M. E., 1992. Reliability of *Escherichia coli* counts for vacuum-packaged ground beef. *J. Food. Prot.*, **55**, no. 4, 266–270; 28 ref.

Hollingworth T. A. Jr., Kaysner C. A., Colburn K. G., Sullivan J. J., Abeyta C. Jr., Walker K. D., Torkelson J. D. Jr., Throm H. R., Wekell M. M., 1991. Chemical and microbiological analysis of vacuum-packed, pasteurized flaked imitation crabmeat. *J. Food Sci. Off. Publ. Inst. Food Technol.* Chicago, IL: The Institute. Jan./Feb. 1991. **56**, no. 1, 164–167.

Jehl-Pietri C., Dupont J., Herve C., Menard D., Munro J., 1991. Occurrence of faecal bacteria, Salmonella and antigens associated with hepatitis A virus in shellfish. *Zentralbl. Hygiene Umweltmedizin,* **192**, no. 3, 230–237; 41 ref.

Jehl-Pietri C., Hugues B., Deloince R., 1990. Viral and bacterial contamination of mussels (*Mytilus edulis*) exposed in an unpolluted marine environment. *Letters Appl. Microbiol.* Oxford: Blackwell Scientific Publications. Sept. 1990. **11**, no. 3, 126–129.

Khalifa K. I., Werner B., Timperi R. Jr., 1986. Non-detection of enteroviruses in shellfish collected from legal shellfish beds in Massachusetts. *J. Food Prot.* Ames, Iowa: International Association of Milk, Food, and Environmental Sanitarians. Dec. 1986. **49**, no. 12, 971–973.

Lambert A. D., Smith J. P., Dodds K. L., 1991. Shelf life extension and microbiological safety of fresh meat—a review. *Food Microbiol.*, **8**, no. 4, 267–297.

McAllister J. S., Stadtherr M. P., Fox T. L., 1988. Evaluation of the 3M petrifilm culture plate method for enumerating aerobic flora and coliforms in poultry processing facilities. *J. Food Prot.* Ames, Iowa: International Association of Milk, Food, and Environmental Sanitarians. Aug. 1988. **51**, no. 8, 658–659, 662.

McMeekin T. A., Ross T., Olley J., 1992. Application of predictive microbiology to assure the quality and safety of fish and fish products. *Int. J. Food Microbiol.*, **15**, no. 1/2, 13–32.

Martinez-Manzanares E., Morinigo M. A., Cornax R., Egea R., Borrego J. J., 1991. Relationship between classical indicators and several pathogenic microorganisms involved in shellfish-borne diseases. *J. Food Protect.*, **54**, no. 9, 711–717; 82 ref.

Park C. E. and Sanders G. W., 1992. Occurrence of thermotolerant campylobacters in fresh vegetables sold at farmers' outdoor markets and supermarkets. *Can. J. Microbiol.*, **38**, no. 4, 313–316; 52 ref.

Pedrosa-Menabrito A. and Regenstein J. M., 1990. Shelf-life extension of fresh fish, a review. III. Fish quality and methods of assessment. *J. Food Qual.*, **13**, no. 3, 209–223.

Prokopowich D. and Blank G., 1991. Microbiological evaluation of vegetable sprouts and seeds. *J. Food Prot.* Ames, Iowa: International Association of Milk, Food, and Environmental Sanitarians. July 1991. **54**, no. 7, 560–562.

Reiber M. A., Hierholzer R. E., Adans M. H., Colberg M., Izat A. L., 1990. Effect of litter condition on microbiological quality of freshly killed and processed broilers. *Poult. Sci.* Champaign, IL: Poultry Science Association. Dec. 1990. **69**, no. 12, 2128–2133.

Reid C. M., 1991. *Escherichia coli* 0157:H7—the "hamburger" but: A literature review. Meat Industry Research Institute of New Zealand, no. 879, iv + 14 p., 32 ref.

Smith H. R., Cheasty T., Roberts D., Thomas A., Rowe B., 1991. Examination of retail chickens and sausages in Britain for vero cytotoxin-producing Escherichia coli. *Appl. Environ. Microbiol.*, **57**, no. 7, 2091–2093; 23 ref.

Smith L. B., Zottola E. A., Fox T. L., Chausse K., 1989. Use of petrifilm to evaluate the microflora of frozen dessert mixes. *J. Food Prot.* Ames, Iowa: International Association of Milk, Food, and Environmental Sanitarians. Aug. 1989. **52**, no. 8, 549–551. charts.

Smith J. J., Ockerman H. W., Plimpton R. F., 1990. The effect of cooking temperature, holding temperature and holding time on proximate analysis, aerobic plate count and coliform count on restructured beef roasts. *J. Food Prot.* Ames, Iowa: International Association of Milk, Food, and Environmental Sanitarians. May 1990. **53**, no. 5, 396–399. charts.

Stantom W. R., 1988. Tropical fermented foods: A scenario for the future. *In:* Robinson R. K., (eds.). *Developments in food microbiology*, vol. 4. Elsevier. ISBN 1 85166 169 7 [see FSTA (1988) 20 12B6].), *Dev. Food Microbiol.*, **4**, 223–243; 68 ref.

Stekelenburg F. K., Zomer W. L. J. M., Mulder S. J., 1990. A medium for the detection of bacteria causing green discolouration of cooked cured meat products. *Appl. Microbiol. Biotech.* Berlin: Springer International. April 1990. **33**, no. 1, 76–77.

Thayer D. W., Lachica R. V., Huhtanen C. N., Wierbicke E., 1986. Use of irradiation to ensure microbiological safety of processed meats. *Food Technol.*, **40**, no. 4, 159–162; 51 ref.

Tokuoka K., Ishitani T., Goto S., Komagata K., 1985. Identification of yeasts isolated from high-sugar foods. *J. Gen. Appl. Microbiol.* Tokyo: Microbiology Research Foundation. **31**, no. 5, 411–427.

Torres J. A. and Karel M., 1985. Microbial stabilization of intermediate moisture food surfaces. III. Effects of surface preservative concentration and surface pH control on microbial stability of an intermediate moisture cheese analog. *J. Food Proc. Preserv.*, **9**, no. 2, 107–119; 18 ref.

Wernars K., Delfgou E., Soentoro P. S., Notermans S., 1991. Successful approach for detection of low numbers of enterotoxigenic *Escherichia coli* in minced meat by using the polymerase chain reaction. *Appl. Environ. Microbiol.*, **57**, no. 7, 1914–1919; 20 ref.

Zhou Y. J., Estes M. K., Jiang X., Metcalf T. G., 1991. Concentration and detection of hepatitis A virus and rotavirus from shellfish by hybridization tests. *Appl. Environm. Microbiol.*, **57**, no. 10, 2963–2968; 34 ref.

PART

MISCELLANEOUS APPLICATIONS OF MICROBIOLOGICAL TESTING

CHAPTER

34

Testing of Starter Cultures for Purity

T. Germain, A. Miclo

The analysis of starter cultures must uncover any contaminants that may be present and often at very low levels among the cultured cells. This analysis must be sensitive and as quantitative as possible as well as selective and able to detect all species considered dangerous.

The use of tested starter cultures obtained from pure cultures of one or several selected strains is becoming increasingly generalized throughout the food industries. Testing starter cultures for purity is becoming increasingly necessary. The example treated in this chapter is from the brewing industry, that is, testing starter cultures of yeast of the genus *Saccharomyces*.

34.1 General Aspects of Testing

When starter cultures come from propagation done in sterile media and the yeasts have been maintained in a state of exponential growth, in the initial phases, at least, the level of contamination is always very low or zero. More often, however, the starter comes from yeasts harvested after having been through some number of fermentations. Harvested yeasts are generally not in the exponential phase of growth, and starter cultures thus obtained are almost always contaminated. Frequent testing of the level of contamination is thus necessary and these tests are largely instrumental in determining the reuse of yeasts.

There are two requirements for testing. First, it must be as rapid as possible in order to make a decision about the reuse of the yeast without having to hold it for such a long time that losses of cell viability occur. Second, it must be quantitative. In order to compensate for the time required for the analysis, it may be advantageous to sample the fermenting wort (during the third or fourth day of the main fermentation, before settling of the yeast in the bottom of the vat) rather than the harvested yeast.

Testing only rarely determines the species or even the genera contaminating a culture. Most producers are satisfied with an overall detection of which some empirical threshold determines acceptance or rejection of the starter culture. Thresholds and levels of contamination are expressed with respect to the number of cells of cultured yeast. Preferably, the threshold should be around $1/10^6$ and should not be any higher than $1/10^4$.

The most significant contaminations are due to yeasts called "wild" yeasts (to distinguish them from cultured yeasts) and to lactic acid bacteria which are considered to be the most dangerous, propagated essentially by the starter cultures. Analysis for acetic acid bacteria—constant contaminants of top fermenting starters and harmful to the viability of the harvested yeast—should not be neglected, however, nor should analysis for gram-negative wort bacteria, particularly enterobacteria which may develop after pitching the starter culture (i.e., during the initial stages of the fermentation when conditions are not yet unfavorable to them). Enterobacteria bring about organoleptic spoilage of the product, a decrease in the rate of fermentation, and an increase in pH favorable to other contaminants. *Obesumbacterium proteus*, in particular, is among the enterobacteria which may survive throughout fermentations and accumulate with the harvested yeast.

The requirement for speed leads us to classify the methods according to this criterion as conventional methods, accelerated conventional methods, and rapid methods. Conventional methods generally use selective solid media. They consist—following the preparation of a yeast suspension of known titer—of spreading or incorporating 10^5–10^7 cultured cells per Petri plate. The media and culture conditions must inhibit the growth of the cultured yeast and allow, with some degree of selection, the growth of certain groups of wild yeasts or bacteria. Conventional methods of detection on solid media are easily quantifiable, the enumeration being only of viable organisms and sensitivity being satisfactory at $1/10^5$–$1/10^7$. Among the rapid methods, immunological methods and methods based on the measurement of a physicochemical parameter will be examined.

34.2 Conventional Methods

34.2.1 Nonselective Microscopic Examination

Direct microscopic examination of the starter culture in water or sodium carbonate can detect an infection if the level of contamination is higher than a threshold around $1/10^3$. This level may be determined by examination of a stained smear. Only contaminants which are morphologically very different from the cultured yeast are detectable by this method. Wild strains of *Saccharomyces*, a few species of which may be very harmful to the product, escape detection.

The sensitivity of microscopic examination may be increased if preceded by a culturing step favoring contaminants over the cultured yeast. This technique is used primarily to increase the speed of detection of lactic acid bacteria on nonselective media. Actidione is added to 10–20 mg/l to the media to inhibit the growth of the yeasts. Beer plus yeast water medium (2:1 proportions) after 3–4 days of partially anaerobic incubation at 25°C gives an enrichment sufficient to observe *Pediococcus* and *Lactobacillus*. BSNB medium (Kretschmer and Kretschmann, 1972) also allows the observation of *Pediococcus* after 3–4 days. Observations made by this method are, however, difficult to quantify. In principle, it may be sufficient to inoculate a liquid medium with an accurately known concentration of the cultured yeast and determine the contaminant level after incubation (e.g., using a stained smear). The real difficulty is in estimating the growth rate of the contaminants.

34.2.2 Analysis for Wild Yeasts

Brewers distinguish contaminating yeasts as belonging either to the genus *Saccharomyces* or to a genus other than *Saccharomyces* based on detection methods using media which favor the growth of one group or the other. An analytical procedure intended to establish the total spectrum of contaminating yeasts must therefore use several complementary media. A critical review of the methods used up until 1982 has been made by Ingledew and Casey (1982). More recent works have sought the development of media that could detect a wider range of wild yeasts covering both *Saccharomyces* and non-*Saccharomyces*.

34.2.2.1 Sporulation Test

The sporulation test used in brewing is based on the fact that the cultured yeasts in use today have almost all lost the ability to sporulate. The sample is first incubated in a rich presporulation medium (e.g., liquid YM) at 25°C for 24–48 hours. The cells are centrifuged in the middle of the growth phase, washed, and placed on sporulation medium, generally an acetate medium, liquid or solid. The presence of asci after 48 hours of incubation at 25°C indicates contamination. Observation may be made easier by staining the spores using the Schaeffer-Fulton method or the Möller method. Media and methods are described by Lodder (1970). The sensitivity of the method is low ($1/10^3$). Since wild yeasts are more thermoresistant than cultured yeasts, the sensitivity may be improved by warming the analyzed suspension to 50°C for 20 minutes prior to culturing in the presporulation medium (Lund and Thygesen, 1957), but the method then becomes difficult to quantify. Independently of the sporulation test, Walsh and Martin (1977 and 1978) have taken advantage of the different thermal sensitivities of *Saccharomyces cerevisiae* and *Saccharomyces uvarum* to detect *S. cerevisiae* in a starter culture of *S. uvarum.* The sensitivity of the test is low. Only sporulating yeasts may be detected and the sporulation rate is highly variable. The test is useful, however, since it does detect wild species of *Saccharomyces.*

34.2.2.2 Lysine Medium

Wild yeasts that are not detectable by the sporulation test are generally able to grow on media containing lysine as sole carbon source. *S. cerevisiae* and *S. uvarum* as well as the majority of contaminating *Saccharomyces* are not able to grow on this medium. This test is suitable for all *Saccharomyces* starter cultures.

The method consists (after washing and diluting the sample in water) of spreading 10^7 cells on lysine medium (Morris and Eddy, 1957). Only "lysine-positive" strains grow. The most often used media are those of Walters and Thiselton (1953), of Morris and Eddy (1957), and of Wickerham which is YCB (yeast carbon base) medium completed by the addition of lysine according to the ASBC method (*Microbiological Controls*, 1962).

34.2.2.3 Actidione Medium

One useful yeast identification criterion is their resistance to actidione (cyclohexamide). At high concentrations, this antifungal agent inhibits the growth of all yeasts. It has thus been used in many conventional methods for the detection and isolation of contaminating bacteria. At relatively low ranges of concentration, however, actidione selectively inhibits the growth of various yeasts to some extent. This has led Harris and Watson (1968) to study the use of this agent for the detection of the relative levels of the various strains in composite starter cultures used in fermentation.

The implementation of this test requires that the actidione sensitivity of the culture strain be well established such that the minimum concentration totally inhibiting growth of the strain during a 48-hour incubation at 28°C is used.

Universal Beer agar or UBA medium published in the ASBC Proceedings of 1968, hopped wort agar, or WLN (Wallerstein Laboratories Nutrient) agar containing 0.1–0.2 mg of actidione per liter are the most often used. The spectrum of wild yeasts detected on this medium is a function of the actidione concentration used which must be determined as precisely as possible before performing the test. This value is fairly close to that which favors non-*Saccharomyces.*

34.2.2.4 Crystal Violet Medium

Because of the poor sensitivity of the sporulation method for the detection of "lysine-negative" wild yeasts of the genus *Saccharomyces*, Kato (1967) sought the selective inhibition of the growth of cultured yeasts by using a stain. At a suitable concentration, crystal violet inhibits the growth of *S. cerevisiae* and *S. uvarum* of which the appearance of colonies on solid medium is delayed. As is the case for actidione medium, the implementation of this test requires that the sensitivity of the strain to the stain be well established. Use of the minimum concentration of crystal violet preventing yeast colony formation for up to 48 hours of incubation at 28°C is recommended.

The most commonly used medium is hopped wort with 2% agar containing 5–20 mg of crystal violet per liter depending on the sensitivity of the cultured yeast. Plates are inoculated by spreading 0.1 ml of yeast suspension titered at 10^6–10^7 cells/ml. The detection sensitivity is in the $1/10^5$ range. Since most yeasts, including *Saccharomyces*, do not utilize lysine, they are detectable on this medium. Lysine and crystal violet media are complementary. In brewing, their joint use for testing purposes is recommended. The EBC recommends in the *Analytica Microbiologica*, the complementary solid media with crystal violet and with actidione or lysine in addition to the sporulation test for the detection of wild yeasts (Martin and Moll, 1984).

34.2.2.5 WLN Medium

The use of WLN medium for the analyses of wild yeasts in starter cultures was proposed by Hall (1971). This nonselective medium is able to detect most contaminants, both yeast and bacterial. Yeasts, however, develop much more quickly than bacteria, and incubation at 25°C for 3 days allows their unique detection. The medium contains bromocresol green. Only *S. cerevisiae* is incapable of reducing the stain and of forming smooth, dark green colonies while wild strains form colonies which are white or faintly colored. *S. uvarum*, like wild yeasts, reduces the stain. WLN agar may thus be used to detect contaminating yeasts in *S. cerevisiae* starters, but is not suitable as such for the analysis of *S. uvarum* starters. It may be used for the latter case when actidione is incorporated.

34.2.2.6 SDM Medium

Schwartz differential medium (SDM) with fuchsin and sulfite, recommended by Brenner and colleagues (1970), allows the growth of most yeast cells with the exception of *S. cerevisiae* and *S. uvarum*. It is not necessary to wash the sample before the analysis. Reading is done after 3 days of incubation at 30°C. The sensitivity is in the range of $1/10^6$. Subsequent work, including that of Seidel (1973), showed that SDM medium did not detect some very important contaminants such as *Saccharomyces bayanus*. In addition, some strains of *S. cerevisiae* and numerous strains of *S. uvarum* may develop on this medium.

34.2.2.7 Lin's Medium

The medium of Lin (1974) is an improvement over SDM medium. It differs by the addition of crystal violet which complements the inhibitory effects of the fuchsin and the sulfite and inhibits the growth of the cultured yeasts to a satisfactory degree. The minimum concentration of crystal violet, between 0.4 and 4 mg/l, must be determined for each cultured yeast. The sensitivity of the method is in the $1/10^6$ range. The detection spectrum of yeasts on Lin's medium is much wider than with crystal violet medium. The few species which do not grow on Lin's medium may be detected on lysine medium.

34.2.2.8 Modified Lin's Medium

Ergosterol and Tween 80 favor the growth of *Saccharomyces* anaerobically. Longley and co-workers (1980) developed a "modified" Lin's medium by adding these two constituents. Anaerobic incubation of a contaminated starter culture on this medium can detect wild *Saccharomyces*. Colonies of non-*Saccharomyces* remain small. This is a more selective medium than Lin's medium which is to be used in conjunction with lysine medium.

34.2.2.9 Copper Sulfate Medium

Lin (1981) proposed a medium in which copper sulfate inhibits the growth of cultured yeasts as well as most *Saccharomyces*. This medium has the advantage over lysine medium of not requiring washing of the starter culture.

The medium MYGP + copper of Taylor and Marsh (1984) has a lower copper sulfate content than the Lin medium (1981). This concentration, around 200 mg/l, is the minimum for inhibiting growth of cultured yeasts. This medium can detect most wild yeasts, both *Saccharomyces* and non-*Saccharomyces*. According to Taylor and Marsh, this is its main feature of interest.

This medium may, however, be made more selective for *Saccharomyces* by adding ergosterol and Tween 80 and incubating anaerobically. The usefulness of such a medium is reported by Thurston (1986a). This author emphasizes the slightly higher resistance to copper sulfate of bottom-fermenting compared to top-fermenting yeasts.

34.2.2.10 Other Methods

Most top-fermenting yeasts require pantothenate. This peculiarity is exploited by Röcken (1983) who proposes a pantothenate-free medium for the detection of a broad spectrum of contaminating *Saccharomyces* and non-*Saccharomyces*.

A new medium called XMACS (xylose, mannitol, adonitol, cellobiose, and sorbitol) has more recently been proposed by De Angelo and Siebert (1987). This group of carbon sources which cultured yeasts cannot assimilate has been retained because it allows the development of most other yeasts, including wild species of *Saccharomyces*. The authors have successfully recovered 84% of the yeast species tested.

Remark: When an analysis is able to isolate *Saccharomyces*, it is often useful to quickly characterize the colonies to make sure that these are not cultured yeasts. Thurston (1986b) proposed a simple test. Wild *Saccharomyces* are, for the most part, able to decarboxylate hydrocinnamic acids and produce strong phenolic odors. The odor produced after 24 hours of incubation of a test colony in hopped wort with such an acid will sharply differentiate a contaminating strain from a cultured strain.

34.2.3 Analysis for Bacteria

The growth of both cultured and wild yeasts is inhibited totally by adding actidione to the media at concentrations of 4–20 mg/l. This concentration does not inhibit the growth of bacteria. Media composition and incubation conditions allow some degree of selection of different groups of bacteria (i.e., lactic acid bacteria, acetic acid bacteria, wort bacteria).

34.2.3.1 Lactic Acid Bacteria

The detection of bacteria of the genera *Lactobacillus* and *Pediococcus* has spawned numerous studies and the search for better media is still ongoing. The detection of the "lactics" using nonselective liquid media has already been mentioned. The wide variety of solid media used are more selective. In general, they can detect lactobacilli in 2 or 3 days and of pediococci in 5 or 6 days of anaerobic incubation under CO_2.

Casey and Ingledew (1981) published a complete review dealing with all of the media developed up to the time of their study (they listed 23), giving their composition and their relative performances. We will mention here only the most widely used.

Media may be rendered very selective for lactic acid bacteria by adding inhibitors of the growth of gram-negative bacteria, such as β-phenylethanol or polymyxin. Phenylethanol allows the selection of *Lactobacillus* and *Pediococcus*, but also slightly inhibits their growth. Polymyxin at the minimum concentration inhibiting the growth of gram-negative bacteria inhibits the growth of pediococci and is usable for the specific detection of lactobacilli. Unless otherwise indicated, the following media do not contain inhibitors, since better recovery and more rapid responses (less inhibition) are often preferred to greater selectivity.

WLD medium (Wallerstein Laboratories Differential medium) differs from WLN medium by containing 4–10 mg of actidione per liter. After incubation at 25°C, lactobacilli are detected in 3 or 4 days, pediococci in 7 or more days. Modifying WLD medium by adding 2% peptonized milk according to Niefind and Spath (1971) gives better yields in a slightly shorter time.

UBA medium (Kozulis and Page, 1968), with added actidione incubated anaerobically at 25°C, gives faster results than WLD medium.

HRM medium (Hsu's rapid medium) is able to detect lactobacilli after 48 hours of aerobic incubation (Hsu et al., 1975b). HLP medium (Hsu's *Lactobacillus pediococcus* medium) is used aerobically in test tubes. The detection of lactic acid bacteria may be done after 24 hours of incubation (Hsu et al., 1975a). The complete formula of HLP medium, which probably includes reducing agents, has not been published.

MRS medium (De Man, Rogosa, Sharpe, 1960), with added actidione incubated anaerobically at 25°C, can detect lactobacilli in 48 hours and pediococci in 7–8 days. This medium, of which the carbon source is glucose, is not selective and allows the growth of other bacteria besides lactic acid bacteria.

Rogosa's medium (Rogosa et al., 1951) is more selective than MRS medium, its pH being lower. With added actidione and incubated anaerobically at 25°C, it can detect lactobacilli in 48 hours and pediococci in 5 days.

Van Keer and associates (1983) proposed a medium called LGM (*Lactobacillus* growth medium), rich in vitamins and in which glucose is the energy source. It allows rapid but nonspecific detection (enumeration possible after 24 hours of semianaerobic incubation). The medium allows the growth of bacteria other than the lactics, including the acetic bacteria.

The type of carbon source is an important parameter in media formulation. Maltose-based media have been proposed for the detection of lactic acid bacteria since the 1960s (Williamson, 1959).

VLB L41 medium by Wackerbauer and Emeis (1969) includes three energy sources: maltose, xylose, and fructose. This medium, which is made selective by the addition of phenylethanol, gives satisfactory results for the detection of lactobacilli and pediococci.

Raka Ray no. 3 medium (Saha et al., 1974) also gives good results. Maltose and fructose are the energy sources. Although it does not contain inhibitors other than actidione for the analysis of starter cultures, this medium does not allow the growth of contaminating facultative anaerobic bacteria, giving it good specificity for lactic acid bacteria with satisfactory recovery.

Boatwright and Kirsop (1976) proposed a sucrose-based medium, which gives better results than MRS medium. The authors suggest making it more selective for the detection of all lactic acid bacteria by adding phenylethanol, and for the detection of lactobacilli alone by adding polymyxin. Many lactic acid bacteria, however, may be detected on media with sucrose as sole carbon source only after a rather long incubation.

Considering that all pediococci are able to grow on glucose and that lactobacilli that are unable to grow on glucose may grow on maltose, Lawrence and Leedham (1979) proposed a "modified" MRs medium with maltose and glucose as energy sources which gave excellent results.

The composition of VLB S7 medium (Emeis, 1969) has not been published. This medium gives good results for lactobacilli and especially for pediococci. Pediococci may be detected in a starter culture in less than 6 days of incubation regardless of the dilution.

NBB medium (Nachweismedium für bierschädliche Bakterien), proposed by Back (1980), containing glucose and maltose as energy sources and malic acid, appears to be very effective for the detection of lactobacilli and especially of pediococci. Comparative studies of different media, mainly VLB S7 with NBB, demonstrated that the development of lactic acid bacteria was more rapid on NBB medium (Dachs, 1981; Wackerbauer and Rinck, 1983; Back et al., 1984).

The contribution of Nakagawa (1964) for the specific detection of pediococci should also be noted. In the Nakagawa medium, D-mannose, the sole carbon source, is utilized by pediococci but not by lactobacilli.

The EBC, in *Analytica Microbiologica*, recommends the use of modified MRS medium and of Raka Ray no. 3 medium (Martin and Moll, 1984).

34.2.3.2 Acetic Acid Bacteria

The isolation of the aerobically respiring acetic acid bacteria is done on low pH culture medium, containing 2–4% ethanol as energy source. The most widely used of these is that of Williamson (1959).

34.2.3.3 Wort Bacteria

The importance of enterobacteria is brewers' starter cultures has at times been neglected. Generally present in wort, they may multiply during the first days of fermentation and survive when the pH is higher than 4.5 and the ethanol concentration is below 1–1.2%. *Obesumbacterium proteus* (a short "*fat*" gram-negative rod), however, resists better than the other coliforms. Furthermore, its ability to flocculate with yeasts leads to rapid enrichment of the starter culture with this species during harvesting.

Detection is done on lactose and bile acid based differential media, such as MacConkey medium and DCL medium (deoxycholate, citrate, lactose) with added actidione (10–20 mg/l). Media are incubated at 30°C. Colonies of coliforms and *Hafnia alvei* appear in 24 hours while those of *Obesumbacterium proteus* are visible only after 48 hours.

34.3 Accelerated Conventional Methods

The first works by Richards (1970), followed by those of Barney and Helbert (1975), led to the association of membrane filtration, staining and observation of microcolonies obtained after short incubation periods (i.e., 10–20 hours aerobically or 2–4 days anaerobically). After incubation, the membranes are heated to 105°C to fix the colonies, immersed in a stain (most often 0.5% safranin), washed with water, and then dried. They are then clarified (using immersion oil for cellulose-based membranes) and examined under the microscope.

Harrison and Webb (1979) proposed a rapid method of detecting yeasts by nondestructive staining of microcolonies, making later identification possible. A mixture of fluorochromes which stain proteins, chitin, and β-polyholosides is added to WLN agar.

These works dealt with the analysis of membranes carrying small yeasts and bacterial loads. Lin (1976a and 1976b) analyzed the influence of high concentrations of cultured yeasts (10^5 cells per membrane) on the development of wild yeasts and contaminating bacteria. The membranes were incubated on the solid media suggested in the conventional methods which inhibit cultured yeasts and detect wild yeasts and bacteria. Taking into consideration a few remarks concerning the significance of the membrane type as well as a few adaptations for the use of lysine medium, the results obtained by Lin suggest the possibility of applying the microcolony method to the analysis of yeasts.

Detection methods based on microcolony observation, although not very rapid, have the advantage of being applicable on a routine basis without implementing overly elaborate techniques, of being quantitative, and of recovering all of the contaminants, given a judicious choice of media.

34.4 Rapid Techniques

Numerous reviews of these methods have been published, including that of Bourgeois (1983), dealing with brewers' starter cultures.

34.4.1 Immunofluorescence

Direct microscopic examination allowing, at best, a detection threshold of $1/10^3$, the use of a specific stain for contaminants must provide a considerable decrease in this value by facilitating the rapid examination of a large number of microscopic fields. The difficulty is finding a stain which labels all cells other than the cultured yeast. Immunological methods, particularly immunofluorescence, appear to provide the answer. Their application in brewing is based on the fundamental works of Campbell (1968, 1971) on the antigenic properties of *Saccharomyces.*

34.4.1.1 Principle of the Method

The basic principle may be taken down into a few steps: immunization of a rabbit against starter culture contaminating yeasts or bacteria, recovery of the γ-globulin fraction produced specifically against these microorganisms (rabbit antiyeast or antibacterial antiserum), labeling of this antiserum by chemical reaction with a fluorescein derivative. When added to a starter culture preparation containing these yeasts or bacteria, the specific γ-globulins of the labeled antiserum will bind to the yeasts and bacteria and the fluorescence observed microscopically with ultraviolet illumination will thus allow their detection.

One variation of the technique consists of not labeling the rabbit antiyeast or antibacterial serum directly but of preparing an antirabbit antiserum in another animal such as a sheep and labeling the sheep antirabbit γ-globulin fraction with a fluorescein derivative such as fluorescein isothiocyanate. This variation provides better fluorescence and requires the preparation of only one fluorescent antibody. The starter culture preparation to be analyzed, containing yeasts or bacteria previously used to immunize a rabbit, is mixed with rabbit antiserum. The antigen–antibody complex formed is detected microscopically after adding sheep fluorescent antirabbit antiserum. When all of the required antisera are available, the determination of a contamination in a starter culture requires only a few hours.

34.4.1.2 Detection of Wild Yeasts

One of the difficulties in applying this method is due to the close serological relatedness of the various species of *Saccharomyces*, the antisera produced generally binding to all *Saccharomyces*, both contaminant and culture. Several methods using "exhaustion" of the antiserum by adsorption have provided solutions.

34.4.1.2.1 Richards-Cowland Method

Developed before 1970 (Richards and Cowland, 1967; Richards, 1968; Cowland, 1968; Richards, 1969) the Richards-Cowland method is well suited to the analysis of *S. cerevisiae* starter cultures.

Saccharomyces belong to two serological types (antigenic subgroups I and II). In order to produce antibodies which react with all *Saccharomyces*, a rabbit antiserum is prepared from species belonging to these two subgroups, *S. bayanus* and *S. cerevisiae* var. *ellipsoideus*. In this mixture, the antiserum antibodies reacting with the cultured yeast strain are eliminated by adsorption with a pure culture of this yeast. This operation must be repeated until the pure culture no longer fluoresces in the presence of the antibody recovered after the previous adsorption plus fluorescent reagent.

The application of this method to *S. cerevisiae* allows the detection of all contaminants belonging to the genus *Saccharomyces* with a sensitivity in the range of $1/10^6$. The execution of the protocol is delicate, however, since in spite of the exhaustion of the rabbit antiserum, cell buds may still fluoresce. The loss of the cell permeability barrier leads to nonspecific fluorescence of the cells.

Applied to *S. uvarum* starters, the method is unsatisfactory. This species includes both antigenic subgroups. The adsorption of the antiserum onto cultured yeasts of subgroup II eliminates all antibodies directed against *S. cerevisiae* and many other species, including *S. diastaticus*. The results are not as poor with *S. uvarum* starters of subgroup I.

Haikara and Enari (1975) proposed other antiserum formulas capable of detecting a very broad spectrum of contaminants of top fermentation starter cultures, covering 85% of wild yeasts, of which almost all are *Saccharomyces*, plus a few other genera such as *Candida* and *Torulopsis*.

Hammond and Jones (1979) observed large variations in the responses to this test in the analysis of *S. cerevisiae* starter cultures coming from successive harvests—variations which parallel tests by conventional tests on selective media do not include. Changes seem to occur in the yeast cell-wall composition which modify the cell antigenic properties. The authors concluded that the method is good for rapidly obtaining evidence of culture contamination but that contamination must be tested by conventional methods.

Immunofluorescent staining does not distinguish living cells from dead cells. Chilver and co-workers (1978) incorporated a stain, methylene blue, which indicates viability.

34.4.1.2.2 Other Methods

The Klaushofer-Dorfwirth (1969) method is for testing bottom fermenting starter cultures using immunofluorescence. Rabbit antiserum is prepared against cultured *S. uvarum.* The antiserum undergoes a treatment making it specific to this species. When the treated serum is used for the analysis of a contaminated starter, all cultured *S. uvarium* cells appear fluorescent while wild yeasts do not. The advantage of this method over that of direct microscopy is the detection of contaminants having the same morphology as the cultured yeast, although the sensitivity is in the same range ($1/10^3$). It may be useful for starter cultures of *S. uvarum* subgroup II.

Bouix and Leveau (1978, 1980) improved this method by using a second fluorescent stain which allows easier observation of contaminants. Evans blue is added after adding the fluorescein isothiocyanate labeled antirabbit antiserum, as in the method of Klaushofer and Dorfwirth. Microscopic observation with ultraviolet light reveals fluorescent green yeast cell walls and reddish-brown centers for the cultured yeast while the contaminating cells are red. The limit of detection is still low at $1/10^3$.

34.4.1.3 Detection of Bacteria

The application of immunofluorescence to the detection of contaminating bacteria is less justifiable than for contaminating yeasts since simpler methods are available.

Dolezil and Kirsop (1976) demonstrated the possibility of distinguishing *Pediococcus* and *Micrococcus* serologically. A commercial antiserum intended for the identification of serological group D strains of *Streptococcus* reacts with both species. This antiserum treated by means of a suitable adsorption may be used to distinguish these genera in starter cultures.

Work by the same authors (Dolezil and Kirsop, 1975) applied the method to the specific detection of *Lactobacillus.* Antisera prepared from a few *Lactobacillus* species (*L. brevis, L. casei* var. *alactosus* and var. *rhamnosus*) contain antibodies specific to the bacteria used as well as less specific antibodies. After suitable adsorption procedures for the nonspecific antibodies, rigorously specific immunosera usable for identifying the *Lactobacillus* species may be obtained. This process is more applicable to specific identifications than to general testing.

34.4.1.4 Other Possibilities

Hope and Tubb (1985) summarized the new possibilities offered by immunological techniques. Fiber-optic based microfluorimeters measures fluorescence emitted by a single cell and eliminate interference associated with nonspecific binding of fluorescent antibodies. Although not immunofluorescence, the enzyme-linked immunosorbent assay (ELISA) now used routinely in medical microbiology may be envisioned for starter culture analysis. This consists of binding a carefully chosen enzyme to the microorganism-directed antibody as a marker rather than using a fluorescent label. The amount of enzyme activity can then be used to assess the contamination. The first results in the brewing industry were for beer analysis (Winnewisser and Donhauser, 1987).

Developments in the production of monoclonal antibodies, combined with progress in the identification of antigenic markers on the yeast cell envelope will undoubtedly increase the specificity of these methods considerably.

34.4.2 Measurements of Physicochemical Parameters

When a selective medium blocks the growth of the cultured yeast, variations of measured physicochemical parameters indicate the presence of a contaminant and may provide an estimate of its concentration.

34.4.2.1 Particle Counters

Yeast is suspended in an electrolyte solution. The suspension is drawn at a known flow rate through an orifice of a diameter in the same range as that of yeast cells. Two electrodes with a voltage across them are placed on either side of the orifice, the current flowing through the electrolyte. The passage of a microbial cell produces a change in the resistance of the circuit which, above a certain threshold, provokes an electrical impulse recorded as a counted event. Thus, a very rapid enumeration of cells in a starter culture is made, but counters may also be used to examine the composition of a heterogenous starter culture made up of cells of different sizes (Claveau et al., 1968; Drake and Tsushiya, 1973). Unfortunately, these counters are poorly applicable to the detection of contaminants.

34.4.2.2 Impedance Measurements

Kilgour and Day (1983) demonstrated the ease of use and the effectiveness of a method based on the measurement of variations of the impedance of top-fermenting starter cultures contaminated by *Obesumbacterium proteus*. The starter culture is prepared in WLN nutrient medium with actidione (20 mg/l) inhibiting yeast growth. Potassium chloride (0.35% w/v) is added to provide sufficient conductivity for the device. The maximum concentration of the yeast is 10^6 cells/ml. Beyond this, the conductivity increases with time, even in the absence of contaminants. When the level of contamination by *O. proteus* reaches 10^6 cells/ml, the conductivity increases sharply. The relationship between the $\log_{10}$ of the initial contamination and the time elapsed before the rise in conductivity is linear. An initial conductivity of 10 cells/ml is detected after 20 hours of incubation. An initial contamination of 10^6–10^7 cells/ml is detectable immediately. This easily automatable procedure is of great interest.

34.4.2.3 Radiometric Detection

A group headed by Bourgeois pioneered attempts to apply methods based on incorporating labeled substances into microorganisms in the brewing industry (Bourgeois et al., 1973; Mafart et al., 1976, 1978, 1981). The method uses the incorporation of ^{14}C-lysine. The performance is quite remarkable, due partly to the intensity of the amino acid absorption phenomenon, but also to the sensitivity of counting by scintillation which can detect quantities of amino acid in the range of 10^{-10}–10^{-11} g in cells retained on filtering membranes. In the case of yeasts, a few dozen cells are detectable in 9 hours. A single cell of *Escherichia coli* is detected in about 6 hours.

Work focused first on testing the sterility of beer, then the authors endeavored to extend the range of possible applications of the technique. Salaun (1978) thus used it to detect lactic acid bacteria in starter cultures and in fermenting wort. Trials were done using mixtures of cultured yeasts and lactic acid bacteria on modified (lysine-free) MRS medium supplemented with 2 mg of actidione per liter and 4% β-phenylethyl alcohol to inhibit growth of yeast and gram-negative bacteria, respectively. Actidione produces a 150-fold reduction in ^{14}C-lysine absorption by yeasts. The absorption of this amino acid by *Lactobacillus* under these conditions is not reduced and is about 30,000 times the amount absorbed by a yeast cell. With this method, the generally desired detection threshold of $1/10^4$ can be reached. Its application to the detection of lactic acid bacteria in starter cultures may certainly be envisaged, but the procedure would likely be quite delicate.

34.5 Conclusion

The microbiologist responsible for testing the quality of starter cultures currently has a set of rather effective techniques at his or her disposal, but no single technique is sufficient. To determine the choice of methods, the objectives must first be well defined: sensitivity, speed of detection, specificity, and so on. The recovery criterion, in terms of broadness of the microbial spectrum detected as much as speed, is often incompatible with selectiveness. This sometimes leads to confusion when comparing publications that recommend one method over another, since these criteria are not always clearly defined.

References

Back W., 1980. Bierschädliche Bakterien. *Brauwelt*, **43**, 1562–1569.

Back W., Dürr P., Anthes S., 1984. Nährböden VLB S7 und NBB; Erfahrungen mit beiden Medien im Jahr 1983. *Monats. Brau.*, **37**, 126–131.

Barney M. C. and Helbert J. R., 1975. A microcolony method for the rapid determination of contaminant microorganisms in beer. *Tech. Quart.*, **12**, 23–27.

Boatwright J. and Kirsop B. H., 1976. Sucrose agar. A growth medium for spoilage organisms. *J. Inst. Brew.*, **82**, 343–346.

Bouix M. and Leveau J. Y., 1978. Application de l'immunofluorescence à la recherche des levures sauvages dans les levains de brasserie. *Bios*, **11**, 35–37.

Bouix M. and Leveau J. Y., 1990. Mise en évidence des levures sauvages dans les levains de brasserie par une méthode de double fluorescence. *Bios*, **2**, 27–35.

Bourgeois C., 1983. Méthodes rapides de contrôle microbiologique en brasserie. *Bios*, **14**, 23–29.

Bourgeois C., Mafart P., Thouvenot D., 1973. Méthode rapide de détection des contaminants dans la bière par marquage radioactif. *Proc. Europ. Brew. Conv.*, 219–230.

Brenner M. W., Karpiscak M., Stern H., Hsu W. P., 1970. A differential medium for detection of wild yeast in the brewery. *J. Am. Soc. Brew. Chem.*, **28**, 79–80.

Campbell I., 1968. Serological identification scheme for the genus *Saccharomyces*. *J. Appl. Bacteriol.*, **31**, 515.

Campbell I., 1971. Antigenic properties of yeasts of various genera. *J. Appl. Bacteriol.*, **34**, 237–242.

Casey P. and Ingledew W. M., 1981. The use and understanding of media used in brewing bacteriology. II: Selective media for the isolation of lactic acid bacteria. *Brew. Digest*, **55**, 38–45.

Chilver M. J., Harrison J., Webb T. J. B., 1978. Use of immunofluorescence and viability stains in quality control. *J. Am. Brew. Chem..*, **36**, 13–18.

Claveau J., Scriban R., Strobbel B., Carpentier Y., 1968. Etude biométrique des levures de brasserie. Cas particulier: Biométrie et microbiologie de quelques levures d'infection. *Brasserie*, **23**, 77–91, 141–159.

Cowland T. W., 1968. Variation in the serological behaviour of yeasts of the genus *Saccharomyces* towards fluorescent antibody. *J. Inst. Brew.*, **74**, 457–464.

Dachs E., 1981. NBB-Nachweismedium für bierschädliche Bakterien. *Brauwelt*, **121**, 1778–1784.

De Angelo J and Siebert K. J., 1987. A new medium for the detection of wild yeast in brewing culture yeast. *J. Am. Soc. Brew. Chem.*, **45**, 135–149,

De Man J. C., Rogosa M., Sharpe M. E., 1960. A medium for the cultivation of *Lactobacilli. J. Appl. Bacteriol.*, **23**, 130–135.

Dolezil L. and Kirsop B. H., 1975. An immunological study of some *Lactobacilli* which cause beer spoilage. *J. Inst. Brew.*, **81**, 281–286.

Dolezil L. and Kirsop B. H., 1976. The detection of *Pediococcus* and *Micrococcus* in breweries, using a serological method. *J. Inst. Brew.*, **82**, 93–95.

Drake J. and Tsuchiya H., 1973. Differential counting in mixed cultures with coulter counters. *Appl. Microbiol.*, **26**, 9–13.

Emeis C. C., 1969. Methoden der brauereibiologischen Betriebskontrolle. III: VLB-S7-Agar zum Nachweis bierschädlicher Pediokokken. *Monats. Brau.*, **22**, 8–11.

Haikara A. and Enari T. M., 1975. The detection of wild yeast contaminants by the immunofluorescence technique. *Proc. Europ. Brew. Conv.*, 363–375.

Hall J. F., 1971. Detection of wild yeasts in the brewery. *J. Inst. Brew.*, **77**, 513–516.

Hammond J. R. M. and Jones M., 1979. The immunofluorescence staining technique for the detection of wild yeast: Practical problems. *J. Inst. Brew.*, **85**, 26–30.

Harris J. O. and Watson W., 1968. The use of controlled levels of actidione for brewing and non-brewing yeast stain differentiation. *J. Inst. Brew.*, **74**, 286–290.

Harrison J. and Webb T. J. B., 1979. Recent advances in the rapid detection of brewery microorganisms and development of a microcolony method. *J. Inst. Brew.*, **85**, 231–234.

Hope C. F. A. and Tubb R. S., 1985. Approaches to rapid microbial monitoring in brewing. *J. Inst. Brew.*, **91**, 12–15.

Hsu W. P., Taparowski J. A., Brenner M. W., 1975a. Schnellzüchtung von brauerei-Milchsäurebakterien. *Brauwissenschaft*, **28**, 157–160.

Hsu W. P., Taparowski J. A., Brenner M. W., 1975b. Two new media for culturing of brewery organisms. *Brew. Digest*, **50**, 52–57.

Ingledew W. M. and Casey P., 1982. The use and understanding of media used in brewing mycology. I: Media for wild yeast. *Brew. Digest*, **57**, 18–22.

Kato S., 1967. A new measurement of infectious wild yeasts in beer by means of crystal violet medium. *Bull. Brew. Sci.*, **13**, 19–24.

Kilgour W. J. and Day A., 1983. The application of new techniques for the rapid determination of microbial contamination in brewing. *Proc. Europ. Brew. Conv.*, 177–184.

Klaushofer H. and Dorfwirth K., 1969. Preparation of a monospecific anti-*Saccharomyces carisbergensis* serum for the biological quality control. Antonie van Leeuwenhoek, 35, suppl. *Yeast Symposium.*

Kozulis J. A. and Page H. E., 1968. A new universal beer agar medium for the enumeration of wort and beer microorganisms. *J. Am. Soc. Brew. Chem.*, **26**, 52–58.

Kretschmer K. F. and Kretschmann W., 1972. Uber einen Universalnährboden zum Nachweis von Bierschädlingen. *Brauwelt*, **88**, 1815–1820.

Lawrence D. R. and Leedham P. A., 1979. The detection of lactic bacteria. *J. Inst. Brew.*, **85**, 119–121.

Lin Y., 1974. Detection of wild yeasts in the brewery. III. A new differential medium. *J. Am. Soc. Brew. Chem.*, **32**, 69–76.

Lin Y., 1976a. Detection of wild yeasts in the brewery. VIII: The use of membrane filtration technique and microcolony method. *Brew. Digest*, **51**, 44–50.

Lin Y., 1976b. Use of various brands of membrane filters for the detection of brewery bacteria. *J. Am. Soc. Brew. Chem.*, **34**, 141–144.

Lin Y., 1981. Formulation and testing of cupric sulphate medium for wild yeast detection. *J. Inst. Brew.*, **87**, 151–154.

Lodder J., 1970. *The yeasts. A taxonomic study.* North Holland Publishing Company.

Longley R. P., Edwards G. R., Mathews S. A., 1980. Improved detection of wild *Saccharomyces. J. Am. Soc. Brew. Chem.*, **38**, 18–22.

Lund A. and Thygesen P., 1957. Detection of spores-forming wild yeasts. *Proc. Europ. Brew. Conv.*, 241–248.

Mafart P., Bourgeois C., Duteurtre B., Moll M., 1976. Radiometric method for control of filtration and pasteurization. *Tech. Quart.*, **13**, 157–160.

Mafart P., Bourgeois C., Duteurtre B., Moll M., 1979. Use of ^{14}C-lysine to detect microbial contamination in liquid foods. *Appl. Environ. Microbiol.*, **35**, 1211–1212.

Mafart P., Cleret J. J., Bourgeois C., 1981. Optimisation of ^{14}C-lysine concentration and specific activity for the radiometric detection of microorganisms. *Europ. J. Appl. Microbiol. Biotechnol.*, **11**, 189–192.

Martin P. A. and Moll A., 1984. *Analytica Microbiologica EBC. Bios*, **15**, 73–82.

1962. Microbiological controls in the brewery. *Proc. Am. Soc. Brew. Chem.*, **20**, 173–175.

Morris E. O. and Eddy A. A., 1957. Method for the measurement of wild yeast infection in pitching yeast. *J. Inst. Brew.*, **63**, 34–35.

Nakagawa A., 1964. A simple method for the detection of beer-sarcinae. *Bull. Brew. Sci.*, **10**, 7–10.

Niefind H. J. and Spath G., 1971. Die Bildung flüchtiger Aromastoffe durch Mikroorganismen. *Proc. Europ. Brew. Conv.*, 459–468.

Richards M., 1968. The incidence and significative of wild *Saccharomyces* contaminants in the brewery. *J. Inst. Brew.*, **74**, 433–435.

Richards M., The rapid detection of brewery contaminants belonging to the genus *Saccharomyces.* Examination of lager yeasts. *J. Inst. Brew.*, **75**, 476–480.

Richards M., 1970. Routine accelerated membrane filter method for examination of ultra-low levels of yeast contaminants in beer. *Wallerstein Lab. Commun.*, **33**, 97–101.

Richards M. and Cowland T. W., 1967. The rapid detection of brewery contaminants belonging to the genus *Saccharomyces* by a serological technique. *J. Inst. Brew.*, **73**, 552–558.

Röcken W., 1983. Fremdhefennachweis in der obergärigen Brauerei mit dem Pantothenat-Agar. *Monats. Brau.*, **36**, 65–69.

Rogosa M., Mitchell J. A., Wiseman R. F., 1951. A selective medium for the isolation and enumeration of oral and fecal *Lactobacilli. J. Bacteriol.*, **62**, 132–133.

Saha R. B., Sondag R. J., Middlkauf J. E., 1974. An improved medium for the selective culturing of lactic acid bacteria. *J. Am. Soc. Brew. Chem.*, **32**, 9–10.

Salaun M., 1978. Détection radiométrique rapide des Lactobacilles dans la bière et le moût en fermentation. Thesis, Ing. INPL, Nancy.

Seidel H., 1973. Differenzierung zwischen Brauerei-Kulturhefen und “wilden Hefen”. *Brauwissenschaft*, **26**, 179–183.

Taylor G. T., Marsh A. S., 1984. MYGP + copper, a new medium that detects both *Saccharomyces* and non-*Saccharomyces* wild yeast in the presence of culture yeast. *J. Inst. Brew.*, **90**, 134–145.

Thurston P. A., 1986a. Detection and control of contaminants in pitching yeast; EBC symposium on brewers' yeast, Helsinki, November 1986, Monograph XII, 84–92.

Thurston P. A., 1986b. The phenolic off-flavour test: a method for confirming the presence of wild yeasts. *J. Inst. Brew.*, **92**, 9–10.

Van Keer C., Van Melkebeke L., Vertriest W., Hoozee G., Van Scooneberghe E., 1983. Growth of *Lactobacillus* species on different media. *J. Inst. Brew.*, **89**, 361–363.

Wackerbauer K. and Emeis C. C., 1969. Uber die bierschäflichen Bakterien der Gattung *Lactobacillus* (Biermilchsäurestäbchen). II: Der Nachweis der bierschädlichen Laktobacillen in der Brauereibetriebskontrolle. *Monats. Brau.*, **22**, 3–8.

Wackerbauer K. and Rinck M., 1983. Über den Nachweis von bierschädlichen Milchsäurebakterien. *Monats. Brau.*, **36**, 392–395.

Walsh R. M. and Martin P. A., 1977. Growth of *Saccharomyces cerevisiae* and *Saccharomyces uvarum* in a temperature gradient incubator. *J. Inst. Brew.*, **83**, 169–172.

Walsh R. M. and Martin P. A., 1978. Detection of wild yeast in *Saccharomyces carlisbergenis. J. Inst. Brew.*, **84**, 79.

Walters L. S. and Thiselton M. R., 1953. Utilization of lysine by yeasts. *J. Inst. Brew.*, **59**, 401–404.

Williamson D. H., 1959. Selective media in the enumeration of bacteria in pitching yeasts. *J. Inst. Brew.*, **65**, 154–164.

Winnewisser W. and Donhauser S., 1987. Enzymimmuntest zum Nachweis von *Pectinatus cerevisiiphilus* in Bier. *Proc. Europ. Brew. Conv.*, 481–488.

CHAPTER

35

Microbiological Monitoring of Factory Equipment, Atmosphere, and Personnel

A. Plusquellec, J. Y. Leveau

35.1 Introduction

The current trend of the food industry is tending towards marketing an ever-growing number of diverse foods which are increasingly formulated. In its final form, a foodstuff is sometimes quite different from its original form, which often provided a natural protection. It has also undergone many handling steps, each of which is likely to introduce its share of contaminants. Furthermore, there is a tendency to require increasingly long preservation times for these foods. Although great progress has been made in this field, the preservation of formulated foods poses serious problems. All of these elements (numerous handling steps, prolonged storage) multiply the health risks associated with our diets. It is therefore absolutely necessary for product safety testing techniques to keep pace with food processing and preparation techniques.

Microbiological testing is too often limited to the finished product. The usefulness of such testing is narrow.

- Bacteriological analysis has a delay in response very often too long to allow intervention before the distribution of the product.
- This type of test serves only to ascertain a result. In the event of defective results, it gives no indication as to the origin of the contamination.

This test must be considered as a verification of manufacturing hygiene. It is essential that the parameters which have an impact on the contamination of the finished product be under control. Contamination depends partly on the quality of the starting material and partly on the input of microorganisms throughout the processing line. Microbiological contamination during manufacturing may be due to various elements:

- the work areas
- the equipment
- the personnel
- the ambient air

The techniques presented here test the microbiological status of these parameters. At the outset, the manufacturing line under consideration must be examined thoroughly in order to identify the critical points where microbiological contamination is possible. Once this is known, a routine testing plan is defined. There must be a sufficiently limited number of samplings for the plan to be carried out frequently. The permanently implemented scheme for monitoring sources of contamination must allow total control of the microbiological quality of the finished product.

35.2 Microorganisms Analyzed

The techniques for the detection of contaminations apply to all microorganisms of interest in food hygiene. The microbial group whose presence is the most worrisome in the finished product will be the object of special analytical attention. In the case of highly handled food products, for example, analysis for staphylococci is very important.

Counts of the total mesophilic flora is a good indication of the overall level of contamination. In some sectors of the food industry (pastry, bread making), molds are the greatest cause for concern, and testing is oriented accordingly. The culture media used are the same as for conventional enumeration.

35.3 Testing of Equipment and Work Areas

Very harmful nests of microbial contamination may be found on the equipment and in the work area. Cleaning and disinfection operations are consequently of paramount importance in the food industries. Cleaning should eliminate soiling which in this industry comes for the most part from organic compounds from starting materials or products in processing. Disinfection with a chemical substance should destroy microorganisms present. Both cleaning and disinfecting must be completed by a rinsing which should eliminate the substances used in the first two steps, without introducing new soiling or new microbes.

It is essential that these operations be performed correctly. The use of improperly treated equipment may have grave consequences. In fact, many fluctuations in product performance during both manufacturing and stocking are attributable to improper cleaning and/or to incomplete or ineffective disinfection. Spoilage of product quality may result from this due to too high a microbial load as well as the possible presence of pathogenic microbes. In addition, poorly performed rinsing may result in the presence of undesirable residues of cleaning and disinfecting substances in the finished product. Furthermore, if the microbiological quality of the water is poor, the rinsing step may be responsible for the recontamination of the equipment, no matter how well the earlier operations are carried out. Microbiological testing of the rinsing water is therefore necessary, as well as of surfaces, such as the equipment, containers, floors, and so on.

The tests carried out first on the surface to be treated provide the means of determining the type(s) of microorganism(s) present and of evaluating the level(s) of contamination. The results determine which treatment procedures to apply. The tests on the treated surface help evaluate the effectiveness of the completed operation and if the results are satisfactory, validate the operation.

There are numerous techniques which may be used. They may be categorized according to their underlying principle (Favero et al., 1968):

- pressing or application techniques;
- techniques based on rinsing;
- techniques based on agar pouring;
- techniques based on swabbing.

Pressing, swabbing, and rinsing techniques involve direct or indirect sampling of microorganisms present on surfaces. Their detection and enumeration is done after amplification by culturing on a suitable medium or possibly by other more rapid methods. The results thus obtained are useful in a qualitative sense, that is, as a function of the culture media used, the type of microorganisms present on an examined surface may be known. Quantitatively, however, these techniques are very imprecise. In fact, the interactions between microorganisms and solid surfaces are very complex, some cells being able to adhere more or less irreversibly. Mechanisms of adhesion depend on several factors (the type of surface, the treatment undergone, etc.). There are currently no means available for detecting simultaneously all of the bacteria adhering to a support. The results are quantitatively obtained by default, the degree of recovery of the microorganisms present on the surface being less than 100%.

If done according to rigorously standardized procedures, the techniques do allow comparisons under proper conditions. It is possible in practice to compare the level of contamination by sampling the same surface at different times. The percentage recovery of microorganisms being a function of the nature of the surface, the comparison of the level of contamination of surfaces of different types is not possible. It is known, for example, that the percentage of microorganisms recovered on a wooden surface is very low compared to that on stainless steel surfaces (Angelotti et al., 1964).

As is usually the case for microbiological techniques, "colony-forming units" or CFUs are counted rather than cells. In actual fact, observation of surfaces by scanning electron microscope reveals clumps of microbial cells rather than isolated cells. Each of these clumps, possibly being part of the biofilm building up on the surface being examined, gives rise to only one colony after its detachment and incubation.

If the examined surfaces are cleansed and disinfected regularly, traces of detergent and disinfectant may be removed at the same time as the microorganisms. These compounds may subsequently inhibit the growth of the microorganisms and thereby bias the results both qualitatively and quantitatively. It is therefore necessary to incorporate into the culture medium substances such as egg lecithin (0.07 to 0.1%), Tween 80 (0.1 to 0.5%), sodium thiosulfate (0.05%), or even histidine (1%) which are able to neutralize the inhibitory action of the disinfectants.

35.3.1 Pressing or Application Techniques

Pressing and application techniques are easy to carry out. They consist of applying an agar-based medium to the surface to be analyzed. They are consequently better suited to flat, easily accessible surfaces.

In order to have proper contact between the aqueous agar and the tested surface, the sites examined must not be greasy as a greasy film can cause inaccuracy. Adding a surface active agent (0.5% Tween 80) may facilitate the recovery of bacteria on the medium from slightly greasy surfaces.

35.3.1.1 Rodac Plates

Petri plates 55 mm in diameter with a grid scored into the bottom portion are often used in order to facilitate reading of the results. They are specially designed (see Figure 35.1) so that the lid is held in place by means of a "trough" on the bottom portion, thereby providing a space between the inside surface of the lid and the top edge of the bottom portion. They are available pre-poured, generally with plate count agar for total flora but they may also be prepared with other media. In this case, the medium is aseptically poured so that a meniscus remains after cooling, which will be the analytical surface.

Once solidified and cooled, the plates are used by applying the agar onto the test surface making sure that the contact is done according to standardized conditions such as contact time (30 seconds to 2 minutes) and application pressure. A 200 g weight may be used for the latter purpose. The plate should not slip or slide around on the examined surface. After replacing the plate cover, the plates are incubated and observed regularly in order to be able to do the count before the colonies become confluent. The results are expressed in number of CFUs/cm^2 of examined surface (see Figures 35.2, 35.3 and 35.4).

Figure 35.1 Cross-section of a "contact" plate.

Figure 35.2 Pouring of the plate.

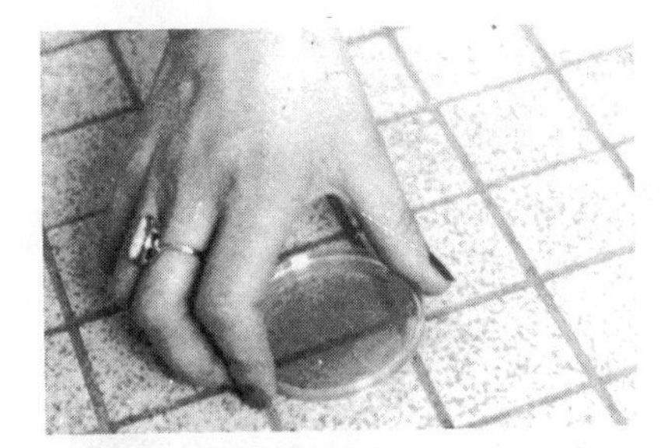

Figure 35.3 Applying to the tested surface.

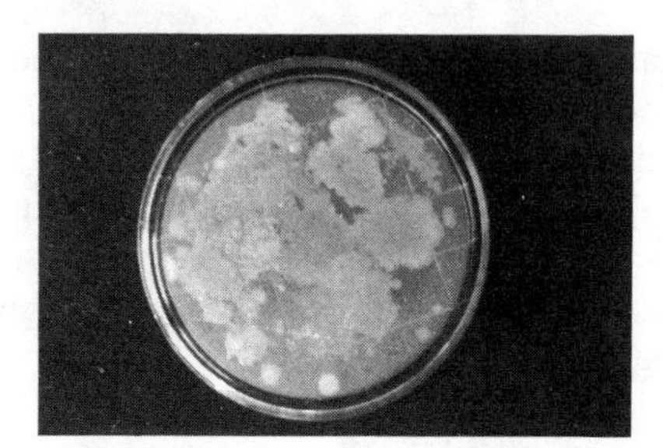

Figure 35.4 Result after incubation.

35.3.1.2 Use of Agar "Sausages"

The agar-based medium is poured into a plastic sheath. The medium is applied to the test surface after exposing a freshly cut agar surface using sterile technique. A slice of the agar-based medium is then removed and placed in a Petri plate (see Figures 35.5 and 35.6). Agar "sausages" may be prepared in the laboratory and thus constitute a very economical method. They are, however, difficult to preserve once cut. They are very convenient for testing in mass production environments.

35.3.1.3 Agar-Coated Slides

Agar-coated slides, thin layers of culture medium on a rigid plastic support, have been available for the past several years. First proposed for rapid tests

Figure 35.5 Exposure of a fresh, sterile surface.

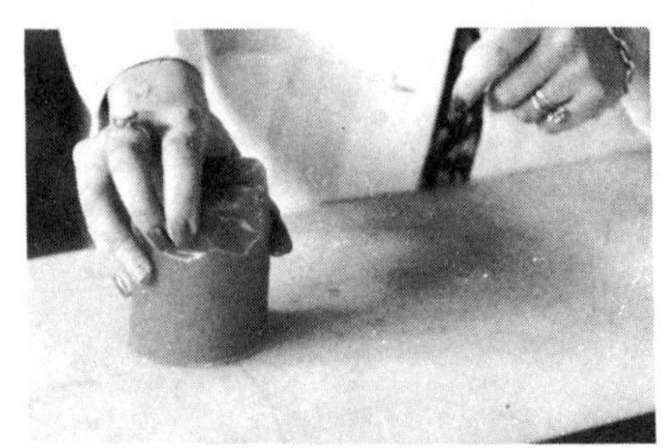

Figure 35.6 Applying to the test surface.

in the field of medical diagnostics—particularly for urine analysis—their use has spread to the food industry for rapid testing of certain types of products such as milk. They may be used for microbiological testing of surfaces either directly, by contact of the agar-based medium with the surface being tested, or indirectly after swabbing, the agar-coated slide being used to enumerate the flora in the liquid in which the germs removed by the swab are suspended. Given their somewhat restrictive form, their direct use is limited to testing surfaces whose profile falls within certain constraints. Flexible agar-coated slides have been commercialized more recently, so-called microbe indicators marketed by a French firm, which are supposed to be applicable to almost all surfaces. Different types of indicators are available for a variety of anticipated microfloral types. Agar-coated slides may also be made in the laboratory (Darbord and Brion, 1976).

35.3.1.4 Petrifilm (Lepage, 1987)

Petrifilm SM (3M Health) is a microbial enumeration system which uses a culture medium (PCA) packed between two flexible films. The dehydrated medium adheres to one of the films. For testing of surfaces, the film is rehydrated with sterile distilled water or a sterile solution of 0.1% lecithin and 0.5% polysorbate 80 (disinfectant neutralizing agent). After 15 minutes, the gel obtained may be applied to the test surface using light finger pressure. The other film serves as a support for a gelling agent and a stain (triphenyltetrazolium chloride) which imparts a red stain to the colonies. During incuba-

Figure 35.7 Applying to the test surface.

tion, the microorganisms develop between the two films, which limits the spreading frequently observed on the solid media in the previously mentioned techniques.

35.3.1.5 Indirect Applications

Via a Velvet Pad

A sterile velvet pad is applied to the test surface and then to the solid medium surface in an ordinary Petri plate.

Via an Adhesive Tape (Veulemans and Jacquemin, 1970)

A strip of adhesive tape removed aseptically is applied to the test surface and then placed on a solid medium surface (i.e., in a Petri plate). The performance of this procedure requires some practice and manual dexterity. Curved surfaces may also be tested by this method (see Figures 35.7 and 35.8).

35.3.2 Techniques Based on Swabbing

The swabbing techniques may be applied to all types of surfaces. It is especially efficient for detecting "microbial nests" which may form in bends, cavities, or curved surfaces (see Figure 35.9). The test surface is swept with a moist, sterile cotton swab. In order to standardize the technique, the test surface is defined

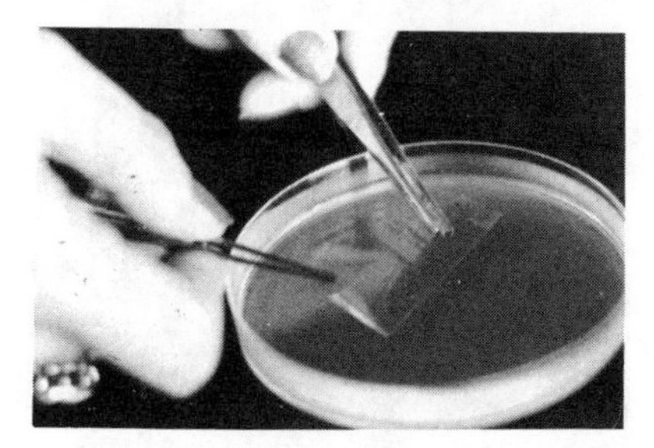

Figure 35.8 Applying to the agar.

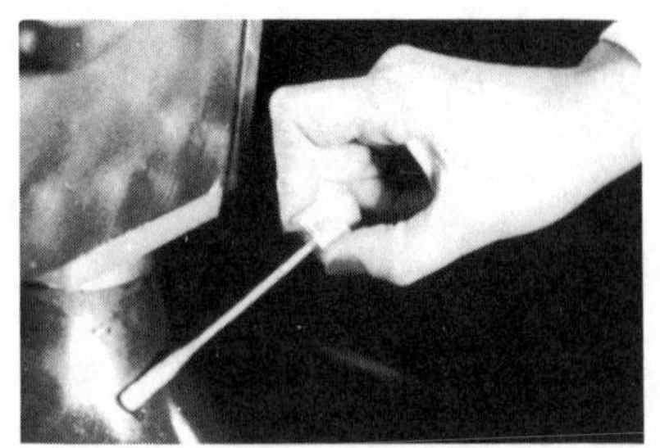

Figure 35.9 Swabbing a nonplanar surface.

with the aid of a sterile guide (see Figure 35.10). The swab is then immersed into a determined volume of sterile dilution medium. This is generally a buffered solution which may also contain disinfectant neutralizing agents (lecithin, Tween 80). The medium of Letheen (bacto Letheen broth, Difco) is suitable for this purpose. The microbes removed from the tested surface by the swab are dispersed into the dilution liquid with the aid of a "Vortex" mixer.

A variety of techniques are then applicable to the microbiological analysis of the suspension thus obtained:

- counting on solid medium inoculated by the conventional pour-plate or spread-plate technique;
- filtering onto a membrane;
- use of a "spiral" plater;
- use of Petrifilm.

Modern microbiological techniques are also applicable for the rapid quantification of the microflora suspended in the dilution liquid. The firm Perstrop offers a device called the Biocounter with reagents (Hygiene Monitoring Kit) which can rapidly determine the level of contamination of a surface by swabbing and ATP-metry. Another device based on the same principle, the "Biotrace Hygiene Monitor," is distributed in France by the firm Chemunex.

Regardless of the technique used for the analysis of the microbial suspension, swabbing is the most significant step determining the analytical result. In

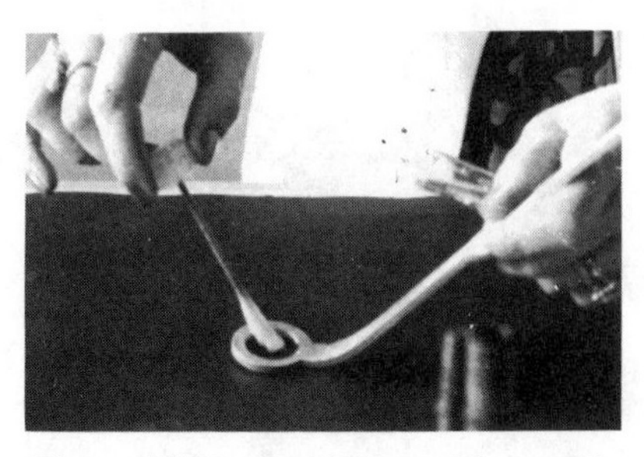

Figure 35.10 Swabbing with a guide.

order for results to be reproducible, it should be done according to as standardized a procedure as possible and by the same person.

As for the previously described application or pressing techniques, results are semiquantitative, the percentage of microorganisms recovered being of the same magnitude. In any event, microbial clumps or portions of biofilm being dispersed by agitating the dilution liquid, the more intense the agitation, the higher are the counts to be expected from conventional microbiological analysis. Under these conditions, comparisons of results obtained by the two different types of techniques should be avoided.

The swabbing technique allows the examination of highly contaminated surfaces because of the possibility of further diluting the microbial suspension obtained after dispersion. For this reason, it is particularly suited to microbiological testing of the surfaces of poultry carcasses. Furthermore, it readily lends itself to analysis for pathogenic bacteria. Swabs are transferred in this case to an enrichment medium (e.g., for *Salmonella* or *Staphylococcus aureus*) from which isolation and characterization procedures for the bacteria of interest are carried out.

35.3.3 Techniques Based on Rinsing

Rinsing techniques are used generally when no alternative exists. They are well suited to the microbiological testing of containers and piping. A sterile rinsing liquid is recirculated such that the test surfaces are contacted after which the liquid is subjected to microbiological analysis.

5.3.3.1 Testing of Bottles

Bottle testing techniques are particularly well suited for routine testing in this case. The sample of bottles to be tested should be representative of a typical washing cycle and will therefore depend on the type of washer being used and whose proper functioning is being tested microbiologically. Given the simplicity of its execution, the analysis is extended to cover lots of 10–12 bottles. As soon as the bottles are removed from the washer they are plugged with sterile cotton batting. Later in the laboratory, a carefully measured and noted volume of rinsing liquid (10 or 15 ml) is aseptically introduced:

Monopotassium phosphate	0.0425 g
Sodium hyposulfite	0.16 g
Tamol M	5 g
Sodium Carbonate	0.800 g
Distilled water	1,000 ml
Sterilize for 15 minutes at 12°C	

It may be necessary to add a surface active agent (Tween 80) to facilitate the suspension of the microbes. Mixing is done according to a standardized procedure (intensity, duration). Microbiological analysis by conventional techniques is then done on the rinsing liquid collected. Total aerobic mesophilic microflora are most often determined. Analysis for microflora capable of causing spoilage during storage may also be done (yeasts and acidophilic bacteria)—in the case of bottled drinks, for example. Results are expressed as numbers of microbes per bottle, based on counts of colonies on Petri plates.

35.3.3.2 Testing of Kegs

The keg testing technique is for testing larger volume containers. The same steps as for bottle testing are followed. A larger volume of sterile rinsing liquid may be introduced (300 ml of rinsing liquid for a 20 l keg), taking care to close the keg with a previously sterilized plug. Microbiological analysis is done by the same conventional technique or using filtering membranes correctly adapted to the low levels of contamination corresponding to a washing properly done.

This technique may also be applied to small containers such as sealers. Care must be taken to work under the best possible conditions in order to avoid contamination by handling.

35.3.4 Techniques Based on Agar Pour Plating

Agar pour plating techniques consist of pouring melted agar-based medium cooled to 45°C directly onto the surface being tested.

For planar surfaces, a testing zone must be defined using sterile glass rods or microscope slides (see Figure 35.11). The tested surface may also serve as the cover of a Petri plate containing liquified medium at 45°C and rapidly inverted to bring the agar into contact with the surface. After cooling in this position, the whole unit is placed in an incubator (Guiteras et al., 1954).

This technique is particularly useful for testing small containers or trays which may then be placed in an incubator (see Figure 35.12). It is obviously difficult to apply to work area surfaces.

The pouring methods are the most quantitative. Each bacterium (or microcolony) present may, in principle, give rise to a colony.

Figure 35.11 Pouring of the agar.

Figure 35.12 Incubation.

Figure 35.13 Test flask after incubation.

35.4 Atmospheric Testing

Atmospheric testing also refers to microbiological testing of the work environment or determination of airborne biocontamination. Air always carries some load of suspended particles. Microorganisms are adsorbed to some of these. For this reason, air is a major contamination factor that must be brought under control.

The simplest technique for determining airborne biocontamination consists of placing open Petri plates in the areas which are being examined, the plates containing the solid medium which corresponds to the category of microbes to be detected. Particles suspended in the air settle on the medium surface and, after exposure, the plates are covered and incubated. Microorganisms, clumps, or spores finding themselves on the agar surface and capable of developing on the medium give rise to one colony each. Preliminary trials must be done in order to determine the exposure time required to obtain a number of colonies between 30 and 300 per plate.

This is again a semiquantitative technique in which results cannot be expressed in terms of CFUs per unit volume of air. It does reflect air movements, convection, and other currents which are very significant in industrial work areas. It also allows comparisons in time and space and is useful for assessing the degree of contamination of work areas and of products by sedimentation of particles in the air as well as the risk of contamination of production lines by air. Furthermore, open Petri plates may be used all along the circuit in order to follow a product throughout its manufacturing and possibly pinpoint the zone of contamination by airborne particles.

Some methods use a sampling device and thus yield a more rigorous determination of airborne contamination (Luquet et al., 1979). Air is drawn by means of a turbine with a stable and low flow rate in order to avoid the destruction of the microorganisms at the points of impact. Bioparticles are suspended by sparging the air into a sterile solution or recovered on gelatin membranes or directly on the surface of an agar-based medium fixed to a plastic film or simply in Petri plates.

The length of sampling time allows the calculation of the volume of air analyzed for a given device. If properly used according to rigorously standardized procedures, these devices make it possible to monitor and control airborne biocontamination trends.

35.5 Testing at the Level of Personnel

Personnel play an important role in the microbiological quality of a finished product, but this role may be of a harmful nature in a variety of ways:

- by disseminating microbes already present, by manual contact, via clothing, shoes, hair, by air movement, sneezing, and so on;
- by introducing new microbes, especially of those having impact on product safety, i.e., staphylococci and fecal microbes;
- indirectly, personnel may be responsible for the proliferation of microorganisms by errors in handling, stocking, and cleaning procedures.

It is especially important for personnel in contact with foods to be particularly sensitized to food hygiene problems. This may be done with the aid of simple procedures similar to those used for product testing. Washing of hands is of major importance in the food industry. In order to test its effectiveness, the simplest method consists of making fingerprints on solid medium (see Figure 35.14). Pressing or application techniques may also be used for testing skin, as may sampling with adhesive tape (Lambion et al., 1972). When analyzing for particular microbial groups on hands (*Staphylococcus aureus*, *Salmonella*, *E. coli*), washing may be done by soaking or swabbing the hands in a sterile diluent which is subsequently taken for enrichment and analysis for the flora

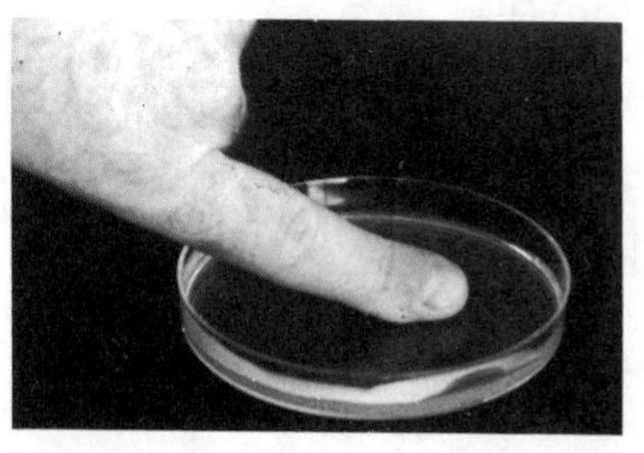

Figure 35.14 Fingerprinting on agar.

of interest. In order to sensitize personnel to the importance of wearing a hair net or cap, the inoculation of a Petri plate with a couple of hairs is surprisingly convincing. In the event that personnel are suspected of carrying dangerous organisms (staphylococci, salmonella, β-hemolytic streptococci), more complete examinations are necessary, which must be carried out by qualified medical personnel (decree* of March 11, 1977).

35.6 Conclusion

The techniques presented here make up a very valuable repertoire of tools for evaluating the microbiological status of surfaces, atmospheres, and personnel in contact with foodstuffs throughout the manufacturing process and all the way to distribution. The implementation of these techniques is generally simple and within the abilities of everyone involved. The interpretation of the results is easy in most cases, although it is nevertheless important to be wary of erroneous interpretations which may be due to the presence of invasive colonies.

Most of these methods remain limited to a narrow range of contamination. Furthermore, the precision of the results yielded is quite relative. This lack of precision may be offset by more rigorous sampling and standardization of operating procedures. Under these conditions, these techniques of contaminations will yield valuable results, mainly for providing answers to simple questions by comparative trials, that is, trends in contamination levels at a specific work post, throughout a production day, comparisons of identical equipment or work areas, verifying the effectiveness of a surface or atmospheric disinfection procedure (Beaucourt et al., 1976). Their well-planned and judicious use should lead to significant reductions in contaminations of finished products. If performed by manufacturing personnel or personnel closely associated with manufacturing, they also constitute an excellent means of sensitization to the notion of microbiological cleanliness.

The use of these techniques following the appearance of a contamination should trace its origin by working upstream through all of the stages of the processing line. However, they should be considered, above all, as techniques for the visualization of microbial contaminations, which allow testing of manufacturing hygiene and provide knowledge about contaminating inputs throughout the processing line, thereby making it possible to control the quality of the finished product.

**Translator's note:* The English word *decree* is used in this chapter to translate the French words *arrêté* and *décret* which corresponds to different administrative levels. An asterisk indicates that the French word used was *arrêté*, otherwise the word was *décret*.

References

Angeloti R., Wilson J. L., Litsky W., Walter W. G., 1964. Comparative evaluation of cotton swab and Rodac method for the recovery of *Bacillus subtilis* spore contamination from stainless steel surface. *Health Lab. Sci.*, **1**, 289–296.

ASPEC-APRIA. *Guide de la bio-contamination.* P. Isoard Cobac.

Beaucourt N., Groux P., Plouvier B., 1976. A propos de la pollution des ambiances en IAA. *Ind. Alim. et Agric.*, **6**, 713–724.

Dabord J. C. and Brion F., 1976. Contrôles microbiologiques des surfaces en milieu hospitalier. *La Nouvelle Presse Médicale*, **21**, 1349–1359.

Favero M. S. et al., 1968. Microbial sampling of surfaces. *J. Appl. Bacteriol.*, **31**, 336–343.

Guiteras A. F., Flett L. H., Shapiro R. L., 1954. A quantitative method for determining the bacterial contamination of dishes. *Appl. Microbiol.*, **2**, 100.

Lambion R., Veulemans A., Kastelyn A., 1972. Le contrôle de la qualité bactériologique dans les industries alimentaires. *Rev. Ferm. Ind. Alim.*, **27**, 243–251.

Lepage C., 1987. Technique nouvelle d'examen microbiologique des surfaces. *Sci. Aliments*, **7**, no. 7, 223–229.

Luquet F. M., Boudier J. F., Mouillet L., Grumelon D., Perre G., 1979. Pollution microbienne de l'air en IAA. *L'alim. et la Vie*, **67**, no. 3.

Veulemans A. and Jacqmin E., 1970. Etude d'une méthode simple pour la détermination du degré de pollution des surfaces, et la comparaison du pouvoir désinfectant de divers produits d'entretien. *Rev. Ferm. Ind. Alim.*, **25**, 58–67.

Etat de santé et Hygiène du personnel appelé à manipuler les denrées animales, ou d'origine animale. Arrêté du 11 mars 1977, *Journal Officiel*, 31 March 1977.

Suggested Readings, Chapters 34 and 35

Anonymous, 1992. Meating the sanitation challenge. *Food Rev.*, **19**, no. 3, 33, 35.

Fliss I., Simard R. E., Ettriki A., 1991. Comparison of three sampling techniques for microbiological analysis of meat surfaces. *J. Food Sci. Off. Publ. Inst. Food Technol.* Chicago, IL: The Institute. Jan./Feb. 1991. **56**, no. 1, 249–250, 252.

Jetton J. P., Bilgili S. F., Conner D. E., Kotrola J. S., Reiber M. A., 1992. Recovery of salmonellae from chilled broiler carcasses as affected by rinse media and enumeration method. *J. Food Prot.*, **55**, no. 5, 329–332; 20 ref.

Kotula A. W. and Emswiler-Rose B. S., 1988. Airborne microorganisms in a pork processing establishment. *J. Food Prot.* Ames, Iowa: International Association of Milk, Food, and Environmental Sanitarians. Dec. 1988. **51**, no. 12, 935–937.

Olenchock S. A., 1990. Endotoxins in various work environments in agriculture. *Dev. Ind. Microbiol.*, **31** (Suppl. 5), 193–197; 24 ref.

Palchak R. B., Cohen R., Jaugstetter J. A threshold for airborne endotoxin associated with industrial-scale production of proteins in Gram-negative bacteria. *Dev. Ind. Microbiol.*, **31** (Suppl. 5), 199–203; 14 ref.

Sonesson A., Larsson L., Schuetz A., Hagmar L., Hallberg T., 1990. Comparison of the Limulus amebocyte lysate test and gas chromatography-mass spectrometry for measuring lipopolysaccharides (endotoxins) in airborne dust from poultry-processing industries. *Appl. Environ. Microbiol.*, **56**, no. 5, 1271–1278; 38 ref.

Suggested Readings, General Interest

Archer D. L., Jackson G. J., Langford C. F., 1991. Control of salmonellosis and similar foodborne infections. *Food Control*, **2**, no. 1, 26–34; 72 ref.

Blanco J., Blanco M., Garabal J., Gonzales E., 1991. Enterotoxins, colonization factors and serotypes of enterotoxigenic *Escherichia coli* from humans and animals. *Microbiologia*, **7**, no. 2, 57–73; 12 ref.

Brackett R. E., 1992. Microbiological safety of chilled foods: Current issues. *Trends Food Sci. Technol.*, no. 4, 81–85; 35 ref.

Cordano A. M. and Virgilio R., 1990. Salmonellosis in Chile; 1971–1985, bacteriological and ecological aspects. *Reyista Latinoamericana Microbiologia*, **32**, no. 2, 137–147; 61 ref.

D'Aoust J. Y. and Stavric S., 1993. Undefined and defined bacterial preparations for the competitive exclusion of *Salmonella* in poultry—a review. *J. Food Prot.*, **56**, no. 2, 173–180; 103 ref.

DeBess E. E., 1993. Foodborne disease outbreaks—a 10 year review (1983–1992) of California data. *Dairy Food Environ. Sanitation*, **13**, no. 5, 286–287; 4 ref.

Doyle M. P., 1992. A new generation of foodborne pathogens. *Dairy Food Environ. Sanitation*, **12**, no. 8, 490, 492–493; 22 ref.

Entis P., 1991. An integrated approach to food microbiology. *Europ. Food Drink Rev.*, Summer, 38, 40, 42; 36 ref.

Eyles M. J., 1990. Rapid methods for the detection of microbial and viral contamination. *CSIRO Food Res. Quart.*, **50**, no. 4, 96–103; 20 ref.

Food & Agricultural Organization, 1992. *Manual of food quality control*. 4. rev. 1. Microbiological analysis, 338p. ISBN 92 5 103189 4.

Fung D. Y. C., 1992. New developments in rapid methods for food microbiology. *Trends Food Sci. Technol.*, **3**, no. 6, 142–144; 39 ref.

Fung D. Y. C. and Mathews R. F. (eds.), 1991. *Instrumental methods for quality assurance in foods*. vi + 310p. ISBN 0 8247 8278 x.

Grant K. A. and Kroll R. G., 1993. Molecular biology techniques for the rapid detection and characterization of foodborne bacteria. *Food Sci. Technol. Today*, **7**, no. 2, 80–88; 48 ref.

Grant I. R. and Patterson M. F., 1991. Effect of irradiation and modified atmosphere packaging on the microbiological safety of minced pork stored under temperature abuse conditions. *Int. J. Food Sci. Technol.*, **26**, no. 5, 521–533; 28 ref.

Halligan A. C., 1990. The emerging pathogens—*Yersinia*, *Aeromonas* and verotoxigenic *E. coli* (VTEC)—a literature survey. *Food Focus*, no. 11, 73p.

Hintlian C. B. and Hotchkiss J. H., 1986. The safety of modified atmosphere packaging: A review. *Food Technol.*, **40**, no. 12, 70–76; 44 ref.

Juven B. J., Meinersmann R. J., Stern N. J., 1991. Antagonistic effects of lactobacilli and pediococci to control intestinal colonization by human enteropathogens in live poultry. *J. Appl. Bacteriol.*, **70**, no. 2, 95–103; 105 ref.

McCann G., Morgan L., Rowe M., 1991. Rapid methods for microbiological testing. *Milk Industry*, **93**, no. 2, 15–16; 2 ref.

Mossel D. A. A., 1989. Adequate protection of the public against food-transmitted diseases of microbial aetiology. *Int. J. Food Microbiol.*, **9**, no. 4, 271–294.

North R., 1991. The investigation of food poisoning. *Letters Appl. Microbiol.*, **12**, no. 5, 145–146; 7 ref.

Oblinger J. L. (ed.), 1988. Bacteria associated with foodborne diseases. *Food Technol.*, **42**, no. 4, 181–200.

Oosterom J., 1991. Epidemiological studies and proposed preventative measures on the fight against human salmonellosis. *Int. J. Food Microbiol.*, **12**, no. 1, 41–52; 24 ref.

Rodrigue D. C., Rowe B., Tauxe R. V., 1990. International increase in Salmonella enteritidis: A new pandemic? *Epidemiology Infection*, **105**, no. 1, 21–27; 26 ref.

Sands R. L., 1991. Industrial perspectives on the microbial contamination of food. *Europ. Food Drink Rev.*, Spring, 71–72, 74–75; 14 ref.

Sinha P. R. and Sinha R. N., 1990. Rapid methods for the detection of coliforms in milk and milk products. *Indian Dairyman*, **42**, no. 10, 427–432; 5 ref.

Smith L. B. and Zottola E. A., 1990. The microbiology of foodborne disease outbreaks: An update. *J. Food Safety*, **11**, no 1, 13–29; 37 ref.

Varadaraj M. C., 1993. Methods for detection and enumeration of foodborne bacterial pathogens: A critical evaluation. *J. Food Sci. Technol.* (India), **30**, no. 1, 1–13.

Waites W. M., 1988. Recent advances in rapid detection and enumeration of microorganisms. *Food Sci. Technol. Today*, **2**, no. 4, 256, 258; 10 ref.

Waites W. M., 1988. Hazardous microorganisms and the hazard analysis critical control point system. *Food Sci. Technol. Today*, **2**, no. 4, 259–261; 10 ref.

Index

AAF (acetylaminofluorene) labeling, 123–124
Absorption or adsorption methods, 82–83
Accelerated spoilage potential tests, 161–163
Accuracy, defined, 145
Acetaldehyde, 203
Acetic acid bacteria, 416
  analysis for, 505
Acetoin production test, 207
Acetylaminofluorene (AAF) labeling, 123–124
Actidione (cycloheximide), 257
Actidione medium, 500
Adenosine triphosphate (ATP), 10–11, 79–81
Adhesive tape, strip of, 521
Aero-anaerobic metabolism, 99
Aerobic metabolism, 99
Afnor reference procedure, 36
Agar-coated slides, 519–520
Agar diffusion techniques, 402
Agar drop culture, 53
Agar pour plate method, 6, 9
Agar pour plating techniques, 524
Agar sausages, 41–42, 519
Agglutination, 110, 112–114
Agitator shafts, 40
Alcoholic fermentation, 255
Alkaline phosphatase, buffer for, 482
Alkaline phosphatase test, 408
AMBIS system, 18
Amino acid degradation, 101–102
Amperometry, 85–86
Amplification, biological, 51
Anaerobic metabolism, 99
Anaerobic sporulating sulfite-reducing bacteria (ASSR), 391–392
Analytical end–point, defined, 145
Anhydrous dairy fat, 414
Animal flesh, *see* Meat
Animal muscle, characteristics of, 422
Antagonistic abilities, 203–204
Antibacterial substances, dairy products analysis for, 401–403
Antibodies, 109
Antifungal additives, assays of, 477
Antigens, 109
API galleries, 103
API tests. combined, 271–272
Application techniques, 518–521
APT medium, 218

Arginine dihydrolase detection, 197–198
Aromatic compounds, 202–203
Arrhenius model, 154–155
Ascomycetes, 250–251
Assimilative capabilities, tests for, 101
ASSR (anaerobic sporulating sulfite-reducing bacteria), 391–392
ATB 32 systems, 103
Atmospheric testing in microbiological monitoring, 525–526
ATP (adenosine triphosphate), 10–11, 79–81
ATPmetry, 234
Automation, 139
Automation of techniques, 139–147
  aim of, 140–141
  analytical aspects of, 144–146
  economic limitations of, 141–143
  limitations of, 141–146
  regulatory constraints on, 143–144
Auxanographic method, 270

Bacteriological quality
  of food products, 23
  of water, regulation of, 386–387
Bacteriophages
  characteristics of, 227–229
  detection and enumeration of, 235–243
  of lactic starters, 227–243
Bactometer, 11, 12
Baird-Parker medium, 364–365
Ballistospore formation, 268
Barnes TTC medium, 304
Basal medium, 354–355
Basidiomycetes, 250, 251, 252
Bea medium, 303–304
Beckers medium, 365
Beef carcasses, sampling, 36
Beer, 415–418
  microbial contaminations of, 415
  microbiological testing on, 416–418
Bergere modified Bryant-Burkey medium, 481
BGLBB (brilliant green lactose-bile broth) medium, 301
Biocounter, 522
Biolog system, 15
Biological amplification, 51
Biotinylated probes, 122–123
Biotrace Hygiene Monitor, 522
"Bone taint," 424
Bottle testing techniques, 523–524
Botulin toxin, detection and typing of, 357–358
Bouix and Leveau method, 508
Breed method, 48–49, 397–398
Brilliant green lactose-bile broth (BGLBB) medium, 301
Brined meats, spoilage of, 425
Brined sausages, spoilage of, 425
Brochothrix thermosphacta, 424–425
Bryant-Burkey medium, Bergere modified, 481
Buffer for alkaline phosphatase, 482
Butter, 412–413
Butter oil, 414
Butyric microbes, enumeration of, 400–401

Campylobacter, 325–332
  analysis in foods, 329–330, 331
  culture conditions and identification, 325–326, 327
  enteritis, 328
  epidemiology, 330–332
  pathogenic potential, 326, 329
  selective media for, 326, 328
Carbohydrate metabolism, examination of, 100–101
Carbon–14, 82
Carbon compounds, assimilation of, 270–271
Carbon substrate metabolism, 196–197
Carcasses
  beef, sampling, 36
  principal spoilages of, 423–425
Casein hydrolysis, 102
Caseinates, 411–412
Caseinolytic flora, enumeration of, 405
Caseins, 411–412
Catalase, 100
Catalase test, 13–14
CCPs (critical control points), 32
Cell count procedure, alternative methods for viable, 6–10
Cell morphological and structural traits, examination of, 98
CFUs (colony-forming units), 28–29

Chalmers's medium, 206–207
Cheeses, 411
  Listeria monocytogenes method of detection in, 378–379
Chemolithotrophs, 99
Chemoorganotrophs, 99
Chemotrophs, 99
Chlamydospore formation, 268
Chloramphenicol, 257
Chromatography, 87–88
Citratase detection, 207
Cleaning, 516
Clostridium, 347–358
  sulfite-reducing, 300
Clostridium botulinum, 353–358
  identification of, 356–357
  indicator value of, 354–355
Clostridium perfringens, 347–353
  detection in foods, 348–349
  enumeration in foods, 349–352
  identification of, 352
  indicator value of, 348
CMM (cooked meat medium), 355
Coagulase, detection of, 366
Coagulase-positive staphylococci, 361–369
  detection and enumeration of, 363–366
  identification of, 366–367
  integration of results for, 367–369
  significance in foods, 362
  taxonomic position of, 361–362
Coenzyme determinations, 79–81
Cold probes, 122, 125
Coliforms, 294–297
  thermotolerant, 389
Colony-forming units (CFUs), 28–29
Commercial quality, 23
Concentrated milk, 409
Conductometry, 86
Cooked meat, regulations and microbiological standards for, 430
Cooked meat medium (CMM), 355
Copper sulfate medium, 502
Copper sulfate technique, 405
Coulter type devices, 66–67
Counting chambers, 48
Cream
  ice, 141
  pasteurized, 413–414
Critical control points (CCPs), 32
Crustaceans
  microbiological testing of, 441–442
  microflora of, 439
Cryptococcaceae, 251, 252
Crystal-violet medium, 500–501
CYA (Czapek yeast autolysate agar), 281–282
Cycloheximide (actidione), 257
Czapek yeast autolysate agar (CYA), 281–282

D_3 medium, 485
Dairy fat, anhydrous, 414
Dairy products
  analysis for antibacterial substances, 401–403
  distributed, microbiological testing of, 406–414
Data acquisition, 160
Decarboxylation-deamination, 101–102
Deep-frozen vegetables, regulation of, 450
Deep putrefaction, 424
DEFT (direct epifluorescent filter technique), 9, 49–51, 398–399
Dehybridization, 132
Dehydrated plant products, 450–451
Deoxycholate agar, 301–302
Detection limit, defined, 145
Detection targets, 74–75
Detection time, 11
Deuteromycetes, 250
Diary products, 411–414
Differential medium for detection of histamine producing germs, 484
Digoxigenin marked UTP derivatives, 123
Diluflo equipment, 39
Dimethyl sulfoxide (DMSO), 37
Dip slides, 56–57
Direct DNA labeling, 123–124
Direct epifluorescent filter technique (DEFT), 9, 49–51, 398–399
Direct immunofluorescence, 114
Direct microscopic examination, 176–177
Disinfection, 516
DMSO (dimethyl sulfoxide), 37
DNA base pair composition, 104
DNA binding, 128–129

DNA/DNA and DNA/RNA homology, 104–105
DNA labeling, direct, 123–124
DNA probe, 15
"Dot–blot" hybridization test, 128–131
Double agar layer method, 230
Drinking water, 385–386
Drop count method, 54–55
Droplette system, 53
Dry milk, 410
Dry sausages, spoilage of, 425

EB enrichment broth, 380
Egg products, pasteurized, regulation of bacteriological quality for, 435
Egg white, unpasteurized, regulation of bacteriological quality for, 435
Eggs, 431–435
  bacteriological analysis of, 432–434
  contamination of, 431–432
  regulation of bacteriological quality for, 434–435
  Salmonella detection in, 433–434
Electrochemistry, 83–86
Electronic time-temperature integrators, 164
ELFA (enzyme-linked fluorescent immunoassay), 16
ELISA (enzyme-linked immunosorbent assay), 16, 115–116
Elliker's medium, 192–193
Enliten procedure, 11
Enteritis, *Campylobacter*, 328
Enterobacteria, 185, 297–298
  enumeration of, 433
Enterotoxigenic staphylococci, characterization of, 362
Enterotoxin, detection of, 353
Enzymatic activity, measurement of, 78–79
Enzyme activities, detection of, 101
Enzyme-linked fluorescent immunoassay (ELFA), 16
Enzyme-linked immunosorbent assay (ELISA), 16, 115–116
Epifluorescent microscopy, 49–51
Equipment, testing of, 516–524
Escherichia coli, 294–297
Esterases, 79
Eumycetes, 250
Exocellular fraction, 273

Factory equipment, microbiological monitoring of, 515–525
Fast acid producers, 202
Fecal contamination indicators, 293–304
  basis for use of, 293–294
  coliforms and *Escherichia coli*, 294–297
  culture media for, 301–304
  enterobacteria, 297–298
  examination for, 294
  streptococci, 298–300
  sulfite-reducing *Clostridium* species, 300
Fecal streptococci, 298–300
  enumeration of, 391
Fermentation, alcoholic, 255
Fermentation test, 195
Fermentative metabolism, examination of, 101
Fermented milk, 410–411
Fermented products
  enumeration of contaminating flora of, 405–406
  spoilage of, 425
Filtration media, concentration onto, 43–44
Filtration membranes, 41–42
Fish, 437–441
  bacteriological quality of, 437–439
  fresh, *see* Fresh fish
  microflora of, 439
  processed, testing of, 440–441
  spoilage processes of, 439
Fisheries products, spoilage processes of, 439
Fishing, contamination subsequent to, 438–439
Fishing waters, contamination of, 438
Flavored sterilized milk, 409
Flesh, animal, *see* Meat
Flow cytometry, 67–68
Fluorimeters, 80
Fluorochromes, 49
Food hygiene, microorganisms analyzed in, 516
Food industry, molds in, 277–279

Food microorganisms
  alternative methods for viable cell count procedure for, 6–10
  estimation of microbial populations and biomass, 10–14
  improvements in sampling and sample preparation, 4–6
  miniaturized microbiological techniques for, 14–15
  novel techniques for, 15–18
  rapid methods and automation in, 3–18
Food products
  bacteriological quality of, 23
  liquid, sampling, 36
  spoilage yeasts in, 259
  Yersinia enterocolitica incidence in, 342–343
Food testing laboratories, 142–143
Foods
  Campylobacter analysis in, 329–330, 331
  Clostridium perfringens detection in, 348–349
  Clostridium perfringens enumeration in, 349–352
  coagulase-positive staphylococci significance in, 362
  detection and typing of botulin toxin in, 357–358
  enumeration of spores and vegetative cells of *Clostridium botulinum* in, 356
  Listeria detection in, 377
  and multiplication of *Listeria monocytogenes*, 377
  Salmonella detection in, 316–321
  Yersinia enterocolitica analysis in, 341–342
Fowells' medium, 268
Fresh fish
  spoilage of, 439
  testing of, 440
Fresh meat, regulations and microbiological standards for, 427–428
Frozen meat, regulations and microbiological standards for, 427–428
Fruits
  characteristics of, 445, 447
  microbiological testing of, 449–450
  processing related contaminations in, 448
  spoilage of, 448
Fungal enumeration method, limitations of, 477
Fungal ergosterol analysis, 479

Galesloot agar, 481
Gardner's medium, 187
Gaseous phase chromatography (GC), 87–88
Gelatin hydrolysis, 102
Gelysate agar, 482
Genetic traits, examination of, 104–105
Gibson-Abdel-Malek test, 195–196
Glucose-yeast extract-peptone medium, 266
Gompertz function, 159
Gorodkowa's medium, 268
Gram-negative rods, 184–185
Gram-positive cocci, 186
Gram-positive rods, 185–186
Gram stain, 183
Gravimetric Diluter, 5
Grinder-blenders, 40

"H" antigens, 311
HACCP (hazard analysis critical control point), 31–32
Halophilic bacteria, detection of, 473
Hams, spoilage of, 425
Hazard analysis critical control point (HACCP), 31–32
Hemagglutination, 110
Heterotrophs, 99
HGMF (hydrophobic grid membrane filters), 55–56, 64–65
High quality pasteurized milk, 407
Histamine producing germs, differential medium for detection of, 484
Horse blood agar, 207–208
Hugh and Leifson medium, 184
Hull microdroplet test, 239–241
Hybridization on colonies, 131
Hybridization parameters, 126–128
Hybridization protocols, 126–132
Hybridization yield, 130–131

Hydrophobic grid membrane filters (HGMF), 55–56, 64–65
Hygiene quality, 23
  indicators of, 25–26
Hypersucrose medium, 484
Hypertonic broth, 195

Ice cream, 414
Ice milk, 414
IMF, *see* Intermediate moisture foods
Immunochemistry, 109
Immunoelectrophoresis, 111
Immunoenzymatic phage detection, 234–235
Immunofluorescence, 110, 114–115
Immunological traits, examination of, 105–106
Immunology, 109
  microbiological applications of, 109–117
Impedance measurements, 233–234, 509–510
Impedance method, 178
Incubation time, 74
Indirect agglutination, 110
Indirect immunofluorescence, 114, 115
Industrial microbiological testing, 23–32
  aims and requirements of, 24
  implementation of, 24–28
  uses of test results from, 28–32
Integrators, electronic time–temperature, 164
Intermediate moisture concept, 469–470
Intermediate moisture foods (IMF), 469–471
  bacteriological analysis of, 473–474
  biochemical tests for, 478–479
  culture media for, 480–485
  microbiological status of, 471
  microbiological testing of, 471–478
  microorganisms likely to develop in, 473
  mycological analysis of, 474–476
  tests of stability of, 472–477
Ion concentrations, 85
Isogrid system, 7–8
ITC broth, 342

Journal of Rapid Methods and Automation, 18

Kauffmann-White scheme, 313–314
Keg testing technique, 524
Killer yeasts, 258
Klaushofer-Dorfwirth method, 508
Kliger medium, 184

Labeled nucleotide incorporation, principles of, 125–127
Laboratories, food testing, 142–143
Lachica medium, 365
Lactic acid bacteria
  analysis for, 503–505
  detection of, 474
Lactic acid isomerism, 196
Lactic microflora, 189–221
  biochemical traits, 195–198
  common techniques of examination, 192–204
  definition, classification, and general properties, 189–192
  examination of different genera, 205–221
  identification, 194–204
  isolation and enumeration, 192–194
  molecular identification, 199–202
  morphology, 194
  physiology, 194–195
  principal species of, 190
  serology, 198–199
  technological properties, 202–204
Lactic starters, bacteriophages of, 227–243
Lactobacillus, 190, 213, 215–221 in beer, 416
Lactococcus, 190
Lactose-gelatin medium, 353
Leuconostoc, 190, 209–211
Limit, defined, 27
Lin's medium, 501–502
  modified, 502
Lipid metabolism, 103
Lipolytic microbes, enumeration of, 403–405
Liquid food products, sampling, 36

Listeria, 373–380
contaminated foods and multiplication of, 377
detection in foods, 377
epidemiology, 375
lysotyping, 375
media used for, 380
pathogenic potential, 375
serotyping, 375
serovarieties of, 376
taxonomy and identification of, 373–374
Listeria monocytogenes
contaminated foods and multiplication of, 377
method of detection in cheeses, 378–379
Litmus milk, 241–243
Litsky medium, 303
L.S. lactose and sulfite-containing medium, 351
Lyophilization, 254
Lysine medium, 500
Lysogenic bacteria, 228

M17 hard agar, 239
M17 medium, 205–206, 482
of Terzaghi and Sandine, 239–240
MacClary's medium, 268
Malt extract agar (MEA), 281
Malt extract medium, 261
Malthus system, 12
Mannitol, lysine, crystal–violet, brilliant green (MLCB) medium, 317–318
Mayeux, Sandine, and Elliker medium, 210
MEA (malt extract agar), 281
Meals, prepared, *see* Prepared meals
Meat, 421–430
endogenous infections of, 422–423
microbiological characteristics of, 422
microbiological testing of, 426–427
processed, regulations and microbiological standards for, 428–429
regulations and microbiological standards for, 427–430
spoilage of, 423–426
Meat products, spoilage of, 425–426
Mechanization, 139
Membrane filtration, 389–390
Mesophilic microflora, 26, 175–179
aerobic, enumeration of, 433
total aerobic, enumeration of, in milk, 396–397
Mesophilic yeasts, 254
Methylene blue method, 27
Microbial populations and biomass, estimation of, 10–14
Microbiological applications of immunology, 109–117
Microbiological monitoring
atmospheric testing in, 525–526
of factory equipment, 515–525
at level of personnel, 526–527
microorganisms analyzed in, 516
pressing and application techniques, 518–521
rinsing techniques, 523–524
swabbing techniques, 521–523
testing of equipment and work areas, 516–524
Microbiological techniques, miniaturized, 14–15
Microbiological testing of distributed dairy products, 406–414
in milk processing, 399–406
Microbiology, 3
predictive, 152–153
Microcalorimetry, 88
Micrococci, tests for distinguishing staphylococci and, 368
Microcolonial enumeration, 53
Microflora, *see* Microorganisms
Microorganisms
electronic counting of, 66–68
enumerating, 47–68
after culturing on filtering membrane, 55–56
in solid medium, 52–53
on solid medium surface, 54–55
estimation by culture, 52–58
evaluation by nonmicrobiological techniques, 73–89
examination of biochemical and physiological traits, 99–101
examination of cell morphological and structural traits, 98

Microorganism (*Continued*)
examination of cultural traits, 97–98
examination of genetic traits, 104–105
examination of immunological traits, 105–106
examination of pathogenic capabilities, 106
examination of sexual traits, 99
food, *see* Food microorganisms
identification of, 97–106
by nucleic acid probe hybridization, 119–136
industrial testing for, *see* Industrial microbiological testing
lactic, *see* Lactic microflora
mesophilic, *see* Mesophilic microflora
microscopic detection of, 47–52
most probable number (MPN) of, 58–66
psychrotrophic, *see* Psychrotrophic microflora
samples of, *see* Samples
Microtiter system, 63–64
Milk, 395–411
concentrated, 409
dry, 410
enumeration of thermoresistant microorganisms in, 399–406
enumeration of total aerobic mesophilic microflora in, 396–397
fermented, 410–411
flavored sterilized, 409
ice, 414
pasteurized, 406–409
raw, 395–399, 406
sterilized, 409
UHT, 409
Milk acidification test, 232–233
Milk processing, microbiological testing in, 399–406
Mineral waters, natural, 387
Minimally processed vegetables, 451–453
MLCB (mannitol, lysine, crystal-violet, brilliant green) medium, 317–318
Modified McBride selective agar (MMA), 380
Moisture, intermediate, *see* Intermediate moisture *entries*
Molds, 277–289
airborne, 284–285
experimental studies and prospects for improved defense against infections of, 288–289
in food industry, 277–279
identification of, 287
methods of analysis of, 279–283
surface, 285
types of analyses performed on, 283–286
undesirable, 278
useful, 278
Molecular hybridization, principle of, 120–121
Monitoring, microbiological, *see* Microbiological monitoring
Morphology, 183
Mossel medium
for *Bacillus cereus*, 485
for enumeration of total contaminants in soft drinks, 483
Most probable number (MPN), 58–66, 177
Motility enrichment, 16–17
MPN (most probable number), 58–66, 177
MRS broth or agar, 193
MRS medium, 218
of De Man and co-workers, 240–241
modified, 216
Multifactorial models in product life span prediction, 157–161
Mycelium, formation of, 266, 267
Mycotoxins, 278, 279
MYGP medium, 261

NAD (nicotinamide adenine dinucleotide), 79
Nalidixic acid medium, 187
Natural mineral waters, 387
Newman's stain, 480
"Nick and patch" technique, 125–126
Nicotinamide adenine dinucleotide (NAD), 79
Nile blue agar, 404
Nitrate-motility medium, 352
Nitrogen compounds, assimilation of, 271

Nitrogenous substances, metabolism of, 101–102
Nucleic acid probe hybridization, identification of microorganisms by, 119–136
Nucleopore polycarbonate membranes, 49

"O" antigens, 310–311
Oat flour medium, 282
Obesumbacterium proteus, 498
OGA (oxytetracycline glucose agar), 262
Omnispec bioactivity monitor system, 12–13
Orange serum broth, 483
Organoleptic quality, 27
Ouchterlony reaction, 111
Oudin reaction, 111
Oxidase, 100
Oxidative metabolism, 253
  examination of, 100–101
Oxygen, availability of, 477–478
Oxyrase, 17
Oxytetracycline glucose agar (OGA), 262

Packaged pasteurized milk, 407
Packaging, role of, 477–478
Paddle massagers, 40
Particle counters, 509
Pasteurized cream, 413–414
Pasteurized milk, 406–409
PCR, *see* Polymerase chain reaction
PDA (potato dextrose agar), 262, 282
Pediococcus, 190, 211–213, 214
  in beer, 416
PEM medium, 348–349
Peroxidase, 100
Peroxidase test, 408–409
Personnel, microbiological monitoring at level of, 526–527
Petrifilm SM, 520
Petrifilm system, 8–9
pH indicator method, 230–231
pH variations, 84–85
Phages, *see* Bacteriophages
Phosphate method, 78
Photochemical labeling, 124
Plant products, dehydrated, 450–451
Plasmid mapping, 201–202
Polymerase chain reaction (PCR), 16
  application of, 133
Polypectate gel medium, 484–485
Potato dextrose agar (PDA), 262, 282
Potentiometry, 83–85
Poultry, regulations and microbiological standards for, 429
Precipitation, 110–111
Predictive microbiology, 152–153
Prepared meals, 455–459
  conditions required for good bacteriological quality of, 456–458
  definition, 455
  microbiological problems posed by, 456
  microbiological testing of, 458–459
  organoleptic aspects of, 456
  regulatory criteria for, 459
  sanitary aspect of, 456
Preserves, 461–467
  definition, 461–462
  role of testing laboratory for, 462
  stability testing for, 462–464
  sterility testing for, 464–467
Pressing techniques, 518–521
Processed fish, testing of, 440–441
Product life span prediction, 151–165
  of manufactured products, 161–164
  modelizing effect of temperature, 153–157
  multifactorial models in, 157–161
  of new products, 164–165
  value and importance of, 151–152
Product spoilage, 27
Protein degradation, 102
PSB broth, 342
Pseudolysogeny, 228
Psychrophilic yeasts, 253
Psychrotrophic microflora, 181–187
  culture media for, 187
  identification of, 183–186
  techniques for examination of, 182–186
PT medium, 485
Pulsed field electrophoretic techniques, 273
Putrefaction, deep, 424

Quantum counters, 80

Radioactive probes, 121–122
Radioimmunology, 110, 116–117
Radiometric methods, 81–83, 510
Random priming, labeling by, 126–127
Rapid lysis method, 231–232
Ratkowsky model, 155–157
Raw milk, 395–399, 406
Recommendation, defined, 27
Redigel system, 9
Redox indicator method, 230–231
Redox potential, 83–84
  measurement of variations in, 77–78
Reductase test, 207
Reference technique, 178
Reference values, choice of, 28–29
Relative light units (RLUs), 11
Resazurin procedure, 77–78
Respiratory metabolism, 99–100
Richards-Cowland method, 507–508
Rinsing techniques, 523–524
RLUs (relative light units), 11
RNA nucleotide sequences, 16S, 105
RNA probe, 15
Rodac plates, 518
Rogosa medium, 216–217
Rothe medium, 302

Sabouraud medium, 261
Saccharomycetaceae, 250, 251
Salmonella, 309–321
  antigenic structure, 310–311
  complementary distinctive traits, 314
  definition, 309
  detection
    in eggs, 433–434
    in foods, 316–321
    sero-enrichment technique for, 113–114
  epidemiology, 315
  hybridization techniques, 320
  identification, 318–319
  Kauffmann-White scheme, 313–314
  pathogenic potential, 315
  principal biochemical traits, 309–310
  taxonomy, 311–313
Salmonella 1–2 test, 17
Salmonella enterica, 312
Samples, 35–44
  collection of, 35–36
  dilution and concentration of, 42–44
  homogenizing, 39–41
  solid, grinding of, 38–41
  storage of, 37
  surface, 41–42
  transportation and preservation of, 37–38
  weighing, 39
Schiemann's CIN medium, 341
Schoolfield model, 155–157
Schwartz differential medium (SDM), 501
Semipreserves, 467–468
Sensititre, 15
Sensitivity, defined, 145
Sero-enrichment technique for *Salmonella* detection, 113–114
Sexual traits, examination of, 99
Shelf life, 27
Shellfish, bacteriological testing of, 442–443
Sherman milk test, 195
Slanetz medium, 303
Slow acid producers, 202
Smoked meats, spoilage of, 426
Soft beverages, analysis of, 418–420
Solid media culture methods, 177
Specification, defined, 27
Spectrometry, 77–81
Spectroscopy, 75–83
Spermophthoraceae, 250, 251
Spiral plate count method, 54
Spiral plating method, 6–7
Spoilage, product, 27
Spoilage potential tests, accelerated, 161–163
Spoilage yeasts in food products, 259
Sporobolomycetaceae, 251, 252
Sporulation test, 499–500
Stabilized powdered Victoria blue, 481
Standard, defined, 27
Standard plate count technique, 52–53
Staphylococci
  coagulase-positive, *see* Coagulase-positive staphylococci
  enterotoxigenic, characterization of, 362–363

tests for distinguishing micrococci and, 368
Staphylococcus aureus, 362
detection of, 474
enumeration of, 433
Staphylococcus hyicus, 362
Staphylococcus intermedius, 362
Starter culture testing for purity, 497–510
accelerated conventional methods for, 505–506
analysis for bacteria, 503–505
conventional methods for, 498–503
general aspects of, 497–498
rapid techniques for, 506–510
Sterilized milk, 409
Stomacher bag, 41
Stomacher instrument, 4, 40
Streptococci, fecal, 298–300
enumeration of, 391
Streptococcus, 190, 205–209
Streptococcus thermophilus strain TJ, 402
Sugar fermentation, 269–270
Sugar-free agar, 482
Sulfite-reducing bacteria, anaerobic sporulating (ASSR), 391–392
Sulfite-reducing *Clostridium* species, 300
Sulfonation, 123
Sulfur dioxide, 256
Surface counts, 54
Swab-catalase test, 13
Swabbing techniques, 521–523

Tap water, 386
Tatini medium, 366
Techniques, automation of, *see* Automation of techniques
Tecra system, 16
Temperature, modelizing effect of, 153–157
Testing thresholds, 29–30
Texturizing ability, 203
Thermocycler, 16
Thermophilic yeasts, 254
Thermoresistant microorganisms, enumeration of, in milk, 399–406
Thermotolerant coliforms, 389
Thomas chamber, 48
"Three tubes per dilution" method, 59–60
Thresholds, testing, 29–30
Time-temperature integration devices, 164
TNase, detection of, 366–367
Tolerance, defined, 29
TPGYC medium, 355
Tributyrin agar, 481
Tributyrin agar count, 404
Turbidometry, 76–77
Tween 80 medium, 217–218, 481

UHT milk, 409
Uncooked ham, spoilage of, 425–426
Unpackaged pasteurized milk, 406
Urea hydrolysis, 102
Uridine analogs, 123

Vegetables, 445–453
biological peculiarities of, 446
characteristics of, 445–447
chemical characteristics of, 447
deep-frozen, regulation of, 450
microbiological testing of, 449–450
minimally processed, 451–453
natural bacterial flora of, 446
processing related contaminations in, 448–449
spoilage of, 447–448
Vegetative reproduction, mode of, 265
Velvet pad, sterile, 521
Vi antigens, 311
Vibrio fetus, 325
Vibrio parahaemolyticus, detection of, 443
Victoria blue agar, 404
VIDAS (Vitek immunodiagnostic assay system), 16
Vitamin-free medium, growth in, 271
Vitek immunodiagnostic assay system (VIDAS), 16
Vitek system, 15
VL-sulfite ferric ammonium citrate agar, 350–351
VRBL agar, 302
VRGB agar, 484

Water, 385–393
bacteriological analysis, 387–392
definition, 385–386

Water, 385–393 (*Continued*)
fishing, contamination of, 438
interpretation of results with, 392–393
regulation of bacteriological quality of, 386–387
Wauter's SSDC medium, 341
Weinzirl technique, 400
Whalley's medium, 262–263
Wild yeasts, 415
Williamson medium, 483
WL medium, 483
WLN medium, 501
Work areas, testing of, 516–524
Wort bacteria, 416
analysis for, 505
Wort media, 261

Xerotolerant yeasts, 262–263
Xylose, mannitol, adonitol, cellobiose, and sorbitol (XMACS) medium, 502

Yeasts, 249–273
biochemical and physiological characteristics of, 269–272
biological characteristics, 252–253
definition and classification of, 250–252
detection and culture techniques for, 260–264
ecological characteristics and consequences of development of, 258–260
enumeration techniques for, 260–263
examination techniques for, 260–273
fine differentiation technique for, 272–273
general characteristics of, 250–257
genetic characteristics of, 257–258
identification techniques for, 264–272
imperfect, 252
isolation and detection of, 263–264
physiological characteristics, 253–257
sexual characteristics of, 268–269
spoilage, in food products, 259
vegetative reproduction characteristics of, 265–268
wild, 415
Yersinia, 335–343
characterization of potentially pathogenic strains, 340
pathogenic potential, 338–340
taxonomy, 335–338
Yersinia enterocolitica, 336–338, 339
analysis in foods, 341–342
incidence in food products, 342–343
pathogenic potential, 339–340
Yersinia pseudotuberculosis, 336, 337
pathogenic potential, 338–339
Yogurt, 410–411